Herbal Medicine Approaches to Antimicrobial Resistance and One Health

Odangowei Inetiminebi Ogidi
https://orcid.org/0000-0002-9961-893X
Bayelsa Medical University, Nigeria

IGI Global
Scientific Publishing
Publishing Tomorrow's Research Today

Vice President of Editorial	Melissa Wagner
Director of Acquisitions	Mikaela Felty
Director of Book Development	Jocelynn Hessler
Production Manager	Mike Brehm
Cover Design	Jose Rosado

Published in the United States of America by
IGI Global Scientific Publishing
701 East Chocolate Avenue
Hershey, PA, 17033, USA
Tel: 717-533-8845 | Fax: 717-533-7115
Website: https://www.igi-global.com E-mail: cust@igi-global.com

Library of Congress Cataloging-in-Publication Data

LCCN: 2026933967 (CIP Data Pending)
ISBN13: 979-8-3373-7810-7.
Isbn13Softcover: 979-8-3373-7811-4
EISBN13: 979-8-3373-7812-1
British Cataloguing in Publication Data
A Cataloguing in Publication record for this book is available from the British Library.

Table of Contents

Detailed Table of Contents

The One Health concept is an integrated, unifying strategy that aims to sustainably balance and optimize the health of people, animals, and ecosystems. The global burden of infectious diseases and the increasing prevalence of drug-resistant strains underscore the need to understand and predict the threats posed by endemic pathogens. The One Health framework holds that human, animal, plant, and environmental health are inherently interconnected and interdependent. The escalation of drug resistance patterns., endemic diseases, vector-borne diseases, are driven by climate change and socioeconomic issues. The next pandemic may emerge gradually, affecting the most vulnerable communities, necessitating immediate action. Notable past outbreaks include the 2003 SARS, 2009 swine flu, 2012 MERS, 2013-2016 Ebola, and Zika virus, caused morbidity and mortality across countries. One Health emphasizes the need for a novel approach to address zoonotic disease transmission, leveraging advanced technology, global collaboration, and data-driven decision-making to improve global disease responses.

Antimicrobial resistance reduces the effectiveness of antibiotics, is a key global health threat, and affects human, animal, and environmental health. This chapter examines antimicrobial resistance within a One Health framework to identify the genetic, biochemical, ecological, and socio-economic drivers that enable resistant pathogens

investigations have demonstrated that traditional remedies employed by indigenous communities often contain bioactive phytochemicals with demonstrable antimicrobial, antiviral and immunomodulatory properties. The chapter is an attempt to explore how traditional knowledge serves as a strategic compass to guide and accelerate the discovery of therapeutic agents for emerging diseases.

Phytochemicals, naturally occurring compounds found in plants, have garnered significant attention in recent years for their potential antimicrobial properties and ability to modify resistance mechanisms. These bioactive substances can disrupt microbial growth and proliferation through various modes of action, including the interference with bacterial cell wall synthesis, inhibition of enzyme activity, and modulation of biofilm formation. Moreover, certain phytochemicals have been shown to enhance the efficacy of conventional antibiotics, offering a promising avenue for combating antibiotic resistance—one of the most pressing global health challenges. Continued exploration of phytochemicals could contribute to the development of novel antimicrobial agents, ultimately enhancing public health outcomes. This chapter elucidates on the recent trends in phytochemicals with antimicrobial and resistance-modifying activities. It also discusses the mode of action as antimicrobial agents and thus addresses the future directions and challenges.

The use of herbal antibiotics has increased as a substitute for synthetic ones, which are ineffective against bacteria that are resistant to drugs. The abuse and overuse of synthetic antibiotics is the root cause of a fast-expanding global health catastrophe. Since people first claimed that medicinal plants had therapeutic qualities thousands of years ago, they have been used. These plants include a variety of phytochemical

to improve public awareness, support rational prescribing, and promote sustainable strategies to mitigate AMR.

Preface

INTRODUCTION

Antimicrobial resistance (AMR) has emerged as one of the most formidable challenges of the 21st century, threatening to reverse decades of progress in modern medicine. Once easily treatable infections are becoming increasingly difficult to manage, placing immense strain on healthcare systems, economies, and global development. The drivers of this crisis are complex and interconnected, spanning human health, animal husbandry, agriculture, environmental contamination, and socio-economic practices. Addressing AMR therefore demands not only scientific innovation but also a holistic, integrative perspective that transcends traditional disciplinary boundaries.

This book, *Herbal Medicine Approaches to Antimicrobial Resistance and One Health*, is conceived as a timely and comprehensive response to this global emergency. It brings together contributions from diverse scholars and researchers to explore how herbal medicine which is rooted in centuries of traditional knowledge can play a transformative role in combating antimicrobial resistance within the One Health framework. The One Health approach recognizes the intrinsic link between human, animal, and environmental health, and this volume embraces that interconnectedness as a guiding principle.

The opening chapters establish the urgency and scale of the AMR crisis, examining its global burden, underlying mechanisms, and key drivers, including genetic mutations, ecological reservoirs, and socio-economic influences. They also highlight the limitations of current therapeutic pipelines and the pressing need for alternative strategies. Within this context, the One Health framework is introduced as a critical lens through which AMR can be better understood and addressed, emphasizing collaboration, data-driven insights, and global coordination.

Subsequent chapters delve into the rich domain of traditional medicine and ethnopharmacology, illustrating how indigenous knowledge systems have long provided effective remedies for infectious diseases. These contributions underscore the importance of preserving biocultural heritage while integrating it with modern scientific methodologies. The exploration of phytochemicals and plant-derived bioactive compounds reveals their diverse antimicrobial mechanisms, including inhibition of microbial growth, disruption of biofilms, and modulation of resistance pathways. Importantly, several chapters demonstrate how these natural compounds can enhance the efficacy of conventional antibiotics, offering synergistic solutions to overcome resistance.

The book further presents compelling case studies of medicinal plants such as *Azadirachta indica*, *Curcuma longa*, *Garcinia kola*, and *Artemisia annua*, highlighting their relevance across human, veterinary, and environmental health sectors. These examples illustrate the practical applicability of herbal medicine within a One Health context, from treating drug-resistant infections to reducing antimicrobial use in agriculture and livestock production. Innovative dimensions are also explored, including the role of artificial intelligence and machine learning in accelerating phytomedicine research. By leveraging computational tools, researchers can identify promising bioactive compounds, predict therapeutic targets, and optimize drug development processes with unprecedented efficiency. In parallel, discussions on environmental reservoirs of resistance, particularly groundwater systems, broaden the scope of intervention strategies to include ecological sustainability.

Recognizing that scientific discovery alone is insufficient, the latter chapters address critical issues of clinical validation, regulatory frameworks, policy integration, and global governance. They examine the challenges of standardizing herbal medicines, conducting rigorous clinical trials, and harmonizing regulatory pathways across regions. The importance of antimicrobial stewardship, behavioral change, and public awareness is also emphasized, highlighting the need for culturally sensitive and evidence-based approaches to reduce inappropriate antibiotic use. Finally, the book situates herbal medicine within the broader agenda of sustainable development, aligning its potential with global efforts to achieve health equity, environmental sustainability, and responsible resource utilization. By linking AMR mitigation strategies with the Sustainable Development Goals, the volume underscores the far-reaching impact of integrating traditional and modern healthcare systems.

This work is not merely an academic compilation; it is a call to action. It advocates for a paradigm shift in how we perceive and address antimicrobial resistance—one that values the synergy between nature and science, tradition and innovation, and local knowledge and global collaboration. The contributors to this volume collectively envision a future where herbal medicine is rigorously validated, responsibly integrated, and widely accessible as part of a comprehensive strategy to combat AMR.

We hope that this book will serve as a valuable resource for researchers, clinicians, policymakers, and students, inspiring further inquiry and fostering interdisciplinary partnerships. More importantly, we hope it will contribute meaningfully to the global effort to safeguard the efficacy of antimicrobial therapies and promote a healthier, more sustainable world.

CHAPTER OVERVIEW

Chapter 1: The One Health Framework and Antimicrobial Resistance: Role of Herbal Medicine

This chapter introduces the One Health framework as a unified approach to tackling antimicrobial resistance. It explores the interconnectedness of human, animal, and environmental health in the spread of infectious diseases. The authors emphasize global collaboration and technology-driven strategies for managing future health threats.

Chapter 2: Mechanisms and Drivers of Antimicrobial Resistance Genetic, biochemical, ecological, and socio-economic factors

This chapter explores the complex mechanisms underlying antimicrobial resistance. It discusses genetic, biochemical, ecological, and socio-economic drivers that contribute to resistance development. The authors highlight predictive modeling and integrated approaches as tools for managing resistance spread.

Chapter 3: Current Therapeutic and Policy Responses to Antimicrobial Resistance

This chapter examines current therapeutic and policy responses to antimicrobial resistance. It highlights advancements in drug development alongside global antimicrobial stewardship efforts. The authors stress the importance of coordinated policies and international collaboration in combating AMR.

Chapter 4: Traditional Knowledge as a Compass: Navigating the Therapeutic Landscape Of Emerging Diseases

This chapter presents traditional knowledge as a valuable resource in addressing emerging infectious diseases. It explores how ethnopharmacology reveals bioactive compounds with therapeutic potential. The authors emphasize integrating indigenous knowledge with modern science to develop new treatments.

Chapter 5: Phytochemicals With Antimicrobial and Resistance-Modifying Activities

This chapter focuses on phytochemicals and their antimicrobial properties. It explains how plant-derived compounds disrupt microbial processes and resistance mechanisms. The authors highlight their potential to enhance existing antibiotics and guide future drug development.

Chapter 6: Herbal Antibiotics for Treating Drug-Resistant Bacteria

This chapter examines herbal antibiotics as alternatives to synthetic drugs. It discusses various plant-derived compounds and their mechanisms against resistant bacteria. The authors emphasize the growing relevance of herbal therapies in addressing drug resistance.

Chapter 7: Selected Medicinal Plants in Focus – Case Studies of Plants with Strong Evidence

This chapter presents case studies of medicinal plants with strong antimicrobial evidence. It explores their mechanisms of action and applications across human and environmental health. The authors highlight their role in supporting sustainable, One Health-based therapeutic strategies.

Chapter 8: Herbal Medicine as Resistance Breakers – Synergy with Antibiotics and Reversing Microbial Resistance

This chapter explores the role of herbal medicine as resistance breakers. It explains how phytochemicals synergize with antibiotics to restore drug effectiveness. The authors highlight clinical benefits and innovative approaches such as nanotechnology integration.

Chapter 9: Veterinary and Agricultural Perspectives of Herbal Medicine Approaches to Antimicrobial Resistance

This chapter focuses on the application of herbal medicine in veterinary and agricultural systems. It explores plant-based alternatives to antibiotics and chemical pesticides. The authors emphasize sustainable practices that support food safety and environmental health.

Chapter 10: Plant-Derived Bioactive Compounds for Mitigating Environmental Antimicrobial Resistance Reservoirs: Focus on Groundwater

This chapter addresses environmental reservoirs of antimicrobial resistance, particularly groundwater. It explores the potential of plant-derived compounds in water treatment and contamination control. The authors highlight the need for sustainable solutions to environmental AMR challenges.

Chapter 11: Natural Defense Mechanisms and Bioactive Compounds in Ipomoea biloba Leaves: Antimicrobial Profiling of the Ethanol Extract

This chapter presents an experimental study on the antimicrobial properties of *Ipomoea biloba*. It analyzes its phytochemical composition and biological activity against pathogens. The authors demonstrate the scientific validation of traditional medicinal plants.

Chapter 12: Herbal Medicine and Antimicrobial Stewardship-Bridging One Health and Traditional Knowledge

This chapter links herbal medicine to antimicrobial stewardship within the One Health framework. It explores mechanisms by which phytochemicals reduce resistance pressure. The authors emphasize evidence-based integration of herbal therapies into healthcare systems.

Chapter 13: Unlocking Nature's Armory: An Ethnobotanical Approach To Discover Novel Anti-Resistance Agents

This chapter explores ethnobotany as a pathway for discovering anti-resistance agents. It highlights the value of indigenous knowledge in identifying bioactive compounds. The authors emphasize systematic approaches to unlocking nature's therapeutic potential.

Chapter 14: Clinical Trials and Regulatory Challenges in Herbal Antimicrobials: Bridging Traditional Wisdom and Modern Standards

This chapter examines the challenges of clinical trials and regulation of herbal antimicrobials. It discusses issues such as standardization, safety, and global regulatory differences. The authors propose strategies for integrating herbal medicine into mainstream healthcare.

Chapter 15: Sustainable Development Goals and Herbal Medicine

This chapter links antimicrobial resistance with the Sustainable Development Goals. It explores how herbal medicine supports health equity and sustainability. The authors highlight global policies and frameworks promoting traditional medicine integration.

Chapter 16: Integrating Behavioral Insights and Herbal Medicine Awareness to Prevent Antimicrobial Resistance: A One Health Perspective from a Moroccan Hospital Setting

This chapter examines behavioral factors influencing antimicrobial resistance. It highlights issues such as self-medication and inappropriate prescribing practices. The authors propose integrating herbal medicine awareness with behavioral interventions to improve outcomes.

Odangowei Inetiminebi Ogidi
Bayelsa Medical University, Nigeria

Chapter 1

The One Health Framework and Antimicrobial Resistance: Role of Herbal Medicine

Sai Priyanka J.
Sri Sathya Sai University for Human Excellence, India

Srusti Shankar Moger
Sri Sathya Sai University for Human Excellence, India

Bhumika Iranna Byalahunasishetter
Sri Sathya Sai University for Human Excellence, India

Ramya Raghavan
http://orcid.org/0000-0002-9953-543X
Sri Sathya Sai University for Human Excellence, India

ABSTRACT

The One Health concept is an integrated, unifying strategy that aims to sustainably balance and optimize the health of people, animals, and ecosystems. The global burden of infectious diseases and the increasing prevalence of drug-resistant strains underscore the need to understand and predict the threats posed by endemic pathogens. The One Health framework holds that human, animal, plant, and environmental health are inherently interconnected and interdependent. The escalation of drug resistance patterns., endemic diseases, vector-borne diseases, are driven by climate change and socioeconomic issues. The next pandemic may emerge gradually, affecting the most vulnerable communities, necessitating immediate action. Notable past outbreaks include the 2003 SARS, 2009 swine flu, 2012

DOI: 10.4018/979-8-3373-7810-7.ch001

MERS, 2013-2016 Ebola, and Zika virus, caused morbidity and mortality across countries. One Health emphasizes the need for a novel approach to address zoonotic disease transmission, leveraging advanced technology, global collaboration, and data-driven decision-making to improve global disease responses.

INTRODUCTION

The global rise of antimicrobial resistance (AMR) poses a critical threat to public health, animal welfare, and environmental sustainability. The One Health approach, which recognizes the interconnectedness of human, animal, and environmental health, has emerged as a holistic framework for addressing complex health challenges like AMR. The One Health framework aims to revolutionize monitoring and mitigation efforts for infectious diseases (Singh et al., 2024).

The conventional use of antibiotics is increasingly challenged by the rise of antimicrobial resistance (AMR), including multidrug-resistant and methicillin-resistant strains. The novel approach should manage the increasing health threats posed by zoonotic diseases and microbes. The WHO-FAO-WOAH Tripartite Collaboration aims to provide sustainable solutions. The One Health High-Level Expert Panel (OHHLEP) emphasizes that the principles of diversity, equity, and inclusion (DEI) are essential for the effective application that balances and optimizes health sectors (Robbiati et al., 2025).

ECONOMIC AND SOCIETAL IMPACT

AMR is a systemic failure that circulates among humans, animals, and the environment, and medicinal plants offer a sustainable, multi-targeted solution. Integrating herbal medicine aligns with global health economic strategies by reducing the financial burden of AMR. The One Health framework integrates human, animal, and environmental health to address complex challenges like neglected tropical diseases (NTDs) and antimicrobial resistance, recognizing their interconnected transmission pathways. NTDs affect over one billion people globally, often in impoverished areas, while AMR exacerbates treatment failures across sectors. Utilizing locally available, drought-tolerant plants (*Phoenix dactylifera or Ziziphus*) reduces healthcare costs and reliance on expensive imported pharmaceuticals, particularly in regions with rich ethnomedicinal traditions (El-Saadony et al., 2025). One Health shifts interventions to holistic strategies aligned with WHO's NTD roadmap 2021-2030. The concept of "One Health" finds profound parallels in traditional medicine systems around the world, particularly in the use of plant secondary metabolites (alkaloids, polyphenols, and terpenes) through a “pleiotropic” (multi-pronged) approach. Quorum Sensing bacteria communicate to coordinate attacks and form resistance. Compounds from plants like *Matricaria aurea* (Golden Chamomile) and *Centella asiatica* disrupt these quorum sensing (QS) signals. These systems inherently support animal and human health via inhibition of bacterial cell wall synthesis, disruption of DNA replication, and the inhibition of ATP (energy) synthesis. Recent global efforts, the National Academy of Medicine Report, and the WHO Global Traditional Medicine Centre, aim at integrating these concepts into national health systems (Soriano, 2025). Our work advocates for equitable collaboration and innovative financial models to support One Health and herbal medicines.

Figure 1. Herbal extracts are being evaluated as potential substitutes for antibiotics to combat antibiotic resistance and improve human health safety

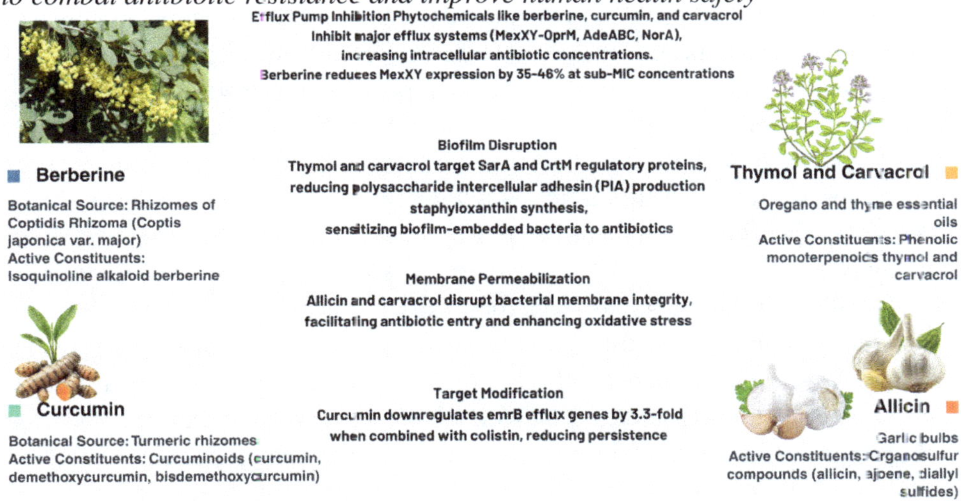

Antimicrobial resistance (AMR) is a global public health issue affecting humans, animals, and the environment. Key drivers of environmental AMR include the misuse of antibiotics and nonantibiotic substances in veterinary care, agriculture, and healthcare. AMR threatens the effectiveness of antibiotics and other antimicrobials, contributing to the emergence of resistant superbugs. The ESKAPE group (Enterococcus faecium, Staphylococcus aureus, Klebsiella pneumoniae, Acinetobacter baumannii, Pseudomonas aeruginosa, Enterobacter spp., and Escherichia coli) comprises the most clinically significant multidrug-resistant pathogens. Drug resistance is characterized by various bacterial evasion strategies, including genetic mutations, horizontal gene transfer, and the development of protective biofilms, all contributing to the rise of dangerous multidrug-resistant (MDR) microorganisms like MRSA.

Environmental hotspots, such as hospitals, wastewater treatment facilities, livestock farms, and pharmaceutical production sites, facilitate the spread of drug-resistant microbes and genetic elements. AMR emphasizes the interconnectedness of human, animal, and environmental health and the need for multidisciplinary collaboration to address antimicrobial resistance (Alem et al., 2025).

Herbal extracts are being evaluated as potential substitutes for antibiotics to combat antibiotic resistance and improve human health safety. *Dorstenia barnimiana* is recognized for its traditional medicinal applications in treating infectious diseases. Dorstenia is rich in bioactive compounds, including phenolics and flavonoids. The ethanol, ethyl acetate, and petroleum ether extract phenolics and flavonoids from D. barnimiana leaves, demonstrated significant antibacterial effects against Staphylococcus aureus, indicating potential for pharmaceutical applications (Dessie et al., 2026).

Herbal medicine is increasingly being integrated into One Health strategies. In vivo and in vitro studies have shown that bioactive compounds from herbal extracts possess effective antimicrobial properties against pathogenic bacteria. Plant secondary metabolites are structurally diverse and exhibit inherent pleiotropic (multi-target) activity against pathogens. This characteristic provides a substantial advantage over conventional single-target antibiotics, as it drastically reduces the pathogen's ability to rapidly evolve single-point resistance. Key findings include the identification of commonly cited plants like *Allium sativum, Azadirachta indica, and Eucalyptus globulus*, for their antibacterial properties. African medicinal plants have a long history of use in traditional medicine, with over 10% of Southern African vegetation used for various ailments and more than 15 species commercially available. Bioactive compounds from these indigenous plants show potential in combating multidrug resistance, with various mechanisms of action against bacteria, including inhibition of cell wall synthesis and disruption of biofilm formation (Moiketsi et al., 2023). Various gelled dosage forms (hydrogels, oleogels, and bigels) were created using deep eutectic solvents (DESs) to enhance the extraction and skin penetration of active compounds. Plant extracts have antibacterial activity against multiple pathogenic bacterial species and pathogenic fungi. Natural products could lead to innovative therapeutic strategies, although further in vivo testing and clinical trials are required for practical application. Future research and development can optimize bioactive compounds as Class 1B adjuvants to salvage existing antibiotics. This strategy maintains the potency of current drug classes while leveraging natural products' ability to attenuate bacterial resistance and virulence.

Table 1. Medicinal plants with proven efficacy with human, veterinary applications (Vladkova et al., 2024)

Botanical Name	Common Name	Part Used	Active Constituents
Coptis chinensis	Goldthread rhizome	Rhizome	Berberine (isoquinoline alkaloid)
Curcuma longa	Turmeric	Rhizome	Curcuminoids (curcumin)
Origanum vulgare	Oregano	Aerial parts	Carvacrol, thymol (phenolic monoterpenoids)
Allium sativum	Garlic	Bulb	Allicin, ajoene (organosulfur compounds)
Thymus vulgaris	Thyme	Leaves	Thymol, carvacrol
Astragalus membranaceus	Astragalus	Root	Polysaccharides, saponins, flavonoids
Atractylodes macrocephala	Bai Zhu	Rhizome	Terpenoids, coumarins, flavonoids
Berberis vulgaris	Barberry	Root bark	Berberine, palmatine

PHENOLIC COMPOUNDS (FLAVONOIDS, TANNINS, AND PHENOLIC ACIDS)

Phenolic compounds, including flavonoids and tannins, disrupt microbial processes through multiple pathways. Curcumin from *Curcuma longa* / Turmeric show Efflux Pump Inhibition activity in yeast. Curcumin suppresses biofilm formation in Pseudomonas aeruginosa by downregulating genes related to quorum sensing. It may diminish toxin production in MRSA.

Polyphenols (PPs)

Polyphenols are defined by their structural possession of phenol units and are subdivided into subclasses, including flavonoids, lignans, phenolic acids, and stilbenes. The hydroxyl groups and electron delocalization facilitate disruptive interactions with microbial cell membranes, proteins, and organelles. PPs can reduce the efficiency of metabolic pathways, cause destructive membrane damage, and critically, amplify their activity through chelation with metal ions. Notable examples include Curcumin, effective against *Staphylococcus aureus*, and Catechin, Tannins derived from *Mangifera indica*.

Quercetin (Flavonoid)

Flavonoids inhibit bacterial fatty acid biosynthesis enzymes and peptidoglycan production, which is essential for the bacterial cell wall. Virulence inhibition by the expression of genes encoding virulence factors and oxidative stress. Quercetin impedes the growth of Candida albicans by increasing intracellular ROS levels.

Catechin and Epigallocatechin Gallate (EGCG) from Green Tea inhibit F1Fo-ATPase and lactate dehydrogenase in *Streptococcus mutans*. EGCG also decreases beta-lactamase activity in K. pneumoniae, restoring susceptibility to beta-lactam antibiotics. Catechin induces ROS generation following membrane permeabilization (Jha et al., 2025).

Ellagic Acid and Tannic Acid interfere with efflux pump activity in bacteria such as *Acinetobacter baumannii*. Ellagic acid helps impede biofilm formation in Candida species.

Phloretin Flavonoids inhibit biofilm development in E. coli by downregulating the curli gene expression, which limits fimbriae formation. Baicalein induces ROS accumulation in Candida species, leading to cell death.

Table 2. Bioactive compounds, botanical sources, and spectrum of antimicrobial activity (Almasri et al., 2025)

	Specific Compound	Botanical Source Examples	MDR Pathogen Targets	Antimicrobial Spectrum
Alkaloid	Berberine (BRB)	Coptis chinensis, Rhazya stricta	A. baumannii (MDR), P. aeruginosa, MRSA, MDR E. coli	Broad-spectrum (Bacteria, Yeast)
Polyphenol (Flavonoid)	Quercetin (QCT)	Fruits, vegetables, seeds, flowers	Uropathogenic MDR E. coli, S. aureus, P. aeruginosa, E. faecalis	Broad-spectrum Bacteria (Gram-positive/Negative)
Terpenoid	(+)-Nootkatone		MDR S. aureus (Biofilms) 8	Bacteria (Biofilm)
Polyphenol (Tannin)	Tannic Acid	Multiple sources (related to Henna/ Clove)	MDR non-albicans Candida	Fungi (Cell wall inhibition)
Alkaloid	Conessine	Holarrhena antidysenterica		General Antimicrobial
Polyphenol (Phenolic Acid)	Rosmarinic Acid (Rs)	Satureja hortensis, Prunella vulgaris	MRSA, E. coli AG100, K. pneumoniae	Resistance Modifier/Bacteria

Alkaloids are nitrogen-containing compounds renowned for their broad-spectrum efficacy. Key examples such as Berberine, Piperine, and Reserpine demonstrate validated activity against difficult pathogens, including Methicillin-Resistant Staphylococcus aureus (MRSA), *Escherichia coli*, and *Pseudomonas aeruginosa*. are nitrogen-containing secondary metabolites that exhibit antibacterial activity, often by interacting with bacterial DNA or enzymes. For example, Berberine from *Berberis vulgaris*, *Corydalis turtschaninovii*, and *Hydrastis canadensis*

Efflux Pump Inhibition: Berberine actively blocks the NorA efflux pump in Methicillin-resistant Staphylococcus aureus (MRSA), which can reverse resistance to antibiotics like ciprofloxacin

Biofilm Inhibition: Berberine inhibits biofilm formation in MDR E. coli by disrupting the synthesis of biofilm-related compounds. Mandate integrated systems modeling requires the use of integrated omics data, including mathematical models and Genome-Scale Metabolic Models (GEMs), for all synergistic combination therapies. One Health hopes to achieve a predictive understanding of drug interactions, ensuring that potential pharmacokinetic (PK/PD) antagonism. For example, Berberine and Ciprofloxacin via host transporter inhibition are identified and mitigated before expensive clinical trials (Robbiati et al., 2025).

Metabolic Interference: Modifies glucose metabolism in *Streptococcus pyogenes* by regulating carbohydrate absorption and transformation. Alkaloids induce the generation of reactive oxygen species (ROS) in bacteria, leading to oxidative damage.

Tomatidine (*Solanum tuberosum* / Potato) is linked to enzyme Inhibition. Tomatidine alkaloid inhibits enzymes involved in ergosterol production in *Candida* and *Saccharomyces* species.

Anguinarine (Apple) inhibits biofilm formation in *Proteus rettgeri*. Piperine from *Piper nigrum* or Black pepper has been identified as a bioactive compound effective against i and *S. aureus*. Alkaloids from *Dicranostigma leptopodum* show membrane disruption. Alkaloids enhance the permeability of *Klebsiella pneumoniae* cells, thereby impeding their development (Robbiati et al., 2025).

Terpenoids derived from isoprene units, particularly oxygenated terpenoids containing alcohol functional groups, exhibit potent antibacterial and antifungal activity. Terpenoids are lipophilic compounds that can easily penetrate and damage bacterial cell membranes. Thymol and Carvacrol from *Thymus vulgaris and Origanum vulgare* compromise the lipid bilayer of bacterial membranes, undermining structural integrity and resulting in cell lysis (Jha et al., 2025)

Carvacrol serves as an effective inhibitor of efflux pumps, particularly in S. aureus. Eugenol (Syzygium aromaticum) disrupts cell membranes, causing the release of cellular contents.

Clove inhibits biofilm formation and eradicates existing biofilms in S. aureus (Jha et al., 2025).

(+)-Nootkatone

Biofilm and Motility Inhibition reduces biofilm thickness, eliminates bacterial cells, and decreases S. aureus motility. Terpinen-4-ol from Melaleuca alternifolia / Tea Tree is effective against S. aureus and Candida. Their presumed mechanism involves compromising the integrity of microbial cell membranes. Specific compounds, including oleanolic acid and (+)-Nootkatone, demonstrate efficacy against biofilms of MDR S. aureus. (Walker et al., 2025).

Cultural and Regional Relevance

Antimicrobial resistance is not solely a clinical issue, but a systemic one propagated through food chains, agriculture, and waste. The role of herbal medicine within this framework combats persistent disparities in outcomes. AMR is driven by social determinants, inequitable healthcare access, and implementation barriers, especially in low- and middle-income countries (LMICs). Despite advancements in precision medicine and pharmacotherapy, these challenges persist (Robbiati et al., 2025). Across continents and cultures, traditional medicine systems have independently developed remarkably similar approaches to managing antimicrobial resistance.

Table 3. Complementary and alternative medicine (CAM) herbal supplements are popular and plants used across continents (Walker et al., 2025) (Jha et al., 2025)

Plant / Herb	Asia (India - Ayurveda, China - TCM)	Africa	Europe	Americas (North/South)	Middle East	Oceania
Bitter Gourd (Momordica charantia)	India China	Used in Nigeria, South Africa	Limited use (migrant communities)	Used among indigenous communities	Used in various systems	Used by Pacific islanders
Fenugreek (Trigonella foenum-graecum)	India, China, Persia	North/East Africa	Mediterranean diet	Growing use in Latin America	Morocco, Egypt	Australia (recent interest)
Cinnamon (Cinnamomum spp.)	India, China	Egypt, Uganda	Culinary/ medicinal use	Used in Traditional Mexican medicine	Widely used	Used in herbal mixtures
Garlic (Allium sativum)	China, India	Across Africa	Central to folk remedies	Native American use	Arabian Peninsula	Used by Aboriginal Australians
Gymnema sylvestre	India	South Africa (imported)	Minimal	Some South Americans use (related plants)	-	Used as a supplement

Interrupting Transmission in the Food Chain and Agriculture

Antibiotic resistance is frequently transmitted from animals to humans via the food chain and agricultural waste. Multidrug-resistant microorganisms and resistance genes can infiltrate agricultural soils through animal manure and sewage, eventually entering the crop microbiome and the human food chain. Natural immunostimulants, such as probiotics and traditional herbal blends, are increasingly used as animal feed additives to manage infections like Salmonella. .(Hu et al., 2025)

- Allicin (Allium sativum / Garlic): Effective against Pseudomonas aeruginosa.
- Azadirachtin (Azadirachta indica / Neem): Klebsiella pneumoniae.
- Plumbagin (Plumbago zeylanica): Oxidative stress in C. albicans, leading to fungal cell death (Ashraf et al., 2023).

Strophanthin (Moringa oleifera): Ion Transport Inhibition of Na+/K+-ATPase by binding to its receptor, altering intracellular ion concentrations (Mapunda & Mramba, 2025). For instance, by-products of *Zanthoxylum bungeanum* and extracts of *Schisandra chinensis* have been shown to improve animal productivity. Plant bioactives' antibacterial properties reduce the need for synthetic antibiotics and thereby lower the environmental burden of resistance residues (Ashraf et al., 2023).

Figure 2. AMR stewardship across the antimicrobial lifecycle (Mapunda & Mramba, 2025)

Overcoming Resistance Evolution

Microorganisms adapt quickly via genetic mutations or horizontal gene transfer. A major failure of antibiotics is the evolution of bacterial resistance to single-molecule drugs. It is logical that different cultures, independently or through ancient trade routes, would discover *Lawsonia inermis, Nigella sativa,* and *Ziziphus spina-christi* that address common issues. The complexity of the herb's alkaloids, phenolics, and terpenoids forces bacteria to undergo lethal mechanisms. A computational screening of 5,230 natural compounds validates ethnopharmacological claims and yields excellent docking scores (Mapunda & Mramba, 2025). The nutrition, vitamins and minerals, which are important for disease prevention and health promotion. The efficacy and safety of the herbal complex *Elaeagnus multiflora* and *Cynanchum wilfordii*, show therapeutic effects for men with lower urinary tract symptoms. It is converging on certain medicinal plants, holistic dietary patterns, and culturally embedded mind-body practices. The recognition of a plant's medicinal properties, including favorable pharmacokinetics and minimal toxicity risks, is a form of deep ecological knowledge. *Coccinia* species have traditionally been used to treat various ailments, such as diabetes, asthma, and jaundice. The shared use of these plants highlights a universal dependence on the health of our ecosystems (Ashraf et al., 2023).

Targeting "Superbugs" Shared by Humans and Animals

Zoonotic pathogens, or diseases transmitted from animals to humans, are a primary concern of the One Health framework. Herbal medicines provide a diverse array of bioactive compounds to combat Methicillin-resistant *Staphylococcus aureus* (MRSA). *Escherichia coli*, and *Salmonella* circulate between humans, livestock, and the environment. MRSA, for example, can be transmitted from livestock to humans (LA-MRSA) (Mapunda & Mramba, 2025). The antimicrobial activity of brown propolis ethanolic extract treats bovine mastitis. It is effective against *Staphylococcus sp.* isolates for intramammary formulations. The extract showed significant antimicrobial and antibiofilm properties, resulting in successful treatment of infected animals without adversely affecting milk quality.

In many regions, herbal medicine is a culturally accepted and accessible form of healthcare. Medicinal plants are generally more accessible, affordable, and accepted, particularly in low-income regions where the burden of infectious disease is highest. Furthermore, phytochemicals can act synergistically with conventional antibiotics. The use of garlic extract combined with amoxicillin may allow lower doses of conventional drugs. Standardizing plant-derived pharmaceuticals could transform healthcare and address the challenges posed by drug-resistant infections (El-Saadony et al., 2025). Traditional medicine is a primary source of healthcare for livestock in Ethiopia, but it faces threats from cultural change, habitat destruction, and over-exploitation of medicinal plants. A recent study documented 33 plant species from 18 families used to treat 20 different livestock diseases in the Bure district, with the most common families being Asteraceae, Solanaceae, Fabaceae, and Acanthaceae. The majority of remedies were prepared from fresh plant materials (84.6%), with leaves being the most commonly used part (57.7%) and oral administration being the predominant application method (67%). The study highlights the importance of the area for ethnoveterinary knowledge, particularly noting a high informant consensus factor for plants used to treat respiratory diseases (ICF = 0.97) and dermatological conditions (ICF = 0.88), indicating a strong agreement among informants on their use (Abdu & Gudesho, 2026).

Table 4. Veterinary applications of herbal alternatives to antibiotic growth promoters

Species	Herbal Formulation	Active Components	Dosage	Effects	Mechanism
Broilers	Astragali radix + Atractylodes macrocephala + Isatis tinctoria + Citri reticulatae pericarpium	Polysaccharides, saponins, flavonoids, terpenoids	0.5-1.5% feed	Enhanced growth performance, immune indices, cecal microbiome	Microbiome modulation, immune organ development
Broilers	Oregano essential oil (84% carvacrol)	Carvacrol, thymol	100-200 mg/kg feed	Improved feed conversion, antioxidant status	Membrane disruption, quorum sensing inhibition
Broilers	Cnidium monnieri + Taraxacum officinale	Coumarins, flavonoids, terpenoids	0.5-1% feed	Coccidiosis alleviation, immune enhancement	Pathogen exclusion, immune modulation
Cattle	Oregano/thyme extracts	Carvacrol, thymol	2.5-50 mg/kg feed	Growth promotion, feed efficiency improvement	Direct antimicrobial, gut microflora modulation
Swine	Garlic extract (allicin-rich)	Allicin, organosulfur compounds	2-10 g/kg feed	Growth enhancement, pathogen control	Membrane disruption, enzyme inhibition

The therapeutic efficacy of *Artemisia annua and Calotropis procera* for treating sub-clinical bovine tropical theileriosis, a significant tick-borne disease in cattle that causes economic losses. The internalization of natural antimicrobial substances into optimal nanocarriers is viewed as the future of "green therapeutics". Herbal formulas, such as *Alisma orientale*, have been shown in animal models to restore gut microbiota balance and reduce the risk of AMR in veterinary environments. The therapeutic efficacy of *Artemisia annua and Calotropis procera* for treating sub-clinical bovine tropical theileriosis, a significant disease affecting cattle. Results indicate that Artemisia annua showed 100% efficacy in improving clinical parameters and hematological indices, while both treatments led to substantial recovery rates and improvements in various health markers without adverse effects on liver and kidney function (Siddiqui et al., 2026).

There is ongoing debate about the long-term efficacy and safety of newer drugs, particularly regarding cardiovascular risks and other side effects. Several herbal extracts demonstrate notable activity against multidrug-resistant (MDR) bacteria, including strains of *Escherichia coli, Klebsiella pneumoniae, Staphylococcus aureus, and* others, often outperforming certain conventional antibiotics in vitro. Garlic (Allium sativum) is effective against MRSA and *Vancomycin-resistant Enterococcus* (VRE), Ziziphus lotus and Ashwagandha (*Withania somnifera*) disrupt the cell membranes of MRSA (Hu et al., 2025). *Cipadessa baccifera* through metabolomics, reductions in blood glucose, insulin, and HbA1c levels, alongside upregulation of GLUT4 expression, suggesting its potential as a valuable supplement for diabetes management.

Figure 3. Medicinal plants and phytochemicals against WHO priority pathogens (Dube et al., 2025) (Hoenders et al., 2024)

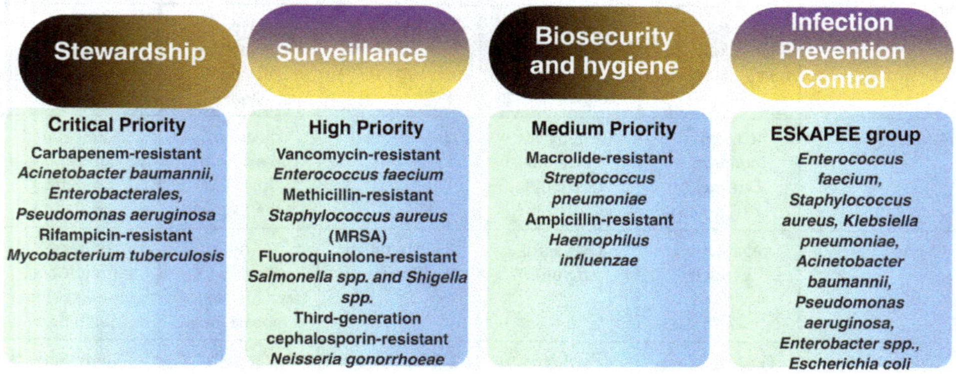

For combating gram-negative resistance use Purslane. *Portulaca oleracea* demonstrates broad-spectrum activity against notorious MDR pathogens like *Acinetobacter baumannii* and *Klebsiella pneumoniae*. Similarly, Henna (*Lawsonia inermis*) has shown efficacy against extended-spectrum beta-lactamase (ESBL) producing strains. Traditional Asian medicine is the most commonly used (59.3%) among all healthcare services. It is an important factor influencing socio-demographic and psycho-social elements. The community's familiarity with traditional medicine and medicinal plants is accessible and culturally appropriate (Febriyanti et al., 2024).

Calotropis procera is effective against MDR *S. aureus, E. coli, and P. aeruginosa*. The antimalarial activity of *Senna alata* has been shown to affect hepatic caspase-3 and NF-κB expression in a murine model.

Allium sativum (garlic) and Coriandrum sativum (coriander): Moderate activity vs MDR S. aureus and biofilms.

Table 5. Plant-based antimicrobials as sustainable solutions, guided by scientific evidence and omics-based standardization (Alem et al., 2025) (Walker et al., 2025)

	Omics Technology	**Pathogen Resistance Mechanism**	**Systems-Level Insights**
Berberine (BRB)	Transcriptomics	MDR Acinetobacter baumannii (AdeABC Efflux)	Confirmed EPI mechanism through reduction in the gene expression of the adeB efflux pump component
Quercetin (QCT)	Proteomics, Real-time PCR (Transcriptomics)	Enterococcus faecalis (Biofilm Formation)	Identified multi-target disruption across crucial pathways: interference with glycolytic, protein folding, and protein translation/elongation machinery
Phyto chemicals	Metabolomics	Pathogen Metabolic Flux and Virulence	Quantification of metabolic flux alterations, enabling the identification of stress responses and disruption of specific virulence metabolite production
Curcumin (Polyphenol)	Omics	Biological responses in specific cell lines/tissues	Elucidation of the broader "network effect" and reprogramming of functional pathways beyond a single target
Antimicrobials	Systems Biology, Mathematical Modeling	AMR Mechanisms in Pathogens	Predictive understanding of bacterial responses achieved by integrating omics data with Genome-Scale Metabolic Models (GEMs)

Omics Technologies in Herbal Medicine Research

Omics technologies are revolutionizing the field of herbal medicine, providing powerful tools for the discovery and development of novel antimicrobials. Their integration into One Health strategies holds great promise for addressing the global challenge of AMR in a holistic, sustainable, and effective manner.

- Artemisinin from Artemisia annua: Genomics and traditional knowledge led to the discovery and scalable production of artemisinin, a key antimalarial agent (Alem et al., 2025) (Walker et al., 2025). Mining plant genomes for biosynthetic gene clusters (BGCs) accelerates the identification of novel antimicrobial peptides and secondary metabolites (Siddiqui et al., 2026). For example, the sequencing of Artemisia annua enabled the identification of genes involved in artemisinin biosynthesis, a key antimalarial compound. (Dube et al., 2025) (Hoenders et al., 2024)
- Mayapple (*Podophyllum hexandrum*): Integrated metabolome and transcriptome analysis identified genes involved in the biosynthesis of podophyllotoxin, a compound with antimicrobial and anticancer properties.
- Synergy with Conventional Antibiotics: Herbal compounds have been shown to act synergistically with antibiotics, reducing the required dose and selective pressure for resistance

The research validates herbal treatments for managing Antimicrobial resistance (AMR). The findings reveal that many AMR patients seek herbal remedies supported by evidence from ethno-pharmacological and phytochemical research. A total of 97 plants were mentioned by participants, with 49 species identified at the National Herbarium, predominantly from the Fabaceae family (Dube et al., 2025). The most frequently cited plants for antimicrobial resistance management included *Zanthoxylum capense, Opuntia ficus-indica, Monotes engleri, and Tamarindus indica*. Ongoing debates persist regarding the evidence base for the safety and effectiveness of traditional medicines in Antimicrobial resistance management, especially when integrated under the One Health umbrella.

Concerns include regulation, quality control of herbal remedies, and lack of standardized protocols, but patients are hesitant to discuss them with health professionals. Improved communication could enhance the dialogue between patients and health professionals about herbal medicines. The studies used both experimental designs and ethnobotanical surveys, with findings indicating that a range of plant species and families are used to treat AMR. Genomics involves the comprehensive analysis of an organism's entire genetic material. In herbal medicine research, the convergence of NGS, dPCR, and CRISPR technologies has transformative potential to ensure transparency in the global food supply. NGS provides comprehensive, untargeted species identification; dPCR delivers precise absolute quantification; CRISPR enables rapid, field-deployable point-of-care detection (Pallavi et al., 2026). Integration with blockchain and IoT enables immutable traceability. The future success of herbal medicine depends on standardization, regulatory harmonization, advances in bioinformatics, and cost reduction to enable widespread deployment across the food industry. The combination of traditional authentication methods with emerging genomic and portable technologies offers a multi-layered approach. (Yoon & Lee, 2025)

GENOMICS: BIOACTIVE COMPOUND BIOSYNTHESIS

The "One Thousand Plant Transcriptomes Initiative" (1KP project) sequenced the transcriptomes of over 1,000 plant species. This massive genomic dataset enabled researchers to identify new thionin sequences (a class of plant AMPs) in Angiosperm families previously not known to contain them, including a large gene family found in Papaver (poppies). Sequencing medicinal plant genomes reveals gene clusters responsible for producing antimicrobial secondary metabolites. For example, the sequencing of Artemisia annua enabled the identification of genes involved in artemisinin biosynthesis, a key antimalarial compound. Research on (+)-nootkatone, a terpenoid, used transcriptional analysis to show that the compound inhibits biofilm formation in MDR S. aureus by downregulating specific biofilm-associated genes. By comparing genomes across species, researchers can annotate and discover sarA, icaA, agrA, RNAIII, and spa. involved in antimicrobial activity.

Table 6. Bioactive compounds found in herbal medicines with antimicrobial properties, which can serve as alternatives or adjuncts to conventional antibiotics (El-Saadony et al., 2025)

Herbal Extract	Active Against MDR Strains	Notable Activity Details
Acacia nilotica (ethanolic)	E. coli, K. pneumoniae, S. aureus	Lowest MIC (9.75-313 µg/ml); potent vs nosocomial/community infections
Syzygium aromaticum (clove)	S. aureus (MDR), non-fermenters, biofilms	Zone of inhibition 20-25 mm; MBC 6.25 mg/ml; >50% biofilm reduction
Cinnamomum zeylanicum (cinnamon)	E. coli, K. pneumoniae, C. albicans	Strong inhibition; effective vs nosocomial strains
Centella asiatica (EtOAc/EtOH)	Gram-positive (S. aureus) and Gram-negative (K. pneumoniae, E. coli)	Highest activity index (93.7-100%); zones 16-21 mm

continued on following page

Table 6. Continued

Herbal Extract	Active Against MDR Strains	Notable Activity Details
Silybum marianum (milk thistle, EtOAc/EtOH)	S. aureus, K. pneumoniae, E. coli	Activity index 75-100%; zones up to 21.8 mm
Camellia sinensis (green tea)	MDR S. aureus, non-fermenters, biofilms	Zone 20-25 mm; strong antibiofilm effects
Pentaclethra macrophylla	Multiple MDR bacteria (62.5% of tested strains)	MIC 64-1024 µg/ml

Genomics-Guided Discovery

Mining plant genomes for biosynthetic gene clusters (BGCs) accelerates the identification of novel antimicrobial peptides and secondary metabolites . A machine learning study based on the Indonesian herbal system "Jamu" used random forest algorithms to screen herbal formulas. The input data included plant constituents, identifying 14 plants as potential natural antibiotic candidates, 10 of which were confirmed to have antibacterial activity in scientific literature (Jha et al., 2025) (Alem et al., 2025).

Integration with Other Omics

Genomics data is often combined with transcriptomics, proteomics, and metabolomics to elucidate complex biosynthetic pathways and optimize the production of bioactive compounds (Patwardhan & Aswar, 2025). The PTPAMP web server integrates compositional models to predict plant-derived AMPs. It classifies peptide sequences based on physicochemical attributes (like amino acid composition) to predict antibacterial, antifungal, or antiviral activity (Teka & Maryo, 2023).

The genomic analysis of Chinese ginseng and opium poppy has led to the identification of genes involved in the biosynthesis of saponins and benzylisoquinoline alkaloids, with antimicrobial properties (Patwardhan & Aswar, 2025).

Proteomics

Proteomics is the large-scale study of proteins, including their expression, structure, and function. In the context of herbal medicine and AMR Advanced computational tools are used to predict potential AMPs from plant genomic data. For instance, in Amaranthus tricolor (red amaranth), a combination of in silico prediction and bottom-up proteomics identified seven novel peptides, including lipid transfer proteins and snakins. Bioactivity screening of one such peptide (Atr-LTP1) confirmed its activity against high-risk, multidrug-resistant pathogens, including *Staphylococcus aureus and Acinetobacter baumannii.*

Table 7. Role of plant bioactive compounds in AMR mitigation in therapeutic and adjunct applications (Dube et al., 2025) (Hoenders et al., 2024)

Compound/ Class	Primary AMR Mitigation Strategy	Mechanism Class (Adjuvant)	Synergistic Partner Example	Effect on Resistance
Berberine (Alkaloid)	Efflux Pump Inhibition (EPI) / QS Quenching	Class 1B (Inhibitor of Passive Resistance)	Ciprofloxacin (CIP), Thioridazine	Breaks MDR in Gram-negatives (P. aeruginosa, A. baumannii) by reducing antibiotic MIC and downregulating resistance genes
Quercetin (Flavonoid)	Biofilm Inhibition / Anti-Virulence Modulation	Class 1B (Biofilm Inhibitor/Quorum Quencher)	Levofloxacin, Meropenem, Ceftazidime	Restores susceptibility by inhibiting biofilm formation and QS gene expression at sub-lethal concentrations
Rosmarinic Acid (Phenolic Acid)	Resistance Modifier / Adjuvant	Class 1E (EPI/ Adjuvant)	Vancomycin, Ampicillin (AMP), Gentamicin	It lowers MICs of conventional antibiotics against MRSA and resistant E. coli AG100
Tannic Acid (Polyphenol)	Cell Wall Integrity Targeting	Class 1B (Membrane/Cell Wall Destabilizer)	Antifungal	Inhibits chitin synthesis, providing a targeted mechanism against MDR Candida strains

Plant-derived compounds, when combined with conventional antibiotics, can produce synergistic effects, reducing the required dosage of synthetic drugs and reversing resistance. Proteomic analysis of *Dendrobium huoshanense* identified crotonylated proteins involved in the biosynthesis of antimicrobial alkaloids and polysaccharides. (Almasri et al., 2025) Proteomics enables the discovery of novel antimicrobial peptides (AMPs) from plant sources. Network pharmacology is optimizing the use of plant-derived compounds (alkaloids, saponins, terpenoids) by identifying optimal combinations that can act synergistically with antibiotics, which can disrupt bacterial membranes or inhibit essential bacterial processes (Anyiam et al., 2025).

- Pomegranate and Amoxicillin: The ethanolic extract of *Punica granatum* (pomegranate) pericarp combined with amoxicillin demonstrated the most effective synergy against *Aggregatibacter actinomycetemcomitans.*
- Garlic and Amoxicillin: *Allium sativum* (garlic) extract combined with amoxicillin ly decreased the minimum inhibitory concentration (MIC) against multidrug-resistant (MDR*) Helicobacter pylori*
- Berberine and Antibiotics: Berberine (from Coptis species) has been effective in treating acute infectious diarrhoea when combined with sulfadimidine and neomycin (Moiketsi et al., 2023). Additionally, berberine can block the NorA efflux pump in MRSA, reversing resistance to ciprofloxacin.

- Resveratrol and Aminoglycosides: Resveratrol enhances the activity of aminoglycoside antibiotics against Staphylococcus aureus
- Fatty Acids and Vancomycin: Unsaturated fatty acids act as adjuvants to vancomycin, rapidly killing Gram-positive bacteria, including vancomycin-tolerant strains, by accumulating membrane-bound cell wall intermediates that disrupt membrane integrity (Ashraf et al., 2023).

Proteomics helps characterize the mechanisms of action of bioactive peptides, their specificity, and their synergistic effects with conventional antibiotics. Quantitative and Functional Proteomics approaches reveal how herbal extracts modulate protein expression in pathogens and host cells, providing insights into resistance mechanisms and potential therapeutic targets (Farey, 2025). Essential oil profiling of *Pogostemon mollis Benth.*, revealing significant antioxidant activity and a diverse range of bioactive compounds. The research identifies 99 bioactive compounds through UPLC-QTOF-MS/MS and 68 compounds in the essential oil via GC–MS, highlighting the therapeutic potential of P. mollis for natural antioxidants and bioactive formulations. The phytochemical composition, antioxidant potential, and essential oil profiling of *Pogostemon mollis Benth.* highlight its diverse pharmacological properties due to various bioactive constituents.UPLC-QTOF-MS/MS analysis identified 99 bioactive compounds, including notable agents such as camptothecin (anticancer), zidovudine (antiviral), luteolin (flavonoid), and nerolidol (terpenoid). GC–MS analysis of the essential oil revealed 68 compounds, with major constituents including lupeol (6.33%), alpha-cyperone (7.15%), and caryophyllene oxide (7.23%), and significant correlations were found between total phenolic and flavonoid contents and antioxidant activities (Momin et al., 2026).

Table 8. Synergistic combinations of herbal extracts with conventional antibiotics (Moiketsi et al., 2023) (Patwardhan & Aswar, 2025) (Li & Shahid, 2025)

Herbal Extract	Antibiotic Partner	Target Pathogen	Mechanism
Daphne genkwa extract	Oxacillin	MRSA	Resistance-modifying activity
Berberine	Imipenem	P. aeruginosa PA012	MexXY-OprM inhibition
Curcumin (100 μM)	Colistin	A. baumannii	AdeABC efflux pump inhibition, ROS increase
Thymol (100 μg/ml)	Vancomycin	MRSA biofilm	sarA-dependent biofilm eradication
Carvacrol (75 μg/ml)	Rifampicin	MRSA	SarA and CrtM targeting
Carvacrol	Tobramycin	E. coli, MRSA	Membrane disruption, efflux inhibition

continued on following page

Table 8. Continued

Herbal Extract	Antibiotic Partner	Target Pathogen	Mechanism
Allicin + Ag NPs	-	MRSA	Membrane disruption, oxidative stress
Curcuma amada extract	Cefixime	S. aureus	Efflux pump modulation
Nigella sativa extract	Cefixime	P. aeruginosa	Multidrug resistance reversal

CHALLENGES

The integration of genomics, proteomics, metabolomics, and microbiome research enables the identification of key genes, proteins, and metabolites involved in antimicrobial activity. Multi-omics provides a holistic understanding of the biosynthesis, regulation, and activity of antimicrobial compounds in medicinal plants. This systems-level approach supports rational drug discovery.

There are challenges that remain in applying omics technologies to herbal medicine research

Multi-omics data are complex and heterogeneous, requiring advanced bioinformatics tools and standardized protocols for integration and interpretation. The mechanistic understanding of how herbal medicines interact with pathogens, the host, and the microbiome to combat AMR. Ensuring reproducibility and efficacy of herbal products through comprehensive profiling. Multi-omics databases (Ginseng Genome Database, Global Pharmacopoeia Genome Database) facilitate data mining and comparative studies. Antimicrobial resistance care needs pooled resources across healthcare sectors for education, infrastructure, and surveillance. However, in many settings, resources for traditional medicine and community-based approaches are limited or poorly coordinated, impacting program sustainability.

FUTURE DIRECTIONS

Integrating Herbal Medicine into Approaches to Antimicrobial Resistance (AMR) and the One Health framework is multifaceted, rooted in the unique ability of plant-derived compounds to circumvent bacterial resistance mechanisms.

- Bitter gourd contains charantin and momordicin, linked with hypoglycemic effects.
- Fenugreek trigonelline modulates insulin sensitivity.

- Cinnamon (cinnamaldehyde, eugenol) enhances insulin receptor activity.
- Garlic's allicin ameliorates insulin resistance.

Herbal phytochemicals restore the efficacy of conventional antibiotics, and their alignment with sustainable, eco-friendly practices (Patwardhan & Aswar, 2025) (Li & Shahid, 2025). Emerging pharmacological studies validate many traditional medicines (Ramya & Ananthakumar, 2025)

Table 9. Herbal medicines against priority resistant pathogens (Walker et al., 2025) (Vladkova et al., 2024) (Patwardhan & Aswar, 2025) (Li & Shahid, 2025)

Pathogen	Effective Herbal Compounds	Example Herbs/ Essential Oils	Notes
Klebsiella pneumoniae, E. coli	Berberine, Kaempferol, Quercetin	Berberis spp., Camellia sinensis	Target carbapenem and colistin-resistant strains
Acinetobacter baumannii, Pseudomonas aeruginosa	Cinnamaldehyde, Carvacrol, Eugenol	Cinnamon, Thyme essential oils	Attacks resistant non-fermenting bacteria
MRSA (Methicillin-resistant S. aureus)	Apigenin, Flavonoids, Terpenoids	Chamomile, Thyme	Alternative for daptomycin and linezolid resistance
Enterococcus species	Various polyphenols and phenolic acids	Burdock, Eucalyptus	Active against vancomycin-resistant strains
Salmonella species	Polyacetylenes, Cinnamaldehyde	Portulaca, Cinnamon	Helpful against azithromycin non-susceptible strains

Bacteriophage and Bacteriocin Combinations

A cocktail of eight phages against avian-pathogenic *E. coli* is used to ensure broader host-range efficacy. *Phages and Cold Atmospheric Plasma* treatment for *Pseudomonas aeruginosa* biofilms involves exposure to cold atmospheric plasma followed by phages, was found to be the most effective. *Bacteriocin-Antibiotic Formulations* comprising two bacteriocins (garvicin KS and micrococcin P1) and penicillin G act synergistically to kill MRSA in skin infections and prevent the development of resistance better than Fucidin cream (Patwardhan & Aswar, 2025) (Li & Shahid, 2025). Academic research centers are increasingly endorsing integrative medicine, supported by phytochemical and ethnopharmacological evidence. There is potential to use the plant "chassis" (Nicotiana benthamiana) to produce clinical-grade antimicrobial peptides at a large scale via transient expression, which could offer a cost-effective solution for mass production.

Table 10. Medicinal plants offer novel nanotechnology targets against Multidrug-Resistant (MDR) pathogens, by efflux pump inhibition, biofilm disruption, synergistic potentiation of existing antibiotics

Mechanism	Description	Herbal Examples
Cell membrane disruption	Causes leakage or breaks membrane integrity	Berberine, Cinnamaldehyde, Eugenol, Carvacrol
Inhibition of nucleic acid synthesis	Blocks DNA topoisomerase or RNA synthesis	Berberine, Sanguinarine
Enzyme inhibition	Disrupts bacterial enzymes essential for metabolism or replication	Flavonoids like Quercetin and Kaempferol
Efflux pump inhibition	Blocks bacterial drug efflux pumps, increasing intracellular antibiotic levels	Flavonoids, Phenolic acids
Anti-biofilm activity	Prevents biofilm formation or disrupts existing biofilms	Chamomile, Cistus species

Nanotechnology and Adjunct Delivery Systems

Nanoparticles and Farnesol composite of silver nanoparticles causes changes in biofilm structure, altering biofilm formation in pathogens like E. coli and Candida albicans. Unlike conventional antibiotics, which typically target a single bacterial pathway (making it easier for bacteria to evolve resistance), medicinal plants offer a complex "cocktail" of bioactive chemicals. Plant extracts attack bacteria through multiple pathways, including membrane disruption, inhibition of DNA replication, and interference with intermediary metabolism (Ananthakumar, Nanjundeshwara, et al., 2025). Biofilms are protective matrices that render bacteria resistant to antibiotics. Plant compounds such as phenols, tannins, and terpenoids can disrupt quorum sensing (bacterial communication) and dismantle existing biofilms, a capability often lacking in synthetic drugs. For example, (+)-nootkatone reduces *Staphylococcus aureus* biofilm thickness and downregulates biofilm-associated genes. Curcumin-Cyclodextrin in Niosomes is tested on cancer membrane-targeted therapy. Curcumin loaded into bacterial cellulose-based hydrogels (using cyclodextrin complexes) or niosomes is used for wound dressings and to target MRSA biofilms. Commiphora molmol (myrrh) and Azadirachta indica (neem) extracts showed synergy with amoxicillin, metronidazole, tetracycline, and azithromycin against periodontopathic bacteria. Antibiotic nanocarriers, Liposome-based nanoparticles, can restore the potency of ceftazidime, imipenem, and cefepime against MDR P. aeruginosa, and of chloramphenicol against MRSA. *Melastoma malabathricum* extract combined with amoxicillin against pathogenic bacteria. Epigallocatechin gallate (EGCG) from green tea restores susceptibility to beta-lactam drugs in Klebsiella pneumoniae and has the potential to decrease beta-lactamase activity via membrane-targeted therapies (Korrapati et al., 2026).

Valorization of Waste and Environmental Sustainability

Agricultural By-products such as onion peels and tomato processing waste, are rich in bioactive compounds. The quercetin and isochlorogenic acid extracts often work via membrane disruption, enzyme inhibition, or efflux pump blockade, offering adjunct potential to reduce resistance pressure. Clinical use requires standardized dosing and further trials due to variability in potency. Native flora leverages plants that have developed resilient bioactive compounds to survive extreme climatic conditions. Sustainable harvesting and utilization of these plants support local ecosystems while providing health solutions. The gap between clinical trial data and official legitimacy is underrepresented in the formal healthcare system (Subedi, 2024). The development of antimicrobial packaging using plant-derived nanoparticles (onion waste) helps prevent contamination in the food industry, further securing the food chain.

Metabolomics involves the comprehensive profiling of small-molecule metabolites in biological samples (P. et al., 2025). Its applications in herbal medicine research show:

- Thymol and EDTA: Thymol combined with EDTA (ethylenediaminetetraacetic acid) shows synergistic antibacterial efficacy against resistant strains.
- Fenugreek and Probiotics: Fenugreek extract combined with the probiotic Bifidobacterium breve was evaluated for in vivo antibacterial efficacy against H. pylori colonization.
- Honey and Propolis: Propolis combined with honey demonstrates a synergistic effect against Escherichia coli and S. aureus.

Metabolomics enables the identification and quantification of antimicrobial compounds in complex herbal extracts using techniques like NMR, LC-MS, and GC-MS . By linking metabolite profiles with antimicrobial activity data, researchers can pinpoint candidate compounds responsible for the observed effects. Metabolomics-based fingerprinting ensures batch-to-batch consistency and helps prevent adulteration. Metabolomics can reveal how plant-derived compounds affect bacterial metabolism and resistance pathways.

GC-MS analysis of endophytic fungi from *Polygonatum sibiricum* identified fatty acid esters and alcohols as major contributors to antibacterial activity against multidrug-resistant pathogens. The use of predictive computational tools and artificial intelligence (AI) is suggested to design and synthesize highly efficient lead drug candidates (Ananthakumar, Nagaraja, et al., 2025). There is potential to use plant systems (Nicotiana benthamiana) as a "chassis" for the large-scale, cost-effective production of clinical-grade antimicrobial peptides.

GREEN SYNTHESIS

Plant-based biological macromolecules and metal-based macromolecules can serve as alternatives or adjuncts to conventional antibiotics, potentially reducing selective pressure for resistance. Nanotechnology helps microbiome-friendly therapeutics with a safety profile of phytocompounds. Nanocarriers promote beneficial host gut microbiota, effectively achieving pathogen eradication and host homeostasis restoration, as an advantage over conventional broad-spectrum antibiotics. Green synthesis of onion peel waste is utilized in the "green synthesis" of metal nanoparticles (silver and gold). These nanoparticles serve as eco-friendly antibacterial agents for wound dressings and antimicrobial coatings, bridging the gap between waste management and advanced biomedical applications.

- MRSA: Extracts from *Caralluma quadrangular* and *Melaleuca alternifolia* (Tea Tree) have been shown to eradicate MRSA biofilms and reduce bacterial loads in skin infections.
- MDR E. coli: Compounds like berberine and extracts from *Rosmarinus officinalis* (Rosemary) are effective against resistant E. coli strains often found in contaminated food.
- Extracts from Golden Chamomile (*Matricaria aurea*) and Miswak (*Salvadora persica*) actively inhibit biofilm development and reduce bacterial adhesion, rendering pathogens like *Pseudomonas aeruginosa, Streptococcus mutans* more vulnerable.
- Compounds like Allicin and Thymol disrupt bacterial cell membranes and interfere with RNA polymerase

There is a need for support to complementary ancient biomedicine through budget allocations, health facilities, human resources, educational institutions, and research output. The research proposes measures to promote a more inclusive and democratic approach to medical pluralism. One Health policies should aim to address the current inequalities and enhance the representation of traditional medical practices. rigorous standards for the chemical characterization, extraction, and quantification of bioactive compounds to ensure consistency and reproducibility across different formulations. Collaboration among pharmacologists, phytochemists, and clinicians is necessary to develop evidence-based, standardized protocols for safely and effectively integrating these therapies.

CLINICAL TRIALS ON HERBAL MEDICINE

Clinical trials can be broadly categorized into three primary domains: Dietary Patterns (whole-food plant-based diets vs. animal-based diets), Phytochemical Isolates (specific compounds such as plant sterols, soy protein, and fiber), and Herbal Medicines (traditional botanical extracts such as Ginkgo biloba, Propolis, and Cannabis). The therapeutic focus is heavily weighted toward cardiometabolic health, though significant research is also evident in neurology, oncology, supportive care, and oral health.

Synergy and Restoration of Antibiotic Efficacy

Herbal medicines are justified not only as alternatives but as potent adjuvants with antibiotics. Efflux Pump Inhibition by herbal compounds like berberine and carvacrol can inhibit efflux pumps, allowing intracellular accumulation of antibacterial agents. The combination of plant extracts with antibiotics can restore the susceptibility of MDR pathogens (Patwardhan & Aswar, 2025) (Li & Shahid, 2025). For instance, Epigallocatechin gallate (EGCG) from green tea restores the susceptibility of Klebsiella pneumoniae. Similarly, berberine reverses ciprofloxacin resistance in MRSA by blocking the NorA efflux pump. Integrating plant-derived antimicrobials reduces the dose of toxic synthetic antibiotics, thereby mitigating adverse side effects and slowing the emergence of further resistance. Microbiome analysis investigates the collective genomes of microorganisms in a specific environment, such as the human gut. In herbal medicine and AMR research (Lakshmaiah & Raghavan, 2026) (Suresh et al., 2024). Research must investigate the effects of extended exposure to medicinal plant extracts on microbial communities to understand how bacteria might develop resistance to these complex combinations. By shifting the microbiome towards a healthier state, herbal interventions may reduce the abundance and transfer of antibiotic resistance genes (ARGs). Biotransformation by Gut microbes metabolizes herbal compounds, sometimes increasing their bioactivity or reducing toxicity (Moiketsi et al., 2023).

Table 11. Clinical trial data on phytotherapy

Herbal compound extract	Clinical condition tested	Intervention	Antimicrobial / clinical Application
Curcumin (local gel / microparticle)	Periodontitis; Chronic kidney disease (inflammatory context)	DRUG: Local delivery of curcumin gel; DRUG: Micro-particle curcumin	Broad-spectrum antibacterial and anti-biofilm activity; inhibits Porphyromonas gingivalis and inflammatory mediators
Tea tree oil (*Melaleuca alternifolia)*	Acne; Blepharitis; Periodontitis	OTHER: Tea tree oil; DRUG: Tea tree oil gel	Potent topical antimicrobial against Cutibacterium acnes, Staphylococcus spp., and ocular pathogens
Thyme (*Zataria multiflora* / thyme oil hydrosol)	Non-alcoholic fatty liver disease (microbiome-linked); Periodontal disease	DIETARY_SUPPLEMENT: Thyme; DRUG: Thyme oil hydrosol	Rich in thymol/carvacrol; disrupts bacterial membranes and oral biofilms
Aloe vera (gel)	Periapical diseases; Oral ulcers; Hyperoxaluria (adjunct)	BIOLOGICAL: Aloe vera gel; DIETARY_ SUPPLEMENT: Aloe vera with cactus	Antimicrobial and wound-healing effects; inhibits oral pathogens and supports mucosal repair
Propolis (nano-propolis)	Traumatic pulp exposure in children	COMBINATION_ PRODUCT: MTA cement + nano-propolis	Strong antibacterial activity against endodontic pathogens; promotes pulp healing
Pomegranate extract	Chronic kidney disease with inflammatory complications	DIETARY_SUPPLEMENT: Flaxseed oil + pomegranate extract	Polyphenol-rich antimicrobial and anti-inflammatory activity; reduces microbial-driven inflammation
Berberine	Diabetes mellitus with obesity	DIETARY_SUPPLEMENT: Berberine	Antibacterial and gut-microbiome-modulating alkaloid; suppresses pathogenic bacteria
Ginger (Zingiber officinale)	Chemotherapy-induced nausea; NAFLD; dyspepsia	DRUG: Ginger; DIETARY_ SUPPLEMENT: Ginger	Antimicrobial against Helicobacter pylori and oral bacteria; adjunct anti-inflammatory
Garlic (*Allium sativum*)	Hypercholesterolemia (supplement arm)	DIETARY_SUPPLEMENT: Fish oil, cinnamon, garlic	Allicin-mediated broad antimicrobial action (bacterial, fungal)
Ginkgo biloba + *Curcumin*	Alzheimer's disease (neuroinflammation context)	DRUG: Ginkgo extract + curcumin	Indirect antimicrobial relevance via anti-inflammatory and oxidative stress pathways
Boswellia serrata (botanical modulator)	Gulf War illness	DIETARY_SUPPLEMENT: Boswellia serrata	Anti-inflammatory with secondary antimicrobial and immune-modulatory effects
Milk thistle (Silybum marianum)	Obesity with metabolic dysregulation	DIETARY_SUPPLEMENT: Milk thistle + Lactobacillus	Hepatoprotective; indirect antimicrobial relevance via microbiome modulation

Ginkgo Biloba herb is a primary focus for cognitive and neurological conditions. Trials assess its efficacy for mild cognitive impairment, dementia, multiple sclerosis, and tinnitus. It is also tested for vitiligo. Green Tea (*Camellia sinensis*) or Green tea extracts are widely studied for obesity and weight management, diabetes/insulin resistance, and oral health (mouthwashes for gingivitis).

Cannabinoids cluster of trials investigates Cannabis and its derivatives for chronic pain, autism, spasticity, and schizophrenia. Interventions vary from "Cannabis oil" to specific CBD formulations. (Wolff et al., 2026) Propolis and Honey, or Bee products with botanical origins, are frequently tested for oral and wound care. Applications include oral mucositis, dentin hypersensitivity, and general wound healing. Flaxseed oil, whole flaxseed, or ground powder is tested for breast cancer, metabolic syndrome, and constipation. Curcumin is in trials for periodontal disease (gel application) and ulcerative colitis. Trials target cognition (Ginkgo), anxiety/stress (Aromatherapy, Cannabis), and depression (St. John's Wort, Saffron) are under study for mental health or neurological conditions (Wolff et al., 2026) (Korrapati et al., 2024). Oncology Support trials investigate interventions for cannabis or cancer-treatment side effects, such as flaxseed for breast cancer or propolis for radiation-induced mucositis.

Most antimicrobial herbal trials are topical or localized. Oral health (periodontitis, pulp exposure, ulcers) dominates due to clear microbial endpoints and lower systemic risk. Trials are classified as drug / dietary supplement / other, not FDA Phase I–III; it demonstrates as a regulatory categorization rather than the absence of human testing. Data on nano-formulations (nano-propolis, micro-curcumin) indicate translational efforts to overcome the bioavailability and potency limitations of herbal antimicrobials. Systemic infections are notably absent as herbal antimicrobials are tested mainly as adjuncts or local alternatives, not stand-alone replacements for antibiotics. The scientific narrative must acknowledge that while many compounds show promise in petri dishes, "clinical evidence remains insufficient" for the vast majority. A robust analysis should not merely list bioactive compounds; it must systematically assess their "clinical readiness" using frameworks such as GRADE (Grading of Recommendations Assessment, Development and Evaluation). As noted in the feedback, with the exception of artemisinin which has established Phase III data, most herbal antimicrobials remain in preclinical stages and require Phase II trials before they can be responsibly recommended for clinical use. Future scholarship must prioritize the discussion of "drug development timelines," highlighting the rigorous path from preclinical isolation to regulatory approval.

Securing the Food Chain and Agriculture

One Health promotes collaborative efforts across disciplines for sustainable outcomes in zoonoses control, food safety, WASH, and AMR stewardship. Key elements include multisectoral surveillance, veterinary public health, and integrated vector management (Bhuvana et al., 2025). Networks like ADAPT in Africa exemplify locally led capacity building for NTDs and AMR. Herbal medicines play a vital role in interrupting the transmission of resistant bacteria through the food supply chain and agricultural systems. Medicinal plants are highly effective against common foodborne pathogens that often harbor resistance, such as *Salmonella, Listeria monocytogenes, E. coli O157:H7*. Lemongrass (*Cymbopogon schoenanthus*), for instance, exhibits activity against these pathogens, thereby aiding food safety preservation.

There is a growing recognition of the effectiveness and sustainability of traditional medicine within veterinary medicine. Beyond human use, plant extracts like Henna (Lawsonia inermis) are effective against plant pathogens (olive knot disease), and reduce the selection pressure for resistance in the environment.

It found that a majority of respondents expressed a strong interest in integrating traditional and modern medical practices (Gurmu et al., 2025). The common use of plants across different cultures for similar ailments suggests a shared biological effectiveness and phytochemical properties. The highest prevalence of use was observed in studies from Africa, Asia and indigenous ethnicity patients covered by both national and private insurance (Hoenders et al., 2024). Herbal remedies for pain management and chronic pain conditions shared knowledge, often passed down through generations, underscore a universal human interaction with the herbal remedies (Raghavan et al., 2025).

Figure 4. Challenges and upcoming solutions in evidence-based phytochemical interventions

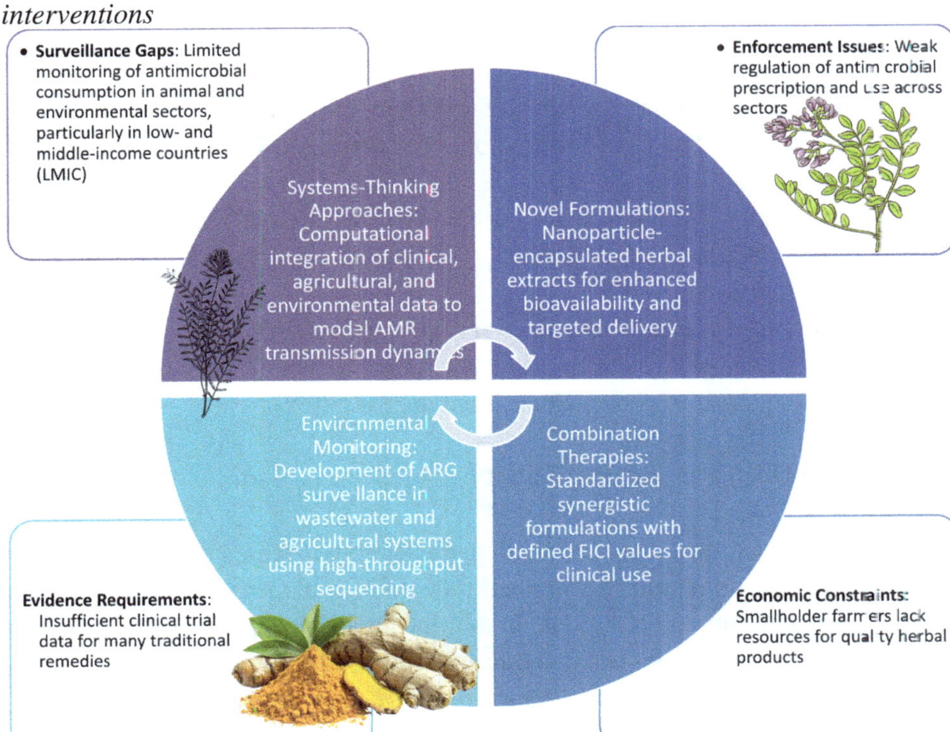

Alignment with One Health and Environmental Sustainability

The "One Health" approach recognizes the interconnection between human, animal, and environmental health. The biogenic synthesis of metal nanoparticles (silver, zinc oxide) using plant extracts is a non-toxic, eco-friendly, and cost-effective alternative to chemical synthesis. These "green" nanoparticles exhibit high antimicrobial activity against multidrug-resistant bacteria and are biocompatible. AMR spreads through animal-derived foods due to antibiotic use in livestock. Herbal immunostimulants (Echinacea, probiotics) are increasingly used as feed additives to manage infections like Salmonella in animals without leaving chemical residues, thus reducing the transmission of resistant genes to humans. The plant-derived fatty acids (Cuphea oil) demonstrate selective antibacterial activity on *Clostridium perfringens* while sparing beneficial gut commensals, supporting microbiome health in a way broad-spectrum antibiotics cannot. Herbal medicine supports this framework by offering sustainable and biodegradable alternatives to synthetic chemicals that pollute ecosystems.

Despite its popularity, safety and contamination of herbal products, and regulatory challenges, which must be addressed for effective integration into healthcare systems (Hoenders et al., 2024).

- Reference Genomes: Many medicinal plants lack high-quality reference genomes, limiting gene discovery and functional annotation.
- Functional Validation: Omics studies often identify candidate genes and compounds, but functional validation in relevant AMR models is frequently lacking.
- Environmental and Phenotypic Variability: The phytochemical composition of medicinal plants is influenced by environmental factors, complicating reproducibility.
- Translational and Regulatory Barriers: Moving from omics discovery to clinical or agricultural application involves regulatory hurdles and the need for standardized quality control (Lakshmaiah et al., 2025).

Addressing Global Health Equity

In developing nations, where the burden of infectious diseases is highest, approximately 85% of the population already relies on traditional medicine. Standardizing these accessible resources offers a viable solution for low-income regions struggling with the high cost of second- or third-line antibiotics. The users of Complementary and Alternative Medicine (CAM), *Prunus africana* for prostate cancer at Korle-Bu Teaching Hospital in Ghana expressed positive views on the legalization. The regulation of CAM treatments through the integration of herbal remedies into the National Health Insurance Scheme is vital. Formal regulation and insurance coverage for CAM therapies are essential for promoting safe and integrated care (Osei et al., 2026). Natural products generally exhibit lower toxicity and fewer side effects compared to synthetic drugs. For example, honey and propolis are effective wound treatments that promote healing, reduce scarring, and prevent infection, offering economic benefits by reducing hospitalization times. Indigenous knowledge regarding 174 medicinal plant species is demonstratedin Namutumba District, Eastern Uganda. Local knowledge of herbal medicine use in traditional healthcare is the most cited factor for the selection species. The decline of medicinal plants, attributed to climate change, poses a significant threat to traditional healthcare systems. The most commonly used plant parts are sourced from bushlands, markets, and wetlands, with Kigelia Africana. Conservation strategies for medicinal plants in the region are lacking, and four species documented in the study are Red-listed, indicating they are vulnerable to extinction (Tugume et al., 2026). Interviews with knowledge holders, practitioners, and key informants reveal the absence of frameworks for access and benefit sharing. The categories of care preferences were professionals living with others and professionals living in the environment. The care preferences are diverse, and individuals with dementia and migration histories are underrepresented in healthcare research (Peters-Nehrenheim et al., 2026). The mistrust, and a lack of understanding regarding the protection of traditional medicine intellectual property should be eradicated. For genuine integration, both allopathic and traditional practitioners need cross-training. Currently, institutional frameworks for joint training are either emerging using nanotechnology and bioengineering (M et al., 2026). Knowledge holders and practitioners need investments for clinical trials or in registering their traditional medicine products and practices. Ethical deployment of ethnopharmacology can help to protect traditional knowledge rights. The WHO has called on member states to advance this integration, and the review recommends establishing a global matrix of academic consortia to strengthen the WHO strategy (Lakshmaiah et al., 2025). Our work advocates for the need for a global redefinition of health that incorporates wholeness and interconnection, drawing from traditional, complementary, and integrative medicine, indigenous knowledge systems, in conventional biomedical models. A few initiatives, such as the National Academy of Medicine Report and the establishment of the WHO Global Traditional Medicine Centre, aim to integrate herbal medicine into national health systems (Soriano, 2025).

CONCLUSION

Antimicrobial resistance affects over 537 million adults globally and is a leading cause of morbidity and mortality. Modern pharmacotherapies and innovations in precision medicine offer promise, yet disparities in outcomes and access persist, rooted in deep social determinants, resource constraints, and implementation gaps. As the world confronts these challenges, there is a renewed interest in traditional medicine systems (Ramya & Ananthakumar, 2025). Ayurveda, Traditional Chinese Medicine, Unani, Indigenous, and African ethnomedicine have developed robust, community-based approaches to metabolic health over millennia. Critically, these systems display remarkable convergence in remedies, principles, and their relationship to the environment, making them key assets for advancing the "One Health" paradigm. Present study highlights the inadequate implementation and enforcement of legal frameworks in ethnopharmacology. Future updates in the protection of rights for holders of traditional medicine knowledge and will avoid the exploitation of traditional medicine resources. We emphasize the importance of societal-level changes, including equitable collaboration and visionary leadership, to facilitate the implementation of new health concepts centered on wholeness and planetary health. The findings suggest that individuals are motivated to use herbal medicines based on their perception of its benefits, regardless of scientific evidence. Future relies on highlighting the importance of understanding benefits for advancing herbal bioactive's research. It calls for the development of innovative financial models to support the sustainable translation of these emergent health concepts into healthcare delivery structures that benefit the common good (Najibi et al., 2025). Multiomics innovations between traditional medicine knowledge holders and the research community, along with regulatory bodies, will progress of traditional medicine research and development. The research in professional engagement and training opportunities, will further One Health research (Gray et al., 2025)

REFERENCES

Abdu, A., & Gudesho, G. (2026). Medicinal plants used to treat livestock diseases in Bure district, illubabor zone, Oromia Regional State, Ethiopia. *BMC Veterinary Research*, *22*(1), 6. DOI: 10.1186/s12917-025-05157-1 PMID: 41491953

Alem, K., Dagnew, M., Gizachew, M., Gelaw, B., & Moges, F. (2025). Environmental Antimicrobial Resistance: Key Drivers, Hotspots, Innovative Strategies, and Challenges in the Fight Against Superbugs. *MicrobiologyOpen*, *14*(5), e70067 DOI: 10.1002/mbo3.70067 PMID: 41085275

Almasri, R. S., Bedir, A. S., & Al Raish, S. M. (2025). Comprehensive Ethnopharmacological Analysis of Medicinal Plants in the UAE: Lawsonia inermis, Nigella sativa, Ziziphus spina-christi, Allium cepa, Allium sativum, Cymbopogon schoenanthus, Matricaria aurea, Phoenix dactylifera, Portulaca oleracea, Reichardia tingitana, Salvadora persica, Solanum lycopersicum, Trigonella foenum-graecum, Withania somnifera, and Ziziphus lotus. *Nutrients*, *17*(3), 411. DOI: 10.3390/nu17030411 PMID: 39940269

Ananthakumar, B., Nagaraja, B. G., Umesh, S. M., & Raghavan, R. (2025). Predictive Medicine. *Predictive Algorithms for Rehabilitation and Assistive Systems*, 291–314. DOI: 10.4018/979-8-3373-0194-5.ch011

Ananthakumar, B., Nanjundeshwara, N., Lakshmaiah, A., & Raghavan, R. (2025). Nanocarriers for Delivering Nucleic Acids in Plants. *Nanocarriers in Plant Science and Agriculture*, 207–234. DOI: 10.4018/979-8-3693-9869-2.ch008

Anyiam, A. F., Muhibi, M. A., Arinze-Anyiam, O. C., & Obeagu, E. I. (2025). Enhancing the bioavailability of Phyllanthus amarus: Implications for traditional and modern medicine in Africa. *Annals of Medicine and Surgery (London)*, *87*(9), 5631–5636. DOI: 10.1097/MS9.0000000000003514 PMID: 40901185

Ashraf, M. V., Pant, S., Khan, M. A. H., Shah, A. A., Siddiqui, S., Jeridi, M., Alhamdi, H. W. S., & Ahmad, S. (2023). Phytochemicals as Antimicrobials: Prospecting Himalayan Medicinal Plants as Source of Alternate Medicine to Combat Antimicrobial Resistance. *Pharmaceuticals (Basel, Switzerland)*, *16*(6), 881. DOI: 10.3390/ph16060881 PMID: 37375828

Bhuvana, G. N., Umesh, S. M., & Raghavan, R. (2025). Expanding the frontiers of Schistosomiasis Treatment: The promise of Nanoparticle-Based Therapies. *Journal of Vector Borne Diseases*, *63*(1), 16–26. Advance online publication. DOI: 10.4103/jvbd.jvbd_180_23 PMID: 40652683

Dessie, Y., Abiyu, E., Wondie, M. A., Awoke, A., & Alehegn, M. (2026). Phytochemical composition, in vitro antioxidant and antibacterial activities of Dorstenia barnimiana Schweinf leaf extracts. *BMC Complementary Medicine and Therapies*, *26*(1), 31. DOI: 10.1186/s12906-026-05250-4 PMID: 41535880

Dube, M., Nyoni, S., Konono, K. C. C., & Chikwambi, Z. (2025). An Ethnobotanical Survey of Medicinal Plants Used to Manage Diabetes in Bulawayo, Zimbabwe. *Journal of Herbal Medicine*, *51*, 101018. DOI: 10.1016/j.hermed.2025.101018

El-Saadony, M. T., Saad, A. M., Mohammed, D. M., Korma, S. A., Alshahrani, M. Y., Ahmed, A. E., Ibrahim, E. H., Salem, H. M., Alkafaas, S. S., Saif, A. M., Elkafas, S. S., Fahmy, M. A., Abd El-Mageed, T. A., Abady, M. M., Assal, H. Y., El-Tarabily, M. K., Mathew, B. T., AbuQamar, S. F., El-Tarabily, K. A., & Ibrahim, S. A. (2025). Medicinal plants: Bioactive compounds, biological activities, combating multidrug-resistant microorganisms, and human health benefits - a comprehensive review. *Frontiers in Immunology*, *16*, 1491777. Advance online publication. DOI: 10.3389/fimmu.2025.1491777 PMID: 40375989

Farey, K. (2025). New Light on Ayurvedic Biology: Protein Bioconformatics Applied to the Integrated Understanding of Ayurveda and Modern Scientific Medicine. *Preprints.Org*, 202505.1446.v1. DOI: 10.20944/preprints202505.1446.v1

Febriyanti, R. M., Saefullah, K., Susanti, R. D., & Lestari, K. (2024). Knowledge, attitude, and utilization of traditional medicine within the plural medical system in West Java, Indonesia. *BMC Complementary Medicine and Therapies*, *24*(1), 1–13. DOI: 10.1186/s12906-024-04368-7 PMID: 38287364

Gray, A. C., Luketic, C. D., Coleman, J., & Roberts, B. (2025). Demographics, practice characteristics, professional satisfaction and professional association engagement of homeopaths in New Zealand: Results from a cross-sectional workforce survey. *European Journal of Integrative Medicine*, *78*, 102500. DOI: 10.1016/j.eujim.2025.102500

Gurmu, A. E., Bitew, B. E., Kifle, Z. D., & Teni, F. S. (2025). Perceptions, practices, and predictors of traditional, complementary, and integrative medicines: A survey of healthcare professionals in Gondar City, northwestern Ethiopia. *European Journal of Integrative Medicine*, *77*, 102488. DOI: 10.1016/j.eujim.2025.102488

Hoenders, R., Ghelman, R., Portella, C., Simmons, S., Locke, A., Cramer, H., Gallego-Perez, D., & Jong, M. (2024). A review of the WHO strategy on traditional, complementary, and integrative medicine from the perspective of academic consortia for integrative medicine and health. *Frontiers in Medicine*, *11*, 1395698. Advance online publication. DOI: 10.3389/fmed.2024.1395698 PMID: 38933107

Hu, W., Zong, J., Gong, J., Xue, R., Lv, Y., Liu, B., Dai, W., Wang, Y., & Chu, C. (2025). Lipase immobilization via dynamic defect-engineered ZIF-8: A bioactivity-enhanced platform for high-efficiency screening of herbal lipase inhibitors. *International Journal of Biological Macromolecules, 335*(Pt 2), 149264. DOI: 10.1016/j.ijbiomac.2025.149264 PMID: 41319765

Jha, S. K., Singh, N., Shanker, O. R., Antil, I., Baghel, J. S., Huddar, V., & Tripathi, R. (2025). A review on integrative approaches in oncology: Bridging ayurvedic medicine and modern cancer therapeutics. *Frontiers in Natural Products, 4*, 1635197. DOI: 10.3389/fntpr.2025.1635197

Korrapati, C., Chowdareddy, S., Suresh, C., & Raghavan, R. (2026). Cancer Cell Membrane-Coated Nanoparticles: Therapeutic Advancements and Hybrid Systems. *Applications of Nanomaterials for Oncology*, 131–160. DOI: 10.4018/979-8-3373-3765-4.ch005

Korrapati, C., Yaraladdi, A. B., Vaddi, H., & Raghavan, R. (2024). Applications of the Saffron European Kitchen Spice Crocus sativus. *Therapeutic and Pharmacological Applications of Ethnobotany*, 98–127. DOI: 10.4018/979-8-3693-1986-4.ch005

Lakshmaiah, A., Korrapati, C., Suresh, C., & Raghavan, R. (2025). Synthetic Biology and AI. *Spatially Variable Genes in Cancer*, 265–290. DOI: 10.4018/979-8-3693-7728-4.ch010

Lakshmaiah, A., & Raghavan, R. (2026). Cardiovascular Protector Activity of Medicinal Mushrooms. *Medicinal Mushrooms: Conservation and Bioprospecting*, 209–218. DOI: 10.1007/978-1-0716-4993-0_28

Li, M., & Shahid, R. (2025). *Antimicrobial Resistance and Modern Therapeutics Approaches*. DOI: 10.3389/978-2-8325-6077-8

M, M. K., P, D. A., J, S. P., Moger, S. S., & Raghavan, R. (2026). Spatial Multi-Omics in Cancers: Integrating Molecular Landscapes for Precision Oncology. *Spatially Resolved Single-Cell and Multiple Omics in Human Diseases*, 217–248. DOI: 10.4018/979-8-3373-4849-0.ch007

Mapunda, P. E., & Mramba, R. P. (2025). Medicinal plant use and conservation in Tanzania: Perspectives from university of Dodoma students. *Discover Social Science and Health, 5*(1), 77. DOI: 10.1007/s44155-025-00232-2

Moiketsi, B. N., Makale, K. P. P., Rantong, G., Rahube, T. O., & Makhzoum, A. (2023). Potential of Selected African Medicinal Plants as Alternative Therapeutics against Multi-Drug-Resistant Bacteria. *Biomedicines, 11*(10), 2605. DOI: 10.3390/biomedicines11102605 PMID: 37892979

Momin, S., Jadhav, M., & Gurav, R. (2026). Exploring phytochemistry, antioxidant potential, essential oil profiling and bioactive profiling of Pogostemon mollis Benth. Through GC–MS and UPLC-QTOF-MS/MS. *Scientific Reports*, *16*(1), 6277. Advance online publication. DOI: 10.1038/s41598-026-35989-2 PMID: 41593157

Najibi, S. M., Sarikhani, Y., Hajimonfarednejad, M., Nimrouzi, M., & Hashempur, M. H. (2025). A scoping review of the barriers and facilitators in the use of traditional, complementary, and integrative medicine: Insights for health policy development. *Journal of Health, Population and Nutrition*, *44*(1), 1–12. DOI: 10.1186/s41043-025-00934-y PMID: 40474255

Osei, B., Laari, L., & Kyei, J. M. (2026). Perspectives on the formalisation of complementary and alternative medicine among current users with prostate cancer: A study at Korle-Bu Teaching Hospital, Ghana. *BMC Complementary Medicine and Therapies*. Advance online publication. DOI: 10.1186/s12906-026-05273-x PMID: 41593602

P., A., D., K. M., M., J., P., S., & Raghavan, R. (2025). The Cancer Exposome. *Cancer Exposomics and Environmental Influences on Carcinogenesis*, 161–192. DOI: 10.4018/979-8-3373-2165-3.ch004

Pallavi, V. N., Chowdareddy, S., Byalahunasishetter, B. I., & Raghavan, R. (2026). CRISPR-Cas9 Technology for Precision Cancer Medicine: Target Discovery, Disease Modeling, and Therapeutics. *CRISPR-Cas9 Technology for Precision Medicine*, 35–72. DOI: 10.4018/979-8-3373-4857-5.ch002

Patwardhan, B., & Aswar, U. (2025). Harnessing the potential of ethnopharmacology for future medicines. *Journal of Ethnopharmacology, 353*(Pt A), 120359. DOI: 10.1016/j.jep.2025.120359

Peters-Nehrenheim, V., Rommerskirch-Manietta, M., Tezcan-Güntekin, H., & Roes, M. (2026). Care preferences of older migrants and minority ethnic groups with various care needs: A scoping review. *PLoS One*, *21*(1), e0341147. DOI: 10.1371/journal.pone.0341147 PMID: 41576142

Raghavan, R., Moger, S. S., Umesh, S., & Bhuvana, G. N. (2025). Revolutionizing Pain Management. *Unveiling Technological Advancements and Interdisciplinary Solutions for Pain Care*, 129–164. DOI: 10.4018/979-8-3693-9501-1.ch005

Ramya, R., & Ananthakumar, B. (2025). Prakriti-Metabolomic Correlations for Precision Medicine and Type 2 Diabetes (T2DM) Prevention. *Traditional Medicine and Modern Medicine*. (world). DOI: 10.1142/S2575900025500065

Robbiati, C., Andriamandroso, A. M., Auerswald, H., González, M. B., Becerra, N. C., Dente, M. G., Dien, N. T., Garnier, J., Onyango, D., Riley, T., Weiszhar, K. L., Winkler, A. S., & Alders, R.Women for One Health Network. (2025). Diversity, equity and inclusion in One Health could crucially support functional health security by fostering prevention, but a change in mindset is needed. *One Health Outlook*, *7*(1), 57. DOI: 10.1186/s42522-025-00175-3 PMID: 41219802

Siddiqui, M. F. M. F., Waghmare, S. P., Chepte, S. D., Ali, S. S., & Shaikh, S. R. (2026). Evaluation of phyto-therapeutic efficacy of Artemisia annua and Calotropis procera against bovine tropical theileriosis. *Tropical Animal Health and Production*, *58*(2), 57. DOI: 10.1007/s11250-026-04867-4 PMID: 41557174

Singh, S., Sharma, P., Pal, N., Sarma, D. K., Tiwari, R., & Kumar, M. (2024). Holistic One Health Surveillance Framework: Synergizing Environmental, Animal, and Human Determinants for Enhanced Infectious Disease Management. *ACS Infectious Diseases*, *10*(3), 808–826. DOI: 10.1021/acsinfecdis.3c00625 PMID: 38415654

Soriano, S. (2025). Wholeness and Interconnection as the Foundation for a New Epistemology of Health. *Global Advances in Integrative Medicine and Health*, *14*, 27536130251328636. DOI: 10.1177/27536130251328636 PMID: 40177304

Subedi, B. (2024). Advancing inclusive and democratic medical pluralism in Nepal. *Journal of Ayurveda and Integrative Medicine*, *15*(6), 100988. DOI: 10.1016/j.jaim.2024.100988 PMID: 39586157

Suresh, G., Raghavan, R., & Narayanan, A. S. (2024). Microbiome Analysis of Metagenome Using RStudio. *Plant Microbiome Engineering*, 323–340. DOI: 10.1007/978-1-0716-4180-4_39

Teka, A., & Maryo, M. (2023). Ethiopian Medicinal Plants Used for Respiratory Tract Disorders: Ethnomedicinal Review. *Evidence-Based Complementary and Alternative Medicine : eCAM*, *2023*(1), 7612804. DOI: 10.1155/2023/7612804 PMID: 36660273

Tugume, P., Ssenku, J. E., Asiimwe, S., Mustafa, A. S., & Walusansa, A. (2026). Indigenous knowledge on the uses and conservation of medicinal plants in Namutumba District, Eastern Uganda. *BMC Complementary Medicine and Therapies*, *26*(1), 93. Advance online publication. DOI: 10.1186/s12906-025-05235-9 PMID: 41484885

Vladkova, T. G., Smani, Y., Martinov, B. L., & Gospodinova, D. N. (2024). Recent Progress in Terrestrial Biota Derived Antibacterial Agents for Medical Applications. *Molecules (Basel, Switzerland)*, *29*(20), 4889. DOI: 10.3390/molecules29204889 PMID: 39459256

Walker, R. J., Kingpriest, P. T., Gong, J., Naisanga, M., Ashraf, M. N., Roberti, J., & Lang, T. (2025). Global perspectives on infectious diseases at risk of escalation and their drivers. *Scientific Reports*, *15*(1), 38630. DOI: 10.1038/s41598-025-22573-3 PMID: 41188418

Wolff, A., Hübner, J., Büntzel, J., & Büntzel, J. (2026). Psychotropic Medicinal Plant Use in Oncology: A Dual-Cohort Analysis and Its Implications for Anesthesia and Perioperative Care. *Applied Sciences (Basel, Switzerland)*, *16*(1), 481. DOI: 10.3390/app16010481

Yoon, S., & Lee, S. (2025). Genomic Data into Traditional Korean Medicine: A Comprehensive Review. *Innovations in Acupuncture and Medicine*, *18*(1), 13. DOI: 10.1186/s44424-025-00019-x

Chapter 2
Mechanisms and Drivers of Antimicrobial Resistance:
Genetic, Biochemical, Ecological, and Socio-Economic Factors

Zahid Ahmad Thoker
http://orcid.org/0009-0002-6897-5835
Lovely Professional University, India

Lovleen Marwaha
http://orcid.org/0000-0002-5175-2590
Lovely Professional University, India

ABSTRACT

Antimicrobial resistance reduces the effectiveness of antibiotics, is a key global health threat, and affects human, animal, and environmental health. This chapter examines antimicrobial resistance within a One Health framework to identify the genetic, biochemical, ecological, and socio-economic drivers that enable resistant pathogens to emerge and spread. It develops through chromosomal mutations, horizontal gene transfer, and mobile genetic elements, while biochemical mechanisms, including enzymatic degradation, target modification, efflux pumps, and biofilm formation, confer multidrug resistance. Transmission is accelerated by ecological reservoirs, contaminated antibiotics, and socio-economic factors, such as misusing antibiotics, having poor infection control practices, and international travel. Case studies of predictive modeling and targeted interventions programmed through integrated molecular and ecological insight are shown.

DOI: 10.4018/979-8-3373-7810-7.ch002

1. INTRODUCTION

Antimicrobial resistance (AMR) is the capacity of microorganisms (such as bacteria, viruses, fungi, and parasites) to resist the impact of a previously applied drug that can treat an infection caused by them (World Health Organization [WHO], 2019). The effect of this phenomenon is the deterioration of the effectiveness of antimicrobial treatment and the occurrence of treatment failures, protracted diseases, more deaths, and increased medical expenses worldwide. AMR occurs in the natural evolutionary process but is increased by abuse and overuse of antimicrobial agents in human healthcare, animal care, food production, and pollution of the environment (Ahmed, 2024; Centers for Disease Control and Prevention [CDC], 2025). The dissemination of AMR across the world by resistant strains and genes has rendered AMR one of the biggest menaces to contemporary medicine, and the projections have put millions of deaths because of AMR in the next few decades, as shown in Table 1 (Naghavi et al., 2024).

The entire global population is affected by AMR, and it has an uneven distribution between low- and middle-income countries since it is associated with insufficient sanitation, inaccessibility to medical services, and unregulated sales of antimicrobials (WHO, 2025). The infections that are resistant cause prolonged hospital admission and more complicated and costly therapy, and in many cases, toxic side effects. Also, resistance reduces the capacity to carry out routine medical care practices safely, such as surgery and chemotherapy (Ahmed, 2024). The AMR is multidimensional and thus requires a multidisciplinary response that incorporates human, animal, and environmental health sectors.

Table 1. Summary of key facts on AMR prevalence and impact worldwide (WHO, 2025; Naghavi et al., 2024; CDC, 2025).

Statistic/Information	Details
Estimated annual deaths attributable to AMR	Over 1.9 million (projected to rise)
The percentage of bacterial infections resistant	Approximately 16% globally
Regions with the highest AMR rates	South-East Asia, Eastern Mediterranean, Africa
Impact on low- and middle-income countries	Disproportionately higher morbidity and mortality
Economic burden	Increased hospital costs and therapy expenses

Historical Background and the Rise of AMR.

The term antimicrobial resistance was first known in the early 20th century, soon after the initial antimicrobial agents had been introduced. The discovery of penicillin by Alexander Fleming in 1928 revolutionized the management of infectious diseases, but the emergence of resistance was noticed early in 1940 when penicillinase-producing bacteria were identified (Davies, 2010). In a 1945 warning, Fleming himself cautioned against the dangers of misuse and the resultant resistance, and emphasized the danger that could happen if antibiotics were highly accessible without proper regulations (Devi, 2024).

A large number of novel classes of antibiotics were developed during the golden age of antibiotic discovery (1950s-1970s), which temporarily balanced out the resistance problem (Muteeb et al., 2023). Throughout that time, however, the pace of the discovery of new antibiotic classes has significantly slowed down, despite the resistance mechanisms spreading quickly across bacterial populations (Hunt, 2024). The initial resistance mechanisms were enzyme degradation of the antibiotics, alterations of drug targets, and decreased uptake of the drugs, and they continue to change and diversify.

Multidrug-resistant organisms have existed in environmental reservoirs of resistance genes, but human activities that cause production and misuse of antibiotics in medicine and agriculture have increased their evolution and spread (Devi, 2024; Muteeb et al., 2023). It is this historical background that highlights the significance of concerted efforts to deal with AMR using stewardship, surveillance, and innovation.

Significance of the One Health Framework.

AMR is a classic One Health problem, which demonstrates the interdependence of the sphere of human and animal health, plants, and the environment (McEwen, 2018). The One Health approach acknowledges that actions by a single industry (e.g., veterinary or agricultural) to utilize antimicrobials and develop antimicrobial resistance have a direct effect on other industries via the transmission of antimicrobial resistance-carrying pathogens and genes across human, animal, and environmental interfaces (Velazquez-Meza et al., 2022).

To illustrate, livestock antibiotics used to prevent diseases and promote animal growth can make animals resistant in their microbiomes and can be transmitted to humans through direct contact, food webs, or even environmental exposure routes like water contamination (WHO, 2025; McEwen, 2018). Environmental reservoirs, comprising soil and wastewater, are also important for maintaining and distributing resistance determinants, and it is essential to focus on a combination of surveillance and mitigation measures (Global AMR Hub, 2024). The collaboration between policymakers, clinicians, veterinarians, environmental scientists, and public health experts is required to create stewardship programs, regulate the use of antimicrobials, improve the containment of infections, and promote sustainable alternatives research. The interdisciplinary approach to studying and fighting AMR, which is represented by the One Health approach, is more effective in its interventions to improve the effectiveness of the interventions and protect the antimicrobial efficacy of future and current generations, as depicted in Figure 1, as well (Velazquez-Meza et al., 2022).

2. GENETIC FACTORS IN AMR

Chromosomal Mutations Leading to Resistance

Chromosomal mutations are one of the key genetic processes through which bacteria may develop resistance to antimicrobials. These mutations occur spontaneously at rather low rates (between 10-6 and 10-10 per cell division) but confer significant adaptive benefits when there is a selective pressure of antibiotics (Davies and Davies, 2010; Bocharova et al., 2023). The mutations change the DNA in bacteria in certain loci and may bring about resistance in multiple biochemical pathways: alteration of the antibiotic target sites, decrease in drug uptake, increase in efflux, and metabolic bypass (Reygaert, 2018).

As an example, the incorporation of mutations in the genes that encode ribosomal RNA or proteins might make the antibiotics such as aminoglycosides and macrolides unable to bind, which leads to resistance (Cesur, n.d.). The mutations that develop in the topoisomerase enzymes gyrA and parC are the common cause of fluoroquinolone resistance as they are the target proteins of the drugs (Cesur, n.d.). Furthermore, the amplification process of genes may result in the multiplication of the number of copies of genes responsible for resistance, thus making levels of resistance higher than merely point mutations (Silva & Khare, 2024).

The clinical implication of the chromosomal mutation is immense, especially due to the fact that the mutation has the capacity to stabilize and maintain resistance in the bacterial population even in the absence of plasmid transfer. Yet, this can be offset by the cost of fitness due to a set of mutations that can sometimes be deadly to bacteria unless it is countered by subsequent mutations (Nair et al., 2024). Table 2 presents the main examples of the resistance mechanisms through chromosomal mutation.

Table 2. Examples of chromosomal mutations conferring antimicrobial resistance (Cesur, n.d.; Silva & Khare, 2024; Reygaert, 2018)

Resistance Mechanism	Mutated Target or Process	Associated Antibiotic Class
Target site modification	16S or 23S rRNA, ribosomal proteins	Aminoglycosides, Macrclides
DNA gyrase and topoisomerase mutations	gyrA, gyrB, parC genes	Fluoroquinolones
Efflux pump gene amplification	norA, sdrM	Multiple, including fluoroquinolones
Reduced permeability mutations	Porin protein genes	Beta-lactams, Carbapenems

Horizontal Gene Transfer Mechanisms.

HGT enables bacteria to obtain resistance genes in other organisms, which is essential in the rapid spread of the resistance characteristic of antimicrobials across different microbial communities (Sultan et al., 2018). There are three major HGT procedures, namely transformation, transduction, and conjugation.

a) Transformation

The process of transformation entails the acquisition of the free DNA fragments present in the environment into the genome of the bacterial cell, where the resistance genes can be incorporated (Frost et al., 2005). Naturally competent bacteria may absorb exogenous DNA by using recombination, which enables the acquisition of resistance without contact with cells. As an illustration, Streptococcus pneumoniae and Neisseria gonorrhoeae exhibit this mechanism conspicuously in that they pick up penicillin resistance genes on environmental DNA pools (Sultan et al., 2018).

b) Transduction

Bacteriophages transduce bacteria and transfer bacterial DNA, including resistance determinants, by accident during viral replication and infection into new host bacteria (Sultan et al., 2018). This translocation facilitates the movement of genes mostly between closely related species, yet plays a very important role in generalizing resistance factors like the beta-lactamase genes in Staphylococcus aureus (Frost et al., 2005).

c) Conjugation

The process of exchanging DNA (typically plasmids) through direct cell-to-cell contact is called conjugation and makes it the most prolific pathway of HGT for transferring antibiotic resistance genes around the world (Sultan et al., 2018). Multidrug resistance has the capability of spreading at the same time due to the presence of many resistance genes on the plasmids. This process is enhanced through conjugative pili and may happen among different species of bacteria, including pathogens and commensals (Frost et al., 2005).

The frequency and spread of conjugation-based dissemination outline the necessity of interventions to address plasmid mobility to reduce the AMR crisis. Figure 1 below shows the three main pathways of transmission, Transduction, and Conjugation through which bacteria quickly learn and transfer antimicrobial resistance genes to different microbial populations.

Figure 1. Plasmids, Integrons, and Transposons in Resistance Spread

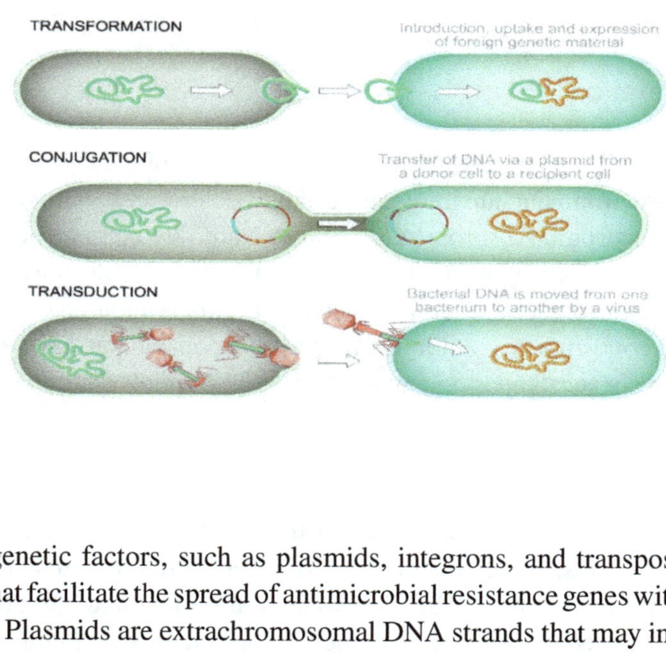

Mobile genetic factors, such as plasmids, integrons, and transposons, are the traffickers that facilitate the spread of antimicrobial resistance genes within bacterial populations. Plasmids are extrachromosomal DNA strands that may independently replicate and commonly carry genes of resistance (Muteeb et al., 2025). Their mating ability increases the processes of horizontal transfer.

Integrons are genetic carriers of gene cassettes, which often contain antibiotic resistance determinants (Sultan et al., 2018). Depending on the sequences of the integrase genes, they are grouped primarily into class 1, 2, and 3 integrons and relate to multidrug resistance in clinical isolates. Transposons mediate resistance genes to move between plasmids and chromosomes through cut-and-paste processes or replicatively, which increases genetic variability and adaptability further (Muteeb et al., 2025). Their interactions maintain resistance gene reservoirs in diverse settings, including in hospitals and wastewater, and agricultural soils; hence, making AMR containment quite challenging (Reygaert, 2018).

Genomic and Metagenomic Understandings of AMR.

Genomic and metagenomic technologies introduced recently have transformed the concept of AMR and enabled the discovery of the complexity and diversity of resistance gene reservoirs in clinical and environmental microbiomes (Bocharova et al., 2023). Whole-genome sequencing can be used to discover mutations, plasmid systems, and resistance determinants, which can be used to conduct epidemiological surveillance of AMR outbreaks and evolution (Muteeb et al., 2025).

Genetic samples obtained without culturing by metagenomics analysis reveal vast resistomes in soil, water, and microbiomes of the gut that are not discovered by traditional methods (Bocharova et al., 2023). Such works educate the effects of anthropogenic interactions on the spread of resistance genes and contribute to the development of specific mitigation approaches in the context of a One Health approach.

3. BIOCHEMICAL AND MOLECULAR MECHANISMS

Antibiotics are targeted by enzymes of the intestinal flora, which break down their chemical structure and render them biologically inactive.

Enzymatic Inactivation of Antibiotics

The intestinal flora enzymes attack the chemical structure of antibiotics and deactivate them biologically. One of the major biochemical processes through which bacteria can avoid the effects of the antimicrobials is the enzymatic inactivation of antibiotics. It is a process in which bacteria produce enzymes that chemically modify or degrade antibiotic molecules, rendering them ineffective (Shaikh et al., 2015). The β-lactamases are among them, and are a diverse and clinically important family of enzymes that cleave the 2-lactam ring of penicillins, cephalosporins, carbapenems, and monobactams, and thus neutralize their antibacterial effect (Eiamphungporn & Bhunya, 2018).

a) Beta-Lactamases and Extended-Spectrum Beta-Lactamases (ESBLs).

The β-lactamases are classified according to their molecular structure and substrate specificity. There are four significant classes (A, B, C, and D), and classes A, C, and D use a serine residue as the active site, and class B comprises metallo-2-lactamases that need zinc ions to catalyze the process (Shaikh et al., 2015). Extensive spectrum β-lactamases (ESBLs) are enzymes that are able to hydrolyze a wide spectrum of β-lactam antibiotics, including third-generation cephalosporins and monobactams; however, these enzymes are typically inhibited by β-lactamase inhibitors, including clavulanic acid (Shaikh et al., 2015).

ESBL-producing bacteria are of significant concern in the world because of their resistance and contribution to hospital- and community-acquired infections, especially in the family of Enterobacteriaceae (Shaikh et al., 2015). Production of β-lactamase is usually plasmid-mediated, and β-lactamase such as New Delhi metallo- 2 3 -lactamase-1 (NDM-1) is associated with horizontal transmission between pathogens (Eiamphungporn and Bhunya, 2018).

The solution to the 2-lactam resistance caused by 2-lactamase is to work with a combination of 2-lactam/2-lactamase inhibitors and create new types of 2-lactamase inhibitors that work with multiple types of 2-lactamases (Eiamphungporn and Bhunya, 2018). Table 3 is a summary of primary β-lactamase groups and their properties.

Table 3. Major classes of β-lactamases and their characteristics (Shaikh et al., 2015; Eiamphungporn & Bhunya, 2018)

β-Lactamase Class	Active Site Feature	Substrate Spectrum	Clinical Importance
Class A	Serine	Penicillins, cephalosporins, carbapenems (some)	Common ESBLs, KPC carbapenemases
Class B	Zinc metalloenzyme	Broad spectrum, including carbapenems	NDM, VIM, IMP metallo-β-lactamases
Class C	Serine (AmpC enzymes)	Cephalosporins, penicillins	Chromosomal and plasmid-mediated
Class D	Serine (OXA-types)	Oxacillinases, variable β-lactam hydrolysis	Carbapenem resistance (OXA-48)

b) Aminoglycoside-Modifying Enzymes.

The variable resistance to the aminoglycoside antibiotics is mainly due to changes in the enzyme that hamper the binding of the antibiotic to the ribosomal targets.

There are three distinct aminoglycoside-modifying enzymes: acetyltransferases (AAC), phosphotransferases (APH), and nucleotidyltransferases (ANT), with distinct chemical changes they make to the antibiotic molecule (Ramirez & Tolmasky, 2010).

These enzymatic changes decrease the binding of aminoglycosides to the bacterial 30S ribosomal subunit, making drugs such as gentamicin, amikacin, and tobramycin inactive (Ramirez and Tolmasky, 2010). The aminoglycoside-modifying enzymes can be easily transmitted between Gram-negative and certain Gram-positive pathogens, and they are frequently plasmid-encoded (Narendrakumar et al., 2023). Some of the strategies to counter this resistance are the prescription of aminoglycosides that are less susceptible to changes, as well as coming up with enzyme inhibitors, but the results have not been successful in clinical practice.

Target Site Modification

Target site modification entails a genetic mutation or enzyme modification of the elements of bacteria that are targeted by antimicrobials, making them less susceptible to drugs. They include the methylation of 23S rRNA to provide resistance to methylisonia (macrolide), mutations in DNA gyrase and topoisomerase IV genes to provide resistance to both fluoroquinolone and penicillin-binding proteins (PBPs), respectively (Reygaert, 2018).

This process typically occurs in the form of a chromosomal mutation or acquisition of mobile genetic elements containing the genes that encode modifying enzymes, and is involved in high-level resistance in a range of pathogens. A good example is the PBP2a of mecA genes in methicillin-resistant Staphylococcus aureus (MRSA), which reduces the 2-lactam affinity (Narendrakumar et al., 2023).

Efflux Pumps and Reduced Permeability

Efflux proteins (membrane proteins) are actively involved in expelling antibiotics in bacterial cells, decreasing the level of drugs in the cell below the therapeutic level (Piddock, 2006). These pumps are categorized into families with the families being ATP-binding cassette (ABC) transporters, major facilitator superfamily (MFS), resistance-nodulation-division (RND), small multidrug resistance (SMR), and multidrug and toxic compound extrusion (MATE) (Poole, 2007).

Efflux also causes multidrug resistance because most pumps have a wide substrate profile, which includes: 8 lactams, tetracyclines, fluoroquinolones, and even macrolides (Poole, 2007). Efflux is complemented by reduced permeability, which is usually achieved by loss or mutation of porin proteins in Gram-negative bacteria, which limits drug entry (Nikaido, 2003). The clinical resistance determinants include overexpression and mutations of the efflux and permeability mechanisms (Poole, 2007).

Mechanisms of Biofilm Formation and Persistence.

Biofilms are organized communities of bacteria entangled in extracellular polymeric substances (EPS) affixed to surfaces, which are a significant method of bacterial persistence and antimicrobial resistance (Hall-Stoodley et al., 2004). In biofilms, the growth rates are low, penetration of drugs is low, and certain resistance genes are expressed by bacteria, hence phenotypic tolerance to antibiotics (Mah & O'Toole, 2001).

A subpopulation of biofilms, the persister cells, is transiently and non-heritably tolerant to drugs, which allows them to be present during the process of antibiotic therapy and helps in the cause of chronic infections and relapses. Infection biofilms are also widely known to be strenuously challenging to eliminate, and in most cases, combination therapy and increased drug levels are necessary (Hall-Stoodley et al., 2004).

4. ECOLOGICAL AND ENVIRONMENTAL DRIVERS

Environmental Reservoirs of AMR.

The environment can also be of importance as a source and a channel of antimicrobial resistance (AMR) genes promoting the maintenance and spread of the resistance out of the clinical and agricultural practices (Berendonk et al., 2015). The natural habitats in which the diverse microbial communities interact and exchange resistance determinants are known as environmental reservoirs and encompass natural habitats like soil, water bodies, and sediments (Martinez, 2015). Knowledge of these reservoirs is also needed to know the entire extent of AMR spread and persistence. The Antimicrobial Resistance Environmental Cycle. Figure 3 also shows a simplified representation of the entry of antibiotics and resistant bacteria into natural environments (both human-originated and natural), which serve as reservoirs of resistance, eventually causing re-exposure.

a) Soil Microbiome and Genes of Resistance.

Soil is a multifaceted biome containing a high microbial diversity, and it is a natural source of a high number of resistance genes. Millions of years ago, antibiotic-producing microorganisms resided in soil, and this has led to the evolution of resistance genes as a mechanism of survival (D'Costa et al., 2011). Nevertheless, human activities like the use of manure, biosolids, and agrochemicals have enhanced the selection pressures, which have made the resistance genes more abundant and diverse in agricultural soils (Heuer et al., 2011).

The occurrence of horizontal gene transfer in soils promotes the dissemination of resistance genes among the environmental bacteria and pathogens that may be potentially (Forsberg et al., 2012). Metagenomic analysis shows that agricultural intensification of soils contains a large number of resistance genes compared to natural or undisturbed soils (Wang et al., 2020). Table 5 presents the factors that affect the soil resistome.

b) **Water Systems and Wastewater Contribution.**

Agricultural runoff, hospital effluents, and wastewater treatment facilities (WWTPs) are the primary sources of contamination of aquatic environments (surface water, groundwater, and sediments). They are major sources of AMR spread (Baquero et al., 2008). The antimicrobials, resistant bacteria, and resistance genes are captured within wastewater systems, and this provides an environment that supports the spread of genetic material and cultivation of the resistant bacteria (Michael et al., 2013).

Research shows that in many instances, WWTPs are not able to fully eliminate resistance genes, and they find their way into natural water systems where they are able to become a part of natural microbiomes (Michael et al., 2013). They are the reservoirs where the dissemination occurs through irrigation, drinking water, and recreational activities (Araújo et al., 2022). There is a high occurrence of resistance genes associated with 2-lactamases, efflux pumps, and other systems in aquatic ecosystems affected by human activities (Michael et al., 2013). Table 6 describes key sources and effects of AMR gene pollution in the water systems.

Agricultural Practices and Antibiotic Pollution.

Agricultural pollution with antibiotics is a major cause of AMR in the environment. Antibiotics are being used in livestock to grow and prevent diseases, and this leads to the selection and excretion of resistant bacteria and antibiotic residues in soils and the water (Van Boeckel et al., 2015). Field application of manure serves as a means of both antibiotics and resistance gene transmission, and in most cases results in the build-up of microbiomes of soil and subsequent horizontal gene transfer events (Heuer et al., 2011).

The antibiotic use in agriculture also has an indirect impact on crop production through either irrigation and application of contaminated organic fertilisers (Martínez,

2009). Selective pressure continues to exist in the environment due to the presence of antibiotics like tetracyclines and sulfonamides even after their use. The regulatory frameworks of different countries differ, and they do not have strict restrictions on non-therapeutic uses of antibiotics, which is why it is necessary to have combined measures of stewardship (Van Boeckel et al., 2017).

Additionally, the soil of agricultural areas with repeated antibiotic exposure not only shows a change in the abundance of the resistance genes but also in the composition of the microbial communities and the very functioning of the ecosystem, which, in turn, could affect the soil fertility and crop yields (Cycoń et al., 2019). The measures that include judicious use of antibiotics, alternative veterinary medicine, and manure management procedures are urgent in order to mitigate environmental AMR pressures.

Wildlife and Zoonotic Transmission Pathways

Wildlife is a reservoir and conduit of AMR genes between natural ecosystems and human and domestic populations of animals. Resistant bacteria are detected in wild species that represent a variety of birds and mammals, as well as aquatic organisms, which indicate environmental pollution and interspecies infection (Vittecoq et al., 2016). The migratory species, especially, are vectors that promote long-range dispersion of the resistant pathogens and genes through geographic boundaries (Ahlstrom et al., 2019).

Zoonotic transmission routes include direct contact with animals, eating contaminated food, or coming into contact with the environment that is shared with wildlife (Allen et al., 2017). Wildlife may obtain human-related sources of resistance genes, including wastewater, agricultural runoff, and landfills, and subsequently, may return resistant strains to humans or domestic animals, which is a complicated epidemiological cycle (Vittecoq et al., 2016).

The ecological interaction and shared habitat between wildlife and human beings require monitoring programs that will include wildlife populations to understand the AMR distribution comprehensively and guide One Health interventions (Allen et al., 2017). Table 4 points out the wildlife species that have been involved in the studies on AMR transmission.

Table 4. Wildlife groups implicated in antimicrobial resistance transmission (Vittecoq et al., 2016; Ahlstrom et al., 2019; Allen et al., 2017)

Wildlife Group	Role in AMR Dynamics	Examples of Resistant Pathogens Detected
Migratory birds	Long-distance dispersal vectors	ESBL-producing Escherichia coli, Salmonella spp.
Mammals (e.g., rodents)	Local reservoirs and transmitters	Methicillin-resistant Staphylococcus aureus (MRSA)
Aquatic organisms	Environmental reservoirs	Multidrug-resistant Vibrio spp., Pseudomonas spp.

5. SOCIO-ECONOMIC AND BEHAVIORAL FACTORS

Antibiotic Misuse and Overuse in Humans.

The misuse and overuse of antibiotics are among the most urgent human practices that contribute to the development and spread of antimicrobial resistance (AMR). The improper use of antibiotics, self-treatment without a doctor's prescription, antibiotic treatment courses, and the management of viral infections with antibiotics are the key causes of resistance development (World Health Organization [WHO], 2023; Ahmed, 2024).

Approximately 1/3 of all antibiotics prescribed to humans are unnecessary or inappropriate, and they are used to treat viral-induced conditions, which antibiotics do not treat (such as colds, influenza, bronchitis, etc.) (Centers for Disease Control and Prevention [CDC], 2025). The pandemic of COVID-19 has been worrying over-prescription of antibiotics, with about 75 percent of hospitalized patients receiving antibiotics even when, in reality, only 8 percent had a confirmed bacterial co-infection, which poses a cause of concern in terms of worsening AMR (WHO, 2024).

Self-medication is widespread, especially in the low- and middle-income countries (LMICs), where antibiotics can be bought without a prescription, which is pushed by low access to healthcare, financial limitations, and the belief in the culture that antibiotics are a fast solution (Nammi et al., 2025). This results in erratic dosing, medication sharing, and discontinuation, encouraging selective pressure in resistant bacteria.

The effects do not just lie in the failed treatment of individuals but also in the high cost of healthcare services, extended hospitalization, and other negative effects on the health systems of the population (Nammi et al., 2025). This complicated issue should be addressed using complex measures, including the reinforcement of stewardship, educating people, establishing more stringent rules on the sale of antibiotics, and enhancing diagnostic abilities.

Veterinary and Livestock Practices that contribute to AMR.

Veterinary and livestock industries also make significant contributions to AMR by the large amount of antibiotics used to prevent diseases, treat diseases, and promote animal growth. In the world, about 70% of the medically significant antibiotics are used in animal farming (Van Boeckel et al., 2015).

The use of antibiotics in livestock facilitates resistance in the microbiomes of the animal, which may be passed on through direct contact, ingestion of contaminated animal food, or indirectly by the environment, through manure runoff, as shown in Table 5 (Tang et al., 2017). This is worsened by the non-veterinary use of antibiotics, particularly in LMICs, where the incorrect prescription of the drugs as well as their dosage can help develop resistance (O'Neill, 2016).

Bans on antibiotic growth promoters and veterinary antibiotic use regulations have been introduced in many countries, but enforcement and implementation are not consistent around the world (Landers et al., 2012). Other interventions like enhanced biosecurity, vaccination, probiotics, and nutritional control are essential so that antibiotic dependency among livestock is minimized (Van Boeckel et al., 2017).

Table 5. Veterinary practices contributing to antimicrobial resistance (Van Boeckel et al., 2015; Tang et al., 2017)

Practice	AMR Contribution
Prophylactic antibiotic use	Sustains resistance selection
Antibiotic growth promoters	Chronic low-dose exposure
Lack of veterinary oversight	Inappropriate antibiotic use
Poor waste management	Environmental contamination

Dissemination of AMR in the Global Trade, Travel, and AMR

Globalization has made the geographical diffusion of resistant bacteria and resistance genes faster through the increase in human movement, international trade, and the importation of food. International travelers may develop resistant bacteria in a foreign country and introduce them into their home country microbial flora, and imported food products could contain resistant pathogens (Tängdén & Cars, 2015).

Healthcare tourism and migration also contribute to the spread of resistant strains to other countries throughout the world, making it harder to control (Kantele and Lääveri, 2021). Moving livestock and animal products across borders through international trade can also introduce new resistance genes (Muloi et al., 2018).

The surveillance systems that incorporate genomic data can be used to track the AMR transmission pathways related to global movement, which highlights the necessity of international policies and monitoring to deal with the challenge (Tängdén and Cars, 2015).

Cultural Practices, Education, and Awareness amongst the people

Poor understanding of AMR by the population, false beliefs regarding the use of antibiotics, cultural beliefs regarding the use of antibiotics, and insufficient education are the behavioral factors that encourage misuse. People link antibiotics and fast recovery and demand to take them even with a viral infection, compelling healthcare professionals to prescribe unnecessary drugs (McCullough et al., 2016).

Patient, caregiver, and healthcare professional educational interventions have been demonstrated to be effective in reducing inappropriate usage of antibiotics (Huttner et al., 2010). Public health messages should be tailored in accordance with certain cultural contexts, especially in communities where conventional beliefs affect health-seeking behaviors.

Awareness about AMR and the importance of using antibiotics responsibly through involvement of the communities, media campaigns, and educational programs at schools are essential elements of stewardship measures. Motivating people through knowledge will lead to behavior change that is required to maintain the effectiveness of antibiotics.

6. INTEGRATED PERSPECTIVES ON AMR DRIVERS

Interaction of Genetic, Biochemical, Ecological, and Socio-Economic Factors

Antimicrobial resistance (AMR) development is a multifactorial process involving complex interactions among genetic, biochemical, ecological, and socio-economic drivers. Studies increasingly illustrate that no single factor acts in isolation; rather, their interplay creates an environment conducive to the emergence, dissemination, and persistence of resistant pathogens across diverse ecosystems, as shown in Figure 4 (WHO, 2023; Ahmed, 2024).

The horizontal gene transfer (HGT) of resistance genes to other bacterial groups takes place through genetic mechanisms, typically including conjugation, transformation, and transduction, and is often assisted by mobile genetic elements, such as plasmids, transposons, and integrons (Bhat et al., 2023). These genetic interactions are commonly prompted by biochemical means like the enzymatic alteration of antibiotics like beta-lactamases and aminoglycoside-modifying enzymes, which confer short-term survival benefits under the impact of antimicrobials (Shaikh et al., 2015).

At the same time, ecological conditions such as environmental reservoirs, such as soil, water systems, and wastewater, can also be regarded as hotspots of gene exchange and horizontal transfer, which even contributes to the spread of resistance (Berendonk et al., 2015). Selective pressure is caused by human actions, especially the misuse, overuse, and poor management of waste, leading to the proliferation of resistant strains in clinical settings as well as in agriculture and the environment (Van Boeckel et al., 2015).

These mechanisms are heavily influenced by socio-economic factors in the sense that they shape practices related to the practices of antibiotic prescription, agricultural policy, and awareness of health in the societies. As an example, poor regulation and ignorance in low-income areas lead to rapid development of resistance due to the high level of antibiotic use (WHO, 2023). The environmental and social aspects are related to one another, and it is important to highlight that the AMR must be controlled through integrated approaches of health (One Health).

Modelling and Predicting AMR Spread in One Health Context

The management of AMR should be done with well-developed modelling systems to combine genetic, ecological, and socio-economic information to anticipate the progression of resistance and implement measures to address it. Over the past, the use of computational models, including agent-based, system dynamics, and genomic epidemiology, that simulate evolution and dissemination pathways of resistance has been adopted (Goryluk-Salmonowicz & Popowska, 2022).

Models that include aspects like regional antibiotic use, movement of resistance genes, environmental contamination, and medical professional activities have proven to predict the possibilities of outbreaks and the hotspots of resistance (Sciencedirect, 2021). As an illustration, combining the environmental data, including the presence of antibiotic residues in wastewater, with bacterial population genomics can be used to determine the points of intervention that are critical (Goryluk-Salmonowicz & Popowska, 2022).

Moreover, predictive models are important components of the One Health framework that can allow policymakers to calculate the effectiveness of interventions, including the banning of non-essential antibiotic use, and to allocate resources appropriately, to conduct surveillance, treat wastewater, and take care of the stewardship (WHO, 2023). Creation of real-time flexible models is also a priority, and there is a focus on joint data sharing between sectors and regions.

Case Studies Illustrating Multi-Factorial Resistance

It is possible to observe real-life case studies that demonstrate the intricate processes that are impacting AMR. The most prominent of them is the development of carbapenem-resistant Enterobacteriaceae (CRE) in health facilities as a result of horizontal transfer of carbapenemase genes through plasmids, which occurs with the help of environmental reservoirs polluted with hospital effluents (Bhat et al., 2023). Examples of the genetic-environmental interface are clinical strains that develop resistance, which is intrinsic to the microbiomes of the environment.

The other example is related to the avian wildlife as a host of multidrug-resistant bacteria that have their origin in livestock farms and wastewater treatment facilities, which exemplifies the ecological and socio-economic interaction (Vittecoq et al., 2016). These strains are resistant and can be transmitted back to human beings as a result of direct contact or contaminated food products, which exemplify the One Health nexus.

The third instance is the application of antibiotics to aquaculture, which causes the development of resistance genes in marine microbiomes. Such genes are passed on to the pathogenic bacteria, which infect people, demonstrating the multi-layered effect of the environment, animal, and human sectors.

7. FUTURE DIRECTIONS AND RESEARCH GAPS

Emerging Molecular Tools for Understanding AMR

The use of new molecular diagnostics has enabled innovation in detecting and comprehending antimicrobial resistance (AMR). Next-generation sequencing (NGS) allows the complete genomic characterization of pathogens, discovery of existing and new resistance genes, mutations, and mobile genetic elements at a level never achieved before (Elbehiry et al., 2025). There are complementary technologies like matrix-assisted laser desorption / ionization-time of flight mass spectrometry (MALDI-TOF MS) that provide high-speed and high-quality pathogen identification and resistance profiling on a proteomic signature basis (Florio et al., 2020).

New diagnostics, such as CRISPR-based diagnostics, are becoming very sensitive to identify the presence of certain resistance genes in clinical and environmental samples with very short turnaround times (Elbehiry et al., 2025). Spectroscopy with machine learning further allows the determination of phenotypes of resistance using

Raman spectroscopy, with promise to perform real-time tests of antibiotic resistance (Novikov et al., 2022).

Molecular diagnostics combined with bioinformatics can support large-scale surveillance and epidemiological monitoring and allow for detecting resistance outbreaks early and inform interventions (Singh et al., 2025). Even with these innovations, access, affordability, and standardization of new diagnostic tools in various health care and environmental contexts, particularly in resource-strained areas, are still problematic.

Innovative Strategies to Mitigate Drivers of Resistance

AMR needs creative approaches that will deal with its origin and transmission. The innovation in drug discovery is aimed at creating new antibiotics possessing new, unique mechanisms that will override the established resistance pathways, including antimicrobial peptides and bacteriophage therapies (Ahmed, 2024). Routine therapies are revitalized through adjunctive therapies such as β-lactamase inhibitors and efflux pump blockers (Shaikh et al., 2015).

Other solutions focus on non-antibiotic strategies such as vaccination, better infection prevention and control, and creation of microbiome-regulating probiotics that decrease pathogen settlement (Tang et al., 2017). Precision farming and phage biocontrol are emerging as a way to use agriculture in a sustainable way to diminish the use of antibiotics in livestock and crops (Van Boeckel et al., 2017).

The environmental interventions are focused on the minimization of the antibiotic pollution by increasing the wastewater treatment technologies that will degrade the remaining antimicrobials and eliminate the resistant bacteria in the effluents (Rizzo et al., 2013). Resistance genes can be monitored in real-time in a variety of environments, which helps to contain the disease early due to the development of nanotechnology and biosensors (Singh et al., 2025).

Policy and Stewardship Recommendations

To control AMR, international policies to promote antibiotic stewardship, surveillance, and regulations are necessary. The WHO Global Action Plan presents a guide to focusing on responsible antimicrobial use in human, animal, and environmental health systems in accordance with the One Health principles (WHO, 2023).

The programs adopted in stewardship should guarantee rational prescriptions, limit the sale of antimicrobial over the counter, and safeguard community and professional education of antimicrobial use (CDC, 2025). Enhancement of laboratory networks to conduct regular susceptibility testing assists in evidence-based treatment and resistance tracking.

Incentives and simplified approval systems, and world funding mechanisms are necessary to renew the antibiotic pipeline (O'Neill, 2016). The environmental laws that stipulate the restriction of the discharge of antibiotics and the encouragement of heightened treatment of wastewater are essential in lowering the environmental AMR reservoirs. The international partnerships and data-sharing systems improve transparency and quick reaction to the emergence of the resistance threat, which emphasizes the necessity of globalized action (GLASS, 2025).

8. CONCLUSION

The development of multidrug-resistant microorganisms is based on complex processes, including genetic changes such as chromosomal mutations and horizontal gene transfer, biochemical processes such as enzyme inactivation of antibiotics and efflux pumps, and ecological conditions such as environmental reservoirs. These mechanisms are aggravated by socio-economic and behavioral factors, such as antibiotic misuse and excess in human and veterinary medicine, which places selective pressure that remains undamaged. Antibiotic residues and resistant bacteria released to the environment further enhance the spread of the gene amongst the microbial communities. The complexity of AMR development is emphasized by these opposite mechanisms and requires the implementation of multivaried interventions that focus on human, animal, and ecosystem health (Irfan et al., 2022; WHO, 2023)

The extensive dissemination of AMR is a threat to the effectiveness of antibiotics, which are important in the treatment of infections in human beings and animals, thereby increasing morbidity, mortality, and health expenditure. Systems that depend on antibiotics have difficulties in sustaining the health and productivity of animals in the long term, with the environmental pollution continuing to transmit resistance genes whose impacts on the ecosystem in the long term remain unknown. One Health integrated strategies are crucial when considering such interdependence and developing holistic surveillance, stewardship, and mitigation measures (Holmes et al., 2016; Berendonk et al., 2015).

To efficiently cope with AMR, it is necessary to formulate a synthesis of knowledge on genetic, biochemical, ecological, and socio-economic levels to address the formation of integrated policies and interventions. Improvements in molecular tools improve the early identification and prediction of the occurrence of an outbreak to implement specific control actions. Aimed at preventing resistance spread, a combination of efforts and strategic approaches in the field of antibiotic stewardship, regulation, education, and environmental control is needed and is coordinated with international models. Reducing health disparities through fair access to diagnostics, new therapeutic options, and clean habitats will be essential to the protection of antimicrobial activity in the generations to come (Ahmed, 2024; WHO, 2023).

REFERENCES

Ahlstrom, C. A.. (2019). Migratory birds as reservoirs and vectors of antimicrobial resistance. *Environmental Microbiology Reports, 11*(4), 420–430. DOI: 10.1111/1758-2229.12751

Ahmed, S. K. (2024). Impacts, challenges, and future avenues of antimicrobial resistance. *Journal of Global Health, 12*(4), 100345. DOI: 10.1016/j.jogh.2024.100345

Allen, H. K., Donato, J., Wang, H. H., Cloud-Hansen, K. A., Davies, J., & Handelsman, J. (2017). Call of the wild: Antibiotic resistance genes in natural environments. *Nature Reviews. Microbiology, 8*(4), 251–259. DOI: 10.1038/nrmicro2312 PMID: 20190823

Araújo, J.. (2022). Antibiotic resistance genes in aquatic environments: From pollution to health threats. *The Science of the Total Environment, 835*, 155505. DOI: 10.1016/j.scitotenv.2022.155505

Baquero, F., Martínez, J. L., & Cantón, R. (2008). Antibiotics and antibiotic resistance in water environments. *Current Opinion in Biotechnology, 19*(3), 260–265. DOI: 10.1016/j.copbio.2008.05.006 PMID: 18534838

Berendonk, T. U., Manaia, C. M., Merlin, C., Fatta-Kassinos, D., Cytryn, E., Walsh, F., Bürgmann, H., Sørum, H., Norström, M., Pons, M.-N., Kreuzinger, N., Huovinen, P., Stefani, S., Schwartz, T., Kisand, V., Baquero, F., & Martinez, J. L. (2015). Tackling antibiotic resistance: The environmental framework. *Nature Reviews. Microbiology, 13*(5), 310–317. DOI: 10.1038/nrmicro3439 PMID: 25817583

Bhat, B. A., Mir, R. A., Qadri, H., Dhiman, R., Almilaibary, A., Alkhanani, M., & Mir, M. A. (2023). Integrons in the development of antimicrobial resistance: Critical review and perspectives. *Frontiers in Microbiology, 14*, 1231938. DOI: 10.3389/fmicb.2023.1231938 PMID: 37720149

Bocharova, Y. A., Tatiana, A. S., & Igor, V. C. (2023). Antimicrobial resistance-associated mutations in chromosomal genes of ESKAPE pathogens. *Clinical microbiology and antimicrobial chemotherapy, 25*(2), 187-201.

Centers for Disease Control and Prevention. (2025). About antimicrobial resistance. https://www.cdc.gov/antimicrobial-resistance/about/index.html

Centers for Disease Control and Prevention. (2025). Antibiotic use and antimicrobial resistance facts. https://www.cdc.gov/antibiotic-use/data-research/facts-stats/index.html

Cesur, S. (n.d.). Antibiotics and the mechanisms of resistance. *International Journal of Antimicrobial Agents.*

Cycoń, M., Mrozik, A., & Piotrowska-Seget, Z. (2019). Antibiotics in the soil environment—Degradation and their impact on microbial activity and diversity. *Frontiers in Microbiology*, *10*, 338. DOI: 10.3389/fmicb.2019.00338 PMID: 30906284

D'Costa, V. M., King, C. E., Kalan, L., Morar, M., Sung, W. W., Schwarz, C., Froese, D., Zazula, G., Calmels, F., Debruyne, R., Golding, G. B., Poinar, H. N., & Wright, G. D. (2011). Antibiotic resistance is ancient. *Nature*, *477*(7365), 457–461. DOI: 10.1038/nature10388 PMID: 21881561

Davies, J., & Davies, D. (2010). Origins and evolution of antibiotic resistance. *Microbiology and Molecular Biology Reviews : MMBR*, *74*(3), 417–433. DOI: 10.1128/MMBR.00016-10 PMID: 20805405

Davies, J., & Davies, D. (2010). Origins and evolution of antibiotic resistance. *Microbiology and Molecular Biology Reviews : MMBR*, *74*(3), 417–433. DOI: 10.1128/MMBR.00016-10 PMID: 20805405

Devi, N. S. (2024). Overview of antimicrobial resistance and mechanisms. *Trends in Microbiology*, *32*(1), 12–23. DOI: 10.1016/j.tim.2023.10.004

Eiamphungporn, W., & Bhunya, A. K. (2018). Tackling the antibiotic resistance caused by Class A β-lactamases. *Current Pharmaceutical Design*, *24*(35), 4194–4206. DOI: 10.2174/1381612825666180712153105 PMID: 30061509

Elbehiry, A., Marzouk, E., Abalkhail, A., Abdelsalam, M. H., Mostafa, M. E., Alasiri, M., Ibrahem, M., Ellethy, A. T., Almuzaini, A., Aljarallah, S. N., Abu-Okail, A., Marzook, N., Alhadyan, S., & Edrees, H. M. (2025). Detection of antimicrobial resistance via state-of-the-art technologies versus conventional methods. *Frontiers in Microbiology*, *16*, 1549044. DOI: 10.3389/fmicb.2025.1549044 PMID: 40071214

Florio, W.. (2020). MALDI-TOF mass spectrometry in clinical microbiology: The use of proteomics for rapid identification and antibiotic resistance detection. *Clinical Proteomics*, *17*, 37. DOI: 10.1186/s12014-020-09271-0

Forsberg, K. J., Reyes, A., Wang, B., Selleck, E. M., Sommer, M. O., & Dantas, G. (2012). The shared antibiotic resistome of soil bacteria and human pathogens. *Science*, *337*(6098), 1107–1111. DOI: 10.1126/science.1220761 PMID: 22936781

Frost, L. S., Leplae, R., Summers, A. O., & Toussaint, A. (2005). Mobile genetic elements: The agents of open source evolution. *Nature Reviews. Microbiology*, *3*(9), 722–732. DOI: 10.1038/nrmicro1235 PMID: 16138100

Global, A. M. R. Hub. (2024). One Health. https://globalamrhub.org/one-health/

Global Antimicrobial Resistance and Use Surveillance System (GLASS). (2025). WHO. https://www.who.int/initiatives/glass

Goryluk-Salmonowicz, A., & Popowska, M. (2022). Factors promoting and limiting antimicrobial resistance in the environment–Existing knowledge gaps. *Frontiers in Microbiology*, *13*, 992268. DOI: 10.3389/fmicb.2022.992268 PMID: 36204635

Hall-Stoodley, L., Costerton, J. W., & Stoodley, P. (2004). Bacterial biofilms: From the natural environment to infectious diseases. *Nature Reviews. Microbiology*, *2*(2), 95–108. DOI: 10.1038/nrmicro821 PMID: 15040259

Heuer, H., Schmitt, H., & Smalla, K. (2011). Antibiotic resistance genes spread due to manure application on agricultural fields. *Current Opinion in Microbiology*, *14*(3), 236–243. DOI: 10.1016/j.mib.2011.04.009 PMID: 21546307

Holmes, A. H., Moore, L. S. P., Sundsfjord, A., Steinbakk, M., Regmi, S., Karkey, A., Guerin, P. J., & Piddock, L. J. V. (2016). Understanding the mechanisms and drivers of antimicrobial resistance. *Lancet*, *387*(10014), 176–187. DOI: 10.1016/S0140-6736(15)00473-0 PMID: 26603922

Hunt, D. (2024). A brief history of antimicrobial resistance. *AMA Journal of Ethics*, *26*(4), E248–E255. DOI: 10.1001/amajethics.2024.248 PMID: 38700525

Huttner, B., Goossens, H., Verheij, T., & Harbarth, S. (2010). Characteristics and outcomes of public campaigns aimed at improving the use of antibiotics in outpatients in high-income countries. *The Lancet. Infectious Diseases*, *10*(1), 17–31. DOI: 10.1016/S1473-3099(09)70305-6 PMID: 20129146

Irfan, M., Almotiri, A., & AlZeyadi, Z. A. (2022). Antimicrobial resistance and its drivers—A review. *Antibiotics (Basel, Switzerland)*, *11*(10), 1362. DOI: 10.3390/antibiotics11101362 PMID: 36290020

Kantele, A., & Lääveri, T. (2021). Travelers as sentinels of antimicrobial resistance in the international spread of multidrug-resistant bacteria. *Frontiers in Microbiology*, *12*, 683729. DOI: 10.3389/fmicb.2021.683729

Landers, T. F., Cohen, B., Wittum, T. E., & Larson, E. L. (2012). A review of antibiotic use in food animals: Perspective, policy, and potential. *Public Health Reports (Washington, D.C.)*, *127*(1), 4–22. DOI: 10.1177/003335491212700103 PMID: 22298919

Mah, T. F., & O'Toole, G. A. (2001). Mechanisms of biofilm resistance to antimicrobial agents. *Trends in Microbiology, 9*(1), 34–39. DOI: 10.1016/S0966-842X(00)01913-2 PMID: 11166241

Martínez, J. L. (2009). Environmental pollution by antibiotics and by antibiotic resistance determinants. *Environmental Pollution, 157*(11), 2893–2902. DOI: 10.1016/j.envpol.2009.05.051 PMID: 19560847

McCullough, A. R., Parekh, S., Rathbone, J., Del Mar, C. B., & Hoffmann, T. C. (2016). A systematic review of the public's knowledge and beliefs about antibiotic resistance. *The Journal of Antimicrobial Chemotherapy, 71*(1), 27–33. DOI: 10.1093/jac/dkv310 PMID: 26459555

McEwen, S. A. (2018). Antimicrobial resistance: A One Health perspective. *Microbiology Spectrum*, 6(2), ARBA-0041-2017. https://doi.org/DOI: 10.1128/microbiolspec.ARBA-0041-2017

Michael, I., Rizzo, L., McArdell, C. S., Manaia, C. M., Merlin, C., Schwartz, T., Dagot, C., & Fatta-Kassinos, D. J. W. R. (2013). Urban wastewater treatment plants as hotspots for the release of antibiotics in the environment: A review. *Water Research, 47*(3), 957–995. DOI: 10.1016/j.watres.2012.11.027 PMID: 23266388

Muloi, D., et al. (2018). Global cross-sectoral zoonotic spillover dynamics. *Science Advances*, 4(7), eaat1934. https://doi.org/DOI: 10.1126/sciadv.aat1934

Muteeb, G., Kazi, R. N. A., Aatif, M., Azhar, A., El Oirdi, M., & Farhan, M. (2025). Antimicrobial resistance: Linking molecular mechanisms to public health impact. *SLAS Discovery : Advancing Life Sciences R & D, 33*, 100232. DOI: 10.1016/j.slasd.2025.100232 PMID: 40216324

Muteeb, G., Rehman, M. T., Shahwan, M., & Aatif, M. (2023). Origin of antibiotics and antibiotic resistance, and their impacts on drug development: A narrative review. *Pharmaceuticals (Basel, Switzerland), 16*(11), 1615. DOI: 10.3390/ph16111615 PMID: 38004480

Naghavi, M., Vollset, S. E., Ikuta, K. S., Swetschinski, L. R., Gray, A. P., Wool, E. E., & Dekker, D. M. (2024). Global burden

Nair, R. R., Andersson, D. I., & Warsi, O. M. (2024). Antibiotic resistance begets more resistance: Chromosomal resistance mutations mitigate fitness costs conferred by multi-resistant clinical plasmids. *Microbiology Spectrum, 12*(5), e04206–e04223. DOI: 10.1128/spectrum.04206-23 PMID: 38534122

Nammi, J., Pasala, R., Andhe, N., Vasam, R., Poruri, A. D., Sherikar, R. R., & Nammi, J. Y. (2025). Antibiotic misuse: An in-depth examination of its global consequences and public health challenges. *Cureus*, *17*(6). Advance online publication. DOI: 10.7759/cureus.85941 PMID: 40656310

Narendrakumar, L., Arumugam, P., & Rangarajan, M. (2023). β-Lactam potentiators to re-sensitize resistant pathogens. *Frontiers in Microbiology*, *14*, 1092556. DOI: 10.3389/fmicb.2022.1092556 PMID: 36970185

Nikaido, H. (2003). Molecular basis of bacterial outer membrane permeability revisited. *Microbiology and Molecular Biology Reviews : MMBR*, *67*(4), 593–656. DOI: 10.1128/MMBR.67.4.593-656.2003 PMID: 14665678

Novikov, A.. (2022). Raman spectroscopy and machine learning for rapid phenotypic detection of antibiotic resistance. *Analytical Chemistry*, *94*(2), 1014–1023. DOI: 10.1021/acs.analchem.1c03884

O'Neill, J. (2016). Tackling drug-resistant infections globally: Final report and recommendations. Review on Antimicrobial Resistance. https://amr-review.org/Publications

Piddock, L. J. (2006). Multidrug-resistance efflux pumps - not just for resistance. *Nature Reviews. Microbiology*, *4*(8), 629–636. DOI: 10.1038/nrmicro1464 PMID: 16845433

Poole, K. (2007). Efflux pumps as antimicrobial resistance mechanisms. *Annals of Medicine*, *39*(3), 162–176. DOI: 10.1080/07853890701195262 PMID: 17457715

Ramirez, M. S., & Tolmasky, M. E. (2010). Aminoglycoside-modifying enzymes. *Drug Resistance Updates*, *13*(6), 151–171. DOI: 10.1016/j.drup.2010.08.003 PMID: 20833577

Reygaert, W. C. (2018). An overview of antimicrobial resistance mechanisms. *Journal of Microbiology & Experimentation*, *6*(2), 1–11. PMID: 31294229

Sciencedirect. (2021). Environmental antimicrobial resistance and its drivers. *Science of The Total Environment*, 772, 145-159. DOI: 10.1016/j.scitotenv.2021.145615

Shaikh, S., Fatima, J., Shakil, S., Rizvi, S. M. D., & Kamal, M. A. (2015). Antibiotic resistance and extended spectrum beta-lactamases: Types, epidemiology and treatment. *Saudi Journal of Biological Sciences*, *22*(1), 90–101. DOI: 10.1016/j.sjbs.2014.08.002 PMID: 25561890

Silva, K. P. T., & Khare, A. (2024). Antibiotic resistance mediated by gene amplifications. *npj Antimicrobials and Resistance*, *2*(1), 35. \

Singh, S.. (2025). Harnessing advanced molecular diagnostics and biosensors to combat antimicrobial resistance. *Current Opinion in Biotechnology*, *78*, 102755. DOI: 10.1016/j.copbio.2024.102755

Sultan, I., Rahman, S., Jan, A. T., Siddiqui, M. T., Mondal, A. H., & Haq, Q. M. R. (2018). Antibiotics, resistome and resistance mechanisms: A bacterial perspective. *Frontiers in Microbiology*, *9*, 2066. DOI: 10.3389/fmicb.2018.02066 PMID: 30298054

Tang, K. L.. (2017). Restricting the use of antibiotics in food-producing animals and its associations with antibiotic resistance: A systematic review and meta-analysis. *The Lancet. Planetary Health*, *1*(8), e316–e327. DOI: 10.1016/S2542-5196(17)30141-9 PMID: 29387833

Tängdén, T., & Cars, O. (2015). Global dissemination of antibiotic resistance: The role of international travellers. *Clinical Microbiology and Infection : The Official Publication of the European Society of Clinical Microbiology and Infectious Diseases*, *21*(10), 969–971. DOI: 10.1016/j.cmi.2015.07.020

Van Boeckel, T. P., Brower, C., Gilbert, M., Grenfell, B. T., Levin, S. A., Robinson, T. P., Teillant, A., & Laxminarayan, R. (2015). Global trends in antimicrobial use in food animals. *Proceedings of the National Academy of Sciences of the United States of America*, *112*(18), 5649–5654. DOI: 10.1073/pnas.1503141112 PMID: 25792457

Van Boeckel, T. P., Glennon, E. E., Chen, D., Gilbert, M., Robinson, T. P., Grenfell, B. T., Levin, S. A., Bonhoeffer, S., & Laxminarayan, R. (2017). Reducing antimicrobial use in food animals. *Science*, *357*(6358), 1350–1352. DOI: 10.1126/science.aao1495 PMID: 28963240

Velazquez-Meza, M. E., Galarde-López, M., Carrillo-Quiróz, B., & Alpuche-Aranda, C. M. (2022). Antimicrobial resistance: One health approach. *Veterinary World*, *15*(3), 743–749. DOI: 10.14202/vetworld.2022.743-749 PMID: 35497962

Vittecoq, M., Godreuil, S., Prugnolle, F., Durand, P., Brazier, L., Renaud, N., Arnal, A., Aberkane, S., Jean-Pierre, H., Gauthier-Clerc, M., Thomas, F., & Renaud, F. (2016). Antimicrobial resistance in wildlife. *Journal of Applied Ecology*, *53*(2), 519–529. DOI: 10.1111/1365-2664.12596 PMID: 27396126

Wang, Q., Wang, P., & Yang, Q. (2020). Antibiotic resistance genes abundance and diversity in soils from organic and conventional vegetable farms. *Environmental Science and Pollution Research International*, *27*(10), 11464–11474. DOI: 10.1007/s11356-020-08032-4

WHO. (2023). Antimicrobial Resistance Fact Sheet. https://www.who.int/news-room/fact-sheets/detail/antimicrobial-resistance

World Health Organization. (2019). Antimicrobial resistance. https://www.who.int/health-topics/antimicrobial-resistance

World Health Organization. (2023). Antimicrobial Resistance Fact Sheet. https://www.who.int/news-room/fact-sheets/detail/antimicrobial-resistance

World Health Organization. (2023). Antimicrobial Resistance Fact Sheet. https://www.who.int/news-room/fact-sheets/detail/antimicrobial-resistance

World Health Organization. (2025). Global antibiotic resistance surveillance report 2025. https://www.who.int/publications/i/item/B09585

Chapter 3
Current Therapeutic and Policy Responses to Antimicrobial Resistance

Chinyere Charity Ezeanya-Bakpa
http://orcid.org/0000-0002-7844-7414
Caleb University, Lagos, Nigeria

Fisayo Emmanuel Lambe
Caleb University, Lagos, Nigeria

Chidozie Vitus Udeogu
Nnamdi Azikiwe University, Awka, Nigeria

ABSTRACT

The escalating threat of antimicrobial resistance (AMR) has prompted a multifaceted approach in both therapeutic and policy domains. Current therapeutic responses emphasize the development of novel antibiotics, alternative therapies that enhance existing drug efficacy while mitigating resistance. In parallel, global health policies are increasingly focused on implementing robust antimicrobial stewardship programs, promoting responsible use of antibiotics in both human health and agriculture, and fostering international collaboration to monitor and combat the spread of resistant pathogens. Collectively, these strategies reflect a comprehensive response to AMR, underscoring the need for innovative solutions and coordinated action across various sectors to safeguard public health effectively. This chapter elucidates on the current therapeutic responses to AMR. This chapter explores the current global advancements in mitigating the crisis of AMR and the impact of policy responses to AMR.

DOI: 10.4018/979-8-3373-7810-7.ch003

INTRODUCTION

Antimicrobial resistance (AMR), remains a challenge and top ten public-health burden globally today (World Health Organisation, 2025). As a result of this global threat, it was estimated in 2019 that 4.95 million mortality worldwide was due to AMR (Murray et al, 2022). Report by the World Bank's report, an anticipated 3.80% decline in world annual gross domestic product (GDP) by 2050 due to the high impact of AMR (Jonas et al, 2017). AMR, describes the capability of microbes with active metabolic activity despite exposure to antibiotics.

Human acquired AMR through their continuous exposure to the environment, agricultural products, animals and healthcare settings (Ezeanya-Bakpa et al., 2021a). AMR evolved from misuse of the antimicrobials, which causes selective pressure, thereby developing resistant against antimicrobial agent (Endale et al., 2023). AMR threat has been found in all classes of micro-organisms leading to public health and socioeconomic problem of a global scale (Prestinaci et al, 2015). People with underlying health conditions are mostly prone to higher infection risk which may not be caused because of their direct disease and its treatments, but may be hospital acquired infection on frequent exposure, invasion steps, and antimicrobial usage (Lyster, 2016).

A. Importance of Addressing AMR in Global Health

The governments, healthcare workers, doctors, pharmacist, pharmaceutical companies and public generally are needed on addressing the issue of AMR in global health. However, various strategies are involved, these include: apposite use of antibiotics, prescription done when it is needed, use of right drugs with corresponding dose, ensuring that complete prescribed regimen is taken as directed. Enhancing infection control measures like separation of an infected patients, can limit the spread and pathogenic AMR bacteria at the clinical sectors. Susceptibility pattern of microbes are importance measure for tracking the spread of resistance and considering treatment plans. Development of novel class of antibiotics like vaccines, phage therapy, immunotherapy are needed to be improved by the government and private sector (Parry and Threlfall, 2008; Burnham et al., 2017). It is very crucial to raise awareness on risk and vital use of antibiotics, also crises about AMR. Educating the mass, healthcare workers, policy makers on future problems with mis-use and the over-use of conventional antibiotics. Similarly, in the field of agriculture, limiting the practice of antibiotics, and appropriate use of livestock antibiotics. These measures can enhance the rapid spread of bacteria that are found in animals and also resistant to humans (Lipsitch and Samore, 2002; McDermott, 2005).

B. Overview of the Current Therapeutics and Policy Response to AMR

Antibiotics, disinfectants, and food preservatives are kinds of known antimicrobial agents that help in reducing the microbial growth, reproduction and destroying. These agents have a unique mechanisms of action, and can either be natural, semi-synthetic, or synthetic. The metabolic and physiological levels of the organism can be altered significantly by the antimicrobials. For instance, cell wall synthesis; inhibiting cell wall synthesis; disrupting bio-metabolic pathways; nucleic acid replication interference and translation (Baquero & Levin, 2021; Smith et al, 2023). During the 21st century, AMR is an established challenge of public health that presents a severe risk thus limiting the ability of a host to prevent various common disease-causing organisms including parasites (CDC, 2025). According to the global report by WHO in respect of AMR, it showed that AMR surveillance provide insight on the spectrum AMR and described the notable gaps in available surveillance systems (CDC, 2025). Additionally, MRSA (methicillin-resistant *Staphylococcus aureus*) is a primary pathogen causative of many infections acquired in the hospital globally, these pathogenic microbes are acquired at the intensive care units and post-surgical cases. About 10 000 mortality is estimated yearly caused by MRSA in the United State (CDC, 2025; Shallcross and Davies, 2014). *Niesseria gonorrhoeae* is another emerging drug-resistant in US, which has presented the first-line therapies in-effective. The United Kingdom also has reported the antimicrobial resistance of gonorrhea pathogenic strain to ceftriaxone and, azithromycin, highlighting issues about the possible health risk associated with sexually transmitted infections (Nardulli et al., 2022, CDC, 2025). Similarly, The outbreak of MDR *Salmonella* strains have been linked up to infectious food products in Pakistan, and about thousands of cases have been reported, was given the sole remaining antibiotic option (Nguyen-Thanh et al., 2023; CDC, 2025). Multiple contributing factors cause the rapid dissemination and advent of AMR, including mis-use as well as the over-use of therapeutics in various fields of life sciences. The unrestricted use of antibiotics in under-developed and average-income nations along with insufficient regulatory oversight, brings about self-medication, poor prescription practices, incomplete course of treatment, contributing to selection and growth of resistant strains of pathogenic bacteria. Antibiotics has been found extensively useful in agricultural sector not only for therapeutic functions but also used as prophylactic agents and hastening of growth, promoting the transmission of drug-resistant bacteria acquired via direct contact, intake of food products and environmental contact to humans (Thomas and Nielsen, 2005). Also, the environment contributes to spread of AMR. For instance, waste from the pharmaceutical companies, untreated industrial effluents, improperly treated hospital sewage or used equipment, and HGT among the bacterial population (Zhu et al. 2022)

UNDERSTANDING ANTIMICROBIAL RESISTANCE

The demonstrated bacterial resistance to different classes of anti-bacterial drugs rely on multiple resistance mechanisms. For example, the survival of methicillin resistance staphylococcus aureus (MSRA) often evolved various drug-resistant. Among the resistance mechanisms is inclusive of the thickening of the cell wall, decrease accumulation, drug target mutation, modification of drugs with the involvement of enzymes, and protective of antibacterial target. Additionally, all the previously mentioned mechanisms of resistance can be conferred to beta-lactam antibacterial (Akujobi et al., 2012; Kyriakidis et al., 2021).

Figure 1. Basic mechanism of antimicrobial resistance

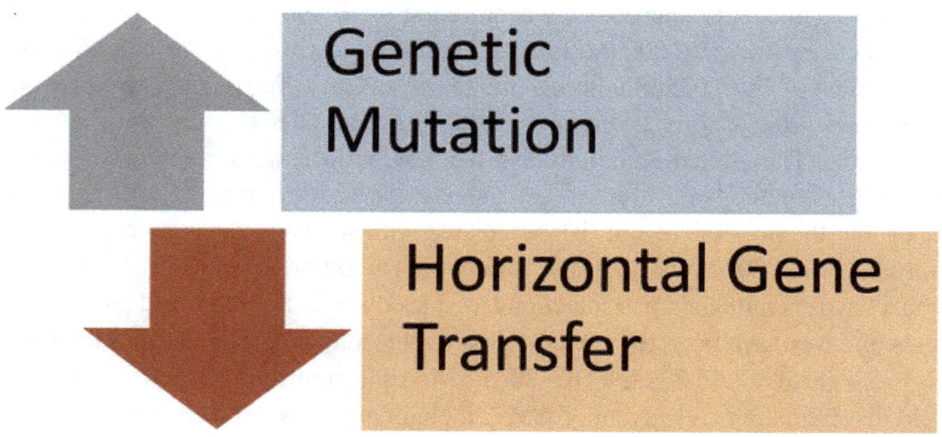

A. Genetic Mutations

The most common mechanisms by which reduce drug potency is through mutation, modification or cellular drug targets shielding, resulting into bacterial disruption. The antibacterial binding ability drastically reduces through changing or modifying the antibacterial target, thus diminishing both its efficacy and value (Walsh, 2000). However, protein activity are favored, by mutations leading to limitation of binding by antibacterial (Smith et al., 2013). Bacteria have shown to change their cellular components such the cell wall structure, membrane and many others that are antibacterial target causing poor susceptibility to antibacterial agent. For instance, the beta lactam bacteria changes cell wall bacterial structure, causing reduction in the site for drug binding that specially designed and directed at the cell wall. DNA mutation usually occur in certain bacteria resulting into alteration of their ribosome structures, the site responsible for protein synthesis. These alterations can lead to antibacterial such as macrolides and tetracycline to bind to their ribosomes, thereby causing the loss of the bacteria (Wang et al, 2023 ; Ezeanya-Bakpa et al, 2024). This kind of resistance mechanism can evolved from either enzymatic changes or mutation in the target site or via replacement of target

B. Horizontal Gene Transfer

Genes that code for resistance are disseminated from cohort to another through heritage. These genes are transmitted via horizontal gene-transfer (HTG) between the bacteria, achieved through the following methods like transformation, conjugation, and transduction (Mellor et al., 2019 ; Collineau et al., 2019 ; Tsigalou et al, 2020). The gene-coding resistance in bacteria have the ability to withstand a particular anti-bacterial or multi-drugs, these genes are often found within mobile genetic elements with sharing potential among bacterial cells (CDC, 2019; Shehreen et al., 2019). HGT occurs in various environment like the gut. The persistence and transfer of AMR genes within the bacterial populations and multi-drug resistance development, are enhanced significantly with the aid of the genetic elements (Bennett, 2008). The tenacity of the resistant genes among microbial species to antimicrobial agents not only by transferring the genes that code for resistance to their young ones but also the capability of sharing these resistant genes via the HGT (Touchon et al., 2017).

The most impactful mechanisms of HGT is conjugation method. This method is achieved through a process by which the bacteria cells have physical interaction facilitating the transmission of genetic material. In the course of this process, there will be formation of sex pilus, enhancing plasmid sharing between the recipient bacterium, thus leading to acquisition of multi AMR. (Wang et al., 2019; Graf et al., 2019). The manner in which a DNA molecule circulates among bacteria species via direct contact, thus the donor and recipient similarly known as F^+ and F^- respectively is termed conjugation approach.

C. Contributing Dynamics to AMR

I. Overuse and Misuse of Antibiotics

Improper directive and the over-use of conventional antibiotics have increased the rate of AMR development in human as well as animal. Research revealed the relationship between the over-use and the upsurge of resistance in bacteria. This challenge persist worldwide, despite the instructions passed by the health authorities, it is becoming glaring, trending, and getting to the critical stage where these challenge is difficult to control. All these parameters have led to the advent and wide-spread of AMR. Comprehensive strategies are required to take care of this global health challenge. In the case of viral infection treatment, such as flu or cold, the perception people is that such infections can be addressed with antibiotics. Also, frequently prescription of antibiotics for treatment of patients, this is very common in developing countries with low diagnostic resources (Chokshi et al, 2019). In the foundation of mis-use of antibiotics, it was revealed that ingestion without a medical doctor's prescription for both humans and animals. Lack of therapeutic instructions and clear antibiotics policies, mostly in developing countries, has been a leading challenge contributing massively to this misuse. Over prescriptions by the health care personnel, pharmacist and veterinarians especially in underdeveloped nations also contribute to problem of antibiotics resistance. At times, the situation AMR become worsen due to poor-quality chain supply of antibiotics (Chaw et al., 2018; Klein et al., 2018; Chokshi et al., 2019; Pires et al., 2024; Yang et al., 2024)

II. Lack of New Antibiotic Development

The challenge of AMR needs urgent attention from the pharmaceutical companies to produce new class and innovative antibiotics (DiMasi et al., 2016). There has been no remarkable improvement in this area despite repeated appeal from the WHO. However, among the most newly 51 developed antibiotics, only about 8 were innovative as therapeutic options for antibiotic-resistant pathogenic bacteria; others are just modification to available drugs. Furthermore, the major concerns now is that, it is high possible that this drugs might lack potency as result to resistance. In view of this, condition like tuberculosis caused by drug-resistant *Mycobacterium tuberculosis* and other infections such as: urinary tract infections (UTI), pneumonia, the management to this disease has been severely compromised due to lack of novel antibiotics, and elderly people have been vulnerably affected by this pathogens (Chang, et al., 2015). Studies showed that the pharmaceutical companies have foreseen that the greatest challenge on producing novel antibiotics result from regulatory hurdles and financial risk, causing investment and innovation scale back. Many of the pharmaceutical companies have focused on manufacturing drugs that target chronic infections instead of acute ones (Chang, et al., 2015; Dutescu and Hillier, 2021; Iskandar et al., 2022; Ezeanya-Bapka et al., 2023; Bo et al, 2024; FAO, 2024).

III. Poor Infection Control Practices

Poor infection practices constitute the emergence of antibiotics resistance, and rapid spreading is due under, over or abuse of the therapeutics. Lack of proper prescription on intake of drugs sped up antimicrobial resistance, is evolved by different factors, such as patients' not following instructions attached to regimen, irrational consumption of drugs by healthcare workers, drugs marketing, unhygienic practices by the healthcare workers, therapeutics use in agriculture, poor quality of therapeutic agents, poor surveillance, and susceptibility examination. However, with patient's knowledge on diagnosis, doctors and other healthcare workers can easily be subjective to patient's demand hence may constitute to the risk of AMR. Some patients could stop treatment plan or forget their prescription or purchase only a small amount of regimen once when they observe they are feeling better. Furthermore, many of the available drugs in pharmacies are without prescription guides, which has massively contributed to the drugs misuse or abuse by patients. Through advertisement, the pharmaceutical companies also play key roles on rapid increase of antibiotics misuse. For instance, use of Ciprofloxacin is best treatment option for patient for at risk patients, lincomycin is encouraged in Philippine as advertised for the treatment of pharyngitis/tonsillitis and, upper respiratory tract infections (RTI) through the use of clindamycin despite knowing-fully that infections are caused by viral infections which do not require the use of antibiotics. Studies revealed that about 30% to 60% take antibiotics than required because doctor's prescriptions varies. Wrong or improper prescription from untrained doctor's and private practitioners can pose a significant health risk.

CURRENT THERAPEUTIC RESPONSES

Figure 2. Overview of current therapeutic responses

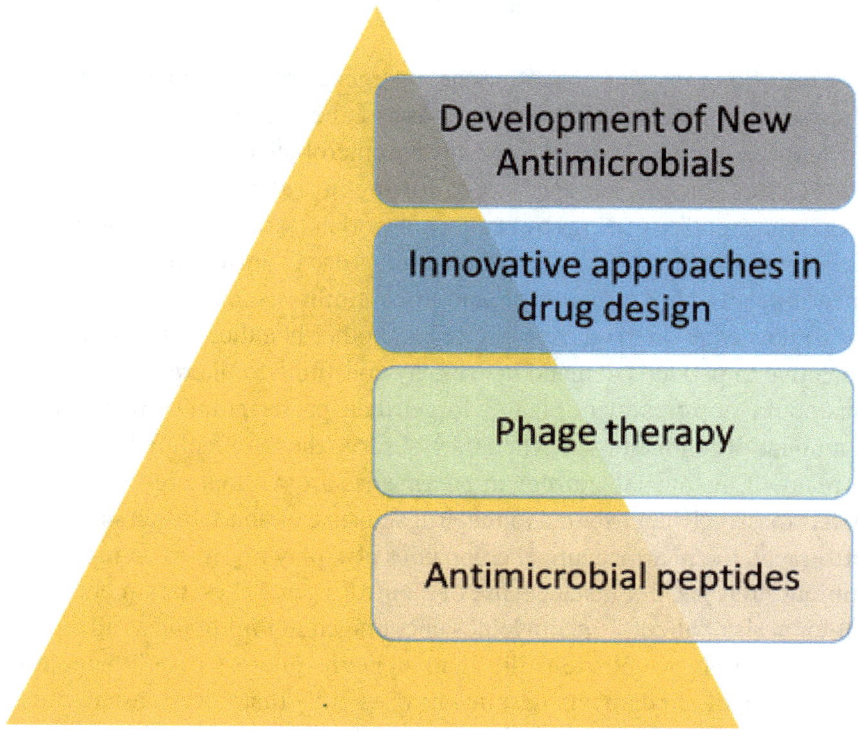

A. Development of New Antimicrobials

Drug discovery and developmental steps are extremely convenient by various computational tools and use of numerous software. In drug discovery, computer programs are required for online screening based on structures and ligand-based as well as with lead optimization (Patel et al., 2022). AI (artificial intelligent) termed as machine learning has improved and transformed many pharmaceutical companies based on drug research and design through computational and data-driven techniques. Drug design is very important since it serves and a template in the discovery of novel therapeutics to address varieties of ailments, thus mostly useful in medicinal chemistry. Findings described the innovative approaches in drug design and development process which include chemical sciences, target protein deposition, DNA-Encoded Laboratories as well as Computer Aided Drug Design (An et al., 2025; Ezeanya-Bakpa et al., 2026). Molecular dynamics in drugs design make use of explicit solution molecular docking simulation, further details is needed to examine the docked postures from Glide (Patton et al., 2021). The term molecular chameleon can be described as the ability of compounds to configure the shapes such that the polar bonds are obtained easily in water but hidden so that non-bonded interactions can be created between molecules around the non-polar environment such as phospholipid membrane. The 3D ligands developing that fit into a target binding site is most important tasks in the drug design. More so, every ligands atoms are equally treated according to structural-based drug design model, thereby avoiding roles that an atom play in ligand for drug design. Hence, making them less effective for testing the vast molecules.

B. Innovative Approaches in Drug Design

Computational methods to drug design, investigating things and production are being examined, utilized, and appreciated (Padole et al., 2022). CADD is described as the method of creating compounds that may be capable of binding to a specific bimolecular target and assuring their binding affinity. The therapeutic potential is improved based on the basic functional groups of the small molecules (SMs)-based chemotherapeutic agents, especially stereo-arrangement and stereo-selectivity, and these promote therapeutic outcomes (Siddiqui et al., 2025). CADD is often used in creating and increasing drug potency processes which are used now, since finding and creating a new drug is always seen as a time-consuming and high cost of processing (Perišić et al., 2025). The basic goal of medicinal chemistry is creating safe and potent molecules that can serve as pharmaceuticals in healthcare. To achieve this, it requires scientist like chemists, biologists, medical doctors, pharmacist and many others necessary professionals. Drugs pharmacokinetics profiling like excretion, absorption, metabolism, toxicity and distribution, are important parameters during drug's development and research in order to understand a drug candidate (Riccardi et al., 2018). The use of upgraded mercury along with graphic program on an interactive space starts the design phase of drug (Kumar, et al., 2024).

Also, the use of AI-driven methods which include AI learning, deep learning, and neutral networks, all these have reduced the time and cost related to associated with the development of convectional drugs. However, this AI-driven methods have significantly be improved by target verification, lead optimization, and drugs reusing (Bharath et al., 2025).

C. Importance of Synergy in Existing Antibiotics

Antibiotic synergistic approach serve as alternative remedy to fight multi-drug resistant bacteria (MDRB), it has the potential to increase the drug potency above what can achieved upon the use of a single antimicrobial agent alone. The effectiveness of available local antibiotics remedies have been significantly limited due to the outbreak and spread of MDRB, giving rise to therapeutic failure and high rate death and illness globally. Antibiotic synergy gives a promising method to fight mechanisms of resistance as well as improve treatment outcomes (Sun et al., 2016; Ezeanya-Bakpa et al., 2025a). One of the key benefit of antibiotic synergy lies on targeting diverse cellular pathways or processes within the bacteria effectively Worthington and Melander (2013). Antibiotics with high therapeutic index in a combined state can disrupt multiple vital functions rapidly, making it difficult for bacteria to withstand the antimicrobials. By focusing of various cellular pathways, antibiotics synergy can improve the bactericidal property and reduce the outbreak of resistance pathogens against individual antibiotics. The combined effort enhances the efficacy of treatment, most especially against multidrug-resistant strains which have intricate mechanisms of resistance. Antibiotic synergy can promote the penetration and accumulation of antibiotics intracellularly with the bacteria (Bolla et al., 2011; Ezeanya-Bakpa et al., 2025b). Findings revealed that some antibiotics can work synergistically in combined form, improve the potential of one drug to effectively infiltrate the cell wall of the bacteria, and enhancing easy penetration of other drug. However, enabling greater exposure of microbes to the antimicrobials, thereby improving their potency against the resistant strains that might have resistant mechanisms to reduce in-flow of antibiotics. Similarly, mixture of certain antibiotics can induce intracellular accumulation of antimicrobial, resulting to presence of high dose within the bacteria and proper disruption of intracellular pathogens (Shang et al., 2019). Antibiotic synergistic has the ability to hinder the resistance mechanism possessed by bacteria (Ezeanya-Bakpa, et al., 2022a). Additionally, enzymes and efflux pumps which is responsible for break down or modification of antibiotic can be inhibited by the mixture of antibiotic synergy, thus acting synergistically. It can also withhold the bacteria from the development of resistance, hence making it a valuable means of fighting MDRB. Antibiotic synergy play a vital role in treating infections related biofilm formation. Pervin and Hassan (2021). Biofilm is described as a complex formation of bacteria, encasing in a self-produced matrix, are very resistant to therapeutics. Antibiotic synergy are mixed to perform complementary actions, disrupting the biofilm architecture or promoting antibiotic penetration via the matrix, total eradication can easily be achieved (Gupta, et al., 2017).

D. Alternatives to Traditional Antibiotics

Among the potential substitutes to available alternative to traditional medicines include bacteriocins, antimicrobial peptides, and phage (Joerger, 2003).

I. Phage Therapy

Phage therapy also called bacterial bacteriophages or involves group of bacterial viruses, to inactivate or disrupt bacterial pathogens. Phages describe as bacteria eaters, offer diverse benefits over available traditional antibiotics, such as easily availability, simultaneous multiplication, diversity, low toxicity level, host range specificity, absence of cross-resistance with antimicrobials, and low environmental impacts. Despite all, challenges like appropriate bacteriophages selection, addressing narrow host spectrum, ability to manage potential immune reactions, and lastly ensuring clinical understanding persist. However, it has proven to be effectively address topical therapies where traditional antibiotics found ineffective. Hence, phage therapy can not completely replace traditional antibiotics (Brives and Pourraz, 2020; Ezeanya-Bakpa et al., 2022b). Bacteriophages offers potential solution to fight the issue of antibiotic-resistant microbes. For example, a newly developed technology by Phico therapeutics known as SAS Project, is designed specifically to achieving about 99.9% disruption onto target bacteria within about 2 minutes. SAS Project uses modified bacteriophages, and break down bacterial viruses carrying antibacterial protein genes. This gene is introduced into the target bacteria, it leads to deactivation of host bacteria Deoxyribonucleic acid (DNA), hence infection is inhibited (Fairhead, 2009). Synergistic effects of phages and traditional antibiotics have been shown to be effectively mostly in tackling multidrug-resist biofilms. It offers a unique mechanisms of both therapies to enhance bacterial inactivation, with phage targeting specific bacteria and antibiotics treating a wide spectrum. The combine use of antibiotics and phage show synergistic effects against difficult-to-treat infections where traditional remedies have been futile (Akturk et al., 2019). The Phage therapy also serve as adjunct-therapy aimed at improving treatment outcome. More so, phage therapeutics can also work alongside probiotics and beneficial microflora, thus sustaining a strong stability of the microbial communities and, enhancing the performance of treatment outcomes (Jassim and Limoges, 2014). Experimental research on phage therapy have utilize model system that included potentially lethal bacterial infections (Soothill, 1992; Merril et al., 1996; Biswas et al., 2002).

II. Antimicrobial Peptides

Antimicrobial peptides (AMPs) production is not restricted to bacteria only but applicable to all living organisms studied so far. AMPs are organic compounds produced by animals, plants, fungi and bacteria. This compound acts on microorganism negatively charged membranes and nucleic acid, also stop protein synthesis, thus target ribosomes one of the intracellular components. For instance, *Polybia paulista* and *Agelaia pallipes,* a social wasps, produce antimicrobial peptides like polybia-CP (ILGTILGLLKSL-NH2) and protonectin respectively, demonstrated antibacterial properties on both classes, multidrug resistant strains, by targeting the bacterial membrane (Wang et al., 2021; Wang et al., 2023). Bacterial species sourced from orthopedic wound was tested against osteoblast cells of human by investigating the cytotoxicity of Dadapin-1 (Campoccia et al., 2023). This compound inhibited the growth of all classes of bacteria significantly according to the result obtained (Boullet et al, 2019; Ezeanya-Bakpa et al., 2021a). *In-vitro* studies showed that synthetic peptides demonstrated wide spectrum of antimicrobial activity towards Gram-positive, Gram- negative, and fungus, also exert synergistic effects, which indicates chances of discovering new class of drugs. AMPs are mostly related to toxicity concerns and pharmatokinetics restrictions as described by Zharkova et al. (2023), results reported on synergistic study described the effects of a number of AMPs, which include protegrin 1 and, REF-ChBac3.4 combine with antiseptic agents such as Sodium hypochlorite, dixydin, protosan and biosulfactants such as amphoteric co-camidopropyl betaine. More research need to be done before considering AMPs-based products on effective development and preservation. Research revealed that predatory *Myobacteria* produced about four AMPs like Myxo_mac104, Coral_AMP411, Stig_213 and So_ce_56_913. *In-vitro* demonstrated an anti-inflammatory and antifungal potential of Coral_AMP411 results demonstrated the anti-inflammatory and anti-fungal properties of Coral_AMP411 (Arakal et al., 2023), and could be regarded as a new class of antibacterial (Chen et al., 2020).

III. Vaccination Strategies

AMR is crisis that spreads globally, resulting to the emergence of development of drug resistant pathogenic microbes, synergistic action of vaccines is required as integrated strategy to combat the risk of AMR via research exploitation (Micoli, et al., 2021). Vaccination as means of addressing the crisis of AMR highlights the main mechanistic differences with respect to available drugs and support the development of novel therapeutic agents to combat drugs resistant microbes. Antibiotics are capable of performing the function killing microbes termed "bactericidal" or inactivate growth activities termed "bacteriostatic", thus grouped based of their mode of action Baquero and Levin (2021). According to report in 2019, about 98.9% deaths per 100 000 evolved due to AMR across all-age in western sub-saharan African, high income countries experienced more of this problem *Klebsiella pneumoniae* was the most pronounced AMR that is responsible for high mortality rate in this region, and about 19.9% of the mortality was caused as result of AMR according the Antimicrobial Resistance Collaborators (2022). High morbidity and mortality should be prioritized by developing vaccines targeting resistant pathogenic microbes (Wang et al., 2021). Pneumococcal and Hib vaccines have been demonstrated to be significantly effective, and capsular polysaccharides (CPS) target against pathogenic *Streptococcus pneumoniae* and *Haemophilus influenza* respectively Henriques-Normark and Normark (2014). From results, pneumococcal conjugated vaccine has been evaluated to be potent at about 86% to towards pneumococcal disease, and Hib vaccine has drastically the Hib cases among children worldwide upon the introduction into the routine infant immunization programs Henriques-Normark and Normark (2014). CPS is mostly targeted and potent bacterial development of vaccines globally. For instance, 12-Valent vaccine is one of the vaccines for bacterial targeting CPS against extra-intestinal pathogenic *E. coli*, bacterial species with high AMR risk, after clinical trials known for its effectiveness against bacterial antigens (World Health Organisation, 2025).

IV. Improving Antibiotic Stewardship

Antibiotic stewardship is a program designed to educate healthcare workers to the right selection, dosage, and required time for antimicrobial agents to improve patients' outcomes as well as to lower the risk of resistant microbes and control its spread. Antibiotics stewardship aimed to focus on ensuring that every healthcare practitioners to place patients on appropriate dose and period of administration for each patient. Moreso, overuse, misuse and abuse of antimicrobials should be avoided. Lastly, ensuring development of resistance is kept at low rate. In 2014, According to Center for disease control, the core elements of required on achieving these goals of antimicrobial stewardship, which mostly applicable in the health centre, irrespective of size, with tailored strategies for both small and critical-access health centre to improve their implementation (Pinto-Ferreira et al., 2022; CDC, 2022)

CHALLENGES AND BARRIERS

Thirty years ago, a cheap of penicillin could cure almost any bacterial infection treated in hospitals. Today, we are fighting infections with drugs that cost many folds more, take longer to work, or don't work at all.

In most countries, the fight against Antimicrobial Resistance (AMR) is not just a scientific battle. It is a battle against an inefficient healthcare system, poor medical infrastructure, pervasive poverty and complex global economics, and deeply ingrained behavioral approach to medicines. In this section, we will analyse the three major barriers: Financial, Global, and Industrial, that are holding countries back from winning this war.

A. Financial Constraints in Research and Development (R&D)

The most disturbing aspect of AMR is that while the bugs are evolving faster than ever, our ability to create or discover new antibiotics is becoming almost impossible.

1. The Immense Cost of Discovery

Developing a single new drug is prohibitively expensive. It is estimated that bringing a new antibiotic from the laboratory to the hospital shelf costs between $1 billion and $2 billion. That is almost the cost a country's total annual budget.

This kind of money is non-existent for R&D in most countries. Scientists are brilliant, but they are often forced to work in laboratories that lack basic consistent infrastructures like electricity and constant running water, let alone the advanced equipment required for genomic sequencing of resistant bacteria (Ezeanya-Bakpa et al., 2021b).

Consequently, a lot of countries relies almost entirely on the Western world to produce new antimicrobials.

2. The Economic Paradox of Antibiotics

Why are big pharmaceutical companies reluctant to spend money on the development of new antibiotics? Because antibiotics are terrible business compared to other drugs. They prefer to develop drugs and medications for other chronic diseases. For instance, a patient takes high blood pressure or diabetes medication every single day for the rest of their life. This is a guaranteed, long-term revenue stream for pharmaceutical companies

When a brilliant new antibictic is developed, medical professionals are told to NOT to prescribe or recommend it. It is kept in reserve as a last resort to prevent bacteria from developing resistance to it.

Therefore, a pharma company spends $2 billion to develop a drug, and then the regulators demands that they only sell a tiny amount of it to prevent AMR. Logically, from a business perspective, it becomes economically unviable to continue investing in antibiotic Research and Development (Miethke *et al.,* 2021).

B. Global Disparities in Access to Medicines

While the world global pharma powers are developing last-resort antibiotics to fight these superbugs, millions in Nigeria are unable to access basic, effective medicines, leading to the national conundrum with AMR. This may not be the case in developed countries where quality of life is far better, and drug regulations are feasibly adhere to (Olagunju *et al.,* 2025).

In Nigeria, anyone with enough of money can walk into a local chemist and purchase powerful antibiotics without a prescription or medical diagnostic report. While a patient in a rural clinic in a remote part of the country, may have a life-threatening bacterial infection, but the clinic has run out of stock of the necessary antibiotic because of a either a broken supply chain or the prohibitive price of the medication.

When the appropriate medicines are unavailable, patients often turn to substandard and fake drugs which directly accelerates resistance. When last-resort antibiotics *are* available, they are often prohibitively expensive (Ozawa *et al.,* 2018).

Case Scenario: A patient in a wound care unit in a hospital has *Klebsiella spp.* infection resistant to almost all drugs except for a very expensive, newer drug like the Ceftazidime-Avibactam combination. One course might cost more than the patient's annual income.

Because of this prohibitive cost, the patient opts for fake and substandard medicines. Fake drugs contain zero active ingredients. The patient dies because the infection is festers and remains untreated.

C. Resistance from Industry and Economic Incentives

Finally, we must look at the resistance from the stakeholders that benefit from the defective status quo. Regulatory responses designed to curb antibiotic use often directly conflict with the economic survival of certain sectors.

1. The Agricultural and Livestock Industry

As mentioned in earlier in this paper, antibiotics are frequently used in farming, not just to treat sick animals, but to prevent illness in crowded, unhygienic conditions, and to make animals grow faster. If the government strictly enforces a ban on using human antibiotics in poultry, poultry farmers fear their mortality rates will skyrocket, making chicken too expensive for the average Nigerians to afford. Pharmaceutical sales agents have no economic incentive to educate the farmers on the reduction of drug use in farming. In the veterinary sector they believe that their livelihood depends on selling large volumes of drugs to these farmer, and not less (Adekanye *et al.,* 2020).

2. Patent Rights and implications

Newly developed antibiotics are expensive, because the company keeps the patent for decades, preventing anyone else from making a cheaper generic version. This keeps prices high. When governments try to stop or suspend these patents for public health emergencies through licensing to allow local manufacturers to produce them cheaply, they face mounting legal and economic resistance from international pharmaceutical bodies and the governments of developed nations where these companies are based (Laxminarayan, *et al.,* 2016).

The challenges facing the response to AMR in Nigeria are legion, but here are a few;

- There are no funds to develop or discover new drugs to discover new drugs locally.
- The global economic model makes discovering new drugs bad for business.
- Economic realities have promoted a defective supply chains leading to a surge in substandard that accelerate resistance.
- Our agricultural sector is economically reliant on the liberal misuse of these drugs.

To overcome these barriers, it is paramount not just to rely on health policies. Therefore, it requires economic restructuring, strengthening of regulatory bodies with better funding and technology, and international collaborations that recognize that a substandard drug sold on the streets, and consumed by humans or animals is a threat to global health security.

POLICY RESPONSES TO ANTIMICROBIAL RESISTANCE

Over the years, several governmental and non-governmental agencies have been involved in the fight against the scourge of Antimicrobial resistance (AMR), albeit without sensitive grassroot and community support, it is difficult to make progress and gain victory over AMR. When microbes change over time and no longer respond to antibiotics, making infections harder to treat and increasing the risk of disease spread, severe illness, and death, AMR becomes the issue. For developing countries, with a high burden of infectious diseases and an overburdened healthcare system, this is a matter of urgent national security and a socio-economic time-bomb that should be stemmed in its tide. To alleviate the AMR problem, policy responses are necessary at both the global and national levels.

A. Global Initiatives

AMR has no international or intra-national borders. A bacterium that develops in a hospital in one country can easily travel to another country or continent via a single traveler. Recognizing that the world is a global village, global health bodies have developed frameworks to guide the Health ministries of nations.

1. World Health Organizat on's Global Action Plan (GAP) on AMR

In 2015, the World Health Assembly adopted a monumental document: the Global Action Plan (GAP) on AMR (World Health Organisation, 2025). Following the launch of AMR-GAP, most countries developed their own National Action Plan on AMR (AMR-NAP) that are country-specific yet outline the global objectives on the AMR challenge.

Generally, GAP outlines five strategic objectives that every country is expected to tailor to their unique needs:

- To improve awareness and understanding of AMR through effective and quality communication, education, and training.
- To strengthen the knowledge and evidence base through the implementation of data, surveillance and research.
- To reduce the incidence of preventable infection through effective sanitation, hygiene, and infection control measures.
- To optimize the use of antimicrobial medicines in human and animal health.
- To develop the economic case for sustainable investment that takes account of the specific needs of all countries, while increasing investment in new medicines, diagnostic tools, vaccines, and other non-pharmaceutical measures.

2. The Quadripartite Collaboration (WHO, FAO, OIE, UNEP)

The global community has realized that that focusing only on doctors and human hospitals have not provided the desired results in the fight against AMR. Antibiotics are used heavily in agriculture (animals) and can leak into the environment (water and soil) (Ohia *et al.,* 2025).

B. National Strategies:

As part of an answer to the global call to action, developed its own National Action Plan on AMR (NCDC, 2017). While the plan is comprehensive on paper, it has not achieved much due to unique challenges in implementation. The political will to enforce has also been very doubtful.

I. Surveillance and Monitoring Programs

The fight against AMR requires painstaking surveillance. This involves collecting data on which bacteria are becoming resistant to which drugs. Data generated from such surveillance will help to formulate policies and regulate the sale, marketing and use of such antibiotics. The information generated will also be used to educate stakeholders in healthcare, environmental and livestock sectors to enhance cooperation (Iregbu *et al.,* 2021).

Current Status: For many years, surveillance was in cohesive. Different hospitals and agencies worked without harmonization and data was rarely shared in the Healthcare space. A central National Health Management Information System database would go a long way enhance data-sharing in healthcare and detect trends of AMR.

Government's Policy Response: The Centre for Disease Control has work to establish a multidisciplinary cohesive national AMR surveillance network that would involve Universities, Hospitals, Laboratories and national surveillance units in the tracking and containment of AMR (NCDC, 2017). The designation of sentinel laboratory sites was adopted and are expected to report and provide data on key trackable pathogens across the country.

However, surveillance does not come cheap. It requires high-complexity medical laboratory technology and equipment, expensive reagents and consumables that are hardly locally available, highly skilled Scientists and laboratory personnel. Many rural areas have no capacity for this level of surveillance and constant power cuts remains a major challenge, leaving huge gaps and information-deficits in our national data.

The Federal Ministry as stakeholders in the One Health Agenda are also encouraged and expected to propel effort geared towards monitoring antibiotic residues in livestock, poultry products consumed in the major markets. Researchers have assayed high levels of antibiotics in livestock products sampled from markets, and the development should be stemmed in its tide (Olatoye *et al.*, 2016).

II. Regulatory Frameworks for Antibiotic Use

Some of the factors that promote the spread of AMR have been identified as poor public awareness, lack of improper use of antibiotics, inefficient management of medical, pharmaceutical and agricultural waste, and weak pharmacovigilance structures, high cost of healthcare and healthcare products which may lead to consideration self-medication.

It is understood that the Policy is moving towards tighter control over antimicrobial stewardship programs in hospitals—sets of rules ensuring doctors prescribe the right drug, at the right dose, for the right amount of time. However, it should not end at that, pharmacies, medical laboratories and other stakeholders in health, agriculture and environment should key into the fight against AMR, which is present crisis, if we must avoid a post-antimicrobial era where routine infections become life-threatening, because previously effective antimicrobials have become ineffectual in treating these infections.

C. Public Awareness Campaigns

Policies made in Abuja would amount to nothing if they are not implemented and understood by all and sundry. Enhancing public awareness is arguably the most important component of the AMR response in Nigeria.

I. Importance of Educating the Public

Public awareness creates a connection between medico-scientific knowledge and community action. The goal is to promote behavioral change in three main groups- The General public, Healthcare providers and Farmers.

The General public who are the end consumers of antimicrobials should be encouraged not to pressure physicians and other health professionals to prescribe or recommend antibiotics, as the causative agent for their illness may be a virus and not caused by bacteria pathogen. Patients are encouraged to always finish their prescribed dosage, even if they feel better. In the public, people share left-over antibiotic pills with family members and friends or stop taking medication once symptoms stops. This allows the strongest bacteria to survive the pharmacological measures and become resistant afterwards.

Healthcare Providers are expected to prescribe antibiotics based on medical diagnostic reports and educate patients on the correct usage. Poor prescription plan and inefficient dispensing of antimicrobials could negate the fight against AMR. Antibiotic Susceptibility testing performed in the medical laboratory serves as a guide for medical prescribers to choose the most viable alternative of antibiotic specific for that patient's treatment needs at that time (Ezeanya-Bakpa et al., 2023).

Veterinarians and Agricultural Workers are expected to maintain healthy farming, promote infection control and preventive measures to reduce the need for antibiotics and educate themselves on the best practices for livestock farming. It is inadvisable to use human antibiotics on animals, due the possibility of zoonotic transfer AMR bacteria.

II. Successful case studies

Successful public health campaigning cannot just be a billboard in a single language. The public must be addressed in several other languages ensuring all ethnic groups are represented.

For instance, in the 90s, the ministry of health leads campaigns that utilize local influencers, radio jingles in local dialects and many other local language spoken to educate people about HIV. Radio dramas portraying the dangers of unprotected proved effective in reaching grassroots populations who may not read newspapers or could afford to own a Television but are frequent radio listeners. This strategy ensured the triumph of the media over the HIV scourge which is evident in the fact the national prevalence rate of the disease among sexually-active young adult dropped to rate less than 2% (National Agency For The Control of AIDS, 2023; Olagunju *et al.,* 2025).

Combating AMR is a complex mix of adopting global standards and navigating realities in the socio-economic milieu. While the GAP provides the map and framework, and the regulatory bodies facilitates national and local strategies, the ultimate win over AMR in tied to enforcement and education.

Empowering regulatory bodies to truly regulate sales is important. However, we must also address the economic factors that make self-medication appealing. Most importantly, we must continue to translate technical AMR jargon into the local languages, ensuring that the average citizen regardless of literacy level understands that by demanding that "mixed drugs" antibiotic wrongly, may be leaving himself or herself defenseless against future illnesses.

FUTURE TRENDS

A. Integration of the "One Health" Approach: Linking Human, Animal, and Environmental Health

It is recommended that we stop treating human health, animal health, and environmental sanitation as different silos (ministries) that rarely talk share information. "One Health" is not just a catchy phrase from the WHO and other world agencies, it is imperative understand that human lives and survival depends on it.

I. The Environment: Integration imperative

Before now, the environmental component has been largely overlooked in the efforts against AMR.

Environmental agencies must work with pharmaceutical companies in industrial areas from dumping their untreated waste-water into rivers. This water contains high concentrations of antibiotics, turning the rivers into "a culture medium" for resistant bacteria.

B. Collaborative Efforts among Stakeholders: Governments, NGOs, and Pharmaceutical Companies

The responsibilities are not cut out for government alone; the pharmaceutical giant in Switzerland, and the local NGO in a rural community all have roles to play.

I. Public-Private Partnerships (PPPs)

In future, robust and enhanced PPPs are vital to fix supply chains and provide efficient service in providing the desired deliverables. The government can partner with local logistics firms and private pharmaceutical distributors to ensure that genuine, high-quality last-resort antibiotics actually reach remote public hospitals at a subsidized rate where they are needed, eliminating the vacuum that substandard and fake-drug peddlers occupy. Clinical laboratories should be encouraged supported to adopt genome-based surveillance and provide clear laboratory information that can translate to pragmatic bedside decisions and promote informed policy-making on AMR (Elbehiry *et al.,* 2025).

II. Engaging Pharmaceutical Industry

International pharma companies must be incentivized. For instance, instead of only looking for high profits, companies can engage in Product Development Alliances (PDAs). In this model, they work with research institutes to develop medical diagnostic tools suited for low-resource settings.

III. Community Engagement via NGOs

Since trust in government is often low. It becomes imperative to engage NGOs, as they are crucial for breaking behavior patterns. They can lead grassroots campaigns that translate "AMR" into community-driven initiative such as organizing clean water projects that reduce the incidence of diarrheal diseases, thereby reducing the *demand* for antibiotics. Several case studies have proven the efficacy of grassroot mobilization and pro-community engagements in combating scourges and enhancing healthcare.

C. Emphasis on Research and Development (R&D) Funding and Local Manufacturing

To be part of the paradigm shift, a country must move away from total dependence on imported medicines and imported research data. Local solutions for local challenges should be the watchword.

I. Changing the Economic Incentive

Funding Input: Direct grants to scientists and universities (through international partners) to undertake foundational research on local medicinal plants that might have antimicrobial properties.

Market Guarantee: The Nigerian government and perhaps ECOWAS, could guarantee that if a company develops a tool or drug designed for the local market, the government will buy a large quantity of it. This makes R&D profitable because the market is guaranteed.

II. Local Manufacturing of Generics

Empowerment of the local pharmaceutical manufacturing industry to produce high-quality generic antibiotics. When drugs are locally produced under strict regulatory supervision, we reduce costs, strengthen our supply chains, reduce the likelihood of imported fakes, and create jobs. R&D should also focus heavily on In

Vitro Diagnostic (IVD) development making quick test kits cheaper and more accessible for diagnostic laboratories and hospitals.

III. Strengthening Genomic Surveillance

Digital health is the future. There is the need to invest heavily in Whole Genome Sequencing (WGS) technology. Instead of just knowing a bacteria is resistant, WGS tells how the resistant genes were acquired and helps trace the spread of superbugs across the country in real-time. This is the ultimate intelligence gathering for the AMR war.

The future directions for AMR therapeutics and policy demands not just political will, but executive bravery. It requires us to acknowledge that promoting sanitation and good hygiene practices in a market in Aba is a medical intervention. It also requires breaking legal and economic barriers to fund R&D differently.

The successful integration of One Health, establish true harmonization between regulatory bodies and the private industry, and fund localized R&D, we will not only protect our citizens but set a blueprint for the rest of the developing world.

CONCLUSION

Antimicrobial resistance (AMR) stands as one of the most pressing public health challenges of our time, threatening the efficacy of antibiotics and other antimicrobial agents that have been pivotal in the treatment of infections. The rise of resistant pathogens not only complicates treatment protocols but also escalates healthcare costs, prolongs hospital stays, and increases morbidity and mortality rates. In response, a multifaceted approach involving both therapeutic innovations and robust policy frameworks has emerged to combat this global crisis.

From a therapeutic standpoint, the current landscape reflects a concerted effort to enhance existing treatments and develop novel antimicrobials. The pharmaceutical industry, recognizing the urgency of AMR, has initiated investment in the research and development (R&D) of new antibiotics and alternative therapies, including bacteriophage therapy, monoclonal antibodies, and immune-modulating drugs. Such innovations are essential, particularly as resistance profiles evolve. In addition to new drugs, emphasis has also been placed on the optimal use of existing antibiotics through stewardship programs. These programs are designed to ensure that antibiotics are prescribed judiciously, minimizing misuse and thereby reducing the selective pressure that drives resistance. Healthcare facilities worldwide are increasingly adopting antimicrobial stewardship guidelines to refine prescribing practices and monitor antibiotic use.

On the policy front, governments and international organizations are striving for coordinated responses to AMR through strategic frameworks. The World Health Organization (WHO) has launched the Global Action Plan on Antimicrobial Resistance, which urges member states to integrate AMR strategies into their national health agendas. Policy initiatives include strengthening regulatory frameworks to oversee antibiotic production and distribution, enhancing surveillance systems to track resistance patterns, and promoting public awareness campaigns to educate both healthcare providers and patients about the perils of antimicrobial mismanagement. Furthermore, collaborations among various stakeholders—governments, academia, industry, and non-governmental organizations—are essential for fostering innovation and ensuring effective implementation of AMR strategies.

Despite these efforts, challenges remain. The complexity of AMR necessitates ongoing research to fully understand the mechanisms of resistance and to explore alternative therapeutic avenues. Moreover, policy implementation often encounters hurdles related to funding, regulatory environments, and healthcare infrastructure, particularly in low- and middle-income countries where the burden of AMR is often greatest. Hence, it is imperative that global efforts continue to evolve, addressing not only the medical and scientific dimensions of AMR but also the socio-economic factors that contribute to its proliferation.

To effectively addressing antimicrobial resistance requires a synthesis of innovative therapeutic approaches and comprehensive policy responses. As the global community confronts this escalating threat, a commitment to stewardship, research, collaboration, and education will be crucial in safeguarding the effectiveness of antimicrobial medications for future generations. Continued vigilance and a united front are essential to combatting AMR and ensuring that effective treatments remain available to manage infectious diseases.

REFERENCES

Adekanye, U. O., Ekiri, A. B., Galipó, E., Muhammad, A. B., Mateus, A., La Ragione, R. M., Wakawa, A., Armson, B., Mijten, E., Alafiatayo, R., Varga, G., & Cook, A. J. C. (2020). Knowledge, Attitudes and Practices of Veterinarians Towards Antimicrobial Resistance and Stewardship in Nigeria. *Antibiotics (Basel, Switzerland)*, *9*(8), 453. DOI: 10.3390/antibiotics9080453 PMID: 32731420

Akturk, E., Oliveira, H., Santos, S. B., Costa, S., Kuyumcu, S., Melo, L. D. R., & Azeredo, J. (2019). Synergistic action of phage and antibiotics: Parameters to enhance the killing efficacy against mono and dual-species biofilms. *Antibiotics (Basel, Switzerland)*, *8*(3), 103. DOI: 10.3390/antibiotics8030103 PMID: 31349628

Akujobi, C. N., Ezeanya, C. C., & Aghanya, N. I. (2012). Detection of Cefotaximase Genes of Beta Lactamase among clinical isolates of *Escherichia coli* in a University Teaching Hospital, Nigeria. *Journal of Medical Sciences*, *12*(7), 244–247. DOI: 10.3923/jms.2012.244.247

An, Q., Huang, L., Wang, C., Wang, D., & Tu, Y. (2025). New strategies to enhance the efficiency and precision of drug discovery. *Frontiers in Pharmacology*, *16*, 1550158. DOI: 10.3389/fphar.2025.1550158 PMID: 40008135

Baquero, F., & Levin, B. R. (2021). Proximate and ultimate causes of the bactericidal action of antibiotics. *Nature Reviews. Microbiology*, *19*(2), 123–132. DOI: 10.1038/s41579-020-00443-1 PMID: 33024310

Bennett, P. M. (2008). Plasmid encoded antibiotic resistance: Acquisition and transfer of antibiotic resistance genes in bacteria. *British Journal of Pharmacology*, *153*(S1), S347–S357. DOI: 10.1038/sj.bjp.0707607 PMID: 18193080

Bharath, E., Thilagavathi, T., Jeeva, T. S., Anbazhagan, D. S., & Vadivel, S. A. (2025). Advances in Artificial Intelligence for drug discovery and development: A review of current trends and applications. *Asian Journal of Pharmaceutical Research and Development*, *13*(2), 67–78. DOI: 10.22270/ajprd.v13i2.1541

Biswas, B., Adhya, S., Washart, P., Paul, B., Trostel, A. N., Powell, B., Carlton, R., & Merril, C. R. (2002). Bacteriophage therapy rescues mice bacteremic from a clinical isolate of vancomycin-resistant Enterococcus faecium. *Infection and Immunity*, *70*(1), 204–210. DOI: 10.1128/IAI.70.1.204-210.2002 PMID: 11748184

Bo, L. T., Sun, H. D., Li, Y. D., Zhu, J. A. T., Wurpel, J. N. D., Lin, H. L., & Chen, Z. S. (2024). Combating antimicrobial resistance: The silent war. *Frontiers in Pharmacology*, *15*, 1347750. DOI: 10.3389/fphar.2024.1347750 PMID: 38420197

Bolla, J. M., Alibert-Franco, S., Handzlik, J., Chevalier, J., Mahamoud, A., Boyer, G., & Pagès, J.-M. (2011). Strategies for bypassing the membrane barrier in multidrug resistant Gram-negative bacteria. *FEBS Letters*, *585*(11), 1682–1690. DOI: 10.1016/j.febslet.2011.04.054 PMID: 21549704

Boullet, H., Bentot, F., Hequet, A., Ganem-Elbaz, C., Bechara, C., Pacreau, E., Launay, P., Sagan, S., Jolivalt, C., Lacombe, C., Moumné, R., & Karoyan, P. (2019). Small antimicrobial peptide with in vivo activity against sepsis. *Molecules (Basel, Switzerland)*, *24*(9), 1702. DOI: 10.3390/molecules24091702 PMID: 31052373

Brives, C., & Pourraz, J. (2020). Phage therapy as a potential solution in the fight against AMR: Obstacles and possible futures. *Palgrave Communications*, *6*(1), 100. DOI: 10.1057/s41599-020-0478-4

Burnham, C. A., Leeds, J., Nordmann, P., O'Grady, J., & Patel, J. (2017). Diagnosing antimicrobial resistance. *Nature Reviews. Microbiology*, *15*(11), 697–703. DOI: 10.1038/nrmicro.2017.103 PMID: 29021600

Campoccia, D., Montanaro, L., Ravaioli, S., Mariani, V., Bottau, G., De Donno, A., & Arciola, C. R. (2023). Antibacterial activity on orthopedic clinical isolates and cytotoxicity of the antimicrobial peptide Dadapin-1. *International Journal of Molecular Sciences*, *24*(1), 779. DOI: 10.3390/ijms24010779 PMID: 36614222

CDC. (2019). Antimicrobial Resistance Facts and Stats. Antimicrobial CDC. Implementation of Antibiotic Stewardship Core Elements at Small and Critical Access Hospitals. Available online: https://www.cdc.gov/antibiotic-use/core-elements/small-critical.html (accessed on 12 November 2022).

Centers for Disease Control and Prevention. (2019). *Antibiotic resistance threats in the United States, 2019*. U.S. Department of Health and Human Services.

Chang, Q., Wang, W., Regev-Yochay, G., Lipsitch, M., & Hanage, W. P. (2015). Antibiotics in agriculture and the risk to human health: How worried should we be? *Evolutionary Applications*, *8*(3), 240–247. DOI: 10.1111/eva.12185 PMID: 25861382

Chaw, P. S., Höpner, J., & Mikolajczyk, R. (2018). The knowledge, attitude and practice of health practitioners towards antibiotic prescribing and resistance in developing countries-A systematic review. *Journal of Clinical Pharmacy and Therapeutics*, *43*(5), 606–613. DOI: 10.1111/jcpt.12730 PMID: 29959783

Chokshi, A., Sifri, Z., Cennimo, D., & Horng, H. (2019). Global Contributors to Antibiotic Resistance. *Journal of Global Infectious Diseases*, *11*(1), 36–42. DOI: 10.4103/jgid.jgid_110_18 PMID: 30814834

Collaborators, A. R. (2022). Global burden of bacterial antimicrobial resistance in 2019: A systematic analysis. *Lancet, 399*(10325), 629–655. DOI: 10.1016/S0140-6736(21)02724-0 PMID: 35065702

Collineau, L., Boerlin, P., Carson, C. A., Chapman, B., Fazil, A., Hetman, B., & Parmley, E. J. (2019). Integrating whole-genome sequencing data into quantitative risk assessment of foodborne antimicrobial resistance: A review of opportunities and challenges. *Frontiers in Microbiology, 10*, 1107. DOI: 10.3389/fmicb.2019.01107 PMID: 31231317

DiMasi, J. A., Grabowski, H. G., & Hansen, R. W. (2016). Innovation in the pharmaceutical industry: New estimates of R&D costs. *Journal of Health Economics, 47*, 20–33. DOI: 10.1016/j.jhealeco.2016.01.012 PMID: 26928437

Dutescu, I. A., & Hillier, S. A. (2021). Encouraging the Development of New Antibiotics: Are Financial Incentives the Right Way Forward? A Systematic Review and Case Study. *Infection and Drug Resistance, 14*, 415–434. DOI: 10.2147/IDR.S287792 PMID: 33574682

Elbehiry, A., Marzouk, E., & Abalkhail, A. (2025). Antimicrobial resistance at a turning point: Microbial drivers, one health, and global futures. *Frontiers in Microbiology, 16*, 1698809. DOI: 10.3389/fmicb.2025.1698809 PMID: 41334162

Endale, H., Mathewos, M., & Abdeta, D. (2023). Potential causes of spread of antimicrobial resistance and preventive measures in one health perspective-a review. *Infection and Drug Resistance, 16*, 7515–7545. DOI: 10.2147/IDR.S428837 PMID: 38089962

Ezeanya-Bakpa, C. C., Adetunji, C. O., Enosabata, I. A., & Olori, E. (2021a). Lycopene for wound infection: In vitro susceptibility of drug-resistant clinical pathogens. *Tropical Journal of Natural Product Research, 5*(1), 49–52. DOI: 10.26538/tjnpr/v5i1.4

Ezeanya-Bakpa, C. C., Agbakoba, N. R., Enweani-Nwokelo, I. B., & Oguejiofor, C. B. (2022a). Phylogeny-based identification of *Mycoplasma genitalium* in a Nigerian population of apparently healthy sexually active female students. *The Pan African Medical Journal, 41*, 71–76. DOI: 10.11604/pamj.2022.41.71.19631 PMID: 35382058

Ezeanya-Bakpa, C. C., Agbakoba, N. R., Oguejiofor, C. B., & Enweani-Nwokelo, I. B. (2021b). Sequence analysis reveals asymptomatic infection with *Mycoplasma hominis* and *Ureaplasma urealyticum* possibly leads to infertility in females: A cross-sectional study. *International Journal of Reproductive Biomedicine, 19*(11), 951–958. DOI: 10.18502/ijrm.v19i11.9910 PMID: 34977452

Ezeanya-Bakpa, C.C., Akinola, K., Adebayo, O.P., Akharaiyi, .F.C. (2023) Antibacterial combination effect of allicin, lycopene and quercetin on antibiotic-resistant bacteria. *Microbes and Infectious Diseases* 5(1) Published First Online Doi: DOI: 10.21608/mid.2023.253557.1701

Ezeanya-Bakpa, C. C., Lambe, F. E., Adebayo-Olajide, T. C., & Esosa-Enoyoze, G. E. (2026). From Lab to Field: Translating Natural Products into Clinical Candidates. In Ogidi, O. (Ed.), *Natural Antimalarial Drug Discovery and Development in the Global South* (pp. 137–168). IGI Global Scientific Publishing., DOI: 10.4018/979-8-3373-5821-5.ch005

Ezeanya-Bakpa, C. C., Lambe, F. E., Ayinde, K., & Adeyeye, N. T. (2025a). Synergistic Antibacterial evaluation of Allicin and *Psidium Guajava* Linnaeus with standard antibiotics. *Tanzania Journal of Health Research, 26*(6), 2881–2893. DOI: 10.4314/thrb.v26i6.13

Ezeanya-Bakpa, C. C., Lambe, F. E., Okwelogu, I. S., Enoyoze, G. E., Dirisu, J., Adediran, B. O., & Alalade, L. I. (2025b). Synergistic antibacterial effect of *Mangifera indica* and *Annona muricata* on antibiotic-resistant wound isolates. *African Journal of Biomedical Research, 28*(1), 174–181. DOI: 10.53555/AJBR.v28i1.5010

Ezeanya-Bakpa, C. C., Shorumu, O., & Ogunsoluju, T. (2022b). Mobile devices and accessories are reservoir for drug-resistant bacteria among university students: A cross-sectional study. *Journal of Medical Pharmaceutical and Allied Sciences, 11*(6), 5494–5499. DOI: 10.55522/jmpas.V11I6.4217

Ezeanya-Bakpa, C. C., Udeogu, C. V., Agbakoba, N. R., Uduchi, I., & Ejike, C. (2024). Probiotics: Mitigating antibiotic resistance in the vaginal microbiome. *Microbes and Infectious Diseases, 5*(3), 1036–1045. DOI: 10.21608/mid.2024.266564.1782

Fairhead, H. (2009). SASP gene delivery: A novel antibacterial approach. *Drug News & Perspectives, 22*(4), 197–203. DOI: 10.1358/dnp.2009.22.4.1367708 PMID: 19536364

Graf, F. E., Palm, M., Warringer, J., & Farewell, A. (2019). Inhibiting conjugation as a tool in the fight against antibiotic resistance. *Drug Development Research, 80*(1), 19–23. DOI: 10.1002/ddr.21457 PMID: 30343487

Gupta, A., Saleh, N. M., Das, R., Landis, R. F., Bigdeli, A., Motamedchaboki, K., & Rotello, V. M. (2017). Synergistic antimicrobial therapy using nanoparticles and antibiotics for the treatment of multidrug-resistant bacterial infection. *Nano Futures, 1*(1), 015004. DOI: 10.1088/2399-1984/aa69fb

Henriques-Normark, B., & Normark, S. (2014). Bacterial vaccines and antibiotic resistance. *Upsala Journal of Medical Sciences*, *119*(2), 205–208. DOI: 10.3109/03009734.2014.903324 PMID: 24694025

Iregbu, K. C., Nwajiobi-Princewill, P. I., Medugu, N., Umeokonkwo, C. D., Uwaezuoke, N. S., Peter, Y. J., Nwafia, I. N., Elikwu, C., Shettima, S. A., Suleiman, M. R., Awopeju, A. T. O., Udoh, U., Adedosu, N., Mohammed, A., Oshun, P., Ekuma, A., Manga, M. M., Osaigbovo, I. I., Ejembi, C. J., & Oduyebo, O. O.Iregbu KC. (2021). Antimicrobial Stewardship Implementation in Nigerian Hospitals: Gaps and Challenges. *African Journal of Clinical and Experimental Microbiology*, *22*(1), 60–66. DOI: 10.4314/ajcem.v22i1.8

Iskandar, K., Murugaiyan, J., Halat, D. H., Hage, S. E., Chibabhai, V., Adukkadukkam, S., Roques, C., Molinier, L., Salameh, P., & Van Dongen, M. (2022). Antibiotic Discovery and Resistance: The Chase and the Race. *Antibiotics (Basel, Switzerland)*, *11*(2), 182. DOI: 10.3390/antibiotics11020182 PMID: 35203785

Jassim, S. A., & Limoges, R. G. (2014). Natural solution to antibiotic resistance: Bacteriophages 'The Living Drugs'. *World Journal of Microbiology & Biotechnology*, *30*(8), 2153–2170. DOI: 10.1007/s11274-014-1655-7 PMID: 24781265

Joerger, R. D. (2003). Alternatives to antibiotics: Bacteriocins, antimicrobial peptides and bacteriophages. *Poultry Science*, *82*(4), 640–647. DOI: 10.1093/ps/82.4.640 PMID: 12710486

Jonas, O. B., Irwin, A., Berthe, F. C. J., Le Gall, F. G., & Marquez, P. V. (2017). Drug-resistant infections: A threat to our economic future (Vol. 2 of 2): Final report. Washington, DC: World Bank Group.

Klein, E. Y., Van Boeckel, T. P., Martinez, E. M., Pant, S., Gandra, S., Levin, S. A., Goossens, H., & Laxminarayan, R. (2018). Global increase and geographic convergence in antibiotic consumption between 2000 and 2015. *Proceedings of the National Academy of Sciences of the United States of America*, *115*(15), E3463–E3470. DOI: 10.1073/pnas.1717295115 PMID: 29581252

Kumar, G., Yadav, S., Mukherjee, A., Hassija, V., & Guizani, M. (2024). Recent advances in quantum computing for drug discovery and development. *IEEE Access : Practical Innovations, Open Solutions*, *6*(3), 64491–64509. DOI: 10.1109/ACCESS.2024.3376408

Kyriakidis, I., Vasileiou, E., Pana, Z. D., & Tragiannidis, A. (2021). Acinetobacter baumannii antibiotic resistance mechanisms. *Pathogens (Basel, Switzerland)*, *10*(3), 373. DOI: 10.3390/pathogens10030373 PMID: 33808905

Laxminarayan, R., Matsoso, P., Pant, S., Brower, C., Røttingen, J. A., Klugman, K., & Davies, S. (2016). Access to effective antimicrobials: A worldwide challenge. *Lancet, 387*(10014), 168–175. DOI: 10.1016/S0140-6736(15)00474-2 PMID: 26603918

Lipsitch, M., & Samore, M. H. (2002). Antimicrobial use and antimicrobial resistance: A population perspective. *Emerging Infectious Diseases, 8*(4), 347–354. DOI: 10.3201/eid0804.010312 PMID: 11971765

Lyster, H. (2016). Antimicrobial stewardship in the immunocompromised patient. In *Antimicrobial Stewardship*. Oxford University Press.

McDermott, P. F. (2005). Antimicrobial resistance in nontyphoidal Salmonellae. In Antimicrobial resistance in bacteria of animal origin (pp. 293-314). DOI: 10.1128/9781555817534.ch17

Mellor, K. C., Petrovska, L., Thomson, N. R., Harris, K., Reid, S. W. J., & Mather, A. E. (2019). Antimicrobial resistance diversity suggestive of distinct salmonella typhimurium sources or selective pressures in food-production animals. *Frontiers in Microbiology, 10*, 708. DOI: 10.3389/fmicb.2019.00708 PMID: 31031720

Merril, C. R., Biswas, B., Carlton, R., Jensen, N. C., Creed, G. J., Zullo, S., & Adhya, S. (1996). Long-circulating bacteriophage as antibacterial agents. *Proceedings of the National Academy of Sciences of the United States of America, 93*(8), 3188–3192. DOI: 10.1073/pnas.93.8.3188 PMID: 8622911

Micoli, F., Bagnoli, F., Rappuoli, R., & Serruto, D. (2021). The role of vaccines in combatting antimicrobial resistance. *Nature Reviews. Microbiology, 19*(5), 287–302. DOI: 10.1038/s41579-020-00506-3 PMID: 33542518

Miethke, M., Pieroni, M., Weber, T., Brönstrup, M., Hammann, P., Halby, L., Arimondo, P. B., Glaser, P., Aigle, B., Bode, H. B., Moreira, R., Li, Y., Luzhetskyy, A., Medema, M. H., Pernodet, J. L., Stadler, M., Tormo, J. R., Genilloud, O., Truman, A. W., & Müller, R. (2021). Towards the sustainable discovery and development of new antibiotics. *Nature Reviews. Chemistry, 5*(10), 726–749. DOI: 10.1038/s41570-021-00313-1

Murray, C. J. L., Ikuta, K. S., Sharara, F., Swetschinski, L., Robles Aguilar, G., Gray, A., Han, C., Bisignano, C., Rao, P., Wool, E., Johnson, S. C., Browne, A. J., Chipeta, M. G., Fell, F., Hackett, S., Haines-Woodhouse, G., Kashef Hamadani, B. H., Kumaran, E. A. P., McManigal, B., & Naghavi, M. (2022). Global burden of bacterial antimicrobial resistance in 2019: A systematic analysis. *Lancet, 399*(10325), 629–655. DOI: 10.1016/S0140-6736(21)02724-0 PMID: 35065702

Nardulli, P., Hall, G. G., Quarta, A., Fruscio, G., Laforgia, M., & Scacco, S. (2022). Antibiotic abuse and antimicrobial resistance in hospital environment: A retrospective observational comparative study. *Medicina (Kaunas, Lithuania)*, *58*(9), 1257. DOI: 10.3390/medicina58091257 PMID: 36143934

National Agency For The Control of AIDS. (2023). The role of the media in Nigeria's fight against HIV. Retrieved from https://naca.gov.ng/the-role-of-media-in-nigerias-fight-against-hiv/ on 1st of December, 2025.

Nguyen-Thanh, L., Wernli, D., Målqvist, M., Graells, T., & Jørgensen, P. S. (2023). Characterizing proximal and distal drivers of antimicrobial resistance: An umbrella review. *Journal of Global Antimicrobial Resistance*, *36*, 50–58. DOI: 10.1016/j.jgar.2023.12.008 PMID: 38128730

Nigeria Centre for Disease Control (NCDC). (2017). National Action Plan on Antimicrobial Resistance 2017-2022: The strategic plan. Retrieved from 77_1511368219.pdf on the 8th of December, 2025

Ohia, C. M., Falodun, O. I., Adebudo, L. I., & Bakarey, A. S. (2025). A one health perspective on multidrug-resistant bacterial infections: Integrated approaches for surveillance, policy and innovation. *Frontiers in Cellular and Infection Microbiology*, *15*, 1614232. DOI: 10.3389/fcimb.2025.1614232 PMID: 40969553

Olagunju, O. J., Ben, E., Olagunju, O., Majolagbe, O. G., Osanyinlusi, O. O., Adewoye, T., Atoyebi, O. F., Ojo, I. O., & Dawha, S. D. (2025). Poorly Regulated Antibiotic Use in Nigeria: A Critical Public Health Concern and Its Impact on Medical Practice. *Cureus*, *17*(6), e85212. DOI: 10.7759/cureus.85212 PMID: 40605881

Olatoye, I. O., Daniel, O. F., & Ishola, S. A. (2016). Screening of antibiotics and chemical analysis of penicillin residue in fresh milk and traditional dairy products in Oyo state, Nigeria. *Veterinary World*, *9*(9), 948–954. DOI: 10.14202/vetworld.2016.948-954 PMID: 27733794

Ozawa, S., Evans, D. R., Bessias, S., Haynie, D. G., Yemeke, T. T., Laing, S. K., & Herrington, J. E. (2018). Prevalence and Estimated Economic Burden of Substandard and Falsified Medicines in Low- and Middle-Income Countries: A Systematic Review and Meta-analysis. *JAMA Network Open*, *1*(4), e181662. DOI: 10.1001/jamanetworkopen.2018.1662 PMID: 30646106

Padole, S. S., Asnani, A. J., Chaple, D. R., & Katre, S. G. (2022). A review of approaches in computer-aided drug design in drug discovery. GSC Biological and Pharmaceutical Sciences, 19(2), 075-083.

Parry, C. M., & Threlfall, E. J. (2008). Antimicrobial resistance in typhoidal and nontyphoidal salmonellae. *Current Opinion in Infectious Diseases*, *21*(5), 531–538. DOI: 10.1097/QCO.0b013e32830f453a PMID: 18725804

Patel, J. R., Joshi, H. V., Shah, U. A., & Patel, J. K. (2022). A review on computational software tools for drug design and discovery. *Indo Global Journal of Pharmaceutical Sciences*, *12*, 53–81. DOI: 10.35652/IGJPS.2022.12006

Patton, E. E., Zon, L. I., & Langenau, D. M. (2021). Zebrafish disease models in drug discovery: From preclinical modelling to clinical trials. *Nature Reviews. Drug Discovery*, *20*(8), 611–628. DOI: 10.1038/s41573-021-00210-8 PMID: 34117457

Perišić, O., Sevim Bayrak, C., & Gunady, M. K. (2025). Machine learning in computer-aided drug design. *Frontiers in Molecular Biosciences*, *12*, 1568437. DOI: 10.3389/fmolb.2025.1568437 PMID: 40070689

Pervin, Z., & Hassan, M. M. (2021). Synergistic therapeutic actions of antimicrobial peptides to treat multidrug-resistant bacterial infection. *Reviews in Medical Microbiology : a Journal of the Pathological Society of Great Britain and Ireland*, *32*(2), 83–89. DOI: 10.1097/MRM.0000000000000239

Pinto Ferreira, J., Battaglia, D., Dorado García, A., Tempelman, K., Bullon, C., Motriuc, N., Caudell, M., Cahill, S., Song, J., & LeJeune, J. (2022). Achieving antimicrobial stewardship on the global scale: Challenges and opportunities. *Microorganisms*, *10*(8), 1599. DOI: 10.3390/microorganisms10081599 PMID: 36014017

Pires, A. J., Pereira, G., Fangueiro, D., Bexiga, R., & Oliveira, M. (2024). When the solution becomes the problem: A review on antimicrobial resistance in dairy cattle. *Future Microbiology*, *19*(10), 903–929. DOI: 10.2217/fmb-2023-0232 PMID: 38661710

Prestinaci, F., Pezzotti, P., & Pantosti, A. (2015). Antimicrobial resistance: A global multifaceted phenomenon. *Pathogens and Global Health*, *109*(7), 309–318. DOI: 10.1179/2047773215Y.0000000030 PMID: 26343252

Riccardi, L., Genna, V., & De Vivo, M. (2018). Metal–ligand interactions in drug design. *Nature Reviews. Chemistry*, *2*(7), 100–112. DOI: 10.1038/s41570-018-0018-6

Shang, D., Liu, Y., Jiang, F., Ji, F., Wang, H., & Han, X. (2019). Synergistic antibacterial activity of designed Trp-containing antibacterial peptides in combination with antibiotics against multidrug-resistant Staphylococcus epidermidis. *Frontiers in Microbiology*, *10*, 2719. DOI: 10.3389/fmicb.2019.02719 PMID: 31824473

Shehreen, S., Chyou, T. Y., Fineran, P. C., & Brown, C. M. (2019). Genome-wide correlation analysis suggests different roles of CRISPR-Cas systems in the acquisition of antibiotic resistance genes in diverse species. Philosophical Transactions of the Royal Society B: Biological Sciences, 374(1772), 20180384.

Siddiqui, B., Yadav, C. S., Akil, M. F., Faiyyaz, M., Khan, A. R., Ahmad, N., & Azad, I. (2025). Artificial Intelligence in Computer-Aided Drug Design (CADD) tools for the finding of potent biologically active small molecules: Traditional to modern approach. *Combinatorial Chemistry & High Throughput Screening*, *28*, 1–50. DOI: 10.2174/0113862073334062241015043343 PMID: 39819404

Smith, T., Wolff, K. A., & Nguyen, L. (2013). Molecular biology of drug resistance in Mycobacterium tuberculosis. *Current Topics in Microbiology and Immunology*, *374*, 53–80. DOI: 10.1007/82_2012_279 PMID: 23179675

Soothill, J. S. (1992). Treatment of experimental infections of mice with bacteriophages. *Journal of Medical Microbiology*, *37*(4), 258–261. DOI: 10.1099/00222615-37-4-258 PMID: 1404324

Sun, W., Weingarten, R. A., Xu, M., Southall, N., Dai, S., Shinn, P., Sanderson, P. E., Williamson, P. R., Frank, K. M., & Zheng, W. (2016). Rapid antimicrobial susceptibility test for identification of new therapeutics and drug combinations against multidrug-resistant bacteria. *Emerging Microbes & Infections*, *5*(1), 1–11. DOI: 10.1038/emi.2016.123 PMID: 27826141

Thomas, C. M., & Nielsen, K. M. (2005). Mechanisms of, and barriers to, horizontal gene transfer between bacteria. *Nature Reviews. Microbiology*, *3*(9), 711–721. DOI: 10.1038/nrmicro1234 PMID: 16138099

Touchon, M., Moura de Sousa, J. A., & Rocha, E. P. (2017). Embracing the enemy: The diversification of microbial gene repertoires by phage-mediated horizontal gene transfer. *Current Opinion in Microbiology*, *38*, 66–73. DOI: 10.1016/j.mib.2017.04.010 PMID: 28527384

Tsigalou, C., Konstantinidis, T., Stavropoulou, E., Bezirtzoglou, E. E., & Tsakris, A. (2020). Potential elimination of human gut resistome by exploiting the benefits of functional foods. *Frontiers in Microbiology*, *11*, 50. DOI: 10.3389/fmicb.2020.00050 PMID: 32117102

Walsh, C. (2000). Molecular mechanisms that confer antibacterial drug resistance. *Nature*, *406*(6797), 775–781. DOI: 10.1038/35021219 PMID: 10963607

Wang, L. M., Cravo Oliveira Hashiguchi, T., & Cecchini, M. (2021). Impact of vaccination on carriage of and infection by antibiotic-resistant bacteria: A systematic review and meta-analysis. *Clinical and Experimental Vaccine Research, 10*(2), 81–92. DOI: 10.7774/cevr.2021.10.2.81 PMID: 34222121

Wang, Y., Yu, Z., Ding, P., Lu, J., Mao, L., Ngiam, L., & Zhang, T. (2023). Antidepressants can induce mutation and enhance persistence toward multiple antibiotics. *Proceedings of the National Academy of Sciences of the United States of America, 120*(5), e2208344120. DOI: 10.1073/pnas.2208344120 PMID: 36689653

World Health Organisation (2025). Nigeria Launches First National Antimicrobial Resistance Survey. Retrieved from Nigeria Launches First National Antimicrobial Resistance Survey | WHO | Regional Office for Africa on 9th of December, 2025

Worthington, R. J., & Melander, C. (2013). Combination approaches to combat multidrug-resistant bacteria. *Trends in Biotechnology, 31*(3), 177–184. DOI: 10.1016/j.tibtech.2012.12.006 PMID: 23333434

Yang, Y. Y., Xie, S. Y., He, F. J., Xu, Y. D., Wang, Z. F., Ihsan, A., & Wang, X. (2024). Recent development and fighting strategies for lincosamide antibiotic resistance. *Clinical Microbiology Reviews, 37*(1), e00161–e23. DOI: 10.1128/cmr.00161-23 PMID: 38634634

Zharkova, M. S., Komlev, A. S., Filatenkova, T. A., Sukhareva, M. S., Vladimirova, E. V., Trulioff, A. S., Orlov, D. S., Dmitriev, A. V., Afinogenova, A. G., Spiridonova, A. A., & Shamova, O. V. (2023). Combined use of antimicrobial peptides with antiseptics against multidrug-resistant bacteria: Pros and cons. *Pharmaceutics, 15*(1), 291. DOI: 10.3390/pharmaceutics15010291 PMID: 36678918

Zhu, Y., Huang, W. E., & Yang, Q. (2022). Clinical perspective of antimicrobial resistance in bacteria. *Infection and Drug Resistance, 15*, 735–746. DOI: 10.2147/IDR.S345574 PMID: 35264857

Chapter 4 Traditional Knowledge as a Compass: Navigating the Therapeutic Landscape of Emerging Diseases

Saritha Thampi
http://orcid.org/0009-0000-8487-5418
CMS College Kottayam, Kerala, India

Soumya Raju
http://orcid.org/0009-0004-1701-6743
CMS College Kottayam, Kerala, India

Rogimon P. Thomas
CMS College Kottayam, Kerala, India

ABSTRACT

Emerging infectious diseases (EIDs) represent a compounding global health crisis that breach the hierarchical barriers secernating wildlife reservoirs from humans. The unprecedented velocity and geographic expansion of pathogens and widespread antimicrobial resistance have severely constrained conventional medical interventions. Simultaneously, traditional knowledge systems accumulated over centuries offer an underexploited therapeutic arsenal to combat diseases. The synergistic integration of traditional medicinal knowledge creates unprecedented opportunities for identifying novel therapeutic pathways and preventive strategies. Ethnopharmacological investigations have demonstrated that traditional remedies employed by indigenous communities often contain bioactive phytochemicals with

DOI: 10.4018/979-8-3373-7310-7.ch004

demonstrable antimicrobial, antiviral and immunomodulatory properties. The chapter is an attempt to explore how traditional knowledge serves as a strategic compass to guide and accelerate the discovery of therapeutic agents for emerging diseases.

INTRODUCTION

The trajectory of human health in the 21st is increasingly defined by a paradox of progress and vulnerability: while biotechnological sophistication has reached unprecedented heights, the global population remains precariously exposed to the swift and often devastating emergence of novel pathogens (Asim *et al.,* 2022). Driven by a complex interplay of anthropogenic factors such as quick urbanization, belligerent land-use change, climate change and increased international trade and travel, emerging infectious diseases (EIDs) represent a compounding global crisis which transcends national borders and biological distinctness (Sahai and Kumar, 2025; Agyarko *et al.,* 2025). The accelerated rate of zoonotic spillover by the breaking of the ecological and physiological barriers defines the biological reality of the Anthropocene (Ellwanger and Chies, 2021).

According to major ecological evaluations, land-use change alone is responsible for over thirty percent of emergent disease occurrences (McFarlane *et al.,* 2013). The surreptitious, creeping pandemic of antimicrobial resistance (AMR) further worsen the scene. The AMR threatens to reverse a century of medical progress by making traditional pharmacological therapies useless, further exacerbates this susceptibility (Oliveira *et al.,* 2024). The shortcomings of the dependence on single-target, high-potency synthetic molecules have become glaringly obvious as the pipeline for novel antibiotics and antivirals gets smaller and the *de novo* drug development process takes decades. Thus, there is a strong scientific and humanitarian need to expand the epistemological maps utilized for preparation and discovery in this setting of pressing therapeutic requirement and systemic instability. An essentially underutilized therapeutic toolkit is provided by traditional knowledge (TK), especially the ethnobotanical and ethnomedical systems amassed throughout millennia of lived human-environment interaction. It is a dynamic, empirically supported compass that may traverse the intricate therapeutic terrain of new diseases rather than just a historical relic or a compilation of folklore.

Traditional knowledge functions as a compass stem from the awareness that, throughout the great majority of human history, Indigenous and local communities have been the world's main guardians of biodiversity and health. These knowledge systems are the culmination of ecological adaptation, trial-and-error experimentation and long-term empirical observation (Kumar *et al.,* 2021). Unlike modern pharmaceutical screening, which often searches enormous, randomized chemical libraries for bioactive hits, ethnopharmacology has a history of safe and effective use in human societies. This method greatly reduces the pool of potential therapies by prioritizing species and formulations that have already passed the generational selection screen. Thus, the chapter is an attempt to explore how traditional knowledge serves as a strategic compass to guide and accelerate the discovery of therapeutic agents for emerging diseases.

THE AGE OF EMERGING DISEASES

The term "emerging infectious diseases" (EIDs) refers to infections that have either just surfaced in a community or that were previously present but are currently quickly expanding in frequency or geographic scope. Diseases that were previously under control but have returned or acquired new traits, including antibiotic resistance, are also included in this category (WHO, 2014). The scope of the phenomenon of emerging infectious diseases is astounding. Pathogens that spread from animal reservoirs are responsible for up to 75% of newly discovered infectious diseases (Sharan *et al.,* 2023). Approximately 60% of newly reported infectious diseases worldwide are now developing zoonoses, with over 30 novel human infections discovered in the past three decades, 75% of which originated in animals (Rahman *et al.,* 2020). Thus, emerging infectious disease events are on the rise for the past few decades and the number of recorded spillover occurrences from zoonoses associated to wildlife has consistently increased. The loss of biodiversity is the main mechanical driver of this trajectory, which is not accidental but rather a direct result of human activity (Adebayo, 2019).

The pathways through which diseases emerge are complex and multifactorial, involving the intricate interplay between pathogen characteristics, host susceptibility and ecological conditions. At the molecular level, zoonotic pathogens must overcome a hierarchical series of barriers to establish themselves in human populations. The process begins with contact between microbes and animals, followed by infection and amplification in the human host. For successful spillover, pathogens must possess cellular and tissue tropism - the ability to recognize and bind to human cell receptors and replicate within human tissues despite the pressures of innate and adaptive immunity. This adaptation is facilitated by the high mutation rates and genetic plasticity of RNA viruses, which can rapidly acquire novel traits through mutation, reassortment and recombination, enabling them to optimize interactions with their new hosts. The kinetics of this process vary dramatically; SARS-CoV-2 became pandemic within three to four months of the initial spillover event, while HIV-1 required approximately 50 years to emerge as a global pandemic after its presumed initial transmission from chimpanzees to humans (Ellwanger and Chies, 2021; Escudero-Pérez *et al.,* 2023).

The contemporary drivers of disease emergence are inextricably linked to human environmental transformation. Urbanization, characterized by rapid agricultural intensification, ecological fragmentation, and land-use change, has fundamentally altered the contact dynamics between wildlife, domestic animals, and human populations. Deforestation of tropical forests, expansion of agricultural frontiers, and the intensification of livestock production systems simultaneously destroy wildlife habitats and create novel opportunities for pathogen spillover. Concurrent with these changes, the global wildlife trade—including the trafficking of bushmeat and exotic species—has become a significant vector for zoonotic pathogen introduction into human populations.

Rapid agricultural intensification, ecological fragmentation and land-use change are hallmarks of urbanization, which has drastically changed the dynamics of interactions between domestic animals, wildlife and human populations. The intensification of livestock production systems, the expansion of agricultural borders and the deforestation of tropical forests all contribute to the destruction of wildlife habitats and the creation of new disease spillover opportunities. Alongside these developments, the international wildlife trade has grown to be a major conduit for the introduction of zoonotic pathogens into human populations. By changing species distributions, vector-borne disease geographic ranges and the creation of novel ecological niches conducive to pathogen transmission, climate change exacerbates these consequences alarmingly (Carbone *et al.,* 2025).

On the otherside, equally troubling as the emergence of novel pathogens is the phenomenon of antimicrobial resistance, which transforms known infections into intractable therapeutic challenges. Antimicrobial resistance occurs when microbes evolve mechanisms protecting them from antimicrobial drugs through spontaneous mutation, horizontal gene transfer, and, critically, increased selective pressure from antimicrobial overuse in both medical and agricultural settings (Reygaert, 2018; Dominic Terkimbi *et al.,* 2025). The World Health Organization has declared that antimicrobial resistance constitutes one of the top global public health threats, with bacterial resistance directly responsible for 1.27 million global deaths in 2019 and contributing to an estimated 4.95 million deaths (WHO, 2023). Without intervention, projections suggest that antimicrobial resistance could result in up to 10 million deaths annually by 2050. This convergence of emerging diseases and antimicrobial resistance creates a particularly perilous scenario wherein novel pathogens may prove resistant to existing therapeutic arsenals.

TRADITIONAL KNOWLEDGE: A LIVING SYSTEM OF CLASSIFICATION, PRACTICE AND ECOLOGY

Traditional ethnobotanical knowledge represents a sophisticated, multidimensional system of understanding, classification and interaction with the botanical world that has evolved through centuries of accumulated observation, cultural practice and ecological integration. This knowledge system, deeply embedded within the worldviews and lifeways of indigenous and traditional communities, constitutes a dynamic living archive of ecological wisdom, botanical classification, practical applications and cultural meaning. Unlike the static categorization of scientific taxonomy, ethnobotanical knowledge remains fluid, responsive to environmental change and intimately connected to the social, spiritual and economic dimensions of community existence (Luković *et al.,* 2023).

Traditional ethnobotanical knowledge encompasses an integrated system of folk taxonomy, vernacular naming and classification systems. Folk taxonomies are generated from social knowledge and emerge from the cultural and ecological contexts in which they develop, functioning to organize plants according to multiple simultaneous criteria including morphological features, habitat characteristics, utilitarian functions, cultural significance and sensory perception. This deep ecological knowledge embedded within folk systems; plants that share morphological characteristics often occupy similar ecological niches and possess related utilitarian properties, creating a taxonomy that simultaneously encodes ecological and practical information (Poncet *et al.,* 2015).

Ethnobotanical knowledge has practical applications in many areas of human need and cultural expression. Approximately 75 to 90 percent of rural inhabitants in developing countries still rely heavily on herbal medicines for healthcare, making ethnomedicinal usage possibly the most well-documented domain (Kibebew, 2001). Because access to pharmaceutical services is still restricted in rural and isolated locations, this reflects both cultural preference and practical necessity. However, traditional ethnobotanical knowledge extends beyond medicine to include techniques for food security, such as identifying and managing wild edible plants that offer vital nutritional resources amid seasonal scarcity or environmental stress. Other realms of knowledge application include building materials, fuel, fiber manufacturing, ceremonial purposes, and spiritual practices.

The convergence of ethnobotanical knowledge with modern scientific inquiry has generated significant synergies in pharmaceutical development and ecological conservation (Pandey and Tripathi, 2017). Historical analysis demonstrates that numerous globally significant pharmaceuticals originated in traditional medicine, including taxol derived from Pacific yew bark, vinblastine from Madagascar periwinkle, quinine from cinchona bark and artemisinin from *Artemisia annua* (Davis and Choisy, 2024). Ethnobotanical approaches to drug discovery accelerate pharmaceutical development by directing researchers toward plants with documented efficacy in traditional contexts. Studies consistently demonstrate that plants used in traditional medicine exhibit higher bioactivity than randomly selected plant species, thereby reducing the time, cost and resource investment required for successful drug discovery. In addition, congeneric medicinal plants are more likely to be used for treating similar indications across geographically distinct cultures, reflecting shared biochemical profiles and phytochemical properties. This concordance enables mining of ethnobotanical patterns to identify taxonomic hotspots around particular indications and to investigate shared therapeutic uses of non-taxonomically related plants that possess shared bioactive compounds (Domingo-Fernández *et al.,* 2023).

Thus, traditional ethnobotanical knowledge, with its deep, place-based understanding of plant-medicine-animal-environment interdependencies, offers a critical conceptual and practical framework for advancing One Health. The holistic paradigm enshrined in One Health when coupled with traditional ethnobotanical knowledge provides not only potential novel therapeutic agents but, more significantly, a preventative, ecological logic for health management.

TRADITIONAL ETHNOBOTANICAL KNOWLEDGE AND ONE HEALTH THINKING

The contemporary One Health paradigm, which recognizes the interconnectedness of human, animal and environmental health, represents a conceptual innovation in Western scientific thinking that indigenous communities have practiced for millennia. One Health is defined as an integrated, unifying approach that aims to sustainably balance and optimize the health of people, animals and ecosystems, acknowledging that their health is closely linked and interdependent (Lu *et al.,* 2016). Yet this theoretical framework has long animated the worldviews embedded within traditional ethnobotanical knowledge systems, which inherently reject artificial divisions between human health, animal welfare and ecological integrity. Integrating traditional ethnobotanical knowledge with One Health principles creates a synergistic framework capable of addressing contemporary global health challenges including zoonotic disease emergence, antimicrobial resistance, food insecurity, and climate change adaptation.

The culturally entrenched subset of ethnobotanical knowledge upholds a holistic viewpoint that views human life and the natural world as interrelated. In order to preserve health at individual, community and ecosystem levels, indigenous societies have long used complex plant-based therapies. The understanding that environmental integrity and animal health are inextricably linked to human health is reflected in this integration. The cosmologies that underlie ethnobotanical knowledge systems place a strong emphasis on interconnection, reciprocity and respect as guiding concepts in human interactions with the natural world. These principles closely align with the transdisciplinary, multisectoral cooperation that One Health demands. Despite these conceptual similarities, indigenous knowledge keepers and researchers have historically been left out of One Health practice, which ignores the fact that indigenous scientists have been working on "One Health" projects from the beginning of time (Mitch Daschuk *et al.,* 2025).

The practical convergence of ethnobotanical knowledge and One Health thinking is particularly evident in addressing antimicrobial resistance, one of the most pressing global health threats. Traditional medicinal plants contain diverse secondary metabolites with demonstrated therapeutic potential against resistant microorganisms. Medicinal plants have been documented for treating zoonotic diseases across regions, with phytochemicals offering mechanisms of action fundamentally distinct from conventional antibiotics, potentially circumventing resistance mechanisms that pathogens develop (Abdallah *et al.,* 2023). Integrating plant-derived antimicrobials into One Health strategies creates opportunities to reduce reliance on synthetic antibiotics while simultaneously preserving biodiversity and supporting traditional knowledge holders economically. This integration must recognize that antimicrobial resistance emerges at the human-animal-environment interface through interconnected drivers across multiple sectors, requiring coordinated responses that transcend conventional disciplinary boundaries (Press, 2025).

ETHNOBOTANY AND ETHNOPHARMACOLOGY AS ENGINES FOR EMERGING INFECTIOUS DISEASES AND ANTIMICROBIAL RESISTANCE

The convergence of ethnobotanical knowledge and ethnopharmacological research represents a critical frontier in addressing two of the most pressing global health threats of the twenty-first century: emerging infectious diseases (EIDs) and antimicrobial resistance (AMR) (Kibibi, 2025). Ethnopharmacology, defined as the study of traditional medicinal knowledge and its application in modern drug discovery, integrates ethnobotanical insights with scientific methodology to identify bioactive phytochemicals capable of therapeutic intervention against pathogens (Khamkar *et al.,* 2015). This integrative approach accelerates drug discovery while simultaneously leveraging centuries of accumulated knowledge about plant efficacy, safety and traditional preparation methods that often optimize bioactive compound extraction and bioavailability.

Ethnopharmacological research targeting emerging viral pathogens demonstrates remarkable therapeutic potential. For diseases including Dengue, Zika, Ebola and Chikungunya causing significant morbidity and mortality with limited conventional therapeutic options, traditional medicinal plants provide promising sources of antiviral compounds (Alagarasu *et al.,* 2022; Atampugbire *et al.,* 2024). Studies from ethno-medicinal systems including Ayurveda, traditional Chinese medicine and African traditional medicine document plants with immunomodulatory, anti-inflammatory, antiviral, and antioxidant properties potentially applicable to emerging pathogens (Anand *et al.,* 2021; Anywar *et al.,* 2021). Bioflavonoids such as luteolin, isolated from multiple dietary and medicinal plants documented in traditional systems, demonstrate activity against herpes simplex virus-1, Dengue, Epstein-Barr virus, Japanese encephalitis, and Chikungunya (Zakaryan *et al.,* 2017). Plant-derived compounds including polyphenols, alkaloids, coumarins, essential oils, flavonoids, terpenoids and tannins exert antiviral effects through multiple mechanisms including inhibition of viral entry, replication and release, generation of reactive oxygen species, modulation of viral gene expression, and enhancement of host immune parameters (Raghu, 2023).

According to Wang *et al.* (2021), the COVID-19 pandemic catalyzed unprecedented ethnopharmacological research, with 231 publications systematically examining traditional medicine interventions between January 2019 and August 2021. Traditional African, Chinese, Indian, and Persian medicines were evaluated for efficacy against SARS-CoV-2, with two high-quality systematic reviews demonstrating that traditional Chinese medicine combined with conventional therapy produced superior outcomes compared to conventional therapy alone, including increased recovery rates and shortened duration of main symptoms. Yet this research also revealed the critical gap in rigorous clinical evidence; most published studies demonstrated low levels of evidence beyond preliminary in vitro and in vivo investigations, emphasizing the need for randomized controlled trials. This evidence gap remains especially acute in low- and middle-income countries where populations depend substantially on traditional medicines due to limited pharmaceutical access.

Antimicrobial resistance, potentially more catastrophic in its consequences presents a challenge equally amenable to ethnopharmacological approaches. Unlike conventional antibiotics that typically target single microbial pathways, phytochemicals from medicinal plants often act on multiple targets within microbial cells through synergistic interactions, substantially reducing the likelihood of resistance development. Certain phytochemicals including polyphenols actively enhance bacterial susceptibility to conventional antibiotics by reversing resistance mechanisms and increasing antibiotic sensitivity (De Rossi *et al.,* 2025). This potentiator function provides a novel strategy for extending the clinical utility of existing pharmaceuticals. Asaaga *et al.* (2024) opined that traditional medicine research targeting the Indian subcontinent documented 69 medicinal plant species used across 29 zoonotic disease conditions, with densest documentation for Dengue, tuberculosis and *Escherichia coli* infections. Yet this ethnomedicinal wealth remains severely undertapped; most existing research remains restricted to localized case studies lacking generalizability to broader populations or landscapes (Guedes *et al.,* 2024).

In order to fully realize the potential of ethnobotany and ethnopharmacology as engines for EID and AMR therapeutics, rigorous clinical trials, respectful collaboration with traditional knowledge holders, fair benefit-sharing arrangements, sustainable sourcing of medicinal plants, and integration of ethnopharmacological research into pharmaceutical development pipelines and public health preparedness strategies are all necessary. In an era of microbial resistance and pathogenic emergence, the integration of traditional botanical knowledge with contemporary scientific methodology offers humanity's most promising path toward the development of diverse, synergistic, and culturally appropriate therapeutic options for addressing infectious diseases (Mickelsson and Simbini, 2025).

ETHNOBOTANY AS A "SHORT-CUT" IN DRUG DISCOVERY

The ethnobotanical approach to pharmaceutical development functions as a strategic shortcut that substantially accelerates drug discovery timelines while reducing developmental costs and early-phase failures. The targeting strategy fundamentally alters the discovery calculus; plants with established traditional therapeutic uses possess significantly higher bioactivity rates compared to randomly selected botanical samples, dramatically improving the probability of identifying bioactive lead compounds (McClatchey *et al.,* 2009). The integration of traditional ethnopharmacological knowledge with modern analytical chemistry creates a synergistic framework that simultaneously preserves the wisdom embedded in traditional medicine while leveraging cutting-edge scientific capability. This convergence enables identification of bioactive phytochemicals with unprecedented efficiency, transforming ancient remedies into validated modern therapeutics while substantially compressing development timelines, reducing costs, and improving success rates relative to conventional pharmaceutical approaches. Thus, the search field for bioactive discovery can be narrowed by ethnobotanical knowledge; instead of randomly screening all plants, species that have been used for infection-related disorders for a long time can be tested. In the context of infectious diseases and antimicrobial resistance (AMR), ethnobotany has been described as an initial step toward choosing species for phytochemical and pharmacological research (Gras *et al.,* 2021). In addition, research on traditional knowledge systems offers tools for interpreting how communities classify illness and remedies, how consensus is formed and how variation across age, gender or specialist roles affects what is considered effective. These insights help avoid "flattening" TK into decontextualized ingredient lists (Reyes-García, 2010). The illustrative plant-based leads and formulations relevant to viral and febrile illness are given below

***Andrographis paniculata* (king of bitters) and viral disease research signals**: *Andrographis paniculata* widely used across Chinese, Indian, Thai and Malaysian traditional medicine systems, demonstrates significant antiviral potential. The plant's primary bioactive compound, andrographolide, exhibits broad-spectrum antiviral activity against dengue virus serotypes, herpes simplex virus-1 (HSV-1), influenza A (H1N1, H3N2, H5N1), Epstein-Barr virus, human papillomavirus, and SARS-CoV-2 (Adiguna *et al.,* 2021). *In vitro* studies document andrographolide concentrations as low as 2.43 µg/mL effectively inhibiting HSV-1 replication, while effectively suppressing viral neuraminidase activity and inflammatory responses (Siridechakorn *et al.,* 2023). A randomized controlled trial examining COVID-19 treatment demonstrated that 180 mg daily of A. paniculata extract shortened olfactory loss symptoms with excellent safety profiles and no hepatic or renal toxicity (Kanokkangsadal *et al.,* 2023). In addition, additional phytochemical constituents like luteolin, quercetin, kaempferol and wogonin also possess synergistic antiviral mechanisms (Nguyen *et al.,* 2023). Perera *et al.* (2022) recorded the evidence suggesting antiviral activity signals against Dengue virus and Chikungunya virus in experimental settings, including inhibition of viral replication. In Siddha formulation research, reports of clinical observations in dengue contexts and also compile *in vitro* findings that estimate strong anti-dengue activity in specific cell-line assays for andrographolide (Maideen, 2023).

***Tinospora cordifolia* (Giloy) as an immunomodulatory and antiviral candidate-promise and caution:** *Tinospora cordifolia* in traditional Indian medicine systems, demonstrates significant immunomodulatory and antiviral properties through its diverse bioactive phytochemical constituents including tinocordiside, berberine, magnoflorine, alkaloids, terpenoids and polysaccharides (Gupta *et al.,* 2024). The plant enhances immune function by stimulating natural killer cells, B cells and T cells while modulating interleukin production and suppressing pro-inflammatory cytokines via NF-κB pathway inhibition (Singh *et al.,* 2024). *In vitro* studies document 61.43% inhibition of herpes simplex virus and 25-49% reduction of chikungunya viral cytopathic effects, while molecular docking analyses reveal six constituent alkaloids and terpenoids with strong binding affinity for SARS-CoV-2 spike proteins and protease targets (Jena *et al.,* 2021). *T. cordifolia* extract increases macrophage phagocytic capacity and CD4+ T cell populations, supporting both innate and adaptive immune responses (Nair *et al.,* 2004).

***Piper sp.* antiviral and antimicrobial potential**: Piper species, particularly *Piper nigrum* (black pepper), *Piper betle* (betel leaf), and *Piper sarmentosum*, demonstrate potent antiviral and antimicrobial properties through diverse bioactive alkaloids and phytochemical constituents. The principal alkaloid piperine from *P. nigrum* exhibits *in silico* binding affinity superior to the antiviral ribavirin against dengue methyl-transferase and Ebola VP35 interferon inhibitory domain proteins, with favorable drug-likeness properties (Tiwari *et al.*, 2020; Nag and Chowdhury, 2020). Piper betle leaf extracts display bactericidal activity against both Gram-positive (*Staphylococcus aureus, Streptococcus species*) and Gram-negative bacteria (*Escherichia coli, Pseudomonas aeruginosa*) including multidrug-resistant strains, with ethyl acetate fractions demonstrating superior antimicrobial potency compared to conventional chlorhexidine (Nayaka *et al.*, 2021). *Piper sarmentosum* exhibits dose-dependent bactericidal effects against methicillin-resistant Staphylococcus aureus (MRSA) at minimal inhibitory concentrations (Fernandez *et al.*, 2012). Additionally, piperine demonstrates synergistic antimicrobial activity when combined with conventional antibiotics against drug-resistant pathogens (Mgbeahuruike *et al.*, 2019). All these research signals support continued investigation of Piper species derivatives as antimicrobial and antiviral therapeutic candidates. Family Piperaceae, employed in Ayurveda, Siddha, Unani and traditional healing practices, constitute a highly valuable resource, owing to their rich diversity of bioactive phytochemicals.

Jwara hara in Ayurveda: In Ayurveda, Jwara denotes a broad spectrum of fever conditions that affect both the body and mind, arising from derangement of doshas and digestive fire (*agni*). Jwara-hara thus refers to the herbs, formulations and therapeutic measures that alleviate fever by correcting doshic imbalance, digesting ama (metabolic toxins), restoring agni and supporting immune responses (Suja *et al.*, 2022). Bindu *et al.* (2025) examined the antiviral properties and mechanisms of action of medicinal plants indicated for "*Jwara*" in the "*Charaka-Sushruta Samhitas.*" Classical texts describe Jwara as one of the most important diseases because it pervades the entire organism, influencing rasa dhatu, prana, manas and the sense organs (Parauha *et al.*, 2016).

The Jwarahara Gana of Charaka is a classical group of ten herbs indicated specifically for fever and related systemic disturbances, including *Hemidesmus indicus,* Sharkara (sugar), *Cissampelos pareira, Rubia cordifolia, Vitis vinifera, Salvadora persica* or related species), *Grewia asiatica, Terminalia chebula, Emblica officinalis* and *Terminalia bellirica.* These herbs are described as cooling, blood purifying, carminative, and immune-supportive; they are said to be useful not only in various types of acute and chronic fevers but also in skin diseases, liver disorders, nervous system problems, and blood-borne pathologies where fever is often a component (Chiplunkar *et al.*, 2018). Formulations based on Jwarahara Gana, sometimes prepared as kashaya (decoction) or syrup, are used to relieve deha-santapa (somatic heat and burning) and manas-santapa (mental distress) that constitute core symptomatology of Jwara. A reverse pharmacology review of Jawarahara Mahakashaya notes that this combination is traditionally credited with antipyretic, anti-inflammatory, hepatoprotective and rasayana effects, providing a potential bridge between classical indications and contemporary pharmacological validation. Modern reviews of plants used for Jwara also highlight their broad antiviral potential, suggesting that traditional fever remedies may harbour candidate molecules for antiviral drug development.

Herbal formulation for emerging diseases: Herbal formulations are increasingly explored as complementary tools for managing emerging infectious diseases, especially viral outbreaks for which specific antivirals or vaccines are limited or slow to develop. These formulations, often derived from Ayurveda, Traditional Chinese Medicine and other ethnomedical systems, combine multiple plants to target viral replication, modulate host immunity and reduce the excessive inflammation that contributes significantly to morbidity and mortality in pandemics. At the same time, the need for rigorous pharmacological, toxicological and clinical evaluation is emphasized, as the complex composition of herbal formulations can lead to unpredictable interactions, variability in potency and risks of contamination or adulteration if not standardized and regulated (Ali *et al.,* 2021). According to Ganjhu *et al.* (2015) the scientific basis for herbal formulations against emerging viral diseases rests on three main pillars: direct antiviral effects, immunomodulation, and anti-inflammatory or organ-protective actions.

Glycyrrhizin from *Glycyrrhiza glabra* has a documented history of use against respiratory and hepatic viral infections (Wahab *et al.,* 2021), while compounds from *Artemisia annua, Lycoris radiata, Lindera aggregata* and related species have shown *in vitro* activity against coronaviruses including SARS-CoV (Li *et al.,* 2005; Sinjini *et al.,* 2021). *Echinacea purpurea* extracts exhibit antiviral activity against influenza, herpes simplex and respiratory syncytial virus, coupled with immunomodulatory and anti-inflammatory effects that may be valuable in multi-herb formulations for respiratory epidemics (Hudson and Vimalanathan, 2011). Moreover, metabolomics-guided research underscores the potential of systematically profiling plant secondary metabolites to identify synergistic combinations and optimize formulations for broad antiviral coverage (Atampugbire *et al.,* 2024).

Immunomodulation is central to the rationale of many traditional formulations for emerging diseases, which aim not only to suppress pathogens but also to strengthen host defences and restore immune homeostasis. Natural products can enhance both innate and adaptive responses, including activation of macrophages and natural killer cells, promotion of antibody production, and support of T-cell function, while simultaneously dampening excessive cytokine release. Several herbal formulations have been tested not only for treatment but also for prevention of emerging infections in high-risk populations. A randomized open-label trial assessed the polyherbal preparations Infuza and Kulzam as prophylactic agents in high-risk individuals during COVID-19 and suggested that their combined use might help prevent infection (Chandra *et al.,* 2022). Sohail *et al.* (2021); Alam *et al.* (2021) and Trivedi and Kumar (2021) reviewed that phytomedicine in the context of COVID-19 discuss how traditional multi-herb prescriptions from Chinese, Japanese, Indian and European systems were repurposed, incorporating plants with known antiviral activity, expectorant and bronchodilatory effects and systemic rasayana or tonic actions. Such formulations often include herbs like *Isatis indigotica, Houttuynia cordata, Nigella sativa* and others that, through molecular docking and experimental studies, have been shown to interact with viral proteases, polymerases or host receptors such as ACE2, thereby offering mechanistic plausibility for their use. Nonetheless, authors consistently caution that extrapolation from *in vitro* and *in silico* findings to clinical practice must be done carefully, with attention to dosage, pharmacokinetics and potential herb–drug interactions (Liu *et al.,* 2022).

FRAMING THE COMPASS: TRADITIONAL KNOWLEDGE AS A GUIDING FRAMEWORK, NOT A PANACEA

Traditional ethnobotanical knowledge represents a profound compass orienting human societies toward sustainable relationships with ecological and social systems, yet contemporary discourse often misframes traditional knowledge either as a panacea requiring no scientific scrutiny or as an inferior system requiring displacement by Western science. Recognizing traditional ethnobotanical knowledge as a directional compass rather than a destination map establishes more realistic and productive relationships between diverse ways of knowing (Ray, 2023). Traditional ethnobotanical knowledge systems possess distinctive characteristics that distinguish them from Western scientific approaches while revealing dimensions often missed by reductionist scientific inquiry. Traditional ethnobotanical knowledge operates through holistic worldviews emphasizing interconnectedness, employs cumulative collective experience and incorporates spiritual and cultural dimensions. Knowledge transmission through oral traditions, narratives and experiential learning creates flexible, continuously updated systems that adapt to environmental changes and incorporate new observations.

The productive path forward requires positioning traditional ethnobotanical knowledge as a guiding compass illuminating research priorities and implementation pathways rather than providing complete solutions requiring no further investigations. In pharmaceutical development, ethnobotanical documentation of traditional uses provides strategic guidance reducing screening timelines and development costs by directing investigation toward botanicals with documented efficacy (Chisamile *et al.,* 2023). This compass metaphor respects traditional ethnobotanical knowledge's genuine value while acknowledging limitations preventing it from functioning as a standalone solution. Thus, traditional ethnobotanical knowledge framed as a guiding compass provides orientation toward sustainability and wellbeing while requiring scientific validation, contextual application, and respectful integration with other ways of understanding the world (Hikuroa *et al.,* 2011; Hao and Xiao, 2020).

ETHNOBOTANICAL KNOWLEDGE IN THE INDIAN PHARMACEUTICAL INDUSTRY

India's pharmaceutical industry represents a distinctive global phenomenon, uniquely positioned at the intersection of ancient ethnobotanical traditions and modern pharmaceutical science. The country's rich ethnobotanical heritage, encompassing between 3,000 and 5,000 medicinal plant species with documented traditional uses spanning millennia, provides an unparalleled resource for pharmaceutical innovation (Kumar *et al.,* 2021; Kacholia, 2022). Approximately 70% of rural Indians continue utilizing plant-based traditional remedies as primary healthcare, while the broader population employs medicinal plants as supplementary treatment alongside modern medicine (Wang *et al.,* 2025). This deep cultural integration of ethnobotanical knowledge with healthcare practices creates a distinctive pharmaceutical ecosystem where traditional knowledge simultaneously informs contemporary drug development and maintains relevance as lived medical practice.

The Indian pharmaceutical industry has leveraged ethnobotanical knowledge to achieve remarkable market expansion. The Indian herbal medicine market was valued at USD 6.3 billion in 2024, projected to reach USD 30 billion by 2035, representing a compound annual growth rate of 15.24% (Gotadki, 2025). The herbal supplements sector specifically generated USD 3,796.2 million in 2024 and is forecast to reach USD 8,796.1 million by 2033, growing at 10% annually. This expansion reflects both domestic consumption and aggressive export penetration; India exported USD 651.17 million of Ayush and herbal products in fiscal 2023-24, with the United States, European Union and Gulf Cooperation Council nations constituting primary export destinations. The Ministry of Ayush established the Ayush Export Promotion Council (Ayushexcil) to systematically expand global market reach for Indian traditional medicines and ethnobotanically-derived products. This market dynamism demonstrates that ethnobotanical knowledge constitutes not merely cultural heritage but economically significant intellectual capital driving pharmaceutical innovation and international commerce.

For more than a millennium, ethnobotanical knowledge has been systematized and passed down through traditional medical systems, including Ayurveda, Siddha and Unani. About 90% of Ayurvedic medicines use plant components. Ayurveda, which originated from Vedic traditions, stresses plant-based remedies to restore physiological equilibrium interrupted by disease. Classical Ayurvedic formulations frequently exhibit empirically validated efficacy; formulations described in ancient texts often exhibit superior therapeutic outcomes compared to isolated individual plant constituents, suggesting synergistic phytochemical interactions (Casanova and Costa, 2017). This recognition is growing in modern pharmaceutical development and has catalyzed systematic phytopharmaceutical development of compounds including Bacoside from *Bacopa monnieri*, Curcumin from *Curcuma longa* and Artemisinin from *Artemisia annua*, each representing ethnobotanically-guided pharmaceuticals achieving global clinical significance (Fatima *et al.,* 2022; Najmi *et al.,* 2022).

INDIAN ETHNOBOTANICAL KNOWLEDGE AND EMERGING DISEASE PREVENTION

Indian traditional medicine systems, particularly Ayurveda, represent a comprehensive repository of botanical knowledge spanning millennia, with significant antiviral and immunomodulatory properties relevant to emerging disease pathways. Recent systematic evaluations have identified at least 22 traditionally utilized Indian medicinal plants exhibiting robust antiviral properties against acute respiratory infections, including COVID-19, SARS, influenza, and respiratory syncytial virus infections, alongside immunomodulatory and thrombolytic mechanisms (Singh *et al.,* 2021). The mechanism of action underlying antiviral efficacy in traditional Indian botanicals operates through multiple interconnected pathways within viral replication cycles. Plants *Glycyrrhiza glabra*, *Curcuma longa* and *Andrographis paniculata* demonstrate particularly pronounced antiviral effects through distinct molecular mechanisms. *Tinospora cordifolia* is extensively documented in Ayurvedic literature as an immunomodulatory rasayana, contains seven bioactive phytochemicals including berberine, tinocordiside, isocolumbin, magnoflorine, tinosporin, cordiofoliside and tinosponone. These compounds exhibit high binding affinity to critical SARS-CoV-2 targets including the main protease (Mpro), receptor binding domain (RBD), surface glycoprotein, and RNA-dependent RNA polymerase. (Khuntia *et al.,* 2022). The neem tree (*Azadirachta indica*), traditionally designated as "the village pharmacy" in South Asian ethnomedicine, contains polyphenolic compounds exhibiting virucidal activity through distinct mechanisms. Aqueous bark extracts of neem function as potent entry inhibitors against Herpes Simplex Virus type-1, blocking viral attachment and preventing glycoprotein-mediated cell-to-cell fusion at even low concentrations (Tiwari *et al.,* 2010).

Withania somnifera integrated into Ayurvedic formulations for centuries as an immune-enhancing adaptogen, contains withanolides capable of direct viral protein binding (Vittal and Vinciguerra, 2025). Recent network pharmacology analyses of Ayurvedic preparations identified *Artemisia annua* as contributing maximally to phytocompound diversity, with documented antiviral activity against SARS-CoV through multiple bioactive constituents (Shaer *et al.,* 2025). Complementary botanical candidates including *Ocimum tenuiflorum*, *Zingiber officinale*, *Punica granatum* and *Allium sativum* demonstrate immunomodulatory effects through stimulation of natural killer cells, T helper cell populations, and interferon-gamma production, with specific efficacy against influenza viruses and emerging coronaviruses (Choudhary and Singh, 2022).

The ethnobotanical integration of complementary mechanisms addresses emerging disease spillover pathways through dual action: direct viral inhibition at molecular targets and systemic immune potentiation reducing susceptibility to zoonotic pathogen establishment. A polyherbal Ayurvedic formulation designated NAOQ19, containing 19 active ingredients derived from 13 traditional herbs, demonstrated clinical efficacy in SARS-CoV-2 management through *in vitro* and *in vivo* validation (Bhardwaj *et al.,* 2023). Thus, integration of traditional knowledge systems with contemporary epidemiological understanding of emerging disease transmission patterns enables development of preventive strategies aligned with ecological and cultural contexts where these botanical resources originate.

The ethnobotanical heritage preserved within indigenous communities of the Western Ghats, particularly among tribal communities inhabiting these biodiversity hotspots, encodes millennia of empirical observation regarding plant-pathogen interactions. Comprehensive documentation of ethnomedicinal uses of endemic Western Ghats species reveals that 126 endemic flowering plant taxa possess 508 documented traditional uses across 39 disease categories, with the majority of species remaining unvalidated by contemporary phytochemical-pharmacological methodologies (Sreekumari *et al.,* 2025). This knowledge asymmetry represents both conservation opportunity and pharmaceutical potential, suggesting that traditional ethnobotanical frameworks offer heuristic guidance for identifying novel antiviral candidates within megadiverse ecosystems most vulnerable to emerging pathogen emergence.

AFRICAN ETHNOBOTANICAL KNOWLEDGE AND DISEASE PREVENTION

Medicinal plants from South Africa have been used in folk medicine since one hundred decades, for the treatment of many diseases. More than ten percent of vegetation finds its applications in traditional medicine systems with over 15 species being partially or fully commercialized. These include the genera Euclea, Ficus, Aloe, Lippia and Artemisia amongst many others. Bioactive compounds from these medicinal plants are used alone or in combination with other antimicrobials as a solution towards multi-drug resistance (Moiketsi, B.N *et al.*, 2023).

ETHNOBOTANICAL KNOWLEDGE AND TRIBALS IN INDIA

India's tribal populations represent one of the world's most significant repositories of ethnobotanical knowledge, with indigenous ethnic groups maintaining sophisticated systems of plant-based medicine refined across millennia of ecological interaction and empirical observation. These tribal communities have developed nuanced understanding of plant pharmacology, disease etiology and therapeutic intervention strategies that remain largely undocumented in formal pharmaceutical literature. The documentation of tribal ethnobotanical knowledge reveals striking patterns of cross-regional therapeutic consensus alongside geographically specific medicinal practices reflecting local ecological adaptation. An ethnobotanical investigation across 17 villages spanning five districts in Mizoram documented 102 medicinal plant species deployed by seven distinct ethnic groups with notable therapeutic consensus regarding disease categories including diabetes, cancer and liver problems and hypertension (Ralte and Singh, 2024). Similarly, comprehensive surveys of tribal communities in Coochbehar district of West Bengal identified 46 plant species from 27 families utilized for medicinal purposes, with traditional knowledge transmission occurring primarily through intergenerational oral communication from medicine men to community members (Datta *et al.,* 2014). Ethnobotanical surveys in the Nilgiris region of Tamil Nadu within the Western Ghats documented the therapeutic applications of wild flora by multiple tribal groups, revealing plant uses spanning 39 distinct disease categories with 508 documented therapeutic applications across 126 endemic flowering plant taxa (Hurkadale and Bidikar, 2023).

The tribal knowledge systems of India encompass not merely isolated botanical remedies, but rather integrated frameworks of disease prevention, health maintenance and therapeutic intervention informed by indigenous disease typology and causation theories. Tribal societies maintain specialized practitioners designated as medicine men, traditional healers, shamans and ritual specialists who possess recognized authority over botanical remedies and therapeutic protocols (Elendu, 2024). This traditional custom encapsulates the profound integration of botanical knowledge with disease prevention philosophy characteristic of tribal ethnomedicinal systems. Among Odisha's tribal communities including Bonda, Didayi, Kandha, Dongria Kandha, Santal, Munda, Bhumij, and numerous other groups, medicinal practices constitute integrated components of healthcare delivery encompassing curative interventions, magico-religious healing practices and disease prevention strategies transmitted through apprenticeship relationships between experienced practitioners and their designated successors (Uniyal *et al.,* 2006).

CONCLUSION

In an era characterized by the rapid emergence and re-emergence of infectious diseases, the pursuit of novel therapeutic solutions can no longer afford to navigate in a single direction. This illuminated the potential of traditional ethnobotanical knowledge systems in pandemic preparedness and emerging disease therapeutics. While conventional pharmaceutical development prioritizes computational high-throughput screening and combinatorial chemistry, indigenous responded to the emerging diseases by deploying ethnobotanical frameworks accumulated over millennia of epidemiological observation. The Shawi people of the Amazonian region, the Irula and Kurumba communities of Tamil Nadu, and countless other indigenous populations intuitively applied One Health principles, recognizing the interconnectedness of human, animal and environmental health, through forest isolation strategies, dietary modifications utilizing traditional foods and deployment of medicinal herbs informed by ancestral knowledge of pathogenic emergence and containment. This parallel response trajectory suggests that traditional knowledge systems function as empirical compasses navigating therapeutic landscapes that Western biomedical frameworks have only recently begun to map through molecular docking studies and network pharmacology.

The integration of ethnopharmacology into mainstream drug discovery represents not merely a retrospective validation of folk remedies, but rather a forward-looking scientific strategy addressing the structural bottlenecks inherent in contemporary pharmaceutical innovation. Ethnopharmacologically-guided bioprospecting circumvents lead by leveraging cross-cultural corroboration of medicinal plant use as predictive filters for bioactivity. When multiple geographically distinct cultures employ identical botanical species for comparable disease presentations, scientific evidence supporting efficacy becomes exponentially more probable. This methodology accelerates candidate identification by reducing the search space from hundreds of thousands of plant species to specifically targeted taxa with documented therapeutic consensus, thereby increasing the probability of successful drug development within compressed timeframes critical for emerging disease response. The seminal discovery of artemisinin from *Artemisia annua*, exemplifies how ethnobotanical guidance transformed intractable antimalarial drug resistance into a validated therapeutic breakthrough. Traditional Chinese medicine practitioners had utilized artemisinin-containing plants for centuries in fever management and parasitic disease treatment, yet systematic ethnopharmacological investigation was required to isolate the bioactive compound, validate its mechanism of action against *Plasmodium* resistance mechanisms and establish clinical protocols. This canonical example demonstrates that traditional knowledge does not constitute static archival information, but rather serves as a heuristic compass directing scientific investigation toward molecular targets and therapeutic pathways that random chemical screening would unlikely identify within equivalent timeframes and resource allocations.

Emerging infectious disease landscapes present unique epidemiological and therapeutic challenges that traditional knowledge systems appear uniquely positioned to address. Zoonotic spillover events occur with increasing frequency driven by anthropogenic land use change, deforestation, wildlife trade and climate-mediated environmental disruptions. These ecological drivers concentrically overlap with regions of highest biodiversity and traditional ethnobotanical knowledge density, particularly the tropics and megadiverse regions of Southeast Asia, the Amazon basin, and the Western Ghats of India. Indigenous communities inhabiting these ecological interfaces have accumulated empirical knowledge of both disease emergence patterns and remedial botanical responses across multiple epidemic generations. A paradigmatic shift toward integrated traditional-biomedical therapeutics necessitates structural transformations in regulatory frameworks, clinical trial methodologies and equitable bioprospecting agreements. India's dual regulatory ecosystem comprising both the AYUSH (Ayurveda, Yoga, Unani, Siddha, Homoeopathy) regulatory pathway and modern pharmaceutical oversight through the Central Drugs Standard Control Organisation provides an emergent model for parallel validation of traditional and contemporary therapeutic modalities.

The mechanistic validation of traditional remedies against emerging pathogens reveals synergistic therapeutic architectures absent from single-compound pharmaceutical models. Ayurvedic polyherbal formulations targeting COVID-19, such as NAOQ19 containing 19 active ingredients across 13 traditional herbs, demonstrated clinical efficacy through mechanistic pathways involving immune stimulation, viral protease inhibition and inflammatory suppression simultaneously. Botanical formulations operate through multiple redundant pathways targeting distinct disease-relevant molecular networks, creating therapeutic resilience against pathogenic mutation and emergence of drug resistance. This systems-level therapeutics approach contrasts sharply with reductionist single-target pharmaceutical models frequently vulnerable to rapid resistance development.

The One Health framework represents the contemporary scientific articulation of principles embedded within indigenous knowledge systems for millennia. One Health acknowledges the interdependent linkages among human, animal and environmental health, utilizing these connections to develop integrated surveillance and disease control strategies. Indigenous communities applied One Health concepts intuitively through sustainable wildlife management, ecological stewardship and recognition of human-animal-environment reciprocity long before formalization of these principles in contemporary public health discourse. The preventive orientation of One Health directly aligns with traditional medicine's emphasis on maintaining physiological balance and preventing disease emergence, rather than reactive symptomatic management following pathogenic establishment. Operationalizing One Health through ethnobotanical integration enables landscape-scale interventions addressing zoonotic spillover drivers coupled with evidence-based botanical prophylaxis informed by traditional knowledge.

To conclude, the therapeutic landscape navigated by emerging infectious disease and antibiotic resistance response demands novel cartography synthesizing ethnobotanical empiricism, contemporary molecular pharmacology and ecological epidemiology. Traditional knowledge systems encoded within indigenous communities represent evolutionary adaptations to recurrent pathogenic challenges, refined across multiple generations and geographic contexts. Rather than dismissing such knowledge as non-scientific or historically superseded, integrating ethnopharmacological frameworks into evidence-based medicine acknowledges the complementary strengths of empirical observation and mechanistic investigation. The compass provided by traditional knowledge does not dictate the destination of therapeutics development, but rather orients navigation through complex pharmacological landscapes where conventional approaches have reached productive limits. As emerging pathogens accelerate spillover events and pandemic recurrence becomes institutionalized within global health planning, the synthesis of traditional knowledge with contemporary biomedical science offers not nostalgic return to premodern medicine, but rather forward-directed innovation toward resilient, equitable, and ecologically-grounded approaches to infectious disease prevention and management.

In the end, traversing the uncharted territory of developing diseases necessitates the use of all available resources and knowledge. Ignoring the customary ethnobotanical knowledge compass is like entering this terrain with an unread critical map. The world may steer toward creative, culturally relevant, environmentally sustainable, and deeply human-centered cures by adopting its guidelines. The first step in creating truly resilient global health systems is realizing that the answers to our most recent problems may reside in the oldest knowledge.

REFERENCES

Abdallah, E. M., Alhatlani, B. Y., de Paula Menezes, R., & Martins, C. H. G. (2023). Back to Nature: Medicinal Plants as Promising Sources for Antibacterial Drugs in the Post-Antibiotic Era. *Plants*, *12*(17), 3077. DOI: 10.3390/plants12173077

Adebayo, O. (2019). Loss of biodiversity: The burgeoning threat to human health. *Annals of Ibadan Postgraduate Medicine*, *17*(1), 1–3.

Adiguna, S. P., Panggabean, J. A., Atikana, A., Untari, F., Izzati, F., Bayu, A., Rosyidah, A., Rahmawati, S. I., & Putra, M. Y. (2021). Antiviral Activities of Andrographolide and Its Derivatives: Mechanism of Action and Delivery System. *Pharmaceuticals (Basel, Switzerland)*, *14*(11), 1102. DOI: 10.3390/ph14111102

Agyarko, R. K. D., Kithinji, D., & Nsarhaza, K. B. (2025). Climate Change and the Rise of Emerging and Re-Emerging Infectious Diseases in Africa: A Literature Review. *International Journal of Environmental Research and Public Health*, *22*(6), 903. DOI: 10.3390/ijerph22060903

Alagarasu, K., Patil, P., Kaushik, M., Chowdhury, D., Joshi, R. K., Hegde, H. V., Kakade, M. B., Hoti, S. L., Cherian, S., & Parashar, D. (2022). *In Vitro* Antiviral Activity of Potential Medicinal Plant Extracts Against Dengue and Chikungunya Viruses. *Frontiers in Cellular and Infection Microbiology*, *12*, 866452. DOI: 10.3389/fcimb.2022.866452

Alam, S., Sarker, M. M. R., Afrin, S., Richi, F. T., Zhao, C., Zhou, J.-R., & Mohamed, I. N. (2021). Traditional Herbal Medicines, Bioactive Metabolites, and Plant Products Against COVID-19: Update on Clinical Trials and Mechanism of Actions. *Frontiers in Pharmacology*, *12*, 671498. DOI: 10.3389/fphar.2021.671498

Ali, S. I., Sheikh, W. M., Rather, M. A., Venkatesalu, V., Muzamil Bashir, S., & Nabi, S. U. (2021). Medicinal plants: Treasure for antiviral drug discovery. *Phytotherapy Research : PTR*, *35*(7), 3447–3483. DOI: 10.1002/ptr.7039

Anand, A. V., Balamuralikrishnan, B., Kaviya, M., Bharathi, K., Parithathvi, A., Arun, M., Senthilkumar, N., Velayuthaprabhu, S., Saradhadevi, M., Al-Dhabi, N. A., Arasu, M. V., Yatoo, M. I., Tiwari, R., & Dhama, K. (2021). Medicinal Plants, Phytochemicals, and Herbs to Combat Viral Pathogens Including SARS-CoV-2. *Molecules (Basel, Switzerland)*, *26*(6), 1775. DOI: 10.3390/molecules26061775

Anywar, G., Akram, M., & Chishti, M. A. (2021). African and Asian Medicinal Plants as a Repository for Prospective Antiviral Metabolites Against HIV-1 and SARS CoV-2: A Mini Review. *Frontiers in Pharmacology*, *12*, 703837. DOI: 10.3389/fphar.2021.703837

Asaaga, F. A., Tomude, E. S., Rahman, M., Shakeer, I., Ghotge, N. S., Burthe, S. J., Schäfer, S. M., Vanak, A. T., Purse, B. V., & Hoti, S. L. (2024). What is the state of the art on traditional medicine interventions for zoonotic diseases in the Indian subcontinent? A scoping review of the peer-reviewed evidence base. *BMC Complementary Medicine and Therapies*, *24*(1), 249. DOI: 10.1186/s12906-024-04553-8

Asim, M., Khan, N. A., Osmonaliev, K., & Sathian, B. (2022). The emergence of novel infectious diseases and the public health impact of mass gathering events: Risks and challenges. *Nepal Journal of Epidemiology*, *12*(4), 1238–1241. DOI: 10.3126/nje.v12i4.50997

Atampugbire, G., Adomako, E. E. A., & Quaye, O. (2024). Medicinal Plants as Effective Antiviral Agents and Their Potential Benefits. *Natural Product Communications*, *19*(9), 1934578X241282923. Advance online publication. DOI: 10.1177/1934578X241282923

Atampugbire, G., Adomako, E. E. A., & Quaye, O. (2024). Medicinal Plants as Effective Antiviral Agents and Their Potential Benefits. *Natural Product Communications*, *19*(9), 1934578X241282923. Advance online publication. DOI: 10.1177/1934578X241282923

Bhardwaj, P., Ganapathy, K., Pathania, M., Naveen, K. H., Charan, J., Dutta, S., Gadepalli, R., Srinivasan, S., Gupta, M. K., Goel, A. D., Midha, N., Kumar, B., Sharma, M., Sharma, P., Banerjee, M., Mitra, P., Misra, S., Subramaniant, G., Dhar, M., & Kanchibhotla, D. (2023). Effectiveness of ayurvedic formulation, NAOQ19 along with standard care in the treatment of mild-moderate COVID-19 patients: A double blind, randomized, placebo-controlled, multicentric trial. *Journal of Ayurveda and Integrative Medicine*, *14*(6), 100778. DOI: 10.1016/j.jaim.2023.100778

Bindhu, A., Nair, S.A., Johnson, A.J., Baby, S. 2025. Plants used in Ayurveda for Jwara or fever: A review of their antiviral studies. *J. Ayurveda Integr Med.* 16(2):101085. doi: 10.1016/j.jaim.2024.101085.

Carbone, G., Boiardi, G., Infantino, C., Cunico, D., & Esposito, S. (2025). Vectors on the Move: How Climate Change Fuels the Spread of Arboviruses in Europe. *Microorganisms*, *13*(9), 2034. DOI: 10.3390/microorganisms13092034

Casanova, L., & Costa, S. (2017). Synergistic Interactions in Natural Products: Therapeutic Potential and Challenges. *Revista Virtual de Química*, *9*(2), 575–595. DOI: 10.21577/1984-6835.20170034

Chandra, K., Das, A. K., Banday, S., Rana, N. A., Arora, M., Jain, S., Islam, F., Agarwal, S., Kashyap, V., Joshi, S., Mueed, A., & Dudeja, M. (2022). Efficacy of polyherbal formulations for prevention of COVID-19 infection in high-risk subjects: A randomized open-label controlled clinical trial. *Phytotherapy Research : PTR*, *36*(9), 3632–3643. DOI: 10.1002/ptr.7531

Chiplunkar, S., Kumar, V. K. C., & Sharashchandra, R. (2018). Preparation of Jwarahara syrup: Treatment for fever. *The Pharma Innovation Journal*, *7*(12), 249–253.

Chisamile, W. A., Sonibare, M. A., Kamanula, J. F. 2023. Ethnobotanical Study of Traditional Medicinal Plants Used for the Treatment of Infectious Diseases by Local Communities in Traditional Authority (T/A) Mbelwa, Mzimba District, Northern Region, Malawi. *J.* 6(1):115-139.

Choudhary, N., & Singh, V. (2022). Multi-scale mechanism of antiviral drug-alike phytoligands from Ayurveda in managing COVID-19 and associated metabolic comorbidities: Insights from network pharmacology. *Molecular Diversity*, *26*(5), 2575–2594. DOI: 10.1007/s11030-021-10352-x

Datta, T., Patra, A. K., & Dastidar, S. G. (2014). Medicinal plants used by tribal population of Coochbehar district, West Bengal, India-an ethnobotanical survey. *Asian Pacific Journal of Tropical Biomedicine*, *4*(Suppl 1), S478–S482. DOI: 10.12980/APJTB.4.2014C1122

Davis, C. C., & Choisy, P. (2024). Medicinal plants meet modern biodiversity science. *Current Biology : CB*, *34*(4), R158–R173. DOI: 10.1016/j.cub.2023.12.038

De Rossi, L., Rocchetti, G., Lucini, L., & Rebecchi, A. (2025). Antimicrobial Potential of Polyphenols: Mechanisms of Action and Microbial Responses—A Narrative Review. *Antioxidants*, *14*(2), 200. DOI: 10.3390/antiox14020200

Domingo-Fernández, D., Gadiya, Y., Mubeen, S., Bollerman, T. J., Healy, M. D., Chanana, S., Sadovsky, R. G., Healey, D., & Colluru, V. (2023). Modern drug discovery using ethnobotany: A large-scale cross-cultural analysis of traditional medicine reveals common therapeutic uses. *iScience*, *26*(9), 107729. DOI: 10.1016/j.isci.2023.107729

Dominic Terkimbi, S., Aja Maduabuchi, P., Chans Mwandah, D., Vandu, C. D., Agbaje, A. B., Paul-Chima, U. O., Dangana, R. S. S., & Mujinya, R. (2025). Molecular Mechanisms of Antimicrobial Resistance in Infectious Diseases: Advances in Therapeutic Strategies and Public Health Interventions. *F1000 Research*, *14*, 783. DOI: 10.12688/f1000research.167514.1

Elendu, C. 2024. The evolution of ancient healing practices: From shamanism to Hippocratic medicine: A review. *Medicine (Baltimore).* 103(28):e39005. doi: . 0000000000039005.DOI: 10.1097/MD

Ellwanger, J. H., & Chies, J. A. B. (2021). Zoonotic spillover: Understanding basic aspects for better prevention. *Genetics and Molecular Biology*, *44*(1, Suppl 1), e20200355. DOI: 10.1590/1678-4685-gmb-2020-0355

Ellwanger, J. H., & Chies, J. A. B. (2021). Zoonotic spillover: Understanding basic aspects for better prevention. *Genetics and Molecular Biology*, *44*(1, Suppl 1), e20200355. DOI: 10.1590/1678-4685-gmb-2020-0355

Escudero-Pérez, B., Lalande, A., Mathieu, C., & Lawrence, P. (2023). Host-Pathogen Interactions Influencing Zoonotic Spillover Potential and Transmission in Humans. *Viruses*, *15*(3), 599. DOI: 10.3390/v15030599

Fatima, U., Roy, S., Ahmad, S., Ali, S., Elkady, W. M., Khan, I., Alsaffar, R. M., Adnan, M., Islam, A., & Hassan, M. I. (2022). Pharmacological attributes of *Bacopa monnieri* extract: Current updates and clinical manifestation. *Frontiers in Nutrition*, *9*, 972379. DOI: 10.3389/fnut.2022.972379

Fernandez, L., Daruliza, K., Sudhakaran, S., & Jegathambigai, R. (2012). Antimicrobial activity of the crude extract of *Piper sarmentosum* against methicilin-resistant Staphylococcus aureus (MRSA), *Escherichia coli, Vibrio cholera* and *Streptococcus pneumoniae. European Review for Medical and Pharmacological Sciences*, *16*(Suppl 3), 105–111.

Ganjhu, R. K., Mudgal, P. P., Maity, H., Dowarha, D., Devadiga, S., Nag, S., & Arunkumar, G. (2015). Herbal plants and plant preparations as remedial approach for viral diseases. *Virusdisease*, *26*(4), 225–236. DOI: 10.1007/s13337-015-0276-6

Gotadki, R. 2025. India Herbal Medicine Market. Market Research Future. https://www.marketresearchfuture.com/reports/india-herbal-medicine-market-44000. Accessed on 10.12.2025.

Gras, A., Parada, M., Vallès, J. and Garnatje, T. 2021. The Role of Traditional Plant Knowledge in the Fight Against Infectious Diseases: A Meta-Analytic Study in the Catalan Linguistic Area. *Front. Pharmacol.* 12:744616. doi: . 744616.DOI: 10.3389/fphar.2021

Guedes, B. N., Krambeck, K., Durazzo, A., Lucarini, M., Santini, A., Oliveira, M. B. P. P., Fathi, F., & Souto, E. B. (2024). Natural antibiotics against antimicrobial resistance: Sources and bioinspired delivery systems. *Brazilian Journal of Microbiology : Publication of the Brazilian Society for Microbiology*, *55*(3), 2753–2766. DOI: 10.1007/s42770-024-01410-1

Gupta, A., Gupta, P., & Bajpai, G. (2024). *Tinospora cordifolia (Giloy)*: An insight on the multifarious pharmacological paradigms of a most promising medicinal ayurvedic herb. *Heliyon*, *10*(4), e26125. DOI: 10.1016/j.heliyon.2024.e26125

Hao, D., & Xiao, P. (2020). Pharmaceutical resource discovery from traditional medicinal plants: Pharmacophylogeny and pharmacophylogenomics. *Chinese Herbal Medicines*, *12*(2), 104–117. DOI: 10.1016/j.chmed.2020.03.002

Hikuroa, D., Morgan, K., Durie, M., Henare, M., & Robust, T. (2011). Integration of Indigenous Knowledge and Science. *The International Journal of Science in Society*, *2*(2), 105–113. DOI: 10.18848/1836-6236/CGP/v02i02/51224

Hudson, J., & Vimalanathan, S. (2011). Echinacea-A Source of Potent Antivirals for Respiratory Virus Infections. *Pharmaceuticals (Basel, Switzerland)*, *4*(7), 1019–1031. DOI: 10.3390/ph4071019

Hurkadale, P.J. and Bidikar, C.M. 2023. Ethno-Medicinal Plants from the North-Central Western Ghats of India for Alternative Health Care. *J Plant Sci Phytopathol*. 7:076-080.

Jena, S., Munusami, P., Mm, B., & Chanda, K. (2021). Computationally approached inhibition potential of *Tinospora cordifolia* towards COVID-19 targets. *Virusdisease*, *32*(1), 65–77. DOI: 10.1007/s13337-021-00666-7

Kacholia, S. 2022. Harnessing the Power of Ethnobotany: The Role of Indian Medicinal Plants: A Review. *International Journal of Recent Research and Review*. XV(1):06-09.

Kanokkangsadal, P., Mingmalairak, C., Mukkasombat, N., Kuropakornpong, P., Worawattananutai, P., Khawcharoenporn, T., Sakpakdeejaroen, I., Davies, N. M., & Itharat, A. (2023). *Andrographis paniculata* extract versus placebo in the treatment of COVID-19: A double-blinded randomized control trial. *Research in Pharmaceutical Sciences*, *18*(6), 592–603. DOI: 10.4103/1735-5362.389947

Khamkar, A. D., Motghare, V. M., & Deshpande, R. (2015). Ethnopharmacology-A Novel approach for Drug Discovery. *Indian Journal of Pharmacy and Pharmacology*, *2*(4), 222. DOI: 10.5958/2393-9087.2015.00007.2

Khuntia, B. K., Sharma, V., Wadhawan, M., Chhabra, V., Kidambi, B., Rathore, S., Agarwal, A., Ram, A., Qazi, S., Ahmad, S., Raza, K., & Sharma, G. (2022). Antiviral Potential of Indian Medicinal Plants Against Influenza and SARS-CoV: A Systematic Review. *Natural Product Communications*, *17*(3), 1934578X221086988. Advance online publication. DOI: 10.1177/1934578X221086988

Kibebew, F. 2001. The status and availability of oral and written knowledge on traditional healthcare in Ethiopia. In: Zewdu, M. and Demissie, A. (Eds.). *Conservation and sustainable use of medicinal plants in Ethiopia*. Proceedings of National Workshop on Biodiversity Conservation and Sustainable Use of Medicinal Plants in Ethiopia. 28 April to 01 May 1998. Addis Ababa: IBCR. 168-175.

Kibibi, J. W. (2025). The Future of Medicinal Plants in Combating Emerging Infectious Diseases. *Newport International Journal of Public Health and Pharmacy*, *6*(2), 57–65. DOI: 10.59298/NIJPP/2025/625765

Kumar, A., Kumar, S., Komal, , Ramchiary, N., & Singh, P. (2021). Role of Traditional Ethnobotanical Knowledge and Indigenous Communities in Achieving Sustainable Development Goals. *Sustainability (Basel)*, *13*(6), 3062. DOI: 10.3390/su13063062

Kumar, M., Rawat, S., Nagar, B., Kumar, A., Pala, N. A., Bhat, J. A., Bussmann, R. W., Cabral-Pinto, M., & Kunwar, R. (2021). Implementation of the Use of Ethnomedicinal Plants for Curing Diseases in the Indian Himalayas and Its Role in Sustainability of Livelihoods and Socioeconomic Development. *International Journal of Environmental Research and Public Health*, *18*(4), 1509. DOI: 10.3390/ijerph18041509

Li, S. Y., Chen, C., Zhang, H. Q., Guo, H. Y., Wang, H., Wang, L., Zhang, X., Hua, S. N., Yu, J., Xiao, P. G., Li, R. S., & Tan, X. (2005). Identification of natural compounds with antiviral activities against SARS-associated coronavirus. *Antiviral Research*, *67*(1), 18–23. DOI: 10.1016/j.antiviral.2005.02.007

Liu, Y. X., Zhou, Y. H., Jiang, C. H., Liu, J., & Chen, D. Q. (2022). Prevention, treatment and potential mechanism of herbal medicine for Corona viruses: A review. *Bioengineered*, *13*(3), 5480–5508. DOI: 10.1080/21655979.2022.2036521

Lu, J., Milinovich, G. J., & Hu, W. (2016). A brief historical overview of emerging infectious disease response in China and the need for a One Health approach in future responses. *One Health*, *2*, 99–102. DOI: 10.1016/j.onehlt.2016.07.001

Luković, M., Stevanović, Z. D., & Veljović, S. (2023). Ethnobotanical knowledge and traditional practice in the frame of urban ecosystem services on the example of the city of Kragujevac (Serbia). *Kragujevac J. Sci*, *45*(45), 315–325. DOI: 10.5937/KgJSci2345315L

Maideen, N. M. P. (2023). Efficacy of Nilavembu Kudineer for Dengue Fever Management – An Overview of Clinical and Preclinical Evidences. *South East Eur J Cardiol*, *6*(1), 109–114. DOI: 10.3889/seejim.2023.6058

McClatchey, W. C., Mahady, G. B., Bennett, B. C., Shiels, L., & Savo, V. (2009). Ethnobotany as a pharmacological research tool and recent developments in CNS-active natural products from ethnobotanical sources. *Pharmacology & Therapeutics*, *123*(2), 239–254. DOI: 10.1016/j.pharmthera.2009.04.002

McFarlane, R. A., Sleigh, A. C., & McMichael, A. J. (2013). Land-use change and emerging infectious disease on an island continent. *International Journal of Environmental Research and Public Health*, *10*(7), 2699–2719. DOI: 10.3390/ijerph10072699

Mgbeahuruike, E. E., Stålnacke, M., Vuorela, H., & Holm, Y. (2019). Antimicrobial and Synergistic Effects of Commercial Piperine and Piperlongumine in Combination with Conventional Antimicrobials. *Antibiotics (Basel, Switzerland)*, *8*(2), 55. DOI: 10.3390/antibiotics8020055

Mickelsson, M., & Simbini, T. (2025). Integrating traditional medicine into antimicrobial resistance education: Community-centric preparedness in Zimbabwe. *PLOS Global Public Health*, *5*(10), e0005245. DOI: 10.1371/journal.pgph.0005245

Mitch Daschuk, C., Dell, A., Chalmers, D., Lisa, S., Larry, L., & Sharon, A. (2025). *Guidance for Canada's Wildfire Emergency Response Efforts: Learning from Volunteer Therapy Dog Program Support During Saskatchewan Crises*. The Palgrave Handbook of Human-Animal Interactions in the Global Context of Climate Change, Disasters, and Other Crises., DOI: 10.1007/978-3-032-00313-3_14

Moiketsi, B.N.; Makale, K.P.P.; Rantong, G.; Rahube, T.O.; Makhzoum. 2023. A.Potential of Selected African Medicinal Plants as Alternative Therapeutics against Multi-Drug ResistantBacteria. *Biomedicines*, 11, 2605.doi:org/.DOI: 10.3390/biomedicines11102605

Nag, A., & Chowdhury, R. R. (2020). Piperine, an alkaloid of black pepper seeds can effectively inhibit the antiviral enzymes of Dengue and Ebola viruses, an in silico molecular docking study. *Virusdisease*, *31*(3), 308–315. DOI: 10.1007/s13337-020-00619-6

Nair, P. K., Rodriguez, S., Ramachandran, R., Alamo, A., Melnick, S. J., Escalon, E., Garcia, P. I.Jr, Wnuk, S. F., & Ramachandran, C. (2004). Immune stimulating properties of a novel polysaccharide from the medicinal plant *Tinospora cordifolia*. *International Immunopharmacology*, *4*(13), 1645–1659. DOI: 10.1016/j.intimp.2004.07.024

Najmi, A., Javed, S. A., Al Bratty, M., & Alhazmi, H. A. (2022). Modern Approaches in the Discovery and Development of Plant-Based Natural Products and Their Analogues as Potential Therapeutic Agents. *Molecules (Basel, Switzerland)*, *27*(2), 349. DOI: 10.3390/molecules27020349

Nayaka, N. M. D. M. W., Sasadara, M. M. V., Sanjaya, D. A., Yuda, P. E. S. K., Dewi, N. L. K. A. A., Cahyaningsih, E., & Hartati, R. (2021). *Piper betle* (L): Recent Review of Antibacterial and Antifungal Properties, Safety Profiles, and Commercial Applications. *Molecules (Basel, Switzerland)*, *26*(8), 2321. DOI: 10.3390/molecules26082321

Nguyen, H. T., Do, V. M., Phan, T. T., & Nguyen Huynh, D. T. (2023). The Potential of Ameliorating COVID-19 and Sequelae From *Andrographis paniculata* via Bioinformatics. *Bioinformatics and Biology Insights*, *17*, 11779322221149622. DOI: 10.1177/11779322221149622

Oliveira, M., Antunes, W., Mota, S., Madureira-Carvalho, Á., Dinis-Oliveira, R. J., & Dias da Silva, D. (2024). An Overview of the Recent Advances in Antimicrobial Resistance. *Microorganisms*, *12*(9), 12091920. DOI: 10.3390/microorganisms

Pandey, A. K., & Tripathi, Y. C. (2017). Ethnobotany and its relevance in contemporary research. *Journal of Medicinal Plants Studies*, *5*(3), 123–129.

Parauha, S., Hullur, M. A., & Prashanth, A. S. (2016). A Litreary Review of Vishama Jwara and its principle of treatment. *Journal of Ayurveda and Integrated Medical Sciences*, *1*(2), 64–73. DOI: 10.21760/jaims.v1i2.3668

Perera, P. K., Diddeniya, J. I. D., & Meedeniya, A. C. B. (2022). *Andrographis paniculata* Usage in the Treatment of Viral Diseases. *Asian J. Biol. Sci*, *15*(2), 77–85. DOI: 10.3923/ajbs.2022.77.85

Poncet, A., Vogl, C. R., & Weckerle, C. S. (2015). Folkbotanical classification: Morphological, ecological and utilitarian characterization of plants in the Napf region, Switzerland. *Journal of Ethnobiology and Ethnomedicine*, *11*(1), 13. DOI: 10.1186/1746-4269-11-13

Press, G. N. (2025). Phytotherapy and One Health: The Role of Medicinal Plants in Combating Antimicrobial Resistance Worldwide. *International Journal of Biochemistry Research & Review*, *34*(6), 86–97. Advance online publication. DOI: 10.9734/ijbcrr/2025/v34i51059

Raghu, K. (2023). Phytochemical profiling and microbial targeting of medicinal plants in the development of alternative antimicrobial therapies. *Microbiology Archives.*, *5*(1), 6–11.

Rahman, M. T., Sobur, M. A., Islam, M. S., Ievy, S., Hossain, M. J., El Zowalaty, M. E., Rahman, A. T., & Ashour, H. M. (2020). Zoonotic Diseases: Etiology, Impact, and Control. *Microorganisms*, *8*(9), 1405. DOI: 10.3390/microorganisms8091405

Ralte, L., & Singh, Y. T. (2024). Ethnobotanical survey of medicinal plants used by various ethnic tribes of Mizoram, India. *PLoS One*, *19*(5), e0302792. DOI: 10.1371/journal.pone.0302792

Ray, S. (2023). Weaving the links: Traditional knowledge into modern science. *Futures*, *145*(January), 103081. DOI: 10.1016/j.futures.2022.103081

Reyes-García, V. (2010). The relevance of traditional knowledge systems for ethnopharmacological research: Theoretical and methodological contributions. *Journal of Ethnobiology and Ethnomedicine*, *6*(1), 32. DOI: 10.1186/1746-4269-6-32

Reygaert, W. C. (2018). An overview of the antimicrobial resistance mechanisms of bacteria. *AIMS Microbiology*, *4*(3), 482–501. DOI: 10.3934/microbiol.2018.3.482

Sahai, S., & Kumar, A. (2025). Climate Change and Emerging Infectious Diseases: A Growing Global Health Threat. *Oman Medical Journal*, *40*(3), e768. DOI: 10.5001/omj.2025.79

Shaer, N. A., Mohamed, A. A., & Schnug, E. (2025). Potential of *Artemisia annua* Bioactives as Antiviral Agents Against SARS-CoV-2 and Other Health Complications. *Pharmaceuticals (Basel, Switzerland)*, *18*(12), 1904. DOI: 10.3390/ph18121904

Sharan, M., Vijay, D., Yadav, J. P., Bedi, J. S., & Dhaka, P. (2023). Surveillance and response strategies for zoonotic diseases: A comprehensive review. *Science in One Health*, *2*, 100050. DOI: 10.1016/j.soh.2023.100050

Singh, J., Saxena, E., Chaudhary, A.R., Kaur, M., Salotra, M., Rasane, P., Kaur, S., Ercisli, S., Durul, M.S., Bozhuyuk, M.R., Urusan, A.H., Ullah, R. 2024. Immunomodulatory properties of Giloy (*Tinospora cordifolia*) leaves and its applications in value-added products. *Heliyon*. 11(1): e40948. doi: . 2024.e40948.DOI: 10.1016/j.heliyon

Singh, R. S., Singh, A., Kaur, H., Batra, G., Sarma, P., Kaur, H., Bhattacharyya, A., Sharma, A. R., Kumar, S., Upadhyay, S., Tiwari, V., Avti, P., Prakash, A., & Medhi, B. (2021). Promising traditional Indian medicinal plants for the management of novel Coronavirus disease: A systematic review. *Phytotherapy Research : PTR*, *35*(8), 4456–4484. DOI: 10.1002/ptr.7150

Sinjini, D., Rachana., Aditi, B., Harshit, D., Karishma, R., Arushi, A., Shreya, K., Kaur, C.I. (2021). Herbs, natural products and bioactive compounds against COVID -19 from South Asia and Africa. *Pharmacologyonline*, *3*, 1893–1922.

Siridechakorn, I., Bhattarakosol, P., Sasivimolrattana, T., Anoma, S., Wongwad, E., Nuengchamnong, N., Kowitdamrong, E., Boonyasuppayakorn, S., & Waranuch, N. (2023). Inhibitory efficiency of *Andrographis paniculata* extract on viral multiplication and nitric oxide production. *Scientific Reports*, *13*(1), 19738. DOI: 10.1038/s41598-023-46249-y

Sohail, M. I., Siddiqui, A., Erum, N., & Kamran, M. (2021). Phytomedicine and the COVID-19 pandemic. *Phytomedicine : International Journal of Phytotherapy and Phytopharmacology*, •••, 693–708. DOI: 10.1016/B978-0-12-824109-7.00005-4

Sreekumari, G. R., Remadevi, R. K. S., Koshy, K. C., & Baby, S. (2025). An assessment of the ethnomedicinal properties of endemic flowering plants of the Western Ghats, India. *Pharmacognosy Reviews/Bioinformatics Trends. Pharmacognosy Reviews*, *19*(38), 260–266. DOI: 10.5530/phrev.20250014

Suja, K., Puranik, P., Sharashchandra, R., & Shetty, S. K. (2022). A randomized controlled comparative clinical trial on Jwarahara effect of Mukkaamukkatukaadi Syrup with Kiratatikta Syrup in Febrile Children. *J. Ayu. Int. Med. Sci*, *7*(7), 40–46. DOI: 10.21760/jaims.7.7.6

Tiwari, A., Mahadik, K.R., Gabhe, S.Y. 2020. Piperine: A comprehensive review of methods of isolation, purification, and biological properties. *Medicine in Drug Discovery*. 7(September(:100027. .DOI: 10.1016/j.medidd.2020.100027

Tiwari, V., Darmani, N. A., Yue, B. Y., & Shukla, D. (2010). In vitro antiviral activity of neem (Azardirachta indica L.) bark extract against herpes simplex virus type-1 infection. *Phytotherapy Research : PTR*, *24*(8), 1132–1140. DOI: 10.1002/ptr.3085

Trivedi, N., & Kumar, D. (2021). Herbal Medicine is the Way of Potential Therapeutic Option for the Treatment of Novel Corona Virus (COVID-19): Recent Updates. *Internal Medicine (Tokyo, Japan)*, *11*, 351.

Uniyal, S. K., Singh, K. N., Jamwal, P., & Lal, B. (2006). Traditional use of medicinal plants among the tribal communities of Chhota Bhangal, Western Himalaya. *Journal of Ethnobiology and Ethnomedicine*, *2*(1), 14. DOI: 10.1186/1746-4269-2-14

Vittal, M., & Vinciguerra, M. (2025). Enhancing healthspan with Ashwagandha (*Withania somnifera*): A comprehensive review of its multifaceted geroprotective benefits. *Biogerontology*, *26*(5), 179. DOI: 10.1007/s10522-025-10320-0

Wahab, S., Annadurai, S., Abullais, S. S., Das, G., Ahmad, W., Ahmad, M. F., Kandasamy, G., Vasudevan, R., Ali, M. S., & Amir, M. (2021). *Glycyrrhiza glabra* (Licorice): A Comprehensive Review on Its Phytochemistry, Biological Activities, Clinical Evidence and Toxicology. *Plants*, *10*(12), 2751. DOI: 10.3390/plants10122751

Wang, J. B., Andrade-Cetto, A., Echeverria, J., Wardle, J., Yen, H. R., & Heinrich, M. (2021). Ethnopharmacological Responses to the Coronavirus Disease 2019 Pandemic. *Frontiers in Pharmacology*, *12*, 798674. DOI: 10.3389/fphar.2021.798674

Wang, X., Anwar, T., Qureshi, H., El-Beltagi, H. S., Sehar, Z., Solieva, D., Azizov, B., Rebouh, N. Y., Abbasov, M. A., Yakubov, F., & Alomran, M. M. (2025). Plant-based traditional remedies and their role in public health: Ethnomedicinal perspectives for a growing population. *Journal of Health, Population and Nutrition*, *44*(1), 300. DOI: 10.1186/s41043-025-01036-5

WHO. 2014. A brief guide to emerging infectious diseases and zoonoses. WHO Regional Office for South-East Asia. https://iris.who.int/handle/10665/204722. ISBN 9789290224587.

WHO. 2023. *Antimicrobial resistance*. World Health Organization. https://www.who.int/news-room/fact-sheets/detail/antimicrobial-resistance. Accessed on 10.12.2025.

Zakaryan, H., Arabyan, E., Oo, A., & Zandi, K. (2017). Flavonoids: Promising natural compounds against viral infections. *Archives of Virology*, *162*(9), 2539–2551. DOI: 10.1007/s00705-017-3417-y

Chapter 5
Phytochemicals With Antimicrobial and Resistance-Modifying Activities

Chinyere Charity Ezeanya-Bakpa
http://orcid.org/0000-0002-7844-7414
Caleb University, Lagos, Nigeria

Ifeoluwa Oluwatosin Olayemi
Caleb University, Lagos, Nigeria

Bridget Omon Adediran
Caleb University, Lagos, Nigeria

ABSTRACT

Phytochemicals, naturally occurring compounds found in plants, have garnered significant attention in recent years for their potential antimicrobial properties and ability to modify resistance mechanisms. These bioactive substances can disrupt microbial growth and proliferation through various modes of action, including the interference with bacterial cell wall synthesis, inhibition of enzyme activity, and modulation of biofilm formation. Moreover, certain phytochemicals have been shown to enhance the efficacy of conventional antibiotics, offering a promising avenue for combating antibiotic resistance—one of the most pressing global health challenges. Continued exploration of phytochemicals could contribute to the development of novel antimicrobial agents, ultimately enhancing public health outcomes. This

DOI: 10.4018/979-8-3373-7810-7.ch005

chapter elucidates on the recent trends in phytochemicals with antimicrobial and resistance-modifying activities. It also discusses the mode of action as antimicrobial agents and thus addresses the future directions and challenges.

INTRODUCTION

Phytochemicals are natural compounds present in plants that are important in sustaining human health. They are also referred to as secondary metabolites, which possess pharmacological effects that go beyond simple nutritional effects. The therapeutic potential of these chemicals, which include antioxidant, anti-inflammatory, antibacterial, anticancer, antidiabetic, and cardioprotective properties help prevent and manage different diseases. These made the compounds to become more widely acknowledged (Rodriguez-Negrete et al., 2024). Their wide area of biological actions makes them valuable drug discovery targets, functional food, and phytomedicine. The recent studies have also emphasized their possible role as drug discovery assistants due to the development of analytical procedures and computational tools, including molecular docking and network pharmacology, which have fast-tracked the discovery of bioactive phytochemicals and their molecular targets (Duda-Madej et al., 2025). Furthermore, the anti-pharmacological aspect of phytochemicals has been advanced to the frontline of promising therapeutic agents due to the integration of traditional medicinal knowledge and modern methods of drug discovery (Rodriguez-Negrete et al., 2024).

Phytochemicals can be classified into a few major groups, which may be grouped based on their chemical structure and their biosynthetic pathways, such as alkaloid, flavonoid, phenolics, terpenoids etc. The alkaloids are compounds containing nitrogen (e.g., morphine, quinine) which may have powerful pharmacological effects, especially in the nervous system and cardiovascular system. Another essential lead in drug discovery is the ability of many alkaloids to have both antimicrobial and resistance-modifying properties (Leclerc & Fournier, 2024). Flavonoids and phenolic compounds contain one or more phenolic rings and are especially known for their antioxidant properties. Flavonoids (e.g., quercetin, catechins) are potent antioxidants that mitigate oxidative stress and modulate inflammation by regulating key cellular signaling pathways such as NF-κB and Nrf2 (Sul & Ra, 2021; Jazvinšćak Jembrek et al., 2021). Phenolic acids and related phenolics contribute to free-radical scavenging and anti-aging effects, as well as antimicrobial and cardioprotective activities (Kumar et al., 2025). Terpenoids (e.g., menthol, artemisinin) are derived from isoprene units ranging from simple monoterpenes to complex triterpenes. Terpenoids exhibit diverse biological activities including antimicrobial, anticancer, and anti-inflammatory effects and several are used in clinically approved drugs such as artemisinin for malaria (Mabou & Yossa, 2021). Saponins are glycosides with amphiphilic properties and are known for immunomodulatory and lipid-lowering effects, contributing to cardiovascular health (Luo et al., 2022). Glycosides, including cardiac glycosides, play key roles in cardiac therapeutics by modulating heart muscle contraction and are used in treating heart diseases (Kumar & Pandey, 2013).

A. Sources of Phytochemicals

Antimicrobial resistance is one of the major worldwide health problems, which has been posing a threat to the efficacy of conventional antibiotics. The growing level of antimicrobial resistance has increased the quest of alternative and complementary antimicrobial agents. Natural products, especially phytochemicals derived from plants, are a promising source of antimicrobial agent based on a variety of mechanisms of action. Phytochemicals usually have multiple microbial targets, as opposed to synthetic antibiotics which usually have only one, which makes them less likely to be able to develop resistance (Rodriguez-Negrete et al., 2024). Phytochemicals can be obtained from different plant-based sources that include:

1. Tea Tree Oil (Melaleuca alternifolia)

Essential oils are volatile aromatic compounds that are extracted out of plants and are abundant in terpenes and phenolic compounds (Bakkali et al., 2008). Tea tree oil is one of the well studied essential volatile oils for its antimicrobial properties that can be extracted from the leaves and small branches of *Melaleuca alternifolia* (Carson et al., 2006; Yasin et al., 2021; Nguyen et al., 2023).

Tea tree oil contains several phytochemicals; however, terpinen-4-ol, γ-terpinene, and α-terpinene are considered the most important bioactive constituents and account for a large proportion of its chemical composition. These compounds are widely recognized as the primary contributors to the antimicrobial activity of tea tree oil (Cox et al., 2000; Yasin et al., 2021). *In vitro* studies have shown that tea tree oil acts mainly by disrupting microbial cell membranes, which leads to leakage of intracellular components, loss of membrane integrity, and the death of microbes (Carson et al., 2006; Yap et al., 2014). For example, when Gram-negative and Gram-positive bacteria are exposed to tea tree oil, the ratio of cell envelope structure, rapid membrane permeabilization, and cell killing are increased in a dose-dependent manner

The antimicrobial activity of tea tree oil is also extended to the most common pathogens, including *Staphylococcus aureus, Escherichia coli*, and the clinical oral microbes like *Porphyromonas gingivalis* and *Streptococcus mutans* among others, which would make it an attractive topical antimicrobial agent and formulation in their proper use (Sharifi-Rad et al., 2017). Nevertheless, tea tree oil should not be ingested and it should be applied topically and diluted because of its evident toxicity and irritation of the skin and there is a high level of safety consideration that should be taken when administering treatment (Cox et al., 2000).

The antimicrobial activity of tea tree oil is predominantly described as mechanistic due to its potential to enter a microbial cell and destabilize its walls, disrupt the enzymatic systems and cause morphological alterations that lead to dysfunction and mortality of the cell (Carson et al., 2006; Yap et al., 2014).

2. Oregano Oil (Origanum vulgare)

Research has shown that essential oils have a strong antifungal and antibacterial action on *Staphylococcus aureus, Escherichia coli, Candida albicans,* and multidrug-resistant strains (Nazzaro et al., 2013; Sharifi-Rad et al., 2017; Ezeanya-Bakpa et al., 2021a). They are expected to be useful in the development of topical antiseptics and food preservative formulations, and in the development of adjuvants to modern antibiotics especially against persistent or resistant pathogens, due to their properties. To be safe and effective in humans however, rigorous clinical validation and formulation are necessary.

3. Garlic (Allium sativum)

Garlic contains organosulfur-containing phytochemicals such as allicin, ajoene (E- and Z-ajoene), vinyldithiins (e.g., 2-vinyl-1,3-dithiin), and diallyl sulfides (e.g., diallyl disulfide and diallyl trisulfide), which are responsible for its antimicrobial activity (El-Saber Batiha et al., 2020; Singh et al.,2025). These compounds are produced enzymatically after when the physical tissue of garlic is disrupted, the precursor allicin is converted to allicin by the enzyme alliinase (Nakamoto et al., 2020; Ezeanya-Bakpa et al., 2023). Allicin specifically is extremely reactive and plays a central role in the antimicrobial activity of garlic. It has a wide-range of antibacterial, antifungal and antiviral activities, as it inhibits enzymes that contain thiol and are vital to the metabolism of the microbes.

4. Grapes (Vitis vinifera L.)

Resveratrol has been shown to inhibit both Gram-positive and Gram-negative bacteria, including methicillin-resistant *Staphylococcus aureus* (MRSA), *Listeria monocytogenes, Escherichia coli, Helicobacter pylori* and possesses antifungal activity against *Candida albicans* (Abbas et al., 2016).

Apart from resveratrol, grapes also include anthocyanins, phenolic acids, proanthocyanidins, and flavonoids (quercetin, catechin, and epicatechin), all of which enhance the antibacterial properties of grape extract. Anthocyanins have bacteriostatic effects by inducing oxidative stress in microbial cells, while flavonoids such as quercetin destabilize bacterial membranes and block vital enzymes. Grape seed extracts are rich in oligomeric proanthocyanidins with potent biological activity. Because they limit bacterial adhesion and colonization, proanthocyanidins from grape seeds are especially effective against oral and gastrointestinal infections.

OVERVIEW OF ANTIMICROBIAL RESISTANCE

The term 'Antimicrobial' denotes 'anti microbes' while 'resistance' means 'withstand'. In the light of these, antimicrobial resistance means the ability of microbes to withstand antimicrobial agents. Antimicrobial Resistance (AMR) usually occurs in bacteria, viruses, fungi, and parasites (Ezeanya-Bakpa et al., 2025a). Thus, resulting in difficult-to-treat infections and increases the chances of disease transmission, and death. Drug resistance leads to antibiotics and other antimicrobial medications becoming ineffective, making infections harder or even impossible to treat. Antimicrobial Resistance (AMR) occurs when microorganisms, such as bacteria, fungi, parasites, and viruses, develop resistance to antimicrobial drugs like antibiotics that are used to treat infections (Sharmila Devi et al., 2024; Ezeanya-Bakpa et al., 2024).

Phytochemicals as Antimicrobial Agents

A. Antimicrobial Activity of Phytochemicals

Phytochemicals exert broad-spectrum antimicrobial activity against bacteria, fungi, and viruses. Their multimodal mechanisms—ranging from membrane disruption to nucleic acid interference—reduce microbial adaptability and resistance development.

I. Activity Against Bacteria

Phytochemicals such as phenolics, flavonoids, alkaloids, and terpenoids demonstrate efficacy against Gram-positive and Gram-negative bacteria, including multidrug-resistant strains:

- **Membrane disruption:** Thymol and carvacrol destabilise lipid bilayers, causing leakage and lysis.
- **DNA and RNA interference:** Alkaloids like berberine and flavonoids such as quercetin intercalate DNA and inhibit topoisomerases.
- **Enzyme inhibition:** Phenolics inhibit energy metabolism and cell wall–associated enzymes.

- **Efflux pump inhibition:** Restores intracellular antibiotic concentration and reverses resistance.

II. Activity Against Fungi

Phytochemicals inhibit pathogenic yeasts and filamentous fungi through:

- **Cell wall biosynthesis inhibition:** Eugenol targets chitin and glucan synthesis.
- **Ergosterol pathway disruption:** Terpenoids compromise membrane integrity.
- **Inhibition of hyphal formation and biofilms:** Flavonoids suppress virulence factors.
- **Oxidative stress induction:** ROS generation leads to fungal apoptosis.

III. Activity Against Viruses

Antiviral actions include:

- **Entry inhibition:** EGCG prevents viral attachment or fusion.
- **Replication suppression:** Quercetin blocks viral polymerases.
- **Protein synthesis inhibition:** Prevents assembly of viral components.
- **Immune modulation:** Upregulation of interferons enhances host antiviral defense.

B. Mechanisms of Antimicrobial Action

Phytochemicals act through multiple mechanisms, often concurrently, reducing the likelihood of inducing resistance. Three core mechanisms are highlighted below.

I. Disruption of Microbial Cell Membranes

Many phytochemicals, particularly terpenoids and phenolics, insert into microbial membranes, causing:

- increased membrane permeability
- leakage of ions and cellular metabolites
- loss of proton motive force
- eventual cell lysis

For example, monoterpenes such as thymol disrupt lipid bilayer integrity, leading to rapid bacterial death.

II. Inhibition of Nucleic Acid Synthesis

Some alkaloids and flavonoids intercalate into DNA or inhibit enzymes such as **DNA gyrase, topoisomerases**, and **RNA polymerases**. This prevents DNA replication and transcription, halting cell division.

III. Interference with Microbial Metabolic Pathways

Phytochemicals can inhibit metabolic enzymes critical for ATP generation, protein biosynthesis, or cell wall formation. Tannins, for example, bind microbial proteins and enzymes, disrupting nutrient acquisition and metabolism.

Additional mechanisms also include:

- inhibition of quorum sensing
- suppression of virulence factors
- disruption of biofilm formation
- metal chelation
- oxidative stress induction

Table 1. Major phytochemical classes and mechanisms

Class	Representative Compounds	Mechanism of Action	Resistance-Modifying Properties
Phenolics	Gallic acid, EGCG, tannins	Membrane disruption, enzyme inhibition, ROS generation	Efflux pump inhibition, metal chelation
Flavonoids	Quercetin, Luteolin, Catechin	DNA intercalation, enzyme inhibition, biofilm disruption	Efflux pump inhibition, synergy with antibiotics
Alkaloids	Berberine, Quinine, Piperine	DNA intercalation, enzyme inhibition, membrane disruption	Inhibit resistance enzymes
Terpenoids	Thymol, Menthol, Carvacrol	Membrane permeabilisation, enzyme inhibition	Enhanced antibiotic uptake

Phytochemicals with Documented Antimicrobial Properties

Phytochemicals are naturally occurring bioactive compounds produced by plants as part of their normal metabolic processes. They are not required for primary physiology such as growth or reproduction, but function as secondary metabolites involved in defence, signalling, and ecological interactions. Over 10,000 phytochemicals have been identified, with many demonstrating therapeutic properties, including antioxidant, anti-inflammatory, anticancer, and antimicrobial effects.

A. Classification of Phytochemicals

Phytochemicals are grouped based on both their chemical structure and biosynthetic origin. For the biosynthetic pathways, they generate the structural diversity that underpins the wide-ranging antimicrobial actions of phytochemicals.

Phenolic compounds - Phenolics or polyphenols contain hydroxylated aromatic rings, including simple phenols, phenolic acids, flavonoids, tannins, and lignans. Representative compounds include gallic acid, caffeic acid, EGCG, and quercetin.

Alkaloids - These alkaloids are derived from amino acids and they have nitrogen present in the heterocyclic ring. The primary precursors of true alkaloids are amino acids. Examples include: berberine, morphine, quinine, and piperine.

Terpenoids - Terpenoids, or isoprenoids, are compounds built from isoprene units (C_5H_8), ranging from monoterpenes (C10) to triterpenes (C30) and beyond. Examples include menthol, carvacrol, thymol, and limonene.

Phytochemicals are derived from several primary biosynthetic pathways in plants, which generate the structural diversity responsible for their varied biological activities:

- **Shikimate pathway**: Produces aromatic amino acids that serve as precursors for phenolic compounds (e.g., flavonoids, phenolic acids, tannins) and aromatic alkaloids. This pathway is central to the formation of hydroxylated aromatic rings critical for antimicrobial and antioxidant activity.
- **Mevalonate (MVA) and methylerythritol phosphate (MEP) pathways**: Generate terpenoids, including monoterpenes, sesquiterpenes, diterpenes, and triterpenes. These pathways determine chain length, cyclic structures, and degree of unsaturation, which influence membrane interaction and enzyme inhibition.
- **Acetate–malonate (polyketide) pathway**: Produces flavonoids, anthraquinones, and other polyketide-derived metabolites. Variations in condensation and cyclization reactions lead to diverse scaffolds with potent antimicrobial and resistance-modifying properties.

Figure 1. Classification of phytochemicals

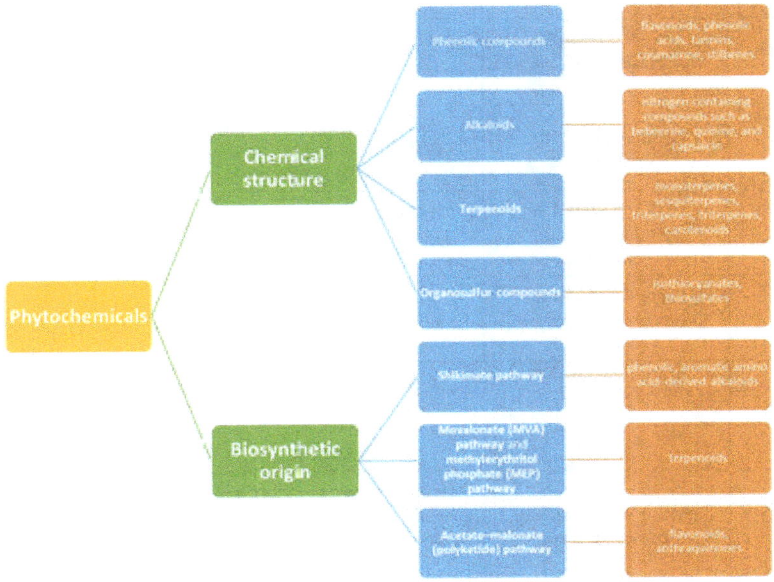

B. Structure–Activity Relationships (SAR) of Phytochemicals

Importantly, their ability to modulate microbial resistance offers new strategies for addressing the global AMR crisis. Continued exploration of phytochemical biosynthesis, SAR, and synergistic interactions will advance their application in modern drug discovery.

Table 2. An overview of Structure–Activity Relationship (SAR) and their effect on antimicrobial activity

Phytochemical Class	Key Functional Groups	SAR Highlights	Effect on Antimicrobial Activity
Phenolics	Hydroxyl, Methoxy, Aromatic ring	More hydroxyls → H-bonding, ROS induction; Methoxylation → lipophilicity	Membrane disruption, enzyme inhibition
Flavonoids	Hydroxylation, C2–C3 double bond, Glycosylation, Prenylation	Hydroxylation → metal chelation; Prenylation → membrane insertion; Glycosylation → reduces permeability	DNA/enzyme targeting, biofilm inhibition
Alkaloids	Planar heterocycles, Quaternary N, Hydroxyl/ methoxy	Planarity → DNA intercalation; Quaternary N → membrane interaction; Chirality → target specificity	DNA replication inhibition, enzyme targeting
Terpenoids	Double bonds, Alcohol/ Aldehyde, Cyclic/Chain structures	Double bonds & cyclic → membrane disruption; Alcohol/aldehyde → enzyme reactivity; Lipophilicity → bilayer integration	Membrane permeabilization, metabolic interference
Tannins & Coumarins	Hydroxyl, Prenylation, Molecular size	Hydroxyl groups → protein binding; Prenylation → enzyme/DNA interactions; Large molecules → protein precipitation	Enzyme inhibition, membrane disruption

Resistance-Modifying Activities of Phytochemicals

Antimicrobial resistance has improved significantly to the extent that research focus shift from discovering new antibiotics to identifying agents that are capable of restoring or enhancing the effectiveness of existing drugs. Most of the mechanisms of antimicrobial resistance include drug inactivation or modification, drug target modification, drug uptake limitation, and drug efflux, which always lead to the reduction of antibiotic concentrations in the cell (Cheema et al., 2024; Ezeanya-Bakpa et al., 2025b). Phytochemicals have been of interest not only due to their direct antimicrobial effect, but also due to their capacity to alter resistance mechanisms of pathogenic microorganisms (Cushnie et al., 2014; Barati & Modarresi Chahardehi, 2024). Most phytochemicals are not conventional bactericidal agents but may be used as resistance modulators, addressing major adaptive responses used by resistant microbes (Cushnie et al., 2014; Barati & Modarresi Chahardehi, 2024). They can inhibit efflux pumps, interfere with biofilm formation, and restore the level of antibiotic sensitivity in resistant microorganisms.

Mechanisms by Which Phytochemicals Modify Resistance

1. Inhibition of Efflux Pumps

Efflux pumps are one of the most common mechanisms by which bacteria develop multidrug resistance. It is an actively functioning membrane-bound transport proteins that expel antibiotics from the microbial cell by reducing intracellular drug concentration below therapeutic level (Li et al., 2015; Du et al., 2018). Phytochemicals have drawn interest recently not just as antibacterial agents but also as efflux pump inhibitors (EPIs) that can restore the effectiveness of antibiotics. Many agents produced from plants have low toxicity and multi-target action, which makes them appealing candidates for resistance modulation in contrast to synthetic EPIs. Efflux pumps are inhibited by phytochemicals via a number of interrelated ways (Cheema et al., 2024). Certain substances prevent substrate recognition or translocation by directly binding to transporter proteins. Others interfere with the ATP supply or proton motive force that drives efflux activity.

Polyphenols such as flavonoids and stilbenes are often effective because they have planar structures and are lipophilic, which enable them to interact with membrane proteins easily (Khameneh et al., 2019; Ezeanya-Bakpa et al., 2021b). While polyphenols such as resveratrol, quercetin and curcumin, as well as alkaloids like berberine have been demonstrated to possess efflux pump inhibitory capacity in resistant strains of *Staphylococcus aureus, Escherichia coli, Klebsiella pneumoniae* and *Pseudomonas aeruginosa* (Tegos et al., 2011; Seukep, et al., 2022). These compounds sensitize bacteria to hitherto resistant antibiotics by disrupting the ATP-dependent or proton-motive force powered transport systems.

In the case of murine infection model with the use of curcumin supplemented with standard antibiotics, the bacterial clearance was improved without the significant rise of toxicity (Praditya et al., 2019). Collectively, these studies confirm that efflux pump inhibition is one of the primary ways that phytochemicals can alter antimicrobial resistance.

2. Disruption of Biofilm Formation

The formation of biofilms is another mechanism of critical resistance, which can be used by microorganisms to survive in adverse conditions and avoid the action of antibiotics (Hall & Mah, 2017). In biofilms, bacteria are enclosed in an extracellular matrix that is self-produced and limits the ability of drugs to penetrate and/or limits metabolic activity and the expression of genetic exchange that also contributes to persistent infections (Flemming & Wingender, 2010; Ezeanya-Bakpa et al., 2026). Associated infections are also identified to be extremely challenging to eliminate, especially the biofilm-related ones, which are prevalent in chronic wounds, medical implants, and urinary tract infections (UTI) (Donlan, 2001). Antibiotics commonly fail to penetrate into biofilm-inspired bacteria, and alternative or complementary methods are necessary (Ciofu et al., 2017). Phytochemicals have emerged as effective anti-biofilm agents with capabilities of disrupting biofilm formation, maturation and maintenance (Alum et al., 2025). One of the main targets is the bacterial communication system quorum sensing that controls biofilm formation and virulence (Alum et al., 2025).

Compounds such as curcumin, resveratrol, and garlic-derived organosulfur molecules have been shown to inhibit biofilm formation and destabilize established biofilms by disrupting extracellular polymeric substances and interfering with quorum sensing pathways (El-Saber Batiha, et al., 2020; Alum et al., 2025). In vitro studies show that the compounds decrease the biofilm biomass and down-regulate the virulence genes in organisms like *S. aureus*, *P. aeruginosa*, and *Candida albicans* (Alum et al., 2025; Praditya et al., 2019). Curcumin that was found to downregulate quorum sensing controlled genes in *Pseudomonas aeruginosa*, which leads to decreased biofilm thickness and a higher sensitivity to antibiotics (Alum et al., 2025). Similarly, resveratrol interferes with the integrity of the extracellular matrix to prevent the synthesis of polysaccharides and make biofilms more vulnerable to antibiotics (Jeyachandran & Sekar, 2025). Although garlic-derived organosulfur compounds, particularly ajoene, have a potent anti-biofilm effect with the ability to inhibit quorum sensing pathways and virulence gene expression in Gram-positive and Gram-negative bacteria (El-Saber Batiha, et al., 2020).

The use of essential oils like oregano oil also destabilizes biofilms because they dissolve lipid substances and disrupt the adhesion of surfaces (Nazzaro et al., 2017). Also, flavonoids, phenolic acids, and terpenoids are capable of inhibiting the production of quorum sensing signals or receptors keeping coordinated biofilm formation at bay (Manso et al., 2021). Phytochemicals compromise biofilm integrity, making microbes more susceptible to antibiotics and host immune-response.

Comparative research shows that phytochemicals have a tremendous effect of increasing the penetration of antibiotics through biofilms (Manso et al., 2021). It was described by Zhang that when resveratrol is used in combination with ciprofloxacin, there was almost full eradication of mature *E. coli* biofilms, but each agent alone demonstrated low effectiveness. Therefore, these observations are also reinforced by *in vivo* studies. Phytochemical-antibiotic combinations were determined to be more effective in wound infection models as compared to antibiotics alone in enhancing healing and decreasing bacterial load (Praditya et al., 2019). These findings put biofilm disruption as one of the key resistance-modifying mechanisms with high translational potential.

3. Restoration of Antibiotic Sensitivity

To re-enable the effects of antibiotics, the reversal of the resistance mechanisms of antibiotics requires restoration of antibiotic sensitivity. Among the most promising resistance-modifying effects of phytochemicals is their capacity to recover the antibiotic tolerance in resistant strains when combined with the conventional antibiotics. Furthermore, phytochemicals may sensitize resistant microorganisms by elevating the membrane permeability. Lysis of the microbial cell membranes enhances the penetration of antibiotics and interferes with the normal functioning of cells.

There are also certain plant-derived compounds that can inhibit resistance-related enzymes, β-lactamases, that break down antibiotics before they reach their pathogenic targets. By neutralizing these enzymes, antibiotics are protected against being inactivated by these enzymes and are able to regain their action. Several researches have indicated that phytochemicals can interact with traditional antibiotics, reducing the minimum inhibitory concentrations (MIC) of drugs, including β-lactams, fluoroquinolones, and aminoglycosides, with resistant bacteria (Akujobi et al., 2012; Praditya et al., 2019). Phytochemicals in these types of combinations are not necessarily primary antimicrobial agents; rather, they increase the action of antibiotics by reversing the resistance mechanisms. This synergy enables to have low doses of antibiotics, less toxicity and slow resistance development. Moreover, this approach also extends the clinical lifespan of existing antibiotics, which is essential, considering the dwindling number of new antimicrobial drugs in the pipeline.

For instant, curcumin and resveratrol have been reported to re-sensitize MRSA strains to oxacillin and ciprofloxacin by altering membrane integrity and inhibiting the resistance-associated genes expression. Similarly, garlic-derived organosulfur compounds enhance antibiotic penetration and inhibit resistance-conferring enzymes, which leads to better antibacterial effects (El-Saber Batiha, et al., 2020; Ezeanya-Bakpa et al., 2022a).

CHALLENGES AND FUTURE DIRECTIONS

The increasing number of evidence supporting phytochemicals as antimicrobial and resistance-modifying agents is promising. However, the compounds face several challenges in translating from laboratory research to clinical setting. These challenges must be address in order to determine the major obstacles encountered when using phytochemicals in combating antimicrobial resistance.

1. Limitations of Current Research

Despite the increasing confirmation or prediction of phytochemicals having antimicrobial potential, there are still a number of limitations that are critical and acting as obstacles to their complete translation into clinically viable resistance-modulating agents.

a. In Vitro Investigations

Majority of research carried out on the antimicrobial potential of phytochemical are limited to *in vitro* studies which is one of the greatest challenges. Though these studies are useful in terms of initial screening and understanding of dynamics, but they fail to adequately mimic the complexity of real biological systems. Biological factors such as host immune responses, interactions between the pathogen and the host, tissue penetration and the impact of the human microbiome are mostly not present in the laboratory models. As a result, the antimicrobial efficacy in controlled conditions might not be reflecting the accurate efficacy in *in vivo* or under clinical settings (Cowan, 1999; Hemaiswarya et al., 2008). Such dependence on *in vitro* data limits the confidence in the therapeutic value of most of the promising phytochemicals.

b. Variability and Lack of Standardization

Inconsistency and non-standardization of phytochemicals composition is another major limitation in research. The chemical composition of plant-based compounds is highly affected by different factors, such as plant species, geographical source, harvesting date, extra solvent and storage, which may cause significant variation in the concentration and types of bioactive metabolites in extracts (Afshar, Najafian, & Radi, 2022; Sun et al., 2025). For instant, extracts from the same plant species may differ significantly in their bioactive constituents, leading to inconsistent antimicrobial outcomes across studies. Additionally, several studies do not characterize their extract properly but often report antimicrobial activity without full phytochemical profiling, which weakens scientific rigor (Heinrich et al. 2022). Therefore, the differences in the concentration, purity, and/or the presence of synergistic or antagonistic compounds may result in variability in the outcome across studies. Also, there are numerous studies where high doses of phytochemicals are used which are difficult to achieve safely in human tissues because of the lack of bioavailability, and high metabolism, or toxicity. Therefore, this raises questions about translational feasibility (Sun et al., 2025). This imprecision complicates finding, and the ability to reproduce or compare results and makes it difficult to establish reliable dose-response relationships, as well as determine the specific compounds responsible in particular antimicrobial effects (Heinrich et al., 2022; Mashele, 2025).

c. Dosage and Translational Feasibility

The issue of dosage and translational viability also serve as limiting factors to applications of current findings. Many studies show only antimicrobial or resistance-modulating effects at comparatively high concentrations of phytochemicals which are not likely to be safely obtained in human tissues. Although these types of concentrations can prove useful in the case of proof-of-concept experiments, this can raise significant doubts upon their clinical applicability. High doses can also put an individual at risk of cytotoxicity or off-target effects especially when taking phytochemicals are administered over a long period of time. Without careful consideration of realistic pharmacological concentrations, the therapeutic potential of these compounds remains theoretical (Gibbons, 2004; Ezeanya-Bakpa et al., 2022b).

d. Limited Understanding of Pharmacokinetics and Metabolism

There is a limited understanding of pharmacokinetics and metabolism of most of the antimicrobial phytochemicals. The data on its absorption, distribution, metabolism, and excretion (ADME) in humans are scarce and most of the compounds exhibit poor bioavailability due to their rapid metabolism or low solubility. These limitations may lead to drastic decrease in in vivo antimicrobial activity, despite its strong activity demonstrated in in vitro. The lack of detailed pharmacokinetic studies interferes with rational formulation strategies and dose optimization, which are essential in developing the drug in a clinical setting (Atanasov et al., 2015).

e. Toxicity and Safety Concerns

The potential toxicity and safety issues are not properly studied as well. Even though Phytochemicals are often considered generally safe due to their natural origin, this assumption is not always corroborated by toxicological data. Certain plant-derived compounds can be genotoxic, hepatotoxic, or endocrine-disrupting, especially at a high or prolong dosage. Moreover, a systematic assessment of the long-term safety of most phytochemicals remains unavailable, which constrains regulatory approval and clinical confidence (Ekor, 2014).

f. Drug–Herb Interactions

Drug-herb interaction is another limitation that is under-researched but clinically relevant in drug discovery. Phytochemicals can either alter the activity of drug-metabolizing enzymes and transporters, which can either increase toxicity or reduce the efficacy of conventional antibiotics. Although synergistic interaction between phytochemicals and antibiotics is widely reported, but there are few studies on antagonistic interaction which leads to an incomplete knowledge regarding the real therapeutic potential of these pairs (Hemaiswarya et al., 2008). It is important to address these gaps before phytochemicals can be integrated safely into antimicrobial treatment regimens.

2. Need For Standardized Methodologies

a. Extraction and Characterization Protocols

It becomes challenging to replicate results across laboratories in the absence of rigorous chemical profiling techniques such as chromatographic fingerprinting (HPLC, LC–MS, GC–MS) and quantitative marker compound analysis. This inconsistency also hides the structure-activity associations and complicates dose standardization, which ultimately jeopardizes the reliance on antimicrobial activity (Pathrikar et al., 2025). Therefore, to guarantee reproducibility, traceability of active ingredients, and conformity with WHO's emphasis on quality-assured evidence for novel antimicrobials, harmonized extraction techniques and obligatory phytochemical characterization are crucial.

b. Standardized Antimicrobial and Resistance-Modifying Assays

WHO AMR frameworks emphasize the fact that standard laboratory procedures are required to facilitate global surveillance and integration of data (WHO, 2023). However, in the majority of studies assessing plant-derived antimicrobials, non-uniform antimicrobial susceptibility testing systems, different MIC determination procedures and disparate techniques have been used to evaluate efflux pump inhibition or biofilm disruption (Duda-Madej et al., 2025). Such methodological heterogeneity renders cross-study comparisons impossible, and it is not possible to find an extract or a compound that would truly restore antibiotic efficacy. It would make the research on phytochemicals more reliable to align the findings with internationally accepted standards and help the WHO to focus on the development of effective antibiotic adjuvants to address the challenges of resistant pathogens.

c. Validated Animal and Preclinical Infection Models

Another requirement of the WHO is the value of the translational research pipelines that move beyond laboratory screening but transition to clinically relevant models. The main sustainability gap in contemporary phytochemical AMR studies is the limited application of tested animal and preclinical infection models that are suitable to human disease. In vitro systems fail to capture important aspects like host immune system responses, tissue penetration, pharmacokinetics and the complexity of biofilm-related infections. Well-validated and standardized in vivo models are thus needed to produce evidence that can be in tandem with WHO priorities towards creating interventions that are not just effective in theory, but can also be applicable in clinical practice.

d. Consensus on Reporting Outcomes

Surveillance and research strategies of WHO AMR emphasize the necessity of coordinated outcomes and endpoints reporting. In phytochemical antimicrobial research, the results include the index of synergy, toxicity level and pharmacological endpoint are not always given consistently or with definite reference points.

This inconsistency makes it more difficult to synthesize the evidence and undermine the validity of arguments about antimicrobial or resistance-modifying potential. Standardized reporting systems would make data more transparent, enable the integration of all data and operate in line with the wider objective of the WHO to develop a coherent and actionable global evidence base against AMR. Therefore, standardization of methods will enhance reproducibility, facilitate meta-analyses and speed up the discovery of phytochemicals with solid translational prospects.

3. Potential for Phytochemical Integration in Conventional Therapies

Phytochemicals offer significant opportunities to enhance conventional antibiotic therapy. By acting as efflux pump inhibitors, biofilm disruptors, and resistance-sensitizing agents, they can potentiate existing antibiotics, reduce required doses, and potentially minimize the selective pressure that drives resistance (Zhang et al., 2014; Sharifi-Rad et al., 2017).

4. Future Research Directions and Clinical Applications If Phytochemical as Antimicrobial and Resistance Modifying Agents

1. **High-throughput screening for synergistic activity:** A high-throughput screening strategy is necessary to identify plant-derived compounds that supplement or enhance the conventional antibiotics to improve their efficacy to tackle multidrug-resistant pathogens. High-throughput screening systems enable the simultaneous screening of many phytochemicals in combination with existing antibiotics, to evaluate their capacity to restore microbial susceptibility. Studies have shown that a number of plant-based molecules can act synergistically with conventional drugs, reducing the effective dose of antibiotics and, in the future, reducing the selective pressure, which leads to resistance (Shahin et al., 2025; Cheesman et al., 2017). This approach enhances the rapid process of identifying promising candidates to undergo further mechanistic and clinical testing.

2. **Mechanistic studies using omics technologies:** It is necessary to have molecular knowledge of how phytochemicals have resistance-modifying action. Genomics, transcriptomics and metabolomics can reveal the interaction of phytochemicals with microbial efflux pumps, biofilm formation disruption, and quorum sensing and metabolic interference. As an example, the RNA sequencing may provide evidence of the differentiated gene expression after exposure to phytochemicals whereas the metabolomic analysis may indicate the disrupted energy metabolism or biosynthetic pathways. The evidence of these mechanistic studies is sufficient to rationally design combination therapies and predict possible off-target effects, which increases the translational potential of plant-derived compounds (Abdelhamid & Yousef, 2023).

3. **Developing innovative methods to improve delivery:** Many phytochemicals are restricted by poor solubility, rapid metabolism or low tissue absorption. Advanced delivery systems, such as nanoparticle encapsulation, liposomal carriers, as well as polymer-based controlled-release systems, increase bioavailability, prevent bioactive compounds degradation, and facilitate delivery to target site of infection. Such innovations are mostly important for topical applications such as wound care and device-associated infections, where local high concentrations can be safely achieved. Besides, improved delivery can not only maximize antimicrobial and antibiofilm effects but also minimize systemic toxicity and improve clinical feasibility (Patra et al., 2018; Abdelhamid & Yousef).

4. **Validated preclinical models:** To translate *in vitro* findings to human infections, preclinical models are required to recapitulate the complexity of biofilm-related and chronic infections. Animal models enable researchers to evaluate the pharmacokinetics, tissue distribution, and combining efficacy with antibiotics in conditions that are approximately human infections. These types of researches are

important to identify the effective dosing regimens, to assess possible toxicity, and to justify further clinical trials (Lewis, 2020).

CONCLUSION

The increasing prevalence of antibiotic resistance constitutes one of the most significant challenges to global health. In recent years, the scientific community has turned its attention to phytochemicals—naturally occurring compounds in plants—that exhibit promising antimicrobial properties and the potential to modify resistance mechanisms. This shift arises from the urgent need for alternative strategies to combat resistant pathogens, which have rendered many conventional antibiotics ineffective.

Phytochemicals, such as flavonoids, terpenes, alkaloids, and phenolic acids, have garnered considerable attention for their multifaceted activities against a range of microbial organisms, including bacteria, fungi, and viruses. These compounds often function through unique mechanisms that can inhibit microbial growth or disrupt critical biological processes. For instance, flavonoids have been shown to interfere with bacterial cell wall synthesis, while terpenes can disrupt membrane integrity. Additionally, many phytochemicals possess the ability to inhibit biofilm formation, a persistent challenge in treating chronic infections.

One of the significant advantages of phytochemicals is their ability to modify resistance mechanisms in pathogens. Several studies have documented the capacity of certain plant-derived compounds to downregulate the expression of efflux pumps and beta-lactamase enzymes, which are critical in mediating antibiotic resistance. By enhancing the susceptibility of resistant strains to existing antibiotics, these compounds may restore the efficacy of treatments that have lost their utility. For instance, the combination of certain phytochemicals with traditional antibiotics has shown synergistic effects, thereby reducing the required dosage of antibiotics and potentially limiting side effects and toxicity.

Furthermore, the utilization of phytochemicals in clinical settings extends beyond direct antimicrobial action; they also play a vital role in immunomodulation and inflammation reduction. By enhancing immune responses and mitigating inflammatory pathways, phytochemicals can contribute to the overall effectiveness of antimicrobial therapies, offering a holistic approach to tackling infections.

However, despite their potential, several challenges must be addressed before phytochemicals can be widely implemented in clinical practice. Research on the pharmacokinetics, bioavailability, and interaction of these compounds with conventional drugs is still in its nascent stages. Moreover, standardization of extraction processes and formulations is crucial to ensure consistent quality and efficacy. As such, rigorous clinical trials are necessary to establish dosages, safety profiles, and the full extent of their antimicrobial capabilities.

In conclusion, while phytochemicals offer a promising avenue for addressing the growing issue of antimicrobial resistance, further research is essential to translate these findings into clinical applications. By understanding and harnessing the potential of these natural compounds, we may develop novel therapeutic strategies that complement and enhance existing antibiotic treatments. As we navigate the challenges of antibiotic resistance, phytochemicals stand out as a beacon of hope, suggesting that nature may provide the solutions we urgently seek in modern medicine.

Phytochemicals represent a promising frontier in the battle against microbial pathogens and antibiotic resistance. These bioactive compounds, derived from various plant sources, have demonstrated significant antimicrobial properties, often exhibiting efficacy against bacteria that are increasingly resistant to conventional treatments. Furthermore, certain phytochemicals possess the ability to modify resistance mechanisms, enhancing the potency of existing antibiotics and potentially restoring their effectiveness. As research continues to unveil the therapeutic potential of these natural substances, it is imperative to integrate phytochemicals into broader antimicrobial strategies, fostering a holistic approach to public health that emphasizes the importance of natural products in combating infectious diseases and mitigating the rising tide of antimicrobial resistance. This multidisciplinary effort will not only advance our understanding of these compounds but also contribute to the development of innovative treatment protocols that prioritize both efficacy and safety.

REFERENCES

Abbas, M., Saeed, F., Anjum, F. M., Afzaal, M., Tufail, T., Bashir, M. S., Ishtiaq, A., Hussain, S., & Suleria, H. A. (2016). Natural Polyphenols: An overview. *International Journal of Food Properties*, *20*(8), 1689–1699. DOI: 10.1080/10942912.2016.1220393

Abdelhamid, A. G., & Yousef, A. E. (2023). Combating Bacterial Biofilms: Current and Emerging Antibiofilm Strategies for Treating Persistent Infections. *Antibiotics (Basel, Switzerland)*, *12*(6), 1005. DOI: 10.3390/antibiotics12061005 PMID: 37370324

Afshar, M., Najafian, S., & Radi, M. (2022). The effect of harvest time on the natural product of *Rosmarinus officinalis* L. from South Iran (Fars province). *Natural Product Research*, *36*(10), 2637–2642. DOI: 10.1080/14786419.2021.1914615 PMID: 33974457

Akujobi, C. N., Ezeanya, C. C., & Aghanya, N. I. (2012). Detection of Cefotaximase Genes of Beta Lactamase among clinical isolates of *Escherichia coli* in a University Teaching Hospital, Nigeria. *Journal of Medical Sciences*, *12*(7), 244–247. DOI: 10.3923/jms.2012.244.247

Alum, E. U., Gulumbe, B. H., Izah, S. C., Uti, D. E., Aja, P. M., Igwenyi, I. O., & Offor, C. E. (2025). Natural product-based inhibitors of quorum sensing: A novel approach to combat antibiotic resistance. *Biochemistry and Biophysics Reports*, *43*, 102111. DOI: 10.1016/j.bbrep.2025.102111 PMID: 40641742

Atanasov, A. G., Waltenberger, B., Pferschy-Wenzig, E. M., Linder, T., Wawrosch, C., Uhrin, P., Temml, V., Wang, L., Schwaiger, S., Heiss, E. H., Rollinger, J. M., Schuster, D., Breuss, J. M., Bochkov, V., Mihovilovic, M. D., Kopp, B., Bauer, R., Dirsch, V. M., & Stuppner, H. (2015). Discovery and resupply of pharmacologically active plant-derived natural products: A review. *Biotechnology Advances*, *33*(8), 1582–1614. DOI: 10.1016/j.biotechadv.2015.08.001 PMID: 26281720

Bakkali, F., Averbeck, S., Averbeck, D., & Idaomar, M. (2008). Biological effects of essential oils—A review. *Food and Chemical Toxicology : An International Journal Published for the British Industrial Biological Research Association*, *46*(2), 446–475. DOI: 10.1016/j.fct.2007.09.106 PMID: 17996351

Barati, M., & Modarresi Chahardehi, A. (2024). *Alkaloids: The Potential of Their Antimicrobial Activities of Medicinal Plants*. IntechOpen., DOI: 10.5772/intechopen.112364

Carson, C. F., Hammer, K. A., & Riley, T. V. (2006). Melaleuca alternifolia (tea tree) oil: A review of antimicrobial and other medicinal properties. *Clinical Microbiology Reviews, 19*(1), 50–62. DOI: 10.1128/CMR.19.1.50-62.2006 PMID: 16418522

Cheema, H. S., Maurya, A., Kumar, S., Pandey, V. K., & Singh, R. M. (2024). Antibiotic potentiation through phytochemical-based efflux pump inhibitors to combat multidrug resistance bacteria. *Medicinal Chemistry (Shariqah, United Arab Emirates), 20*(6), 557–575. DOI: 10.2174/0115734064263586231022135644 PMID: 37907487

Cheesman, M. J., Ilanko, A., Blonk, B., & Cock, I. E. (2017). Developing New Antimicrobial Therapies: Are Synergistic Combinations of Plant Extracts/Compounds with Conventional Antibiotics the Solution? *Pharmacognosy Reviews, 11*(22), 57–72. DOI: 10.4103/phrev.phrev_21_17 PMID: 28989242

Ciofu, O., Rojo-Molinero, E., Macià, M. D., & Oliver, A. (2017). Antibiotic treatment of biofilm infections. *APMIS : Acta Pathologica, Microbiologica, et Immunologica Scandinavica, 125*(4), 304–319. DOI: 10.1111/apm.12673 PMID: 28407419

Cowan, M. M. (1999). Plant products as antimicrobial agents. *Clinical Microbiology Reviews, 12*(4), 564–582. DOI: 10.1128/CMR.12.4.564 PMID: 10515903

Cox, S. D., Mann, C. M., Markham, J. L., Bell, H. C., Gustafson, J. E., Warmington, J. R., & Wyllie, S. G. (2000). The mode of antimicrobial action of the essential oil of *Melaleuca alternifolia. Journal of Applied Microbiology, 88*(1), 170–175. DOI: 10.1046/j.1365-2672.2000.00943.x PMID: 10735256

Cushnie, T. P. T., Cushnie, B., & Lamb, A. J. (2014). Alkaloids: An overview of their antibacterial, antibiotic-enhancing and antivirulence activities. *International Journal of Antimicrobial Agents, 44*(5), 377–386. DOI: 10.1016/j.ijantimicag.2014.06.001 PMID: 25130096

Donlan, R. M. (2001). Biofilms and Device-Associated Infections. *Emerging Infectious Diseases, 7*(2), 277–281. DOI: 10.3201/eid0702.010226 PMID: 11294723

Du, D., Wang-Kan, X., Neuberger, A., van Veen, H. W., Pos, K. M., Piddock, L. J. V., & Luisi, B. F. (2018). Multidrug efflux pumps: Structure, function and regulation. *Nature Reviews. Microbiology, 16*(9), 523–539. DOI: 10.1038/s41579-018-0048-6 PMID: 30002505

Duda-Madej, A., Viscardi, S., Niezgódka, P., Szewczyk, W., & Wińska, K. (2025). The Impact of Plant-Derived Polyphenols on Combating Efflux-Mediated Antibiotic Resistance. *International Journal of Molecular Sciences, 26*(9), 4030. DOI: 10.3390/ijms26094030 PMID: 40362268

Ekor, M. (2014). The growing use of herbal medicines: Issues relating to adverse reactions and challenges in monitoring safety. *Frontiers in Pharmacology*, *4*. Advance online publication. DOI: 10.3389/fphar.2013.00177 PMID: 24454289

Ezeanya-Bakpa, C. C., Adetunji, C. O., Enosabata, I. A., & Olori, E. (2021a). Lycopene for wound infection: In vitro susceptibility of drug-resistant clinical pathogens. *Tropical Journal of Natural Product Research*, *5*(1), 49–52. DOI: 10.26538/tjnpr/v5i1.4

Ezeanya-Bakpa, C. C., Agbakoba, N. R., Enweani-Nwokelo, I. B., & Oguejiofor, C. B. (2022a). Phylogeny-based identification of *Mycoplasma genitalium* in a Nigerian population of apparently healthy sexually active female students. *The Pan African Medical Journal*, *41*, 71–76. DOI: 10.11604/pamj.2022.41.71.19631 PMID: 35382058

Ezeanya-Bakpa, C. C., Agbakoba, N. R., Oguejiofor, C. B., & Enweani-Nwokelo, I. B. (2021b). Sequence analysis reveals asymptomatic infection with *Mycoplasma hominis* and *Ureaplasma urealyticum* possibly leads to infertility in females: A cross-sectional study. *International Journal of Reproductive Biomedicine*, *19*(11), 951–958. DOI: 10.18502/ijrm.v19i11.9910 PMID: 34977452

Ezeanya-Bakpa, C.C., Akinola, K., Adebayo, O.P., Akharaiyi, .F.C. (2023) Antibacterial combination effect of allicin, lycopene and quercetin on antibiotic-resistant bacteria. *Microbes and Infectious Diseases* 5(1) Published First Online Doi: DOI: 10.21608/mid.2023.253557.1701

Ezeanya-Bakpa, C. C., Lambe, F. E., Adebayo-Olajide, T. C., & Esosa-Enoyoze, G. E. (2026). From Lab to Field: Translating Natural Products into Clinical Candidates. In Ogidi, O. (Ed.), *Natural Antimalarial Drug Discovery and Development in the Global South* (pp. 137–168). IGI Global Scientific Publishing., DOI: 10.4018/979-8-3373-5821-5.ch005

Ezeanya-Bakpa, C. C., Lambe, F. E., Ayinde, K., & Adeyeye, N. T. (2025a). Synergistic Antibacterial evaluation of Allicin and *Psidium Guajava* Linnaeus with standard antibiotics. *Tanzania Journal of Health Research*, *26*(6), 2881–2893. DOI: 10.4314/thrb.v26i6.13

Ezeanya-Bakpa, C. C., Lambe, F. E., Okwelogu, I. S., Enoyoze, G. E., Dirisu, J., Adediran, B. O., & Alalade, L. I. (2025b). Synergistic antibacterial effect of *Mangifera indica* and *Annona muricata* on antibiotic-resistant wound isolates. *African Journal of Biomedical Research*, *28*(1), 174–181. DOI: 10.53555/AJBR.v28i1.5010

Ezeanya-Bakpa, C. C., Shorumu, O., & Ogunsoluju, T. (2022b). Mobile devices and accessories are reservoir for drug-resistant bacteria among university students: A cross-sectional study. *Journal of Medical Pharmaceutical and Allied Sciences, 11*(6), 5494–5499. DOI: 10.55522/jmpas.V11I6.4217

Ezeanya-Bakpa, C. C., Udeogu, C. V., Agbakoba, N. R., Uduchi, I., & Ejike, C. (2024). Probiotics: Mitigating antibiotic resistance in the vaginal microbiome. *Microbes and Infectious Diseases, 5*(3), 1036–1045. DOI: 10.21608/mid.2024.266564.1782

Flemming, H.-C., & Wingender, J. (2010). The biofilm matrix. *Nature Reviews. Microbiology, 8*(9), 623–633. DOI: 10.1038/nrmicro2415 PMID: 20676145

Gibbons, S. (2004). Anti-staphylococcal plant natural products. *Natural Product Reports, 21*(2), 263–277. DOI: 10.1039/b212695h PMID: 15042149

Hall, C. W., & Mah, T.-F. (2017). Molecular mechanisms of biofilm-based antibiotic resistance and tolerance. *FEMS Microbiology Reviews, 41*(3), 276–301. DOI: 10.1093/femsre/fux010 PMID: 28369412

Heinrich, M., Jalil, B., Abdel-Tawab, M., Echeverria, J., Kulić, Ž., McGaw, L. J., Pezzuto, J. M., Potterat, O., & Wang, J. B. (2022). Best Practice in the chemical characterisation of extracts used in pharmacological and toxicological research-The ConPhyMP-Guidelines. *Frontiers in Pharmacology, 13*, 953205. DOI: 10.3389/fphar.2022.953205 PMID: 36176427

Hemaiswarya, S., Kruthiventi, A. K., & Doble, M. (2008). Synergism between natural products and antibiotics against infectious diseases. *Phytomedicine : International Journal of Phytotherapy and Phytopharmacology, 15*(8), 639–652. DOI: 10.1016/j.phymed.2008.06.008 PMID: 18599280

Jazvinšćak Jembrek, M., Oršolić, N., Mandić, L., Sadžak, A., & Šegota, S. (2021). Anti-Oxidative, Anti-Inflammatory and Anti-Apoptotic Effects of Flavonols: Targeting Nrf2, NF-κB and p53 Pathways in Neurodegeneration. *Antioxidants, 10*(10), 1628. DOI: 10.3390/antiox10101628 PMID: 34679762

Jeyachandran, S., & Sekar, S. (2025). Natural anti-biofilm agents: A comprehensive review and future perspectives. *Current Microbiology, 83*(2), 115. Advance online publication. DOI: 10.1007/s00284-025-04684-2 PMID: 41461973

Khameneh, B., Iranshahy, M., Soheili, V., & Fazly Bazzaz, B. S. (2019). Review on plant antimicrobials: A mechanistic viewpoint. *Antimicrobial Resistance and Infection Control, 8*(1), 118. DOI: 10.1186/s13756-019-0559-6 PMID: 31346459

Kumar, A., Saranyadevi, S., Thirumalaisamy, S. K., Dapana Durage, T. T., Jaiswal, S. G., Kavitake, D., & Wei, S. (2025). Phenolic acids in fermented foods: Microbial biotransformation, antioxidant mechanisms, and functional health implications. *Frontiers in Molecular Biosciences*, *12*, 1678673. Advance online publication DOI: 10.3389/fmolb.2025.1678673 PMID: 41195418

Kumar, S., & Pandey, A. K. (2013). Chemistry and biological activities of flavonoids: An overview. *TheScientificWorldJournal*, *162750*(1), 162750. Advance online publication. DOI: 10.1155/2013/162750 PMID: 24470791

Leclerc, A., & Fournier, P. (2024). The Role of Alkaloids in Natural Products Chemistry. International Journal of Herbal Medicine 12(5), 07-10. DOI: https://doi.org/DOI: 10.22271/flora.2024.v12.i5a.945

Lewis, K. (2020). The science of antibiotic discovery. *Cell*, *181*(1), 29–45. DOI: 10.1016/j.cell.2020.02.056 PMID: 32197064

Li, X. Z., Plésiat, P., & Nikaido, H. (2015). The challenge of efflux-mediated antibiotic resistance in Gram-negative bacteria. *Clinical Microbiology Reviews*, *28*(2), 337–418. DOI: 10.1128/CMR.00117-14 PMID: 25788514

Luo, H., Chen, J., Su, C., & Zha, L. (2022). Advances in the Bioactivities of Phytochemical Saponins in the Prevention and Treatment of Atherosclerosis. *Nutrients*, *14*(23), 4998. DOI: 10.3390/nu14234998 PMID: 36501028

Mabou, F. D., & Yossa, I. B. N. (2021). TERPENES: Structural classification and biological activities. *IOSR Journal of Pharmacy and Biological Sciences*, *16*, 25–40. DOI: 10.9790/3008-1603012540

Manso, T., Lores, M., & de Miguel, T. (2021). Antimicrobial Activity of Polyphenols and Natural Polyphenolic Extracts on Clinical Isolates. *Antibiotics (Basel, Switzerland)*, *11*(1), 46. DOI: 10.3390/antibiotics11010046 PMID: 35052923

Mashele, S. S. (2025). Phytochemicals as Multifunctional Agents: Antimicrobial, Enzyme Inhibitory, and Wound-Healing Potentials in the Era of Drug Resistance. *Preprints*. DOI: 10.20944/preprints202510.1183.v1

Nakamoto, M., Kunimura, K., Suzuki, J. I., & Kodera, Y. (2020). Antimicrobial properties of hydrophobic compounds in garlic: Allicin, vinyldithiin, ajoene and diallyl polysulfides. *Experimental and Therapeutic Medicine*, *19*(2), 1550–1553. DOI: 10.3892/etm.2019.8388 PMID: 32010337

Nazzaro, F., Fratianni, F., Coppola, R., & Feo, V. D. (2017). Essential Oils and Antifungal Activity. *Pharmaceuticals (Basel, Switzerland)*, *10*(4), 86. DOI: 10.3390/ph10040086 PMID: 29099084

Nazzaro, F., Fratianni, F., De Martino, L., Coppola, R., & De Feo, V. (2013). Effect of essential oils on pathogenic bacteria. *Pharmaceuticals (Basel, Switzerland)*, *6*(12), 1451–1474. DOI: 10.3390/ph6121451 PMID: 24287491

Nguyen, L., DeVico, B., Mannan, M., Chang, M., Rada Santacruz, C., Siragusa, C., Everhart, S., & Fazen, C. H. (2023). Tea tree essential oil kills escherichia coli and Staphylococcus epidermidis persisters. *Biomolecules*, *13*(9), 1404. DOI: 10.3390/biom13091404 PMID: 37759804

Pathrikar, S. Y., Mukhmale, V. S., Deshmukh, N. B., & Deshmukh, S. P. (2025). Standardization extraction and phytochemical screening for herbal drugs. *GSC Advanced Research and Reviews, 22(01),* 084-091 I: https://doi.org/DOI: 10.30574/gscarr.2025.22.1.0518

Patra, J. K., Das, G., Fraceto, L. F., Campos, E. V. R., Rodriguez-Torres, M. D. P., Acosta-Torres, L. S., Diaz-Torres, L. A., Grillo, R., Swamy, M. K., Sharma, S., Habtemariam, S., & Shin, H. S. (2018). Nano based drug delivery systems: Recent developments and future prospects. *Journal of Nanobiotechnology*, *16*(1), 71. DOI: 10.1186/s12951-018-0392-8 PMID: 30231877

Praditya, D., Kirchhoff, L., Brüning, J., Rachmawati, H., Steinmann, J., & Steinmann, E. (2019). Anti-infective Properties of the Golden Spice Curcumin. *Frontiers in Microbiology*, *10*, 912. DOI: 10.3389/fmicb.2019.00912 PMID: 31130924

El-Saber Batiha, G., Magdy Beshbishy, A., G. Wasef, L., Elewa, Y. H. A., A. Al-Sagan, A., Abd El-Hack, M. E., Taha, A. E., M. Abd-Elhakim, Y., & Prasad Devkota, H. 2020. Chemical Constituents and Pharmacological Activities of Garlic Allium sativum L.Review, A. *Nutrients*, *12*(3), 872. DOI: 10.3390/nu12030872

Rodríguez-Negrete, E. V., Morales-González, Á., Madrigal-Santillán, E. O., Sánchez-Reyes, K., Álvarez-González, I., Madrigal-Bujaidar, E., Valadez-Vega, C., Chamorro-Cevallos, G., Garcia-Melo, L. F., & Morales-González, J. A. (2024). Phytochemicals and their usefulness in the maintenance of health. *Plants*, *13*(4), 523. DOI: 10.3390/plants13040523 PMID: 38498532

Rodríguez-Negrete, E. V., Morales-González, Á., Madrigal-Santillán, E. O., Sánchez-Reyes, K., Álvarez-González, I., Madrigal-Bujaidar, E., Valadez-Vega, C., Chamorro-Cevallos, G., Garcia-Melo, L. F., & Morales-González, J. A. (2024). Phytochemicals and their usefulness in the maintenance of health. *Plants*, *13*(4), 523. DOI: 10.3390/plants13040523 PMID: 38498532

Rodríguez-Negrete, E. V., Morales-González, Á., Madrigal-Santillán, E. O., Sánchez-Reyes, K., Álvarez-González, I., Madrigal-Bujaidar, E., Valadez-Vega, C., Chamorro-Cevallos, G., Garcia-Melo, L. F., & Morales-González, J. A. (2024). Phytochemicals and their usefulness in the maintenance of health. *Plants*, *13*(4), 523. DOI: 10.3390/plants13040523 PMID: 38498532

Seukep, A. J., Fokoua-Maxime, C. D., Mbuntcha, H. G., Chen, G., Assob, J. C., Tenniswood, M., Sarker, S. D., Kuete, V., & Ming-Quan, G. (2022). Bacterial drug efflux pump inhibitors from plants. *Antimicrobial Resistance*, 487–532. DOI: 10.1007/978-981-16-3120-7_16

Shahin, H. H., Baroudi, M., Dabboussi, F., Ismail, B., Salma, R., Osman, M., & El Omari, K. (2025). Synergistic Antibacterial Effects of Plant Extracts and Essential Oils Against Drug-Resistant Bacteria of Clinical Interest. *Pathogens (Basel, Switzerland)*, *14*(4), 348. DOI: 10.3390/pathogens14040348 PMID: 40333114

Sharifi-Rad, J., Sureda, A., Tenore, G. C., Daglia, M., Sharifi-Rad, M., Valussi, M., Tundis, R., Sharifi-Rad, M., Loizzo, M. R., Ademiluyi, A. O., Sharifi-Rad, R., Ayatollahi, S. A., & Iriti, M. (2017). Biological Activities of Essential Oils: From Plant Chemoecology to Traditional Healing Systems. *Molecules (Basel, Switzerland)*, *22*(1), 70. DOI: 10.3390/molecules22010070 PMID: 28045446

Sharmila Devi, N., Mythili, R., Tijo Cherian, R., Dineshkumar, G. K., Sivaraman, R., Jayakumar, M., Prathaban, M., & Duraimurugan, V. (2024). Overview of antimicrobial resistance and mechanisms: The relative status of the past and current. *The Microbe*, *100083*(3), 100083. DOI: 10.1016/j.microb.2024.100083

Singh, N., Gusain, A., Nigam, M., & Mishra, A. P. (2025). The pharmacological and therapeutic versatility of *Allium* species: A comprehensive exploration of bioactive constituents and biological activities. *Discover Applied Sciences*, *7*(4), 349. DOI: 10.1007/s42452-025-06800-0

Sul, O. J., & Ra, S. W. (2021). Quercetin Prevents LPS-Induced Oxidative Stress and Inflammation by Modulating NOX2/ROS/NF-kB in Lung Epithelial Cells. *Molecules (Basel, Switzerland)*, *26*(22), 6949. DOI: 10.3390/molecules26226949 PMID: 34834040

Sun, S., Yu, Y., Jo, Y., Han, J. H., Xue, Y., Cho, M., Bae, S.-J., Ryu, D., Park, W., Ha, K.-T., & Zhuang, S. (2025). Impact of extraction techniques on phytochemical composition and bioactivity of natural product mixtures. *Frontiers in Pharmacology*, *16*, 1615338. Advance online publication. DOI: 10.3389/fphar.2025.1615338 PMID: 40808686

Tegos, G. P., Haynes, M., Strouse, J. J., Khan, M. M., Bologa, C. G., Oprea, T. I., & Sklar, L. A. (2011). Microbial efflux pump inhibition: Tactics and strategies. *Current Pharmaceutical Design*, *17*(13), 1291–1302. DOI: 10.2174/138161211795703726 PMID: 21470111

World Health Organization. (2023). *Global action plan on antimicrobial resistance*. WHO. https://www.who.int/publications/i/item/9789241509763

Yap, P. S. X., Yiap, B. C., Ping, H. C., & Lim, S. H. E. (2014). Essential oils, a new horizon in combating bacterial antibiotic resistance. *The Open Microbiology Journal*, *8*(1), 6–14. DOI: 10.2174/1874285801408010006 PMID: 24627729

Yasin, M., Younis, A., Ramzan, F., Javed, T., Shabbir, R., Noushahi, H. A., Skalicky, M., Ondrisik, P., Brestic, M., Hassan, S., & Sabagh, E. L. (2021). Extraction of essential oil from River Tea Tree (Melaleuca bracteata F. Muell.): Antioxidant and antimicrobial properties. *Sustainability (Basel)*, *13*(9), 4827. DOI: 10.3390/su13094827

Zhang, X. L., Guo, Y. S., Wang, C. H., Li, G. Q., Xu, J. J., Chung, H. Y., Ye, W. C., Li, Y. L., & Wang, G. C. (2014). Phenolic compounds from *Origanum vulgare* and their antioxidant and antiviral activities. *Food Chemistry*, *152*, 300–306. DOI: 10.1016/j.foodchem.2013.11.153 PMID: 24444941

Chapter 6
Herbal Antibiotics for Treating Drug-Resistant Bacteria

K. Gayathri
http://orcid.org/0009-0007-3786-2009
Department of Microbiology, Sacred Heart College, Tirupattur, India

P. Saranraj
http://orcid.org/0000-0002-5052-6532
Department of Microbiology, Sacred Heart College, Tirupattur, India

M. Kokila
Department of Microbiology, Sacred Heart College, Tirupattur, India

Mabel Joshaline C.
http://orcid.org/0000-0002-6091-1485
Arul Anandar College, Madurai, India

M. Shyamala
NMSSVN College, Nagamakai, India

Sri Rajalakshmi R.
M.R. Government Arts College, Mannargudi, India

ABSTRACT

The use of herbal antibiotics has increased as a substitute for synthetic ones, which are ineffective against bacteria that are resistant to drugs. The abuse and overuse of synthetic antibiotics is the root cause of a fast-expanding global health catastrophe. Since people first claimed that medicinal plants had therapeutic qualities thousands of years ago, they have been used. These plants include a variety of phytochemical substances, such as phenolics, terpenoids, alkaloids, flavonoids, tannins, and essential oils. Numerous of these substances have shown broad-spectrum antibacterial action. By interfering with the bacterial cell wall and membrane, preventing the production of proteins and nucleic acids, blocking the efflux pump, and blocking quorum-sensing pathways, these phytochemicals have antimicrobial activity that increases bacterial virulence and resistance.

DOI: 10.4018/979-8-3373-7810-7.ch006

1. INTRODUCTION

With the emergence of mass antibiotic manufacturing, human beings had been relying on natural mechanisms of defence and herbs as a means of curing bacterial infections. In the long run, though, extensive, and usually random, application of antibiotics has imposed a strong selective pressure on the population of bacteria, and many of them have acquired resistance to a variety of antimicrobial agents. Consequently, the formerly very effective antibiotics are progressively being rendered useless as therapeutic agents. This frightening trend has led to increased efforts by countries all over the world to discover new categories of antibiotics that are safe to humans and effective against the resistant pathogens. Even though antibiotics have indisputable advantages in the context of the treatment of infectious diseases in humans and animals, the concept is also linked to several negative effects, such as toxicity, changes in normal microbiota, and drug-drug interactions. The increasing prevalence of Multidrug-Resistant (MDR) and Extensively Drug-Resistant (XDR) bacteria is one of the gravest challenges to successfully manage the bacterial infections nowadays (Mundy et al., 2016).

The difficulty of antimicrobial resistance is aggravated by the fact that the novel antibiotics are developed very slowly and are quite expensive. The process of drug discovery and approval is resource intensive and hard to accomplish in a short time and respond to the changing crisis of resistance. Pharmaceutical industry has been trying to find other measures such as developing new antimicrobial agents and non-traditional treatment methods. The causes of antibiotic resistance are a combination of several factors that are largely interdependent, with the most prominent one being the misuse and excessive use of antibiotics, both in the medical and veterinary practice and agricultural activities. The development of resistance is caused by inappropriate prescribing practices and the absence of effective infection control measures, the limited number of alternative therapies, and the genetic differences among people (Liang et al., 2022).

The transfer of resistant bacteria is further enhanced by environmental pollution with traces of antibiotics. Even though antibiotics are the key to protection against infections, they can stimulate the development of resistance and the emergence of negative clinical results in case of their abuses. The resistance to most of the major classes of antibacterial drugs has increased drastically within the last 10 years. WHO reported unprecedented rates of hospital-acquired infections by Gram-negative bacteria, which are resistant to a variety of antibiotic classes. Worryingly, almost two-thirds of nations across the globe report resistance rates of over 50 percent in *Escherichia coli* with third-generation cephalosporins (Bhattacharya & Chandar, 2018).

Over the past years the production of new antimicrobial drugs has become more difficult and costly. Herbal medicines, on the other hand, have several bioactive compounds working in a synergistic manner against bacterial pathogens, even though they have been in use since time immemorial. In developing countries, to this date, some two-thirds to three-quarters of the population continues to use plant-based medicines to treat bacterial infections (Stefanović, 2018). Their efficacy has been shown historically; e.g., Coptis chinensis Franch. was administered as Traditional Chinese Medicine more than 1000 years ago to acute bacterial dysentery. In addition, plant-derived compounds have been useful in modern pharmacology wherein a vast number of these compounds have been used as a template in modern antibacterial drugs. An example is the creation of an oral formulation of Salvia miltiorrhiza Bunge which can be used to treat acne due to *Cutibacterium acnes*. Combined with regular antibiotics, herbals can improve the antibacterial effect and contribute to the recovery of the susceptibility of resistant strains. Notably, plant-based drugs are usually linked to a smaller number of side effects and a reduced possibility of resistance emergence in comparison to synthetic drugs (Ionescu, 2018).

The case of Tuberculosis (TB) treatment is a vivid illustration of the advantages as well as the limitations of antimicrobial therapy. The establishment of the effective anti-TB drugs made TB not a deadly disease, but a curable one. This development, unfortunately, was spoilt by the development of drug-resistant strains of *Mycobacterium tuberculosis*. Multidrug-resistant TB (MDR-TB) is characterized by resistance to no fewer than 2 major first-line medications, isoniazid and rifampicin, and extensively drug-resistant TB (XDR-TB) also has resistance to fluoroquinolones and injectable second-line medications. These kinds of TB are both deadly and have a few treatment alternatives. The last four decades have seen very few new anti-TB drugs approved, which adds to the fears surrounding the increasing burden of incurable TB (Chassagne et al., 2021).

Recent findings have proven that the development of new TB strains resistant to drugs, has further lowered the success rates of the treatment. However, a well planned combination treatment with several anti-TB drugs have proven to be more effective even in areas that depict high resistance. The problems met by clinicians who treat XDR-TB now are like the ones that were experienced before the introduction of antibiotics when it was believed that host immunity, nutrition, and environment were the factors to rely on. Positively, the recent studies and the accruing clinical experience suggest that in most cases, MDR and XDR TB can be treated successfully once the choice of drugs is reasonable and the available treatment options are used wisely (Khare et al., 2021).

2. PLANT-DERIVED ANTIMICROBIAL AGENTS

Phytochemicals can be applied in an antimicrobial setting as they are promiscuous and can react with a variety of molecular targets. Phytochemicals have the additional advantage of being potentially active against a variety of molecular targets, and as such, it is extremely challenging to develop resistance mechanisms against phytochemicals by bacteria. Phytochemical in medicinal plants is diverse and consists of alkaloids, coumarins, essential oils, flavonoids, lectins, phenolic compounds, polypeptides, polyacetylene terpenoids, tannins and so on. Most of these chemical classes also have compounds that might exhibit bactericidal or bacteriostatic activity against bacteria that have known Multidrug-Resistant (MDR) profiles (Mehboob et al., 2021). In addition, these phytochemicals can offer a point of attack to the development of new antibiotics that can help in the treatment of the infections caused by MDR bacteria. The rate of antimicrobial resistance in the world is on the rise and is complicating the process of treating most of the infectious diseases (Dassanayake et al., 2021).

The emergence of Multi-Drug-Resistant (MDR) bacteria in clinical setting is posing an urgent health concern in the population because of the widespread use of antibiotics and the capacity to generate biofilms. The management of infections by biofilm-forming bacteria is especially a challenging task because of the resistance which biofilms have to the action of the usual antibiotics. Existing research literature records numerous papers that prove the incapability of antibiotics to remove biofilms (Liang et al.,2022). The necessity of the innovative approaches to eliminate the failure of antibiotics is high. There are numerous applications that plant-based natural products are likely to be used in treating bacteria and yeast-induced infections. The numerous types of secondary metabolites generated by plants have been observed to disrupt biofilm formation, as they very differently affect biofilm formation as compared to conventional antibiotics (Kitpipit et al., 2023).

These pathways include quorum sensing inhibition, bacteria movement inhibition, bacterial adhesion impediment, Reactive Oxygen Species (ROS) generation and response, and so on. Combination of various phytochemicals and antibiotics has been reported to exhibit additive or synergistic effects whether used to regulate bio film development. Biofilms are multifaceted microbial communities that colonize both biotic and abiotic surfaces through a protective matrix which is made up of protein and polysaccharide components (Harkin et al., 2022).

3. ANTIMICROBIAL HERBS AND THEIR COMPONENTS

Medicinal plants are an important source of bioactive compounds that have strong antimicrobial properties. The alkaloids are one of the most important groups of antimicrobial agents found in nature. Alkaloids are organic substances containing nitrogen and are common in medicinal plants and they are the focus of their pharmacological effects. Besides their use in traditional medicine, alkaloids have been important in drug discovery in the modern era, including the discovery of a number of valuable antimicrobials like quinine, derivatives of quinolone and synthetics of nitroimidazoles based on azomycin. The main bioactive substance in it is the isoquinoline alkaloid berberine with a broad -spectrum antibacterial and antifungal effect (Tirant et al., 2024). The accumulation of berberine in the bacteria cells disrupts the DNA replication process and transcription. It is also a target of bacterial multidrug resistance (MDR) efflux pumps, especially the NorA in Gram-positive bacterium and MDR pumps in Gram-negative bacteria. Berberine has a greater antimicrobial effect when used with efflux pump inhibitors. Berberine impairs the cell division process in Gram-positive bacteria by interfering with the FtsZ protein, which is one of the major cytokinesis regulators. Berberine hydrochloride has demonstrated moderate to high potency against *Staphylococcus aureus, Escherichia coli* and *Staphylococcus aureus*, including MDR strains. Besides, berberine has antifungal properties against dermatophytes, including *Trichophyton mentagrotes, T. rubrum, Microsporum canis*, and *M. gypseum* (Matei & Vi an, 2025).

The alkaloids obtained by the marine are also of interest in terms of antimicrobial properties. Magnoflorine is a naturally occurring alkaloid that has been shown to have an inhibitory effect on Candida species with Minimum Inhibitory Concentration (MIC) level of about 50 mg/mL. Experimental works have demonstrated that the magic known as magnoflorine interferes with the cell wall of the fungus named *Candida albicans* by altering α-glucosidase an enzyme which plays a vital role in glucose metabolism as well as in the cell wall production of fungi. In addition to that, magnoflorine inhibits biofilm formation, which is a most important virulence factor of fungi. The toxicity tests have shown that magnoflorine has low cytotoxicity to human cell lines, which implies that it can be used as a safe antifungal (Husain et al., 2022).

Though these alkaloid rich herbal medicines have a potential of therapeutic value they need to be used with caution because they may be toxic. There are a variety of alkaloids that are found in Coppt s chinensis Franch. such as berberine, coptisine, palmatine, and epiberberine that have been reported to have dose-dependent toxicity. It has been found that extracts that have a higher concentration of alkaloids are more cytotoxic compared to whole-plant extracts. The experimental work demonstrates that the berberine demonstrates maximum cytotoxicity in hepatic (HepG2) and adipocyte (3T3-L1) cell models, and palmatine is less toxic. The LD 50 of berberine, coptisine, palmatine, and epiberberine are estimated to be 713.57, 852.12, 1533.68 and 1360 mg/kg, respectively (Anand et al., 2019). It has also been reported that berberine inhibited human ether-a-go-go-related gene (hERG) potassium channels, which may cause long QT interval and cardiac arrhythmia. Moreover, the Coptis chinensis alkaloids have inhibitory performance against acetylcholinesterase, and may cause mitochondrial dysfunction, which is a concern about the toxicity of the organs in excess or chronic use (Bhardwaj, 2016).

3.1 Antimicrobial Activity

The activity of the antimicrobials was tested on standardized Clinical and Laboratory Standards Institute (CLSI) guidelines. The disc diffusion technique was utilized as a pre-screening assay with the calculation of Minimum Inhibitory Concentrations (MICs) being done with the broth microdilution technique. Antibacterial agents (ampicillin, erythromycin, linezolid, ampicillin-sulbactam, cefuroxime, amikacin, ceftazidime, and doxycycline) were made in series of two folds in Mueller-Hinton broth whereas antifungal assays were done in RPMI medium. Bacterial and fungal inoculate were diluted to the ultimate concentrations of 1 x 10 6 and 1 x 10 3 CFU/mL, respectively. The microtiter plates were incubated at 37 °C and growth of the microbes was evaluated both visually and spectrophotometrically at 530nm. According to the CLSI guidelines, the lowest concentration of the antimicrobial agent that inhibited the visible growth was considered the MIC (Ramesh et al., 2024).

3.2 Herbal Actions

The formulations of herbs implemented in treatment of urinary tract infection (UTIs) are usually multi-functional to improve treatment. These are diuretic and aquaretic properties, antimicrobial and antiseptic properties, de-mucosal tissues, antispasmodic properties to relieve pain and immune-stimulating properties. Immunomodulators and tonic herbs are the most important supportive agents that help in increasing the host practice and make the tissues strong, which minimizes the chances of repeated infections (El-Saadony et al., 2025).

3.3 Antiseptic Herbs

Herbs with antiseptic and antimicrobial properties are particularly valuable in supporting recovery from UTIs. Hydroquinone derivatives, primarily released from arbutin-containing plants, exert antiseptic effects within the urinary tract by converting into active antimicrobial compounds in the presence of bacteria. Volatile oils present in many medicinal plants contribute to both specific and nonspecific antimicrobial activity. Herbs with notable urinary tract specificity include *Agathosma betulina* (buchu) and *Chimaphila umbellata* (pipsissewa), each containing distinct bioactive compounds that target urinary pathogens. Other plants rich in volatile oils, such as *Thymus vulgaris*, *Juniperus communis*, and *Apium graveolens*, also exhibit antimicrobial activity against uropathogens. *Achillea millefolium* (yarrow) combines antimicrobial and anti-inflammatory effects, while *Usnea* species contain usnic acid, a potent antibacterial compound. *Vaccinium macrocarpon* (cranberry) prevents bacterial adhesion to the bladder epithelium through mechanisms involving fructose and proanthocyanidins, making it effective for both prevention and management of UTIs (Bhardwaj et al., 2023).

3.4 Tonic Herbs

Tonic herbs deal with strengthening of connective tissues and organs resilience by offering structural and functional support to the urinary system. *Centella asiatica* (Gotu Kala) is a connective tissue tonic, fascia and ligament of the urinary tract, and a nervous system tonic, which reduces neurogenic irritation of the UTIs. *Parietaria diffusa* (pellitory of the wall) is a tropostorative of the kidney and urinary system that is associated with tissue regeneration and alleviation of inflammation via stainless steel mechanisms involving flavonoids. The herb can be used both in acute diseases, as in the case of urinary irritation and kidney stones, and in chronic ones, in recurrent UTIs or in interstitial cystitis (Cock et al., 2017; Seukep et al., 2022).

4. INDICATIONS SUITABLE FOR TREATMENT WITH HERBAL MEDICINES

Clinical observations and the history behind the therapeutic application of herbal medicines can readily prove its usage especially in the treatment of non-life-threatening illnesses. The traditional and modern literature suggests that herbal products have been useful in curing the Upper Respiratory Tract Infections (URI) and gynaecological, gastrointestinal, Urinary Tract Infections (UTIs), and psychosomatic issues like anxiety and depression. The fact that they have been in use over a long period of time underscores the possibility of phytotherapy being used as an alternative or complementary method of conventional treatment (Khan et al., 2018).

The psychosomatic disorders are normally treated with psychotherapy coupled with synthetic psychotropic drugs, which comprise antidepressants, anxiolytics, sedatives, and analgesics. Despite its effectiveness, the drugs have also been linked to side effects of weight gain, sexual dysfunction, tremors, headaches, insomnia, and mental impairment. Besides, some groups of psychotropic medications are associated with dependency and abuse risks, especially benzodiazepines. These issues are of particular concern in older patients (70 years and above) who are often a comorbidity burden, and more susceptible to drug-drug interactions. According to epidemiological evidence, it is believed that close to 23 percent of people who are aged 70 years and over experience psychosomatic disorders and about 40 percent of them will need medical care. Herbal medicines in this case provide a promising source of treatment with a better safety profile and less likelihood of dependence (Theuretzbacher, 2025).

Although they have advantages, the use of herbal therapies is not accompanied by suitable representation in official treatment schemes since they are not covered by mainstream pharmacological models. However, they are also very important in self-medication practices mostly due to their high benefit to risk ratio. The herbal medicines are also popular in the treatment of gynecological disorders like Premenstrual Syndrome (PMS) and menopausal syndrome. Although Hormone Replacement Therapy (HRT) is the standard therapy in vasomotor symptoms in menopause, it is associated with side effects such as edema, joint pain, psychiatric disorders, thrombosis and high risk of breast cancer. Herbal interventions, on the contrary, have fewer side effects and improved patient adherence, which decreases the rate of therapy discontinuation (Chadha et al., 2021).

The current practice of using antibiotics to treat infectious diseases, especially UTIs has brought serious concerns because of side effects associated with the gastro-intestinal system, as well as the fast development of resistant bacteria to antibiotics. Therefore, minimisation of unnecessary antibiotic exposure has become an issue of concern in medical practice. On the same note, Lower Urinary Tract Symptoms (LUTS) associated with Benign Prostatic Hyperplasia (BPH) are normally treated with synthetic medications, which tend to produce side effects that include erectile dysfunction, hypotension, dizziness, gynecomastia and decrease in libido. These adverse effects often lead to the lack of treatment compliance. Herbal medicine is also safer to use in long-term management of such conditions (Woo et al., 2023).

4.1 Importance of Phyto therapeutics in Aquaculture

The Phyto therapeutics have been receiving growing interest in the aquaculture and animal production systems because of their multipurpose biological functions and non-toxicity. The use of herbal supplements in animal feed is determined by age, species, and production purposes. Medicinal plants possess individual bioactive compounds that have different physiological effects, which lead to an increase in growth performance and disease resistance (Angelini, 2024).

4.1.1 Herbal Drugs as Growth Promoters

Phytogenic feed supplements become the most common natural alternative to antibiotic growth promotion. They mostly work by influencing intestinal microbiota leading to increased gut health and nutrient absorption. Phytogenic compounds contribute to optimal development in genetic potential of animals by maintaining microbial balance in the gastrointestinal tract, which indirectly increases the activity of digestive enzymes and the bioavailability of nutrients, thus facilitating optimal growth. Some researchers have indicated that herbal extracts of various varieties of plants like *Salvia officinalis* (sage), *Thymus vulgaris* (thyme), and *Rosmarinus officinalis* (rosemary) enhance the digestibility of the food and growth performance of poultry. Bioactive compounds such as carvacrol, cinnamaldehyde, and capsaicin are reported to stimulate the appetite, increase secretions in the digestive tracts, and have antibacterial activity, all of which lead to better feed efficiency (Gupta et al., 2017). Though essential oils have been shown to have varied effects on growth performance depending on the age and dosage, evidence has also been indicated that essential oils can affect lipid metabolism and fat digestion when they stimulate digestive enzymes (Maramba-Lazarte et al., 2020).

4.1.2 Herbal Drugs as Antibacterial Agents

The occurrence of various phytochemicals and their interactions with one another is what causes the antibacterial quality of herbal medicines. These are flavonoids, alkaloids, tannins, saponins, glycosides, phenolic acids, coumarins, steroids as well as anthraquinones. These compounds have antibacterial action focusing on their concentration, chemical structure, and interaction with microbial cells (Riaz et al., 2024). The herbal medicines have immunomodulatory, anti-inflammatory, antioxidative, and antimicrobial effects and all these make the herbal medicines antibacterial. Most of the plant-based compounds have the ability to boost the humoral and cell-mediated immune response and decrease the immunosuppression caused by stress. Also, their antioxidative property prevents the formation of free radicals during the metabolic processes, and the anti-inflammatory effect is mediated by selective inhibition of the enzymes, including cyclooxygenase-2, 5-lipoxygenase, and glutathione S-transferase, and, consequently, the production of prostaglandins (Porras et al., 2020).

4.1.3 Herbal Drugs as Antiviral Agents

Many medicinal plants have shown strong antiviral effects against most human and animal viruses. Preliminary research that followed the World War II showed that extracts of various plant species had the ability to inhibit the growth of influenza virus. This has been followed by massive screening programs worldwide which since then have identified many antiviral agents derived out of plants in in vitro and in vivo screening. Grape, apple and strawberry juices are some of the fruit juices that when extracted have exhibited an inhibitory effect against the viruses such as Herpes Simplex Virus (HSV), poliovirus and coxsackievirus. Research of medicinal plants of British Columbia had reported amplified antiviral properties in intestinal coronavirus and Respiratory Syncytial Virus (RSV) (Fadhil, 2025). Moreover, aqueous extracts of *Sanicula europaea* were observed to suppress influenza RNA-replication, and ethanol-based extraction abolished the antiviral effect which implied that water-soluble compounds have a central role (Marghani et al., 2019). The marine algae are another useful source of antiviral compounds. Cyanovirin-N is a blue-green algae protein that has shown excellent anti-HIV efficacy of preventing viral fusion and infection. Seaweed polysaccharides sulphated have demonstrated a general antiviral activity, and the antiviral activity was directly related to the level of sulphation. Algae antiviral potential is not adequately studied, considering that little chemical characterisation studies have been conducted (Ríos & Recio, 2005).

5. MECHANISMS OF ACTION OF HERBAL ANTIBIOTICS

Herbal antibiotics have multi-targeted and various modes of action which make them different compared to conventional antibiotics which tend to be directed at only one specific cellular target. This multi-modal practice minimizes the chances of resistance development and increases the ability to combat Multidrug-Resistant (MDR) pathogens. The phytochemicals that can react are alkaloids, flavonoids, terpenoids, phenolics and the essential oils, and they react with bacterial structures, metabolic pathways, and host defence systems, thus leading to both direct antimicrobial effects and indirect immunomodulatory effects (Kopel et al., 2022).

5.1 Cell Wall and Membrane Disruption

Alteration of membranes and cell walls in bacteria and the cytoplasmic membranes is one of the main processes through which herbal antibiotics work. Phytochemicals (terpenoids, essential oils: thymol, carvacrol, eugenol), saponins interfere with lipid bi-layers and change the membrane permeability and integrity. This causes the leakage of vital intracellular materials, such as ions, nucleotides and proteins, which eventually causes cell lysis. In Gram-positive bacteria, herbal compounds disrupt peptidoglycan monomerization and cross-linking and in Gram-negative bacterium disrupt the outer membrane by binding lipopolysaccharides. Membrane targeting is non-specific and thus bacteria have no means of developing resistance thus rendering this mechanism quite effective against MDR strains (Álvarez-Martínez et al., 2021).

5.2 Inhibition of Protein and Nucleic Acid Synthesis

There are several herbal antibiotics that prevent bacteria growth by disrupting both protein synthesis and nucleic acid metabolism. Alkaloids, including berberine and quinine, attach to ribosomes subunits and interfere with the translation process and prevention of peptide chain elongation. Flavonoids and polyphenols may suppress DNA gyrase and topoisomerase IV which are enzymes necessary in the replication and transcription of DNA. Part of phytochemicals can also be able to interfere with DNA strands making the latter structurally distorted, hence preventing replication and gene expression. Herbal antibiotics focus on several different stages of macro-molecule synthesis, and they can prevent bacterial reproduction and survival, even in those pathogens that are resistant to antibiotics (Khameneh et al., 2019).

5.3 Efflux Pump Inhibition

One of the significant antibiotic resistance mechanisms is efflux pumps, which enables the bacteria to get rid of the antimicrobial and decrease the concentration of drugs in the cells. Numerous herbal compounds are Efflux Pump Inhibitors (EPIs) which revive the performance of the traditional antibiotics. Flavonoids, alkaloids, and phenolic acids block efflux pump proteins in *Staphylococcus aureus* and Gram-negative NorA and AcrAB-TolC, respectively. Herbal antibiotics make intracellular accumulation of antimicrobial agents high, reduce the susceptibility of bacteria, and reverse multidrug resistance by blocking these transport systems. This mechanism brings out the importance of the herbal compounds as resistance-modifying agents in combination therapy (Balkrishna et al., 2024).

5.4 Quorum Sensing and Biofilm Inhibition

The bacterial communication system is known as quorum sensing (QS) and it controls the production of virulence factors and the formation of biofilms. QS signalling is disrupted by herbal antibiotics which prevent the production or the reception of autoinducer molecules. These compounds (ajoene (garlic), curcumin, cinnamaldehyde) disrupt QS regulatory proteins, inhibit the production of toxins and pathogenicity (Qassadi et al., 2023). Moreover, herbal agents inhibit biofilm because of the inhibition of bacterial adhesion and Extracellular Polymeric Substance (EPS) generation. They are also able to disturb the mature biofilms, thereby increasing the vulnerability of the bacteria to host immune system and antibiotics. This action is especially important in overcoming chronic and device-associated infections as biofilms are an important part of both (Mickymaray, 2019).

5.5 Modulation of Host Immune Responses

In addition to the direct effects which are antibacterial, herbal antibiotics increase the host defense mechanisms via immunomodulatory effects. Innate immune responses are stimulated by many medicinal plant compounds which activate the activity of macrophages, neutrophils and natural killer cells. Phytochemicals including polysaccharides and flavonoid products, increase phagocytosis, cytokine secretion, and antibody functions. Also, the herbal antibiotics possess anti-inflammatory and anti-oxidant effects, thereby minimizing tissue destruction due to excessive immune response during an infection. Herbal antibiotics have a combined therapeutic benefit and this is strengthened by boosting host immunity and limiting bacterial growth, especially in immunocompromised patients and in chronic infections (Marghani et al., 2019).

Figure 1: demonstrates herbal antibiotics are the emerging alternative and complementary medicine to the conventional antimicrobials, especially in a period where drug resistance is increasing. As opposed to synthetic antibiotics that in most cases only attack one bacterial pathway, herbal antibiotics have variety of bioactive phytochemicals which are able to attack many cellular and molecular targets simultaneously. These compounds disrupt the bacterial cell structure, block key metabolic functions, disrupt resistance mechanisms including efflux pumps and biofilm formation and alter host immune responses. Comprehension of the action of herbal antibiotics is essential to not only justify their therapeutic possibilities, but also to maximise their clinical application, and to combine traditionally accepted medicine with the current policies to deal with the multidrug resistant bacterial infections.

Figure 1. Mechanisms of action of herbal antibiotics against drug-resistant bacteria

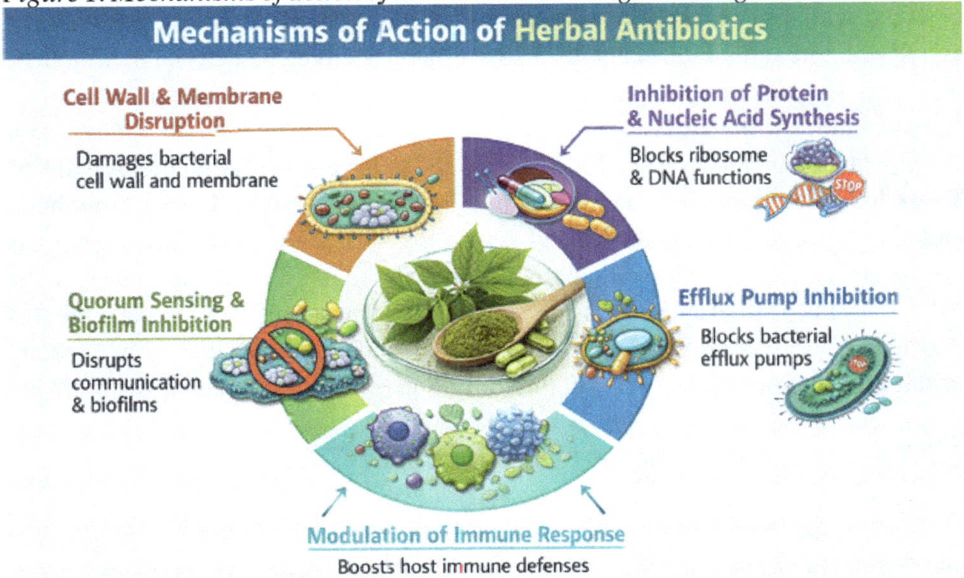

6. HERBAL ANTIBIOTICS AGAINST MAJOR DRUG-RESISTANT PATHOGENS

The quick acquisition of multidrug-resistant (MDR) bacterial pathogens has strongly decreased the efficacy of the conventional antibiotics, which has revived fresh interest in using herbal antibiotics as an alternative or complementary therapeutic agent. Plants that are used medicinally have a variety of bioactive compounds, which can address the problem of resistant bacteria in a variety of ways such as disrupting membranes, inhibiting enzymes, disrupting efflux pumps, disrupting quorum sensing, and modulating the immune system. All these multifactorial actions increase their effectiveness against key drug-resistant pathogens and minimize their chances of developing resistance (Gupta and Birdi, 2017).

6.1 Methicillin-Resistant *Staphylococcus aureus* (MRSA)

Methicillin-resistant *Staphylococcus aureus* (MRSA) is also a significant cause of hospital and community-acquired infections such as skin and soft tissue infections, pneumonia, endocarditis and sepsis. The MRSA resistance is mainly explained by the fact that it has a gene named mecA that codes a modified penicillin-binding protein (PBP2a) with a low affinity to β-lactam antibiotics. There has been a great deal of activity of herbal antibiotics against MRSA that has targeted bacterial membranes, inhibited virulence factors and biofilm formation. Berberine, allicin, curcumin, Epigallocatechin Gallate (EGCG), and essential oils of *Thymus vulgaris, Origanum vulgare* and *Cinnamomum* sp., have been found to have strong anti-MRSA activity. These toxins interfere with membrane integrity, block DNA and protein synthesis and also inactivate resistance-related genes. Besides, some of the herbal extracts are synergistic to the traditional antibiotics, reinstating their functioning against the MRSA strains. Herbal antibiotics have an additional therapeutic effect that increases their efficacy in chronic and device-related MRSA infections because of the capacity of the herbal antibiotics to prevent the formation of biofilms (Vaou et al., 2021).

6.2 Vancomycin-Resistant Enterococci (VRE)

Enterococci resistant to vancomycin (VRE) especially *Enterococcus faecium* and *Enterococcus faecalis* are severe clinical problems since few treatment options exist. VRE resistance is modulated by the effect on the precursors of cell walls, which decreases the binding of vancomycin. Herbal antibiotics have potential alternatives because they focus on bacterial membranes, metabolic pathways, and not the cell wall synthesis. Medicinal plants like *Allium sativum* (garlic), *Azadirachta indica* (neem), *Punica granatum* (pomegranate), and *Terminalia chebula* have been found to be of great activity against VRE strains, as extracts. Phenolic substances and tannins disrupt the integrity of membrane, prevent ATP production, and hinder the ability to attach. Additionally, some of the herbal compounds regulate the host immune reactions, which are improved to clear the bacteria. The low-toxicity and wide-range activity of these phytochemicals justify their possible usage as complementary therapy to the treatment of VRE infections (Wink, 2015).

6.3 Extended-Spectrum β-Lactamase (ESBL)-Producing Bacteria

Due to their broad spectrum of infections that include UTIs, bloodstream infections, and pneumonia, extended-spectrum β-lactamase (ESBL)-producing bacteria, including *Escherichia coli* and *Klebsiella pneumoniae*, are the causes of numerous infections. ESBL enzymes break down 2-lactam antibiotics making them useless. Herbal antibiotics oppose these pathogens with the mechanisms that are independent of β-lactamase activity. Plant phenolic acids, including *Camellia sinensis* (green tea), *Ocimum sanctum* (holy basil) and *Syzygium aromaticum* (clove) have an inhibitory effect on ESBL-producing bacteria. These substances cause disruption in the replication process of bacteria, slow down efflux pumps, and decrease resistance related enzyme activities. There are also other herbal extracts that inhibit the transfer of plasmid-mediated resistance genes, curbing the diffusion of ESBL characteristics. They are effective in UTIs, so they are also relevant in minimizing antibiotic dependence in clinical experience (Cowan, 1999).

6.4 Carbapenem-Resistant Enterobacteriaceae (CRE)

Carbapenem-Resistant Enterobacteriaceae (CRE) are considered among the greatest threats to human health worldwide because they are resistant to last-resort antibiotics. The common mechanisms of resistance are production of carbapenemases, loss of porins, and overexpression of efflux pumps. Herbal antibiotics offer the multi-targeting effect that can defeat these resistance mechanisms. Plant bioactive compounds like *Curcuma longa* (turmeric), *Glycyrrhiza glabra* (licorice) and *Nigella sativa* have been shown to be inhibitory against CRE strains. These are compounds that destabilize bacterial membranes, prevent the activity of carbapenemases and disrupt quorum sensing. Moreover, combinations of herbal extracts with carbapenems have been demonstrated to have a stronger antibacterial effect than either of the two, implying the use of such combinations as resistance-modifying agents. These combination therapies have the potential to increase clinical use of current antibiotics (Anand et al., 2019).

6.5 Multidrug-Resistant *Pseudomonas* and *Acinetobacter*

Pseudomonas aeruginosa and *Acinetobacter baumannii* are multidrug-resistant pathogens that are known to cause intense hospital-acquired infections, especially among immunocompromised patients. Their resistance is due to various mechanisms such as overexpression of enzyme, biofilm development, and enzyme degradation of antibiotics. Their inherent resistance and flexibility make it especially hard to treat these pathogens. The use of herbal antibiotics on such organisms has proven to be promising as they inhibit biofilms and quorum sensing. Phytochemicals like quercetin and catechins, essential oils of Eucalyptus, Tea tree and Cymbopogon species prevent biofilm formation and decrease virulence factor production. Moreover, the herbal compounds can also increase the penetration of antibiotics by interfering with the membranes of bacteria and suppressing efflux pumps. The properties make the herbal antibiotics useful as adjuncts in dealing with the MDR Pseudomonas and Acinetobacter infection (Pane, 2024).

Table 1: Demonstrated the emergence of drug-resistant bacterial pathogens, in a relatively short period of time, has made the use of conventional antibiotics very ineffective, which has become a substantial worldwide health challenge. In this regard, herbal antibiotics have received a growing interest because they have broad-spectrum effects, multi-target effects, and a reduced likelihood of causing resistance. Many of these medicinal plants and their bioactive constituents have proven to be highly effective against most drug-resistant pathogens, such as: methicillin-resistant *Staphylococcus aureus* (MRSA), Vancomycin-Resistant Enterococci (VRE), Extended-Spectrum β-Lactamase (ESBL)- producing bacteria, Carbapenem-Resistant Enterobacteriaceae (CRE) and multidrug-resistant Pseudomonas and Acinetobacter species. Key herbal sources, their active phytochemicals, mechanisms of action postulated and their therapeutic relevance against these clinically important resistant pathogens are summarized in the following table giving an idea about the possibility of herbal antibiotics in use as complementary or alternative therapies in the management of antimicrobial resistance.

Table 1. Herbal antibiotics against major drug-resistant pathogens

Drug-Resistant Pathogen	Common Clinical Infections	Effective Herbal Sources / Compounds	Proposed Mechanisms of Action	Therapeutic Significance
Methicillin-Resistant *Staphylococcus aureus* (MRSA)	Skin and soft tissue infections, wound infections, pneumonia, bloodstream infections	*Curcuma longa* (curcumin), *Allium sativum* (allicin), *Camellia sinensis* (epigallocatechin gallate), *Azadirachta indica* (nimbidin)	Disruption of cell membrane integrity, inhibition of protein synthesis, suppression of biofilm formation, efflux pump inhibition	Reduces virulence and restores antibiotic sensitivity; effective against biofilm-associated MRSA infections
Vancomycin-Resistant Enterococci (VRE)	Urinary tract infections, endocarditis, intra-abdominal infections	*Berberis vulgaris* (berberine), *Punica granatum* (ellagic acid), *Terminalia chebula* (tannins)	DNA intercalation, inhibition of cell wall synthesis, modulation of quorum sensing	Acts as resistance-modifying agents and improves host immune response
ESBL-Producing Bacteria (*E. coli, Klebsiella pneumoniae*)	UTIs, sepsis, gastrointestinal and respiratory infections	*Syzygium aromaticum* (eugenol), *Cinnamomum verum* (cinnamaldehyde), *Ocimum sanctum* (ursolic acid)	Inhibition of β-lactamase activity, membrane destabilization, efflux pump suppression	Enhances susceptibility to β-lactam antibiotics; suitable for combination therapy

continued on following page

Table 1. Continued

Drug-Resistant Pathogen	Common Clinical Infections	Effective Herbal Sources / Compounds	Proposed Mechanisms of Action	Therapeutic Significance
Carbapenem-Resistant Enterobacteriaceae (CRE)	Hospital-acquired infections, ventilator-associated pneumonia, septicemia	*Glycyrrhiza glabra* (glycyrrhizin), *Andrographis paniculata* (andrographolide), *Nigella sativa* (thymoquinone)	Inhibition of carbapenemase enzymes, oxidative stress induction, immune modulation	Potential alternative in limited-option infections; lowers mortality risk when used adjunctively
Multidrug-Resistant *Pseudomonas* and *Acinetobacter*	Burn wound infections, pneumonia, UTIs, device-associated infections	*Thymus vulgaris* (thymol), *Origanum vulgare* (carvacrol), *Azadirachta indica* (azadirachtin)	Quorum sensing inhibition, biofilm disruption, membrane permeability alteration	Effective against biofilm-forming nosocomial pathogens; reduces chronic infection persistence

7. SYNERGISTIC EFFECTS OF HERBAL ANTIBIOTICS AND CONVENTIONAL DRUGS

The recent trend of antimicrobial resistance has contributed to the growing curiosity in combination therapies as a way of combining herbal antibiotics with traditional antimicrobial agents. In comparison to monotherapy, synergistic interactions take advantage of the complementary properties of action, increasing antibacterial efficacy and decreasing the doses of drugs needed and the associated toxicity. Herbal antibiotics have numerous bioactive agents that are able to target bacterial membrane, resistance enzymes, efflux pumps, and virulence factors, thus rendering drug-resistant pathogens susceptible to the available antibiotics. These integrative methods have a bright future in terms of offering a lasting solution to the lifespan of traditional antimicrobials and clinical outcomes (Zouine et al., 2024).

7.1 Herb–Antibiotic Combinations

Combination of herb-antibiotics has shown improved antibacterial effect against most multidrug-resistant (MDR) pathogens. Phytochemicals which include flavonoids, alkaloids, phenolics and terpenoids can interact with antibiotics in synergy whereby they contribute to uptake of the drug, damaging bacterial membranes or preventing enzymes which confer resistance. As an example, plants compounds such as epigallocatechin gallate (EGCG), berberine, curcumin, and allicin have been reported to boost the efficacy of -lactams and aminoglycosides, fluoroquinolones, and glycopeptides. Out of several studies, herbal extracts are found to reinstate the vulnerability of the resistant bacteria like MRSA, ESBL-producing *Escherichia coli* and carbapenem-resistant *Klebsiella pneumoniae* when combined with conventional antibiotics. Also, the combinations have the ability of decreasing minimum inhibitory concentration (MIC) of antibiotics thereby decreasing therapeutic doses to reduce adverse effects. The selective pressure of bacteria is also slowed down by the multitarget activity of herbal antibiotics, which slows the development of resistance (Berida et al., 2024).

7.2 Reversal of Antibiotic Resistance

The fact that herbal antibiotic can reverse mechanisms of antibiotic resistance is among the most promising of herbal antibiotics. Most phytochemicals are resistance-modifying agents as they either prevent bacterial efflux pumps, destroy resistance enzymes, or stem out resistance gene expression. Antimicrobial efficacy against MDR pathogens is restored by efflux pump inhibition, that maximally increases intracellular accumulation of antibiotics. Flavonoids and alkaloids are herbal compounds which have been reported to inhibit plasmid-mediated resistance and biofilm-associated tolerance. The herbal antibiotics prevent virulence and formation of protective biofilms that shield bacteria against antibiotics by blocking quorum sensing pathways. Herbal compounds also affect the activity of 2-lactamase and carbapenemase in a few instances and thus, restore the efficacy of 2-lactam antibiotics. These reversal effects make the herbal antibiotics effective adjuncts of chronic and repeat infections (Al-Mahdi et al., 2024).

7.3 Challenges in Standardization and Validation

The fact that herbal antibiotic can reverse mechanisms of antibiotic resistance is among the most promising of herbal antibiotics. Most phytochemicals are resistance-modifying agents as they either prevent bacterial efflux pumps, destroy resistance enzymes, or stem out resistance gene expression. Antimicrobial efficacy against MDR pathogens is restored by efflux pump inhibition, that maximally increases intracellular accumulation of antibiotics. Flavonoids and alkaloids are herbal compounds which have been reported to inhibit plasmid-mediated resistance and biofilm-associated tolerance. The herbal antibiotics prevent virulence and formation of protective biofilms that shield bacteria against antibiotics by blocking quorum sensing pathways. Herbal compounds also affect the activity of 2-lactamase and carbapenemase in a few instances and thus, restore the efficacy of 2-lactam antibiotics. These reversal effects make the herbal antibiotics effective adjuncts of chronic and repeat infections (Pauzi et al., 2024).

8. HERBAL ANTIBIOTICS IN ONE HEALTH PERSPECTIVE

One Health concept acknowledges the complexity of human, animal and environmental health as a solution to the global issues like Antimicrobial Resistance (AMR). The extensive use of antibiotics in the human medical field, veterinary sector, agriculture, and aquaculture has increased the occurrence and spread of drug-resistant pathogens in the ecology. Herbal antibiotics, in this case, provide a comprehensive and sustainable solution that is consistent with the One Health principles through safer application of antimicrobials, decreased pollution of the environment, and healthier ecosystems. Their wide-spectrum activity, biodegradability, and multi-target make them especially useful in combined AMR mitigation efforts (Álvarez-Martínez et al., 2024).

8.1. Human, Animal, and Environmental Health Linkages

Antimicrobial resistance is not an issue only in clinical environments but spreads among human beings, livestock, wildlife, food systems, and the environment. The bacteria and resistance genes which are resistant are spread by direct contact, food chain, water sources, and soil creating a looping process of exposure. It can be stopped with the help of herbal antibiotics which minimize the use of synthetic antibiotics in human and veterinary medicine. Phyto therapeutic prophylaxis and treatment in livestock and aquaculture has reduced the use of antibiotics and hence the chances of selection increases on resistant strains. Environmentally, the herbal antibiotics are usually biodegradable and less persistent as compared to the synthetic drugs, hence causing less ecological disturbance. In comparison to traditional antibiotics which build up in water bodies and the soil, herbal components decay naturally thereby minimizing the transmission of resistance genes in the microbiomes of the environment. The immunomodulatory and anti-inflammatory effects of herbal antibiotics have also been demonstrated to aid host defenses in human beings without subjecting microbial communities to excessive selective pressure. A combination of these effects indicates that herbal antibiotics can bridge the human, animal, and environmental health in the One Health framework (Izah et al., 2023).

8.2 Role in Sustainable Antimicrobial Strategies

The presence of herbal antibiotics is essential in promoting the concept of sustainable antimicrobial approaches since the latter provide viable alternatives that are cheap and eco-friendly. They have a multi-component nature and this is aimed at attacking pathogens in many pathways at a time making them less prone to resistance development. Also, most of the herbal antibiotics may be locally produced and sourced, which will lead to self-sufficiency and diminish the reliance on imported pharmaceutical products (especially in low- and middle-income nations). Phytogenic compounds are finding application in agriculture and aquaculture in the replacement of antibiotic growth promoters, which play a significant role in AMR, as growth promoters, immunostimulants, and disease-preventive agents. Herbal antibiotics may be employed to augment the efficacy of antibiotics and reduce the length of treatment in the medical care of human beings. Herbal-drug combinations are beneficial by reducing the dose and frequency of antibiotics used, which preserves the effectiveness of the available antimicrobials. These qualities make herbal antibiotics central figures of sustainable, long-term measures to counter AMR without jeopardizing productivity and society health (Arip et al., 2022).

8.3 Implications for Public Health Policy

Policy implications of incorporation of herbal antibiotics in the systems of the people. To acknowledge the role of the herbal medicines as complementary adjuncts in the AMR control, the regulatory frameworks should be established based on the evidence of quality, safety and efficacy. The recommended policies in the field of public health include research, standardization, and clinical validation of herbal antibiotics and the rational use of herbal antibiotics in human and veterinary medicine. Antibiotic misuse could be eliminated by including herbal antibiotics into national AMR action plans which may help to uphold preventative healthcare strategies. It is necessary to conduct educational programs targeting medical workers, veterinarians and farmers which will help to ensure responsible usage and avoid misinformation. Moreover, the conservation of medicinal plants and sustainable harvesting policies are also the keys to preserving biodiversity and guaranteeing the long-term supply. Through coordination, education, and sustainability, the public health policies can use herbal antibiotics as an effective One Health intervention to the antimicrobial resistance crisis experienced globally (Herman & Herman, 2023).

9. CONCLUSION

The high rate of development and dissemination of drug-resistant bacterial pathogens are a major danger to the health of the population, which requires new and sustainable approaches to treatment that are not based on traditional antibiotics. Herbal antibiotics in this regard provide an alternative and complementary solution to the Multidrug-Resistant (MDR) and extensively drug-resistant (XDR) microorganisms. Medicinal plants represent highly concentrated reservoirs of bioactive Phyto transformative phytochemicals such as alkaloids, flavonoids, phenolics, terpenoids, tannins, and essential oils that have an antimicrobial effect with multi- and overlapping mechanisms. Such contain cell wall and membrane disruption, protein and nucleic acid synthesis, efflux pump inhibition, quorum sensing disruption, and biofilm formation inhibition. These multi-targeted measures have a high value of mitigating the chances of developing resistance in comparison with single-target synthetic antibiotics. Moreover, herbal antibiotics show significant synergetic action with orthodox antimicrobial agents, reversing antibiotic resistance in the orthodox antibiotic-resistant strains, and allowing a reduction in dosage, which reduces toxicity and adverse effects. Their other immunomodulatory, anti-inflammatory, and anti-oxidant effects boost host defensive mechanisms and result in better treatment outcomes especially in infections that are chronic and reoccurring. In addition to human medicine, the use of herbal antibiotics in veterinary medicine, agriculture and aquaculture is also in line with the One Health concept and presents non-toxic and sustainable solutions to the problem of antimicrobial resistance. Although they have great potential, there are still issues like standardization, quality assurance, toxicity testing, dose- optimization and large-scale clinical validation. It is necessary to begin to address these limitations by developing sophisticated phytochemical studies and innovative drug delivery systems along with evidence-based clinical trials. On the whole, the use of herbal antibiotics in modern antimicrobial practices is a promising and sustainable solution to the conservation of the efficacy of antibiotics and managing the increasing global epidemic of antimicrobial resistance.

REFERENCES

Al-Mahdi, A., Ali, A., Alabed, A. A. A., Chen, H., Baobaid, M. F., & Ruhi, S.. (2024). Antibacterial activity of herbal essential oils against Gram-positive and Gram-negative bacteria with a potential for multidrug resistance. *Journal of Angiotherapy, 8*(2). Advance online publication. DOI: 10.25163/angiotherapy.829517

Álvarez-Martínez, F. J., Barrajón-Catalán, E., Herranz-López, M., & Micol, V. (2021). Antibacterial plant compounds, extracts and essential oils: An updated review on their effects and putative mechanisms of action. *Phytomedicine : International Journal of Phytotherapy and Phytopharmacology, 90*, 153626. DOI: 10.1016/j.phymed.2021.153626 PMID: 34301463

Álvarez-Martínez, F. J., Díaz-Puertas, R., Barrajón-Catalán, E., & Micol, V. (2024). Plant-derived natural products for the treatment of bacterial infections. *Handbook of Experimental Pharmacology*. Advance online publication. DOI: 10.1007/164_2024_706 PMID: 38418668

Anand, U., Jacobo-Herrera, N., Altemimi, A. B., & Lakhssassi, N. (2019). A comprehensive review on medicinal plants as antimicrobial therapeutics: Potential avenues of biocompatible drug discovery. *Metabolites, 9*(11), 258. DOI: 10.3390/metabo9110258 PMID: 31683833

Angelini, P. (2024). Plant-derived antimicrobials and their crucial role in combating antimicrobial resistance. *Antibiotics (Basel, Switzerland), 13*(8), 746. DOI: 10.3390/antibiotics13080746 PMID: 39200046

Arip, M., Selvaraja, M., Mogana, R., Tan, L. F., Leong, M. Y., & Tan, P. L.. (2022). Review on plant-based management in combating antimicrobial resistance: Mechanistic perspective. *Frontiers in Pharmacology, 13*, 879495. DOI: 10.3389/fphar.2022.879495 PMID: 36249774

Balkrishna, A., Sharma, N., Srivastava, D., Kukreti, A., Srivastava, S., & Arya, V. (2024). Exploring the safety, efficacy, and bioactivity of herbal medicines: Bridging traditional wisdom and modern science in healthcare. *Future Integrative Medicine, 3*(1), 35–50. DOI: 10.14218/FIM.2023.00086

Berida, T., Adekunle, Y. A., Dada-Adegbola, H., Kdimy, A., Roy, S., & Sarker, S. D. (2024). Plant antibacterials: The challenges and opportunities. *Heliyon, 10*(10), e31145. DOI: 10.1016/j.heliyon.2024.e31145 PMID: 38803958

Bhardwaj, A. K., Kashyap, N. K., Hait, M., Bera, S. K., & Dewangan, H. (2023). *Physicochemical characterization of rhizome of Curcuma caesia Roxb*. ES Food & Agroforestry., DOI: 10.30919/esfaf813

Bhardwaj, M. (2016). Potential of herbal drug and antibiotic combination therapy: A new approach to treat multidrug resistant bacteria. *Pharmaceutica Analytica Acta*, *7*(11). Advance online publication. DOI: 10.4172/2153-2435.1000523

Bhattacharya, D., & Chandar, B. (2018). An overview of New Delhi metallo-beta lactamase-1 and extended spectrum beta lactamase producing bacteria: Need for an alternate. *Journal of Antimicrobial Agents*, *4*(1). Advance online publication. DOI: 10.4172/2472-1212.1000155

Chadha, J., Gupta, M., Nagpal, N., Sharma, M., Adarsh, T., & Joshi, V.. (2021). Antibacterial potential of indigenous plant extracts against multidrug-resistant bacterial strains isolated from New Delhi region. *GSC Biological and Pharmaceutical Sciences*, *14*(2), 185–195. DOI: 10.30574/gscbps.2021.14.2.0053

Chassagne, F., Samarakoon, T., Porras, G., Lyles, J. T., Dettweiler, M., Marquez, L., Salam, A. M., Shabih, S., Farrokhi, D. R., & Quave, C. L. (2021). A systematic review of plants with antibacterial activities: A taxonomic and phylogenetic perspective. *Frontiers in Pharmacology*, *11*, 586548. DOI: 10.3389/fphar.2020.586548 PMID: 33488385

Cock, I. E., Cheesman, M. J., Ilanko, A., & Blonk, B. (2017). Developing new antimicrobial therapies: Are synergistic combinations of plant extracts/compounds with conventional antibiotics the solution? *Pharmacognosy Reviews*, *11*(22), 57–72. DOI: 10.4103/phrev.phrev_21_17 PMID: 28989242

Cowan, M. M. (1999). Plant products as antimicrobial agents. *Clinical Microbiology Reviews*, *12*(4), 564–582. DOI: 10.1128/CMR.12.4.564 PMID: 10515903

Dassanayake, M. K., Khoo, T.-J., & An, J. (2021). Antibiotic resistance modifying ability of phytoextracts in anthrax biological agent *Bacillus anthracis* and emerging superbugs: A review of synergistic mechanisms. *Annals of Clinical Microbiology and Antimicrobials*, *20*(1), 1–14. DOI: 10.1186/s12941-021-00485-0 PMID: 34856999

El-Saadony, M. T., Saad, A. M., Mohammed, D. M., Korma, S. A., Alshahrani, M. Y., Ahmed, A. E., Ibrahim, E. H., Salem, H. M., Alkafaas, S. S., Saif, A. M., Elkafas, S. S., Fahmy, M. A., Abd El-Mageed, T. A., Abady, M. M., Assal, H. Y., El-Tarabily, M. K., Mathew, B. T., AbuQamar, S. F., El-Tarabily, K. A., & Ibrahim, S. A. (2025). Medicinal plants: Bioactive compounds, biological activities, combating multidrug-resistant microorganisms, and human health benefits—A comprehensive review. *Frontiers in Immunology*, *16*, 1491777. DOI: 10.3389/fimmu.2025.1491777 PMID: 40375989

Fadhil, N. (2025). Innovations in the treatment of antibiotic-resistant bacterial infections: From drugs to alternative approaches. *Journal of University of Babylon for Pure and Applied Sciences, 33*, 193–205. DOI: 10.29196/jubpas.v33i3.5985

Gupta, P., & Birdi, T. J. (2017). Development of botanicals to combat antibiotic resistance. *Journal of Ayurveda and Integrative Medicine*, *8*(4), 266–275. DOI: 10.1016/j.jaim.2017.05.004 PMID: 28869082

Harkin, C. (2022). Systematic review and meta-analysis of plant-derived antimicrobials in WHO priority pathogens. *International Journal of Complementary & Alternative Medicine*, *14*(2), 122–137. DOI: 10.15406/ijcam.2022.15.00600

Herman, A., & Herman, A. P. (2023). Herbal products and their active constituents used alone and in combination with antibiotics against multidrug-resistant bacteria. *Planta Medica*, *89*(2), 168–182. DOI: 10.1055/a-1890-5559 PMID: 35995069

Husain, S. A., Ahmad, S., Abass, S., Parveen, R., Irfan, M., & Jan, B. (2022). Synergy-based extracts of medicinal plants: Future antimicrobials to combat multidrug resistance. *Current Pharmaceutical Biotechnology*, *23*(13), 1527–1540. DOI: 10.2174/1389201023666220126115656 PMID: 35081888

Ionescu, M. (2018). *Are herbal products an alternative to antibiotics?* InTechOpen. https://doi.org/DOI: 10.5772/intechopen.72110

Izah, S. C., Joshua, M. T., Torru, K. E., Ngun, C. T., Ogwu, M. C., & Hait, M. (2023). Antimicrobial resistance and the role of herbal medicine: Challenges, opportunities, and future prospects. In *Reference series in phytochemistry* (p. 1). Springer., DOI: 10.1007/978-3-031-21973-3_71-1

Khameneh, B., Iranshahy, M., Soheili, V., & Bazzaz, B. S. F. (2019). Review on plant antimicrobials: A mechanistic viewpoint. *Antimicrobial Resistance and Infection Control*, *8*(1), 118. DOI: 10.1186/s13756-019-0559-6 PMID: 31346459

Khan, M. F., Tang, H., Lyles, J. T., Pineau, R. M., Mashwani, Z., & Quave, C. L. (2018). Antibacterial properties of medicinal plants from Pakistan against multidrug-resistant ESKAPE pathogens. *Frontiers in Pharmacology*, *9*, 815. DOI: 10.3389/fphar.2018.00815 PMID: 30116190

Khare, T., Anand, U., Dey, A., Assaraf, Y. G., Chen, Z., Liu, Z., & Kumar, V. (2021). Exploring phytochemicals for combating antibiotic resistance in microbial pathogens. *Frontiers in Pharmacology*, *12*, 720726. DOI: 10.3389/fphar.2021.720726 PMID: 34366872

Kitpipit, W., Scholfield, C. N., Sangkanu, S., Nissapatorn, V., Pereira, M. de L., Paul, A. K., & Mitsuwan, W. (2023). Virulence factors and quorum sensing as targets of new therapeutic options by plant-derived compounds against bacterial infections caused by human and animal pathogens. *Veterinary World, 16*, 1346–1355. DOI: 10.14202/vetworld.2023.1346-1355 PMID: 37577190

Kopel, J., McDonald, J., & Hamood, A. N. (2022). An assessment of the in vitro models and clinical trials related to the antimicrobial activities of phytochemicals. *Antibiotics (Basel, Switzerland), 11*(12), 1838. DOI: 10.3390/antibiotics11121838 PMID: 36551494

Liang, J., Huang, X., & Ma, G. (2022). Antimicrobial activities and mechanisms of extract and components of herbs in East Asia. *RSC Advances, 12*(45), 29197–29213. DOI: 10.1039/D2RA02389J PMID: 36320733

Maramba-Lazarte, C. C., Cavinta, L. L., & Sara, M. (2020). Antibacterial activity of Guyabano, Ulasimang Bato, Sambong, and Tsaang Gubat leaf extracts against common drug-resistant bacteria. *Acta Medica Philippina, 54*(1). Advance online publication. DOI: 10.47895/amp.v54i1.1087

Marghani, A. A. E., Algariri, K., & Ridzuan, P. M. (2019). Prospective influence of phytotherapy on resistant bacterial tonsillitis. *Journal of Pure & Applied Microbiology, 13*(1), 167–176. DOI: 10.22207/JPAM.13.1.17

Matei, A. T., & Vi an, A. I. (2025). Mechanism, efficacy, and safety of natural antibiotics. *Antibiotics (Basel, Switzerland), 14*(10), 981. DOI: 10.3390/antibiotics14100981 PMID: 41148673

Mehboob, A., Rafique, A., & Abbas, T. (2021). Bioceutical properties of culinary herbs, spices and their possible outcomes with standard antibiotics. *Food and Nutrition Sciences, 12*(3), 262–280. DOI: 10.4236/fns.2021.123021

Mickymaray, S. (2019). Efficacy and mechanism of traditional medicinal plants and bioactive compounds against clinically important pathogens. *Antibiotics (Basel, Switzerland), 8*(4), 257. DOI: 10.3390/antibiotics8040257 PMID: 31835403

Mundy, L., Pendry, B., & Rahman, M. M. (2016). Antimicrobial resistance and synergy in herbal medicine. *Journal of Herbal Medicine, 6*(2), 53–58. DOI: 10.1016/j.hermed.2016.03.001

Pane, Y. S. (2024). Effectiveness of traditional herbal extracts against multidrug-resistant bacteria: A review. *bioRxiv*. https://doi.org/DOI: 10.1101/2024.11.03.621775

Pauzi, R. Y., Ihtiaringtyas, S., & Wahyudin, W. (2024)... *Medical and Health Journal, 3*(2), 128. DOI: 10.20884/1.mhj.2024.3.2.10459

Porras, G., Chassagne, F., Lyles, J. T., Marquez, L., Dettweiler, M., Salam, A. M., Samarakoon, T., Shabih, S., Farrokhi, D. R., & Quave, C. L. (2020). Ethnobotany and the role of plant natural products in antibiotic drug discovery. *Chemical Reviews*, *121*(6), 3495–3560. DOI: 10.1021/acs.chemrev.0c00922 PMID: 33164487

Qassadi, F., Zhu, Z., & Monaghan, T. (2023). Plant-derived products with therapeutic potential against gastrointestinal bacteria. *Pathogens (Basel, Switzerland)*, *12*(2), 333. DOI: 10.3390/pathogens12020333 PMID: 36839605

Ramesh, S. P., Karajgi, S., Singh, R., Naik, P., Sahoo, U., Sharma, M. C., Nainwal, P., & Wal, A. (2024). Herbal antibiotics: A review of traditional medicinal plants with antibacterial properties and their potential for future treatment. *Anti-Infective Agents*, *23*(2), e080724231696. Advance online publication. DOI: 10.2174/01221 13525302796240528080758

Riaz, N., Zubair, F., Ali, A., Fiaz, S., Batool, S., & Batool, T.. (2024). Plant-based antimicrobials against multi drug resistant bacterial strains (MDRS). *Polish Journal of Environmental Studies*. Advance online publication. DOI: 10.15244/pjoes/190564

Ríos, J. L., & Recio, M. C. (2005). Medicinal plants and antimicrobial activity. *Journal of Ethnopharmacology*, *100*(1-2), 80–84. DOI: 10.1016/j.jep.2005.04.025 PMID: 15964727

Seukep, A. J., Nembu, N. E., Mbuntcha, H. G., & Kuete, V. (2022). Bacterial drug resistance towards natural products. In *Advances in botanical research* (p. 21). Elsevier., DOI: 10.1016/bs.abr.2022.08.002

Stefanović, O. D. (2018). Synergistic activity of antibiotics and bioactive plant extracts: A study against Gram-positive and Gram-negative bacteria. In *InTechOpen*. https://doi.org/DOI: 10.5772/intechopen.72026

Theuretzbacher, U. (2025). The global resistance problem and the clinical antibacterial pipeline. *Nature Reviews. Microbiology*, *23*(8), 491–504. DOI: 10.1038/s41579-025-01169-8 PMID: 40210708

Tirant, M., Tirant, H., & Wollina, U. (2024). Herbal bioactive compounds for skin infections and inflammatory conditions. *Open Access Macedonian Journal of Medical Sciences*, *12*, 1–10. DOI: 10.3889/oamjms.2024.11888

Vaou, N., Stavropoulou, E., Voidarou, C., Tsigalou, C., & Bezirtzoglou, E. (2021). Towards advances in medicinal plant antimicrobial activity: A review study on challenges and future perspectives. *Microorganisms*, *9*(10), 2041. DOI: 10.3390/microorganisms9102041 PMID: 34683362

Vaou, N., Stavropoulou, E., Voidarou, C., Tsigalou, C., & Bezirtzoglou, E. (2021). Towards advances in medicinal plant antimicrobial activity: A review study on challenges and future perspectives. *Microorganisms*, *9*(10), 2041. DOI: 10.3390/microorganisms9102041 PMID: 34683362

Wink, M. (2015). Modes of action of herbal medicines and plant secondary metabolites. *Medicines (Basel, Switzerland)*, *2*(3), 251–286. DOI: 10.3390/medicines2030251 PMID: 28930211

Woo, S., Marquez, L., Crandall, W. J., Risener, C. J., & Quave, C. L. (2023). Recent advances in the discovery of plant-derived antimicrobial natural products to combat antimicrobial resistant pathogens: Insights from 2018–2022. *Natural Product Reports*, *40*(7), 1271–1306. DOI: 10.1039/D2NP00090C PMID: 37439502

Zouine, N., Ghachtouli, N. E., Elabed, S., & Koraichi, S. I. (2024). A comprehensive review on medicinal plant extracts as antibacterial agents: Factors, mechanism insights and future prospects. *Scientific African*, *26*, e02395. DOI: 10.1016/j.sciaf.2024.e02395

Chapter 7
Selected Medicinal Plants in Focus:
Case Studies of Plants With Strong Evidence

Rahul Kumar
http://orcid.org/0009-0000-1323-2220
Department of Microbiology, Graphic Era (Deemed to be) University, Dehradun, India

Mohammad Aves
http://orcid.org/0009-0008-9927-4410
Department of Computer Science and Engineering, Graphic Era (Deemed to be) University, Dehradun, India

ABSTRACT

This chapter explores four medicinal plants, Azadirachta indica (Neem), Garcinia kola (Bitter Kola), Curcuma longa (turmeric), and Artemisia annua (Sweet Wormwood), with strong evidence of antimicrobial activity and relevance to the One Health Framework. Each plant contains unique bioactive compounds that target resistant pathogens through mechanisms such as membrane disruption, quorum sensing inhibition, efflux pump suppression, and biofilm reduction. Neem and turmeric show broad spectrum antibacterial effects, Garcinia kola exhibits anti-inflammatory and resistance modulatory activities, and Artemisia annua has potent antiparasitic and antiviral effects. Human medicine and its applications in livestock health, agriculture, and environmental management highlight sustainable strategies for reducing antimicrobial misuse. Integrating these plants into a single health approach can strengthen global efforts to mitigate antimicrobial resistance through eco-friendly, evidence based herbal therapeutics.

DOI: 10.4018/979-8-3373-7810-7.ch007

INTRODUCTION

Antimicrobial resistance (AMR) has emerged as a major threat to global public health, undermining decades of progress in the prevention and treatment of infectious diseases (Salam et al., 2023; Kumar et al., 2025). The early twentieth century brought about a medical revolution through antibiotic discovery, which enabled doctors to cure bacterial infections that had previously caused high death rates. The medical field and agriculture, animal husbandry, and fish farming sectors have created resistant pathogens because they use antimicrobials without proper control in their practices. Microorganisms have developed diverse resistance mechanisms, which include enzymatic drug degradation and target modification and efflux pump activation and biofilm formation, resulting in the worldwide distribution of multidrug resistant (MDR) infections (Elshobary et al., 2025). AMR creates an epidemiological impact that spreads across the entire world, yet low and middle income countries bear the most severe impact because their surveillance systems do not function properly and their antibiotic distribution remains unregulated while their infection control methods remain weak. Pathogens that develop resistance, including methicillin resistant *Staphylococcus aureus* and extended spectrum β-lactamase producing Enterobacteriaceae and drug resistant Plasmodium species, create major difficulties for healthcare providers to treat (Abebe and Birhanu, 2023). The World Health Organization recognizes AMR as a One Health issue, as resistant microbes circulate between humans, animals, food systems, and the environment through complex transmission pathways. People have used medicinal plants to treat illnesses since ancient times, which establishes their essential role as the primary medicinal resources from which contemporary pharmacology emerged (Napagoda and Wijesundara, 2022).

Scientists are currently studying plant-based antimicrobials because these substances contain diverse chemical structures and can combat multiple targets while developing resistance at a lower rate. The One Health framework presents a complete system for combining herbal medicine practices with antimicrobial resistance control methods. This chapter explores the medicinal plants *Azadirachta indica, Garcinia kola, Curcuma longa* and *Artemisia annua* which show proven effectiveness against resistant pathogens through experimental research and practical use. The chapter presents phytochemical information about the substances which demonstrate antimicrobial effects and their various uses in human and animal health and environmental protection to show their potential as sustainable evidence based solutions for One Health programs.

Azadirachta indica (Neem)

Azadirachta indica (Neem) is a medicinal tree which belongs to the Meliaceae family and grows throughout the tropical and subtropical zones of Asia and Africa and the Americas (Kadir et al., 2024). The Indian subcontinent is the native home of neem which Indian people have used for more than 2000 years in traditional medicine practices including Ayurveda and Unani and folk medical traditions to cure infections and skin diseases and inflammatory disorders and parasitic diseases. The entire plant contains medicinal properties which extend to its leaves and bark and seeds and flowers and oil. Phytochemical studies reveal the existence of multiple bioactive compounds in neem which include limonoids that contain azadirachtin and nimbidin and nimbolide and nimbin and flavonoids and tannins and polysaccharides which together produce antimicrobial effects (Kumari and Singh, 2024). Neem shows effectiveness against various types of bacteria and fungi and viruses and parasites through numerous laboratory and animal research studies which include the examination of methicillin resistant *Staphylococcus aureus* and extended spectrum β-lactamase producing Enterobacteriaceae which are resistant to multiple drugs. Neem demonstrates its antibacterial properties through two main methods which include breaking down microbial cell walls and membranes and stopping the process of microbial essential pheromone production which leads to biofilm development and pathogenesis. Neem extracts and their individual components show combined effects with traditional antibiotics which lead to improved drug performance and decreased resistance. The neem plant serves multiple purposes in which extends beyond its use in human medicine to include its application in veterinary disease control, production of environmentally safe biopesticides, and establishment of water treatment systems and environmental cleaning methods. The various characteristics of *Azadirachta indica* make it an effective sustainable plant-based solution that helps fight antimicrobial resistance and protects human and animal and environmental health (Wylie and Merrell, 2022).

Botanical Description and Distribution

The tree named *Azadirachta indica A. Juss.* or neem grows rapidly as an evergreen tree which belongs to the Meliaceae family. The tree typically reaches a height range of 15 to 20 meters and displays a straight trunk which supports a wide spreading crown and a dark brown to grey bark that develops rough fissures (Bakewell-Stone 2024). The leaves grow in an alternate pattern with a pinnate structure that includes 8 to 19 serrated leaflets which have a lanceolate shape and produce a bitter taste. Neem produces small flowers which have a pleasant scent and display white to pale yellow colors that grow in axillary panicles while the flowers contain abundant nectar which attracts pollinators (Malakar and Mandal 2024). The fruit develops into a smooth olive shaped drupe which holds one seed that develops inside a solid endocarp from which neem oil gets extracted. The plant demonstrates excellent drought resistance and it flourishes in various soil conditions that include sandy loamy and saline soils which make it suitable for dry environments of arid and semi-arid regions. The Indian subcontinent serves as the original homeland of neem which has been cultivated throughout India Pakistan Bangladesh and Sri Lanka for more than a thousand years (Sujarwo et al. 2016). *A. indica* has spread throughout tropical and subtropical areas in Africa Southeast Asia Australia Central and South America and the Middle East because of its valuable medicinal uses and agricultural applications and environmental benefits (Kriticos et al. 2003). The plant can thrive in extreme heat conditions while handling limited water resources and poor soil environments which has enabled it to spread across different regions of the world. The neem tree operates as a crucial multiple use tree species which provides essential ecological and medicinal benefits together with socioeconomic advantages for areas that deal with infectious disease outbreaks and agricultural insect problems and environmental sustainability challenges.

Morphological Features and Global Occurrence

The Neem tree *Azadirachta indica* grows as a fast growing evergreen or semi evergreen tree which reaches medium to large height with its straight trunk and broad crown that creates dense shade. The tree reaches its maximum height of 15 to 20 meters under ideal conditions but sometimes grows as high as 30 meters. The bark exhibits a thick and rough appearance that displays fissures between its dark gray and brown color range while it releases a bitter resinous sap (Alamgir, 2017). The leaves grow in an alternate pattern and form imparipinnate structure which contains 8 to 19 leaflets that have lanceolate and ovate shapes with serrated edges and pointed tips and display deep green shiny surface and extreme bitterness (Sinha et al., 2024). The neem tree produces small white to cream colored flowers which emit fragrance and grow from axillary or terminal panicles to attract pollinators. The fruit develops into a smooth ovoid drupe which starts as green and changes to yellow when it reaches ripeness. The seed exists inside a hard endocarp which contains oil bearing kernels that surround it. Neem originated from the Indian subcontinent yet its worldwide presence has expanded because of its exceptional adaptability and various functional uses (Arunachalam et al., 2025). The plant grows successfully in tropical and subtropical regions because it can withstand temperatures from freezing point to 45 °C while enduring extended periods without rainfall. Neem develops successfully in various soil conditions which include sandy and loamy and alkaline and moderately saline environments. The neem tree exists throughout South and

Southeast Asia and sub-Saharan Africa and the Middle East and Australia and parts of Central and South America because of its deliberate cultivation and its natural spread (Witt et al., 2018). *A. indica* has developed into a crucial species because of its ability to withstand extreme environmental conditions and its resistance to pests which are used for afforestation and agroforestry and medicinal applications and sustainable environmental management throughout the world.

Traditional Medicinal Uses in Ayurveda and Folk Medicine

The scientific study of *Azadirachta indica* (Neem) demonstrates its value to traditional healing practices which include Ayurveda because its healing powers treat all types of diseases. Classical Ayurvedic texts describe neem as having tikta (bitter) and kashaya (astringent) properties, which produce cooling effects and detoxification effects and blood purification effects (Ashalatha, 2025). The medical properties of neem leaves and bark and flowers serve to treat skin disorders which include eczema and acne and psoriasis and leprosy and to treat fever and ulcers and inflammatory conditions. Traditional medicine uses neem bark decoctions to treat gastrointestinal infections and malaria and metabolic disorders while neem leaf pastes serve as topical applications which medical professionals use to protect wounds and burns and ulcers from infections while they accelerate the healing process (Acharya, 2025). In rural South Asian and African and Southeast Asian communities traditional medicine uses neem as a treatment which protects people from diseases. People in India use neem twigs to protect their teeth because they believe it helps them avoid dental caries and gingivitis and periodontal infections. The traditional use of neem oil, which people extract from seeds, functions as an antiseptic and antifungal and antiparasitic treatment for skin infections and lice infestation and scabies (Singh et al., 2021). People use neem leaf infusions to drink or take baths which treat their fever and parasitic diseases and microbial infections. People use neem leaf infusions to treat fever and parasitic diseases and microbial infections through drinking and bathing. Traditional veterinary medicine incorporates neem for treating livestock infections and wound management and controlling external parasites. The traditional medicinal use of neem shows its extensive ability to kill microbes and provides solid evidence to evaluate antimicrobial resistance in modern scientific studies and research (Latif and Nawaz, 2025).

Phytochemical Constituents

The tree produces a wide array of different chemical compounds which create multiple biological effects and antibacterial properties. The plant contains more than 300 different bioactive substances which exist in its various parts that include leaves and bark and seeds and fruits and flowers (Altemimi et al., 2017). The main components of neem extract exist as limonoids and triterpenoids which serve as the main active medicinal ingredients. The most common research on azadirachtin centers on its abilities to fight against germs and parasites and insects. Other vital limonoids include nimbolide nimbin nimbidin salannin and gedunin which possess antimicrobial and anti-inflammatory and immunomodulatory properties (Nagini et al., 2021). Neem contains high concentrations of flavonoids which include quercetin and kaempferol together with phenolic compounds and tannins and saponins and polysaccharides which boost its capacity to fight oxidation and microbial growth. Quercetin demonstrated the ability to stop bacterial and fungal pathogens from growing. Neem seed oil contains fatty acids which include oleic and palmitic and stearic and linoleic acids that give the oil its ability to fight germs and protect skin (Ni Putu Ratna et al., 2024). Various factors such as geographical distribution, plant section, growth stage, and extraction technique determine the phytochemical composition of neem, which in turn affects its biological properties. Neem contains multiple bioactive compounds that function through distinct mechanisms, which create an antimicrobial effect that targets various pathogens, making it especially useful for treatment of multidrug-resistant infections. Neem has been used in traditional medicine for centuries because of its complex chemical composition, which scientists now study for potential use in contemporary health and antimicrobial research (Table 1).

Table 1. Major bioactive compounds of Azadirachta indica (Neem) and their biological significance

Bioactive Compound	Chemical Class	Primary Plant Source	Biological Activities	Antimicrobial / AMR Relevance	References
Azadirachtin	Limonoid (Tetranortriterpenoid)	Seeds (kernel), seed oil	Antibacterial, antiparasitic, insecticidal, antiviral	Disrupts microbial growth and metabolism; interferes with cell signaling; reduces virulence and resistance development	(Almadiy et al., 2025)
Nimbidin	Triterpenoid	Seed oil	Antibacterial, antifungal, anti-inflammatory	Damages microbial cell membranes; inhibits growth of *Staphylococcus aureus*, *Escherichia coli*, and *Candida* spp.	(Acharya, 2025)
Nimbolide	Limonoid	Leaves, flowers	Antibacterial, antifungal, anticancer, antioxidant	Inhibits biofilm formation; effective against MDR bacteria including MRSA	(Jubair et al., 2021)
Nimbin	Limonoid	Seed oil, bark	Antimicrobial, antipyretic, anti-inflammatory	Suppresses bacterial growth and enhances host immune response	(Nagini et al., 2024)
Salannin	Limonoid	Seeds	Antimicrobial, insect repellent	Contributes to inhibition of microbial proliferation and resistance modulation	(Qasim et al., 2024)
Gedunin	Limonoid	Bark, seeds	Antimalarial, antibacterial, antifungal	Inhibits microbial stress-response proteins; active against resistant parasites	(Al-Khayri et al., 2023)
Quercetin	Flavonoid	Leaves, bark	Antibacterial, antiviral, antioxidant	Inhibits nucleic acid synthesis; enhances antibiotic efficacy	(Davidova et al., 2024)
Kaempferol	Flavonoid	Leaves	Antimicrobial, anti-inflammatory	Alters membrane permeability; reduces bacterial virulence	(Periferakis et al., 2022)
Catechin	Polyphenol	Leaves	Antioxidant, antimicrobial	Inhibits bacterial enzymes and biofilm formation	(Bernal-Mercado et al., 2018)

continued on following page

Table 1. Continued

Bioactive Compound	Chemical Class	Primary Plant Source	Biological Activities	Antimicrobial / AMR Relevance	References
Tannins	Polyphenolic compounds	Bark, leaves	Astringent, antimicrobial	Precipitate microbial proteins; inhibit enzyme activity	(Ferdes, 2018)
Polysaccharides	Complex carbohydrates	Bark, leaves	Immunomodulatory, antimicrobial	Enhance host defense mechanisms against infections	(Porwal et al., 2021)
Fatty acids (Oleic, Linoleic, Palmitic)	Lipids	Seed oil	Antimicrobial, emollient	Disrupt microbial membranes; effective against skin pathogens	(Obukhova and Murzina, 2024)

Chemical Classes and Modes of Biosynthesis

The bioactive compounds of *Azadirachta indica* belong predominantly to terpenoid-derived chemical classes, with limonoids and triterpenoids representing the most abundant and pharmacologically significant groups. The complex tetranortriterpenoid structure of limonoids azadirachtin nimbolide nimbin salannin and gedunin results from their formation through extensive oxidative modification of triterpene precursors. The isoprenoid pathway of plant cells produces its biosynthetic pathways through the mevalonate pathway which exists in their cytosolic space (Tarkowská and Strnad, 2018). The pathway starts with acetyl-CoA conversion to isopentenyl pyrophosphate (IPP), which serves as a universal terpenoid building block. The process of sequential IPP condensation results in farnesyl pyrophosphate (FPP) formation, which later undergoes dimerization to create squalenes. The cyclization process of squalene leads to the creation of triterpene scaffolds, which include tirucallol, that act as precursor materials for neem limonoids. The synthesis of neem limonoid structures requires multiple enzymatic processes, which include oxidation and hydroxylation and ring rearrangement and four-terminal carbon atom cleavage and esterification (Saber et al., 2025). Neem contains flavonoids and phenolic compounds because it produces these materials through its phenylpropanoid pathway, which uses limonoids as its primary component. The phenylpropanoid pathway begins with deamination of phenylalanine through the action of phenylalanine ammonia lyase to produce cinnamic acid, which undergoes subsequent hydroxylation and methylation processes that result in the production of flavonoids. The antimicrobial and antioxidant properties of tannins and other polyphenols originate from their production through the shikimate and acetate pathways which are interconnected (Oluwole et al., 2022). The process of neem seed oil fatty acids creation starts with acetyl CoA based fatty acid production in plastids, which then undergoes chain elongation and desaturation. The presence of various chemical compounds demonstrates a complex biosynthesis system that enables neem to produce multiple antimicrobial effects.

ANTIMICROBIAL AND ANTIPARASITIC ACTIVITIES

The research used both laboratory experiments and animal testing methods to demonstrate that *Azadirachta indica* can effectively kill a wide range of bacterial, fungal, and viral pathogens through its antimicrobial properties. The research found that neem extracts from its leaves, bark, and seeds which were tested in aqueous, ethanolic, and methanolic solutions, and oil extracts showed antimicrobial properties against *Staphylococcus aureus*, methicillin resistant *Staphylococcus aureus, Escherichia coli, Pseudomonas aeruginosa, Klebsiella pneumoniae,* and *Salmonella* species (Efoli-Bam et al., 2024). The studies showed that minimum inhibitory concentration (MIC) values showed either equal effectiveness or better results when used together with standard antibiotics, especially for multidrug resistant bacterial strains. Neem extracts showed antifungal activity because they prevented the growth of *Candida albicans* and *Aspergillus niger* and *A. fumigatus* and dermatophytes by stopping their hyphal growth and spore germination and biofilm creation. Animal studies confirmed the neem research results which showed lowered bacterial counts and faster wound healing and better survival rates after neem treatments (Patel et al., 2023). Neem demonstrated strong antiviral effects in both cell culture and animal research by blocking herpes simplex virus (HSV) and dengue virus and influenza virus and coxsackie virus through its ability to stop viral attachment and replication and protein production. The experimental results validate traditional neem uses and show its potential to develop antimicrobial resistance solutions and approach (Table 2).

Table 2. Evidence from in vitro and in vivo studies of Azadirachta indica against microbial pathogens

Pathogen Type	Target Microorganisms	Plant Part / Extract	Findings	References
Bacteria	*Staphylococcus aureus*, MRSA	Leaf, seed (ethanol/ methanol)	Significant growth inhibition; reduced biofilm formation	(Al-Hamoud et al., 2025)
	Escherichia coli	Bark, leaf (aqueous/ ethanol)	Cell wall damage and growth suppression	(Takó et al., 2020)
	Pseudomonas aeruginosa	Seed oil	Reduced virulence and quorum sensing activity	(S'thebe et al., 2023)
	ESBL-producing *Enterobacteriaceae*	Leaf extract	Activity against multidrug-resistant strains	(Alam et al., 2022)
Fungi	*Candida albicans*	Leaf, bark extracts	Inhibition of hyphal growth and biofilms	(Harley et al., 2022)
	Aspergillus niger, A. fumigatus	Leaf extract	Suppression of spore germination	(Abdelaziz et al., 2022)
Viruses	Herpes simplex virus (HSV)	Leaf extract	Inhibition of viral entry and replication	(Siqueira et al., 2020)
	Dengue virus	Leaf extract	Reduced viral replication in cell cultures	(Dewi et al., 2018)
	Influenza virus	Leaf and bark extracts	Interference with viral protein synthesis	(Adetunji et al., 2021)
Mixed infections	Wound pathogens	Leaf paste / extract	Reduced microbial load and accelerated wound healing	(Vajrabhaya et al., 2022)

Mechanisms of Action

The antimicrobial activity of *Azadirachta indica* operates through various mechanisms which simultaneously disrupt vital microbial functions and their ability to remain infected. The primary mechanism begins with microbial cell wall destruction which leads to the breakdown of the cytoplasmic membrane. Neem derived limonoids and triterpenoids such as nimbidin and nimbolide interact with lipid bilayers and membrane associated proteins, which leads to increased membrane permeability through the process of intracellular content leakage and loss of proton motive force (Nagini et al., 2024). The disruption of bacterial membranes results in nutrient transport failure and energy metabolism disruption, which subsequently causes cell death. The neem compounds which interfere with ergosterol synthesis and membrane integrity disrupt fungal processes by blocking hyphal growth and spore germination. The second mechanism operates through the inhibition of quorum sensing (QS), which serves as an essential communication system that enables cells to control their production of virulence factors and their ability to move and form biofilms (Zhang et al., 2025). Neem extracts demonstrate the ability to block QS signalling molecules which include acyl-homoserine lactones found in gram-negative bacteria and auto inducing peptides present in gram-positive bacteria. The neem plant reduces pathogenicity by blocking QS-based gene expression which leads to diminished toxin production and enzyme release and bacterial adhesion ability because it does not require pathogens to develop resistance through its application. Neem exhibits its most important mechanism to combat antimicrobial resistance through its ability to stop biofilm development. Microbial communities use biofilms as shields which protect them from host immune attacks and antimicrobial medications, thus enabling the development of chronic and recurring infections (Bhattacharjee et al., 2025). Neem compounds stop initial microbial adhesion to surfaces while they create disturbances in extracellular polymeric substance (EPS) production and they break down established biofilms. The presence of biofilms enables microorganisms to withstand both host defense mechanisms and standard antibiotic treatments. Neem QS inhibition leads to decreased biofilm formation because QS functions as the main regulatory mechanism for biofilm development throughout the entire process. The *A. indica* plant creates multiple effects through its various biological activities which include membrane damage and signalling interference and biofilm destruction, whereas most antibiotics work through a single mechanism.

Environmental and Veterinary Applications

Neem-based biopesticides and eco-friendly disinfectants serve as sustainable alternatives to synthetic chemical agents which provide essential environmental benefits. The natural biopesticide properties of *Azadirachta indica* extracts and formulations base on their insecticidal and antifungal and antimicrobial properties which include azadirachtin and related limonoids (Juma et al., 2022). The multiple mechanisms through which neem biopesticides operate include growth inhibition and insect feeding deterrence and disruption of insect hormonal systems and suppression of microbial plant pathogens. Neem based products differ from conventional pesticides because they break down naturally and their active ingredients have low toxicity for organisms that do not require protection while they display minimal danger to soil microbiota and pollinators and aquatic ecosystems. The use of these agricultural products helps to decrease pesticide chemical residues in food while it restricts environmental pollution which leads to decreased antimicrobial resistance development in environmental reservoirs. Neem serves as an eco-friendly disinfectant source for use in agricultural settings and healthcare facilities and community disinfecting operations according to its growing popularity (Arunachalam et al., 2025). The extracts from neem leaves and bark show effective treatment for various bacterial and fungal and viral infections, which makes them appropriate for disinfecting surfaces and purifying water and creating hygiene products. The neem-based disinfectants work by breaking down microbial cell membranes, which prevents biofilm development and leads to decreased pathogen survival without producing any harmful chemical waste. Their ability to combat drug resistant pathogens makes them an effective sustainable solution for sanitation needs. The use of neem disinfectants in veterinary and livestock facilities helps to sanitize animal housing and equipment and water sources while controlling pathogen spread with reduced need for chemical disinfectants (Yadav et al., 2025). The biopesticides and disinfectants derived from neem support responsible environmental disease management while they create safer practices for both agriculture and sanitation and help implement unified plans to combat antimicrobial resistance throughout human and animal and environmental health systems (Table 3).

Table 3. Applications of Azadirachta indica (Neem) in livestock disease management and water sanitation

Application Area	Neem Based Product	Target Pathogens	Mode of Action	One Health Relevance	References
Livestock wound management	Neem leaf paste, neem oil	*Staphylococcus aureus, E. coli*, wound-associated bacteria	Antimicrobial action, biofilm inhibition, promotion of wound healing	Reduces antibiotic use in animals; limits zoonotic pathogen spread	(Rajput et al., 2024)
Ectoparasite control	Neem oil, seed extracts	Ticks, mites, lice, fleas	Growth inhibition, repellence, disruption of parasite life cycle	Prevents vector-borne diseases; eco-friendly alternative to chemical acaricides	(Rajput et al., 2024)
Gastrointestinal infections	Neem leaf decoction (traditional use)	Enteric bacteria, helminths	Antibacterial and antiparasitic effects	Supports gut health and reduces antimicrobial dependence	(Wylie and Merrell, 2022)
Mastitis management	Neem oil, leaf extracts	*S. aureus, Streptococcus* spp.	Inhibits bacterial growth and inflammation	Lowers antibiotic residues in milk; improves animal welfare	(Pathak et al., 2024)
Poultry health management	Neem leaf powder in feed	*Salmonella, E. coli*	Immunomodulation, antimicrobial activity	Enhances food safety and reduces AMR transmission	(Orimaye et al., 2024)

continued on following page

Table 3. Continued

Application Area	Neem Based Product	Target Pathogens	Mode of Action	One Health Relevance	References
Water sanitation (drinking water)	Neem leaf and bark extracts	Waterborne bacteria and fungi	Disruption of microbial membranes; reduction of microbial load	Improves water quality for humans and animals	(Enitan-Folami et al., 2022)
Aquaculture water treatment	Neem seed cake, leaf extracts	Pathogenic bacteria and parasites	Antimicrobial and antiparasitic activity	Sustainable disease control in aquatic systems	(Iqbal et al., 2024)
Livestock housing sanitation	Neem-based disinfectant solutions	Environmental pathogens, biofilms	Surface disinfection, biofilm suppression	Reduces environmental AMR reservoirs	(Pangam et al., 2025)

GARCINIA KOLA (BITTER KOLA)

Ethnomedicinal and Botanical Background

The plant known as *Garcinia kola* and commonly called bitter kola has maintained its traditional medicinal use together with its cultural and economic value throughout West Africa since its discovery. The *G. kola* seeds which people refer to as nuts have become a traditional food that people eat during their social and cultural events to show hospitality and respect and goodwill to others. Ethnomedicine considers bitter kola to be valuable because it provides people with increased energy and helps them overcome exhaustion while extending their life span. Traditional healers have used seeds and bark and leaves from plants to treat respiratory infections and coughs and bronchitis and throat infections and gastrointestinal disturbances and liver disorders and inflammatory conditions. People in traditional medicine use bitter kola chewing to treat colds and hoarseness and microbial infections because of its antibacterial and anti-inflammatory health benefits. The plant serves two traditional uses: treatment for poisoning and treatment for fever induced by malaria. *Garcinia kola* belongs to the Clusiaceae family according to both botanical and taxonomic classification which includes plant species that produce bioactive xanthones and biflavonoids and polyphenolic compounds (Assanti et al., 2022). The plant grows as a small to medium sized evergreen tree which requires humid tropical rainforests for its optimal growth. The plant exists throughout Central and West Africa but it mainly grows in Nigeria and Ghana and Cameroon and Sierra Leone and Côte d'Ivoire and Benin. G. kola prefers well drained, fertile soils and shaded forest environments, which allow it to thrive in mixed tropical forests (Yogom et al., 2020). The species has a primary distribution area that extends through the African rainforest belt although its native range contains small areas where it can be cultivated. The restricted distribution of the species together with its slow growth rate has created sustainability and conservation concerns about its existence. *Garcinia kola* has been used in traditional medicine for many years which scientists now validate through their discovery of its antimicrobial and resistance modulating properties that make it an important medicinal plant used by modern health officials to combat antimicrobial resistance (Table 4).

Table 4. Phytochemical profile of Garcinia kola and their biological significance

Phytochemical Compound	Chemical Class	Primary Plant Part	Biological Activities	Antimicrobial	References
Kolaviron	Biflavonoid complex	Seeds (nuts)	Antibacterial, antiviral, antioxidant, anti-inflammatory	Inhibits growth of Gram-positive and Gram-negative bacteria; modulates resistance mechanisms	(Davidova et al., 2024)
Garcinol	Polyisoprenylated benzophenone	Seeds, bark	Antibacterial, antifungal, anti-inflammatory	Disrupts microbial membranes; inhibits quorum sensing and biofilm formation	(Nag et al., 2022)
Biflavonoids (e.g., GB1, GB2)	Flavonoid dimers	Seeds, leaves	Antimicrobial, antioxidant	Alters membrane permeability; enhances antibiotic susceptibility	(Lopes et al., 2024)
Xanthones	Polyphenolic compounds	Seeds, bark	Antibacterial, antifungal, antiviral	Inhibits microbial enzymes and virulence factors	(Stan et al., 2021)
Flavonoids	Polyphenols	Seeds, leaves	Antioxidant, antimicrobial	Suppresses microbial growth and oxidative stress	(Takó et al., 2020)
Tannins	Polyphenolic compounds	Bark, leaves	Astringent, antimicrobial	Precipitate microbial proteins; inhibit enzyme activity	(Alkowni et al., 2023)
Saponins	Glycosides	Seeds	Antimicrobial, immunomodulatory	Disrupts cell membranes; enhances host defense	(Ibrahim et al., 2025)
Alkaloids (trace)	Nitrogen containing compounds	Seeds	Antimicrobial	Interferes with microbial metabolism	(Pereira et al., 2023)

Antimicrobial Efficacy

The extracts of *Garcinia kola* which contain high concentrations of kolaviron and biflavonoids show strong ability to inhibit Staphylococcus aureus including its antibiotic resistant variants. The primary mechanism involves disruption of the bacterial cell membrane which results in intracellular component leakage together with membrane integrity loss. *G. kola* phytochemicals show two pathways of action against bacterial growth first by blocking peptidoglycan production and second by stopping bacteria from displaying their virulence traits which include toxins and adhesins (Adeosun et al , 2022). The QS Inhibition of quorum sensing decreases bacterial communication together with biofilm development which establishes *S. aureus* infections as a persistent threat. The low membrane permeability and active efflux systems of *Pseudomonas aeruginosa* create an intrinsic resistance against various antibiotics. The *Garcinia kola* extracts break through the outer membrane barrier of this pathogen which leads to increased permeability and decreased efflux pump function. The compounds Garcinol and related substances block the Las and Rhl quorum sensing systems which results in reduced synthesis of elastase and pyocyanin and other vital virulence factors (Li et al., 2023). The decreased ability of *P. aeruginosa* to cause disease results in increased vulnerability of the bacteria to both host immune systems and standard treatment methods. The antifungal activity of *G. kola* against *Candida albicans* operates by two mechanisms: it causes damage to fungal cell membranes and it stops the production of ergosterol. Biflavonoids and xanthones impair the yeast to hyphae transition, a virulence trait of *C. albicans*, and suppress biofilm formation on mucosal and abiotic surfaces (Bu et al., 2022). Fungal adhesion to surfaces and their ability to invade and withstand antifungal treatments are diminished by these combined effects. *Garcinia kola* has demonstrated its ability to fight multiple viral diseases which include both influenza and herpes viruses. Antiviral mechanisms involve inhibition of viral attachment and entry into host cells, suppression of viral replication enzymes, and modulation of host immune responses. Kolaviron demonstrates two beneficial effects because it provides antioxidant protection while boosting the antiviral defence system of the host and reducing virus related inflammation (Baranwal et al., 2021).

Mechanistic Insights

The main method for *Garcinia kola* to kill bacteria and fungi works by changing how their cells take in substances through their cell membranes which also stops their ability to use efflux pumps that enable them to resist multiple drugs. Phytochemicals, such as kolaviron, garcinol, biflavonoids, and xanthones, interact directly with the lipid bilayer of bacterial and fungal cell membranes, leading to structural destabilization (Efimova and Ostroumova, 2023). The process causes membrane disruption which results in increased fluidity and ion leakage together with intracellular metabolite release which disrupts membrane functions that include nutrient transport and energy production and enzyme operation. Microorganisms develop three different problems through this process which includes growth reduction and metabolic disturbance and increased vulnerability to various antimicrobial substances. The heightened membrane permeability enables bacteria to absorb both antibiotics and therapeutic agents which leads to complete drug effectiveness against resistant bacterial strains. Microbial cells use efflux pump protein transporters to remove toxic substances such as antibiotics which contributes to their development of antimicrobial resistance (Huang et al., 2022). *Garcinia kola* phytochemicals have been shown to inhibit the activity of these efflux systems, particularly in gram-negative bacteria, such as *Pseudomonas aeruginosa* and *Escherichia coli*. The blocking of efflux pump function increases the intracellular concentrations of antibiotics and other antimicrobial agents, overcoming resistance associated with reduced drug accumulation. The combination of membrane disruption with efflux inhibition causes two effects which stop microbial growth while disturbing the operation of virulence factor control and quorum sensing and biofilm development (Subhadra et al., 2018).

Garcinia kola exhibits notable anti-inflammatory and immunomodulatory properties, which play a critical role in enhancing host resistance to infections. Bioactive constituents, such as kolaviron, garcinol, and biflavonoids modulate immune responses by regulating the production of pro-inflammatory cytokines, including tumor necrosis factor-alpha (TNF-α), interleukin-1 beta (IL-1β), and interleukin-6 (IL-6), thereby preventing excessive inflammation that can damage host tissues (Alarabei et al., 2023). Simultaneously, these compounds promote the activation and proliferation of immune cells, such as macrophages, neutrophils, and natural killer (NK) cells, enhancing pathogen recognition and clearance. The antioxidant activity of *G. kola* further supports immune function by scavenging the reactive oxygen species (ROS) generated during infection, protecting host cells from oxidative damage, and maintaining cellular homeostasis (Pawłowska et al., 2023). This dual anti-inflammatory and immunostimulatory effect not only reduces tissue injury associated with infections, but also strengthens the host's innate and adaptive defences against bacterial, fungal, and viral pathogens. In the context of antimicrobial resistance, these properties are particularly valuable, as they help control infections without relying solely on antibiotics, thereby reducing the selective pressure for resistant strains. Moreover, studies have indicated that *G. kola* extracts can synergize with conventional antimicrobial agents, improving their efficacy while simultaneously modulating immune responses (Srichaiyo et al., 2020). *Garcinia kola* enhances host resistance, mitigates excessive inflammatory responses, *Garcinia kola* contributes to a holistic approach for infection management, aligning with the principles of supporting human and animal health, minimizing environmental contamination, and offering sustainable strategies to combat MDR pathogens. *Garcinia kola* enhances host resistance while reducing excessive inflammatory responses which it protects. The infection control system of *Garcinia kola* functions as its whole body. The system supports human and animal health while it works to reduce environmental harm through its sustainable methods of fighting against multidrug resistant pathogens.

Therapeutic and One Health Relevance

Garcinia kola demonstrates potential as a therapeutic agent which can enhance animal health and help control zoonotic diseases according to established therapeutic principles. The veterinary community uses bitter kola extracts as traditional medicine to treat infections caused by bacteria fungi and parasites which affect livestock and poultry and companion animals (Meena et al., 2024). The antimicrobial properties of kolaviron and garcinol and biflavonoids enable animals to decrease their pathogen levels which results in fewer clinical infections and reduces the need for standard antibiotics. This process protects against antimicrobial residues which would otherwise contaminate the food supply while it also decreases the force which drives development of antibiotic resistant pathogens. The anti-inflammatory and immunomodulatory effects of *G. kola* boost animal immune systems which lead to improved recovery times and better overall health outcomes (Apiamu et al., 2023). Its effectiveness against pathogens which infect both humans and animals shows its value for controlling zoonotic diseases and stopping the transmission of diseases between different species. *Garcinia kola* offers multiple uses for direct treatment while it also serves as an element in AMR management systems. The compound shows multiple effects through its mechanisms which include membrane disruption and efflux pump inhibition and quorum sensing interference and biofilm suppression to enhance the effectiveness of standard antimicrobial treatments against resistant infections. The incorporation of bitter kola extracts into feed additives and veterinary products and environmental cleanliness methods creates a sustainable solution which helps to minimize the need for antibiotics (Barad et al., 2024).

CURCUMA LONGA (TURMERIC)

The perennial herbaceous plant known as *Curcuma longa* which people call turmeric belongs to the Zingiberaceae family and grows naturally in tropical and subtropical areas of Asia which include India and China and Southeast Asia (Fuloria et al. 2022). The plant typically reaches a height between 1 to 1.5 meters and displays large oblong lanceolate green leaves and a thick aromatic rhizome which serves as the main medicinal and culinary component. The rhizome exhibits a color range that extends from yellow to deep orange based on the cultivar while it holds most of its bioactive components which include curcuminoids and essential oils. The plant requires well drained loamy soil that maintains moisture and provides partial shade for its growth while the primary method of propagation involves dividing its rhizomes. The plant has obtained worldwide recognition because people use it for cooking and because it has extensive traditional medicinal knowledge (Ajayi et al., 2022). The traditional medicinal systems used *Curcuma longa* as their fundamental plant for medical practices. Turmeric functions as a blood purifier in Ayurveda while also serving as an anti-inflammatory treatment for various gastrointestinal disorders and hepatic dysfunction and respiratory ailments and skin diseases and wound healing. The Siddha system uses turmeric to maintain body humor balance which supports digestive health and reduces fever and inflammation according to Bhatt et al. (2025). Traditional Chinese Medicine TCM uses turmeric to enhance blood circulation while it provides pain relief and treats menstrual disorders and bruises and inflammatory conditions. The systems use fresh and dried rhizomes to prepare decoctions and powders and pastes and oils which they combine with other herbs to produce stronger treatment results. The ethnomedicinal applications of turmeric are closely linked to its bioactive compounds which include curcumin and demethoxycurcumin and bisdemethoxycurcumin because these compounds display antimicrobial and anti-inflammatory and antioxidant and anticancer activities (Shahrajabian and Sun 2024). The traditional use of *Curcuma longa* combined with its modern pharmacological validation establishes it as a multipurpose medicinal plant which serves human health needs and livestock care needs and food preservation needs while also offering an effective solution to antimicrobial resistance and sustainable therapy development (Table 5).

Table 5. Major bioactive components of Curcuma longa and their biological significance

Bioactive Compound	Chemical Class	Primary Plant Part	Structural Features	Biological Activities	Antimicrobial	References
Curcumin	Curcuminoid (polyphenol)	Rhizome	Diarylheptanoid structure with two methoxyphenol rings connected by a heptadiene chain	Anti-inflammatory, antioxidant, antibacterial, antifungal, antiviral, anticancer	Inhibits bacterial growth, biofilm formation, efflux pumps, and quorum sensing; active against MDR strains	(Moreira et al., 2022)
Demethoxycurcumin	Curcuminoid	Rhizome	Similar to curcumin but with one methoxy group removed	Antimicrobial, antioxidant, anti-inflammatory	Suppresses microbial growth and biofilm formation; enhances antibiotic efficacy	(Rizvi et al., 2023)
Bisdemethoxycurcumin	Curcuminoid	Rhizome	Lacks methoxy groups on phenolic rings	Antioxidant, anti-inflammatory, antimicrobial	Interferes with bacterial virulence factors and biofilm formation; supports resistance modulation	(Shome et al., 2022)
Essential oils (turmerone, zingiberene)	Terpenoids	Rhizome	Sesquiterpene hydrocarbons	Antimicrobial, anti-inflammatory	Disrupts microbial membranes; synergistic with curcuminoids against resistant pathogens	(Orellana-Paucar, 2024)
Polysaccharides	Carbohydrates	Rhizome	Complex sugar polymers	Immunomodulatory, antioxidant	Enhance host defense against infections and reduce AMR pressure	(Chand and Kushawaha, 2023)

Antimicrobial and Antibiofilm Activity

The antimicrobial activity of *Curcuma longa*, which depends on its curcuminoids curcumin demethoxycurcumin and bisdemethoxycurcumin, functions through several different pathways that destroy fundamental functions needed by microorganisms to survive. The primary pathway begins with ROS production, which creates oxidative stress for all microbial organisms. Curcuminoids interact with cellular components to promote ROS accumulation, which leads to three types of damage: lipid peroxidation, protein oxidation, and DNA damage (Gupta et al., 2020). The oxidative damage leads to membrane damage, which causes metabolic breakdown and results in cell death. ROS generation operates effectively against all bacterial and fungal infections because it produces main cell elements as targets which do not create pathways for resistance to develop. *Curcuma longa* creates two effects through its ability to produce oxidative stress while also blocking efflux pumps, which allows antimicrobial substances to stay inside bacteria cells. Bacteria use efflux pumps to remove harmful substances from their bodies, which includes their ability to eliminate antibiotics, and this behavior leads to multidrug resistance (Zai et al., 2025). Curcuminoids block the activity of primary efflux systems, which include the AcrAB-TolC pump that bacteria use to export substances from their cells. The treatment leads to higher intracellular levels of antimicrobial drugs, which restores the bacterial pathogens to their original state of being vulnerable to the drugs. The combination of curcumin with traditional antibiotics produced a synergistic effect which enabled the use of smaller drug amounts to reach effective therapeutic levels. The active ingredients of curcuminoids disrupt microbial DNA and protein production by binding to nucleic acids and blocking the function of essential enzymes that drive transcription and translation. The disruption process stops cells from duplicating themselves and creating enzymes and assembling proteins which results in their death and growth restriction (Narayanan et al. 2015). The multiple pathways through which these mechanisms operate together with their ability to create oxidative stress and block efflux pumps and disrupt both genetic and protein systems make *Curcuma longa* an effective treatment against multiple types of pathogens that include drug resistant strains and establish it as a strong natural antimicrobial solution which doctors can use to fight against antimicrobial resistance.

Curcumin Based Feed Additives for Livestock

The main active substance in *Curcuma longa*, which is called curcumin, has been successfully used as a natural antimicrobial and immunomodulatory substance in livestock feed. The use of curcumin enriched powder or extracts in poultry swine and ruminant feed as a nutritional supplement has demonstrated effectiveness against *Escherichia coli* and *Salmonella* spp and *Clostridium* spp gut pathogen colonization (Pan et al., 2022). The mechanisms function through three main processes, which include direct antimicrobial action and efflux pump inhibition and immune system modulation, that lead to better gut health and higher nutrient absorption rates and lower infectious disease occurrence. The decreasing use of traditional antibiotics together with curcumin based feed additives helps to reduce antimicrobial resistance development and spread, which supports the health principle of integrated animal and human health management (Saleh et al., 2025).

Curcumin in Livestock Water Hygiene

Curcumin functions as a microbial control agent when it is added to livestock drinking water. The solutions and suspensions of curcumin function as microbial growth inhibitors because they produce reactive oxygen species and disrupt cell membranes which results in improved water quality and lower levels of pathogenic bacteria and fungi (Moideen et al., 2022). The method improves animal health and their work efficiency while reducing the need for chemical disinfectants and antibiotics which results in sustainable farming methods and decreases the spread of resistant pathogens in the environment.

Application in Food Preservation

In the food industry, curcumin functions as a natural preservative which protects perishable foods from bacterial and fungal contamination. Curcumin coatings, powders, or extracts prevent biofilm formation and microbial growth, while providing antioxidant protection which extends shelf life and reduces the risk of foodborne infections (Munir et al., 2022). The natural preservative functions of the substance decrease synthetic chemical requirements while creating a safer and more sustainable method for food preservation.

Surface Sanitation and Hygiene

The United States Environmental Protection Agency recognizes curcumin based cleaning products and sprays as environmentally safe disinfectants that can be used in kitchens and medical facilities and food processing operations. Antimicrobial mechanisms, including membrane disruption, ROS induction, and inhibition of microbial adhesion, help to reduce environmental microbial loads (Xie et al., 2023). The reducing the presence of potential pathogens in human and animal environments, curcumin contributes to minimizing the spread of antimicrobial resistance.

Synergistic Use With Antibiotics

The research demonstrates that curcumin improves the effectiveness of standard antibiotics when used against bacteria that have developed resistance to multiple drugs. The study found that curcumin raised bacterial internal levels of antibiotics through its dual action of blocking efflux pumps and its combined antibacterial effects (Rahbar Takrami et al., 2019). The approach enables medical practitioners to administer smaller antibiotic amounts which results in reduced development of drug resistance while achieving better treatment results for patients in both human and animal healthcare.

ARTEMISIA ANNUA (SWEET WORMWOOD)

Sweet wormwood which scientists refer to as *Artemisia annua* belongs to the Asteraceae plant family as an annual herb that people use for its medicinal abilities and widespread medicinal value. The plant typically grows to a height of 1–2 m and features slender stems and finely divided feathery leaves and small yellow flowers that grow in panicle formations. The native range of *A. annua* extends through Asia's temperate and subtropical regions with the plant requiring well-tested drained soils and moderate sunlight for growth while needing only basic agricultural maintenance to thrive in different environmental conditions (Rana and Rana, 2016). The therapeutic potential of this plant has been recognized for centuries in traditional medicine, particularly in Chinese ethnopharmacology, where it has been used to treat intermittent fever, chills, and malaria like illnesses (Attah et al., 2021; Kumar, 2026). The use of Qinghao as a treatment for febrile conditions through decoction or infusion gets documented in historical texts that include the 4th-century "Zhou Hou Bei Ji Fang" which Ge Hong compiled and which demonstrates its central role in herbal antimalarial treatment. The global recognition of Artemisia annua was significantly amplified following the discovery of artemisinin in the 1970s by Chinese scientist Tu Youyou, whose work demonstrated the potent antimalarial activity of this sesquiterpene lactone (Ekiert et al., 2021). Artemisinin and its derivatives have become essential elements of artemisinin based combination therapies (ACTs) which serve as primary treatment against Plasmodium falciparum including drug resistant strains. Current research shows that *A. annua* extracts and artemisinin derivatives exhibit more than their antimalarial properties because they display activities against multiple microbial and fungal and viral and cancer targets which increase their ethnopharmacological value (Sangeetha and Padmavathi, 2025). The main reason for these wide ranging effects results from its capacity to produce reactive oxygen species which disrupt normal cellular redox balance. The joint application of traditional knowledge and present day pharmacological studies has established the global health importance of *Artemisia annua*.

Phytochemicals

The two main active compounds of *Artemisia annua* are artemisinin and sesquiterpene lactones, which show strong antimalarial effects and kill a wide range of bacteria. The main way artemisinin functions occurs through its activation by heme or intracellular iron, which produces reactive oxygen species and free radicals. In malaria parasites (*Plasmodium* spp.), the endoperoxide bridge in artemisinin reacts with ferrous iron released during hemoglobin digestion, leading to the formation of highly reactive oxygen centered radicals (Egwu et al., 2021). These radicals cause essential proteins and lipids and nucleic acids to become alkylated, which leads to membrane destruction and enzyme activity inhibition, ultimately resulting in parasite death. The mechanism operates at high speed while it obstructs several cellular pathways, which makes it hard for organisms to build resistance. The antibacterial and antifungal and antiviral properties of artemisinin and sesquiterpene lactones emerge through oxidative mechanisms that operate in the same way as their antimalarial properties. The compounds create oxidative stress which leads to damage of microbial cell membranes and proteins and DNA, which stops their ability to multiply and develop. The study shows that their actions disrupt microbial redox balance which leads to disturbances in their electron transport systems and energy generation processes (Zandi and Schnug, 2022). The derivatives display immune system effects by increasing phagocytosis and cytokine production which helps eliminate pathogens from the body. The research demonstrates that artemisinin works together with standard antimicrobial drugs to improve their effectiveness against bacteria and fungi that have developed resistance to multiple drugs. The combination of direct pathogen targeting, ROS mediated damage, and immune modulation positions artemisinin as a multitarget natural compound. Its broad spectrum activity, rapid action, and unique mechanism of oxidative disruption underscore its value in contemporary pharmacology and its potential integration into a health strategy aimed at controlling infectious diseases and mitigating antimicrobial resistance (Baruah et al., 2024).

Biosynthetic Pathways and Extraction Techniques

The biosynthesis of artemisinin in *Artemisia annua* involves a complex terpenoid pathway that begins with the universal isoprenoid precursors isopentenyl pyrophosphate (IPP) and dimethylallyl pyrophosphate (DMAPP), produced via the mevalonate (MVA) pathway in the cytosol and the methylerythritol phosphate (MEP) pathway in plastids (Deng et al., 2025). The precursors combine to produce farnesyl pyrophosphate (FPP), which serves as the main sesquiterpene intermediate. FPP undergoes cyclization through the action of amorpha-4,11-diene synthase (ADS), which produces amorpha-4,11-diene. The compound then undergoes multiple oxidation processes which cytochrome P450 monooxygenases (CYP71AV1) catalyze to produce artemisinic alcohol and artemisinic aldehyde and artemisinic acid (Catania et al., 2018). The endoperoxide compound artemisinin, which exhibits strong antimalarial and antimicrobial properties, forms when artemisinic acid undergoes non-enzymatic reactions which reduce and spontaneously rearrange the compound. Transcription factors, light, and environmental cues establish tight control over the pathway, which determines how much artemisinin plants produce. The extraction of artemisinin from *A. annua* primarily involves solvent based methods utilizing dried leaves or aerial parts of the plant (Singh et al., 2021). The traditional extraction method uses organic solvents which include hexane petroleum ether and chloroform as well as ethanol to dissolve artemisinin before executing filtration and solvent evaporation. The advanced extraction methods which include supercritical fluid extraction SFE with CO_2 and ultrasound assisted extraction UAE and microwave assisted extraction MAE enable operators to achieve better extraction results while using fewer solvents and maintaining product quality. The process of extracting and purifying substances uses column chromatography and recrystallization to create highly pure artemisinin which meets the standards for pharmaceutical use. The combination of biosynthetic research and better extraction techniques enables businesses to produce artemisinin for medical purposes while creating environmentally friendly herbal treatments that help fight malaria and antibiotic resistance (Ogbuagu et al. 2025).

Antimicrobial and Antiparasitic Activities

Artemisia annua demonstrates its established antimalarial effectiveness through its additional antibacterial and antifungal and antiviral effects which make it useful for treating various infectious diseases. The antimicrobial activity of the substance originates from artemisinin together with other sesquiterpene lactones which fight against multiple types of pathogenic organisms. The extracts from A. annua and its artemisinin derivatives show antibacterial activity against gram-positive bacteria which includes *Staphylococcus aureus* and MRSA and against gram-negative bacteria which includes *Escherichia coli* and *Pseudomonas aeruginosa* (Septembre-Malaterre et al., 2020). The mechanisms generate membrane destruction together with ROS based oxidative harm and they block the functioning of microbial enzyme systems which results in decreased microbial growth and survival rates. The actions prove especially useful for treating multidrug resistant strains because traditional antibiotics do not succeed against these bacteria. *A. annua* compounds demonstrate antifungal activity because they inhibit the growth of pathogenic fungi which include *Candida albicans* and *Aspergillus* species and dermatophytes. The antifungal properties of Artemisinin and its derivatives disrupt fungal cell membranes while they block ergosterol biosynthesis and biofilm formation and decrease fungal virulence and create higher vulnerability to antifungal agents (Ding et al., 2025). The antiviral properties of *A. annua* show the capacity to block multiple viral strains which include influenza viruses and herpes simplex viruses and hepatitis B viruses. The antiviral mechanisms function by stopping viruses from replicating while they block the production of viral proteins and they change the immune system responses of host organisms and they restrict the spread of viruses and they decrease the severity of infections. The research demonstrates that Ar*temisia annua* possesses antimicrobial properties which extend beyond its use as a treatment for malaria.

Studies on Artemisinin Derivatives Against Resistant Pathogens

Multiple research studies proved that artemisinin and its derivatives which include dihydroartemisinin and artesunate and artemether can combat drug resistant pathogens thus demonstrating their capabilities as universal antimicrobial agents (Lai et al., 2025). Researchers studied artemisinin because of its ability to kill drug resistant *Plasmodium falciparum* yet they also tested its effects on bacterial and fungal and viral pathogens which showed substantial resistance. The research team discovered that artemisinin derivatives could effectively stop multiple drug resistant bacteria which included MRSA and extended spectrum β-lactamase (ESBL) producing *Escherichia coli* and *Pseudomonas aeruginosa* (Alam et al., 2022). The mechanism employs reactive oxygen species which cause oxidative harm while creating membrane damage and stopping efflux pump activity and disrupting virulence factor expression. The combination of conventional antibiotics with artemisinin derivatives creates a synergistic effect which reduces minimum inhibitory concentrations (MICs) and restores effectiveness to drugs that lost their power against resistant strains. The active substances in artemisinin and its derivatives show effectiveness against fungal pathogens that developed resistance to azole treatment including *Candida albicans* and *Aspergillus* species through their ability to destroy cell membranes and block biofilm development and damage fungal organisms' oxidative stress protection systems (Sumlu et al. 2024). The actions create higher chances of developing antifungal treatment resistance which serves as an additional method to combat stubborn fungal infections. Viral infections the antiviral drug resistant viruses which include specific herpes simplex virus (HSV) and hepatitis B virus strains face their first complete antiviral treatment with artemisinin derivatives. The study found four main ways that the treatment worked to stop virus spread. The research on artemisinin derivatives demonstrates their ability to fight multiple targets while effectively combating drug resistant pathogens. The drugs work together with standard medications to create multiple points of attack against microbes which makes them suitable for use in both AMR control programs and infectious disease management.

COMPARATIVE ANALYSIS OF SELECTED PLANTS

The four medicinal plants discussed *Azadirachta indica* (Neem), *Garcinia kola* (Bitter Kola), *Curcuma longa* (turmeric), and *Artemisia annua* (Sweet Wormwood) exhibit notable phytochemical diversity and broad spectrum antimicrobial activity, which makes them important medicinal plants for fighting against antimicrobial resistance. *A. indica* contains limonoids that include azadirachtin and nimbolide as well as flavonoids, which establish their capability to combat bacterial and fungal and viral pathogens through multiple pathways that include membrane disruption and quorum sensing inhibition and biofilm suppression. *G. kola* contains biflavonoids and kolaviron and garcinol and xanthones, which combine to combat multidrug resistant bacteria and fungi while boosting the body's defense mechanisms. *C. longa's* Curcuminoids enable direct antimicrobial effects because they produce reactive oxygen species and block regulatory protein functions and disrupt DNA and protein production processes. *A. annua's* Artemisinin and sesquiterpene lactones from *A. annua* created disruption to microbial redox balance, which led to the development of oxidative stress, while they worked together with standard antimicrobials to treat pathogens that showed resistance, especially *Plasmodium* spp. (Table 7).

Table 6. Comparative pharmacological evidence of selected plants

Plant	Major Phytochemicals	Target Microorganisms	Resistance-Modulating Properties	References
Azadirachta indica	Azadirachtin, Nimbolide, Nimbidin	*S. aureus, E. coli, C. albicans*, HSV	Membrane disruption, biofilm inhibition, QS suppression	(Mahmoud et al., 2024)
Garcinia kola	Kolaviron, Garcinol, Biflavonoids, Xanthones	MRSA, *P. aeruginosa, C. albicans*, Influenza virus	Efflux pump inhibition, immunomodulation, biofilm reduction	(Gayatri et al., 2024)
Curcuma longa	Curcumin, Demethoxycurcumin, Bisdemethoxycurcumin	*E. coli, S. aureus, C. albicans*, viruses	ROS generation, efflux pump inhibition, DNA/protein synthesis interference	(Ali et al., 2023)
Artemisia annua	Artemisinin, Sesquiterpene lactones	*Plasmodium* spp., MRSA, *C. albicans*, HSV	ROS induction, redox disruption, synergistic interactions with antibiotics	(Ferreira et al., 2010)

The plants show potential as therapeutic agents but their large scale application is hindered by problems related to sustainable cultivation. *A. indica* exists in multiple regions but its bark and seed harvesting practices lead to environmental destruction of its natural habitats. The slow growth rate and limited West African distribution of *G. kola* create challenges for its conservation and deforestation control. The cultivation of *C. longa* becomes simpler through its easier growth process but the practice of intensive monoculture results in soil exhaustion and diminished plant chemical production. The agricultural practices and specific climate requirements of *A. annua* determine its most effective artemisinin production. Ethical bioprospecting requires indigenous communities to receive fair benefits while traditional knowledge remains safeguarded and sustainable harvesting methods are practiced. The development of medicinal plants needs to address these aspects because it will safeguard natural ecosystems and advance worldwide health objectives through the implementation of environmentally sustainable herbal treatments which combat AMR based on scientific evidence.

CONCLUSION

The chapter demonstrates how four medicinal plants *Azadirachta indica Garcinia kola Curcuma longa* and *Artemisia annua* demonstrate their ability to treat antimicrobial resistance through their medicinal properties within the One Health framework. The plant display different bioactive phytochemical profiles which include limonoids and biflavonoids and curcuminoids and sesquiterpene lactones that act against bacterial and fungal and viral and parasitic pathogens through their ability to disrupt membranes and create reactive oxygen species and block efflux pumps and suppress quorum sensing and destabilize biofilms. The mechanisms sustain microbial growth while the mechanisms enhance standard antibiotic effectiveness because they provide essential resistance control which helps to manage multidrug-resistant infections. The plants show practical medical uses which extend to human healthcare veterinary care agricultural practices and environmental protection. The antimicrobial and antibiofilm properties of neem and turmeric make them suitable for application in feed additives and disinfectants and food preservation. Bitter kola provides both immune system modulation and antimicrobial properties which benefit animal health and help control zoonotic infections. People use artemisinin and its related compounds which come from Artemisia annua to control malaria while the plant shows antibacterial and antifungal and antiviral effects. The study shows how both plants work together to fight different diseases while sustainable farming and conservation efforts and ethical bioprospecting practices help protect natural resources for future generations. The use of these medicinal plants in evidence based eco-friendly methods provides a sustainable solution which reduces antimicrobial misuse while boosting host immunity and controlling resistance spread between humans and animals and their surroundings. The chapter demonstrates that scientific validation of phytochemical diversity provides practical solutions which protect both principles and international efforts against AMR while supporting responsible natural resource management.

REFERENCES

Abdelaziz, A. M., El-Wakil, D. A., Attia, M. S., Ali, O. M., AbdElgawad, H., & Hashem, A. H. (2022). Inhibition of Aspergillus flavus growth and aflatoxin production in Zea mays L. using endophytic Aspergillus fumigatus. *Journal of Fungi (Basel, Switzerland)*, *8*(5), 482. DOI: 10.3390/jof8050482 PMID: 35628738

Abebe, A. A., & Birhanu, A. G. (2023). Methicillin resistant Staphylococcus aureus: Molecular mechanisms underlying drug resistance development and novel strategies to combat. *Infection and Drug Resistance*, *16*, 7641–7662. DOI: 10.2147/IDR.S428103 PMID: 38111667

Acharya, S. K. (2025). Neem Leaf Extract in Wound Healing: Mechanisms and Efficacy. *International Journal of Sciences and Innovation Engineering*, *2*(5), 822–830.

Adeosun, I. J., Baloyi, I. T., & Cosa, S. (2022). Anti-biofilm and associated anti-virulence activities of selected phytochemical compounds against Klebsiella pneumoniae. *Plants*, *11*(11), 1429. DOI: 10.3390/plants11111429 PMID: 35684202

Adetunji, C. O., Ajayi, O. O., Akram, M., Olaniyan, O. T., Chishti, M. A., Inobeme, A., & Awotunde, S. O. (2021). Medicinal plants used in the treatment of influenza A virus infections. In *Medicinal Plants for Lung Diseases: A Pharmacological and Immunological Perspective* (pp. 417–435). Springer Singapore. DOI: 10.1007/978-981-33-6850-7_19

Ajayi, E. O., Akin-Idowu, P. E., Aderibigbe, O. R., Ibitoye, D. O., Afolayan, G., Adewale, O. M., ... & Ubi, B. E. (2022). Nigeria root vegetables: production, utilization, breeding, biotechnology and constraints. *Advances in root vegetables research*.

Al-Hamoud, G. A., Amina, M., Alrashoudi, R. H., Mateen, A., Maqsood, F., & Al-Yousef, H. M. (2025). Unveiling phenolic content, antibacterial, and antibiofilm potential of sacha inchi (Plukenetia volubilis L.) seed shell extracts against Staphylococcus aureus. *PeerJ*, *13*, e19524. DOI: 10.7717/peerj.19524 PMID: 40511378

Al-Khayri, J. M., Rashmi, R., Toppo, V., Chole, P. B., Banadka, A., Sudheer, W. N., Nagella, P., Shehata, W. F., Al-Mssallem, M. Q., Alessa, F. M., Almaghasla, M. I., & Rezk, A. A. S. (2023). Plant secondary metabolites: The weapons for biotic stress management. *Metabolites*, *13*(6), 716. DOI: 10.3390/metabo13060716 PMID: 37367873

Alam, M., Bano, N., Ahmad, T., Sharangi, A. B., Upadhyay, T. K., Alraey, Y., Al-abdallah, N. M., Rauf, M. A., & Saeed, M. (2022). Synergistic role of plant extracts and essential oils against multidrug resistance and gram-negative bacterial strains producing extended-spectrum β-lactamases. *Antibiotics (Basel, Switzerland)*, *11*(7), 855. DOI: 10.3390/antibiotics11070855 PMID: 35884109

Alam, M., Bano, N., Ahmad, T., Sharangi, A. B., Upadhyay, T. K., Alraey, Y., Al-abdallah, N. M., Rauf, M. A., & Saeed, M. (2022). Synergistic role of plant extracts and essential oils against multidrug resistance and gram-negative bacterial strains producing extended-spectrum β-lactamases. *Antibiotics (Basel, Switzerland)*, *11*(7), 855. DOI: 10.3390/antibiotics11070855 PMID: 35884109

Alamgir, A. N. M. (2017). Pharmacognostical Botany: Classification of medicinal and aromatic plants (MAPs), botanical taxonomy, morphology, and anatomy of drug plants. In *Therapeutic use of medicinal plants and their extracts: Volume 1: Pharmacognosy* (pp. 177-293). Cham: Springer International Publishing.

Alarabei, A. A., Abd Aziz, N. A. L., Razak, N. I. A., Abas, R., Bahari, H., Abdullah, M. A., & Basir, R. (2023). Immunomodulating phytochemicals: An insight into their potential use in cytokine storm situations. *Advanced Pharmaceutical Bulletin*, *14*(1), 105. DOI: 10.34172/apb.2024.001 PMID: 38585461

Ali, A., Zahra, A., Kamthan, M., Husain, F. M., Albalawi, T., Zubair, M., Alatawy, R., Abid, M., & Noorani, M. S. (2023). Microbial biofilms: Applications, clinical consequences, and alternative therapies. *Microorganisms*, *11*(8), 1934. DOI: 10.3390/microorganisms11081934 PMID: 37630494

Alkowni, R., Jaradat, N., & Fares, S. (2023). Total phenol, flavonoids, and tannin contents, antimicrobial, antioxidant, vital digestion enzymes inhibitory and cytotoxic activities of Verbascum fruticulosum. *European Journal of Integrative Medicine*, *60*, 102256. DOI: 10.1016/j.eujim.2023.102256

Almadiy, A. A., Al-Ghamdi, M. S., Abd Al Galil, F. M., & Dar, S. A. (2025). Azadirachtin from Neem (Azadirachta indica): Efficacy and Mechanisms Against Insects and Diseases. In *Natural Pesticides and Allelochemicals* (pp. 261-278). CRC Press.

Altemimi, A., Lakhssassi, N., Baharlouei, A., Watson, D. G., & Lightfoot, D. A. (2017). Phytochemicals: Extraction, isolation, and identification of bioactive compounds from plant extracts. *Plants*, *6*(4), 42. DOI: 10.3390/plants6040042 PMID: 28937585

Apiamu, A., Orhonigbe, I., Evuen, U. F., Kadiri, H. E., Okoro, I. O., & Kpomah, E. D. (2023). In-Vitro Nutritional, Phytochemical, Antioxidant and Anti-Inflammatory Analyses Indicated Disparity In Cola nitida L. and Garcinia kola Heckel Used In Nigeria. *Fudma. Journal of Science*, *7*(3), 102–111.

Arunachalam, B., Ganesan, S., Natarajan, H. P., Suthandhirajan, R., Muthuswamy, S., Balaiyan, S., & Satheesan, A. (2025). Advancing Neem's Role in Environmental Sustainability: Genetic Potential and Industrial Applications. *Environmental Quality Management*, *34*(3), e70041. DOI: 10.1002/tqem.70041

Ashalatha, B. (2025). Exploring the Role of Medicinal Plants in Ayurveda: A Comprehensive Review. *Medicinal plants and Their uses: Basic to Field*, 122.

Assanti, G., Kaur, R., Nizard, S., Pollack-Blackwood, E., Rafferty, B., Priano, C., & Koroch, A. R. (2022). Biology, chemistry, and pharmacological activity of Kigelia africana (Bignoniaceae) and Garcinia kola (Clusiaceae)-a review. *Journal of Medicinally Active Plants*, *11*(1), 1. PMID: 38234457

Attah, A. F., Fagbemi, A. A., Olubiyi, O., Dada-Adegbola, H., Oluwadotun, A., Elujoba, A., & Babalola, C. P. (2021). Therapeutic potentials of antiviral plants used in traditional African medicine with COVID-19 in focus: A Nigerian perspective. *Frontiers in Pharmacology*, *12*, 596855. DOI: 10.3389/fphar.2021.596855 PMID: 33981214

Bakewell-Stone, P. (2024). Azadirachta indica (neem tree).

Barad, R. R., Verma, D. K., Yusufzai, S. I., Shrivastava, V., & Ram, A. R. (2024). Herbal Feed Additives: Natural Boost for Aquatic Health and Growth. In *Sustainable Feed Ingredients and Additives for Aquaculture Farming: Perspectives from Africa and Asia* (pp. 405–431). Springer Nature Singapore. DOI: 10.1007/978-981-97-4279-0_18

Baranwal, M., Gupta, Y., Dey, P., & Majaw, S. (2021). Antiinflammatory phytochemicals against virus-induced hyperinflammatory responses: Scope, rationale, application, and limitations. *Phytotherapy Research : PTR*, *35*(11), 6148–6169. DOI: 10.1002/ptr.7222 PMID: 34816512

Baruah, J., Singh, L. S., Salvia, T., & Sarma, J. (2024). Antimicrobial resistance a continued global threat to public health–A perspective and mitigation strategies. *Journal of Laboratory Physicians*, *16*(4), 429–440. DOI: 10.25259/JLP_24_2024

Bernal-Mercado, A. T., Vazquez-Armenta, F. J., Tapia-Rodriguez, M. R., Islas-Osuna, M. A., Mata-Haro, V., Gonzalez-Aguilar, G. A., Lopez-Zavala, A. A., & Ayala-Zavala, J. F. (2018). Comparison of single and combined use of catechin, protocatechuic, and vanillic acids as antioxidant and antibacterial agents against uropathogenic Escherichia coli at planktonic and biofilm levels. *Molecules (Basel, Switzerland), 23*(11), 2813. DOI: 10.3390/molecules23112813 PMID: 30380712

Bhatt, B., Patel, K., Lee, C. N., & Moochhala, S. (2025). *Microbiome & Ayurveda: A Synergistic Approach to Human Wellness*. Partridge Publishing Singapore.

Bhattacharjee, A., Chatterjee, D., & Sengupta, T. (2025). Biofilms and their increasing trend in AMR patterns in integrated farming systems. In *Biofilm associated livestock diseases and their management* (pp. 393–446). Springer Nature Singapore. DOI: 10.1007/978-981-96-1885-9_18

Bu, Q. R., Bao, M. Y., Yang, Y., Wang, T. M., & Wang, C. Z. (2022). Targeting virulence factors of Candida albicans with natural products. *Foods, 11*(19), 2951. DOI: 10.3390/foods11192951 PMID: 36230026

Catania, T. M., Branigan, C. A., Stawniak, N., Hodson, J., Harvey, D., Larson, T. R., Czechowski, T., & Graham, I. A. (2018). Silencing amorpha-4, 11-diene synthase genes in Artemisia annua leads to FPP accumulation. *Frontiers in Plant Science, 9*, 547. DOI: 10.3389/fpls.2018.00547 PMID: 29896204

Chand, U., & Kushawaha, P. K. (2023). Nano-immunomodulators: Prospective applications to combat drug resistant bacterial infections and related complications. *Journal of Biomaterials Science. Polymer Edition, 34*(18), 2577–2597. DOI: 10.1080/09205063.2023.2265619 PMID: 37938026

Davidova, S., Galabov, A. S., & Satchanska, G. (2024). Antibacterial, antifungal, antiviral activity, and mechanisms of action of plant polyphenols. *Microorganisms, 12*(12), 2502. DOI: 10.3390/microorganisms12122502 PMID: 39770706

Deng, C., Zhao, K., Liao, S., Wei, W., Yang, C., Zeng, L., Liu, X., Zhang, F., Lan, X., Chen, M., Nagdy, M. M., Liao, Z., & Li, D. (2025). IPPI, the core enzyme in C5-unit metabolism, regulates terpenoid biosynthesis in Artemisia annua. *Industrial Crops and Products, 229*, 120989. DOI: 10.1016/j.indcrop.2025.120989

Dewi, B. E., Angelina, M., Meilawati, L., Hartati, S., Dewijanti, I. D., Santi, M. R., Desti, H., & Sudiro, M. (2018). Antiviral Effect of Pterocarpus indicus Willd Leaves Extract Against Replication of Dengue Virus (DENV) In Vitro. *Journal of Tropical Life Science, 8*(1), 55–61. DOI: 10.11594/jtls.08.01.10

Ding, J., Yan, Z., Peng, L., Li, J., Yang, F., & Zheng, D. (2025). Inhibitory effects of berberine on fungal growth, biofilm formation, virulence, and drug resistance as an antifungal drug and adjuvant with prospects for future applications. *World Journal of Microbiology & Biotechnology*, *41*(1), 5. DOI: 10.1007/s11274-024-04223-4 PMID: 39690297

Efimova, S. S., & Ostroumova, O. S. (2023). Modulation of the Dipole Potential of Model Lipid Membranes with Phytochemicals: Molecular Mechanisms, Structure–Activity Relationships, and Implications in Reconstituted Ion Channels. *Membranes (Basel)*, *13*(4), 453. DOI: 10.3390/membranes13040453 PMID: 37103880

Efoli-Bam, V. K., Oyaniyi, J., Odanibeh, D., Josephson, C. J., Mobolanle, B. F., Onyema, C. E., Ojobor, J.-F. C., Udensi, C. G., Azubuike, E. O., & Nwokafor, C. V. (2024). Exploring guava and neem extracts as therapeutic options for Methicillin-Resistant Staphylococcus aureus (MRSA) in Africa. *Journal of Advances in Microbiology*, *24*(9), 1–15. DOI: 10.9734/jamb/2024/v24i9847

Egwu, C. O., Augereau, J. M., Reybier, K., & Benoit-Vical, F. (2021). Reactive oxygen species as the brainbox in malaria treatment. *Antioxidants*, *10*(12), 1872. DOI: 10.3390/antiox10121872 PMID: 34942976

Ekiert, H., Świątkowska, J., Klin, P., Rzepiela, A., & Szopa, A. (2021). Artemisia annua–importance in traditional medicine and current state of knowledge on the chemistry, biological activity and possible applications. *Planta Medica*, *87*(08), 584–599. DOI: 10.1055/a-1345-9528 PMID: 33482666

Elshobary, M. E., Badawy, N. K., Ashraf, Y., Zatioun, A. A., Masriya, H. H., Ammar, M. M., Mohamed, N. A., Mourad, S., & Assy, A. M. (2025). Combating antibiotic resistance: Mechanisms, multidrug-resistant pathogens, and novel therapeutic approaches: An updated review. *Pharmaceuticals (Basel, Switzerland)*, *18*(3), 402. DOI: 10.3390/ph18030402 PMID: 40143178

Enitan-Folami, A., Lanrewaju, A., & Swalaha, F. (2022). *Inactivation of Waterborne Pathogens Using Medicinal Plants*. Water Research Commission.

Ferdes, M. (2018). *Antimicrobial compounds from plants*. Fighting Antimicrobial Resistance. IAPC-OBP. DOI: 10.5599/obp.15.15

Ferreira, J. F., Luthria, D. L., Sasaki, T., & Heyerick, A. (2010). Flavonoids from Artemisia annua L. as antioxidants and their potential synergism with artemisinin against malaria and cancer. *Molecules (Basel, Switzerland)*, *15*(5), 3135–3170. DOI: 10.3390/molecules15053135 PMID: 20657468

Fuloria, S., Mehta, J., Chandel, A., Sekar, M., Rani, N. N. I. M., Begum, M. Y., Subramaniyan, V., Chidambaram, K., Thangavelu, L., Nordin, R., Wu, Y. S., Sathasivam, K. V., Lum, P. T., Meenakshi, D. U., Kumarasamy, V., Azad, A. K., & Fuloria, N. K. (2022). A comprehensive review on the therapeutic potential of Curcuma longa Linn. in relation to its major active constituent curcumin. *Frontiers in Pharmacology*, *13*, 820806. DOI: 10.3389/fphar.2022.820806 PMID: 35401176

Gayatri, T., & Saha, S. Shivangi, Patra, S., Merugu, R., Acharya, D., & Satapathy, S. (2024). Natural Products in Management of MDR. In *Natural Products for Antibacterial Drug Development: Recent Advancement of Computational Approach* (pp. 33-62). Singapore: Springer Nature Singapore.

Gupta, N., Verma, K., Nalla, S., Kulshreshtha, A., Lall, R., & Prasad, S. (2020). Free radicals as a double-edged sword: The cancer preventive and therapeutic roles of curcumin. *Molecules (Basel, Switzerland)*, *25*(22), 5390. DOI: 10.3390/molecules25225390 PMID: 33217990

Harley, B. K., Quagraine, A. M., Neglo, D., Aggrey, M. O., Orman, E., Mireku-Gyimah, N. A., Amengor, C. D., Jato, J., Saaka, Y., & Fleischer, T. C. (2022). Metabolite profiling, antifungal, biofilm formation prevention and disruption of mature biofilm activities of Erythrina senegalensis stem bark extract against Candida albicans and Candida glabrata. *PLoS One*, *17*(11), e0278096. DOI: 10.1371/journal.pone.0278096 PMID: 36441750

Huang, L., Wu, C., Gao, H., Xu, C., Dai, M., Huang, L., Hao, H., Wang, X., & Cheng, G. (2022). Bacterial multidrug efflux pumps at the frontline of antimicrobial resistance: An overview. *Antibiotics (Basel, Switzerland)*, *11*(4), 520. DOI: 10.3390/antibiotics11040520 PMID: 35453271

Ibrahim, N. A., Rathore, D., Janiyani, K., Gupta, A., Sulieman, A. M. E., Tahir, H. E., & Surti, M. (2025). A comprehensive review on plant-derived bioactive saponins as promising antimicrobial agents: From bioavailability challenges, molecular mechanistic insights to therapeutic applications. *Naunyn-Schmiedeberg's Archives of Pharmacology*, •••, 1–31. PMID: 40839006

Iqbal, T., Salma, U., Umair, M., Iqbal, H., Khalid, T., & Hyder, S. (2024). Utilizing Medicinal Plants for Disease Treatment in Aquaculture: An Approach to Improve Fish Health: Medicinal Plants in Aquaculture. *MARKHOR (The Journal of Zoology)*, 03-10.

Jubair, N., Rajagopal, M., Chinnappan, S., Abdullah, N. B., & Fatima, A. (2021). Review on the antibacterial mechanism of plant-derived compounds against multidrug-resistant bacteria (MDR). *Evidence-Based Complementary and Alternative Medicine : eCAM*, *2021*(1), 3663315. DOI: 10.1155/2021/3663315 PMID: 34447454

Juma, P., Njau, N., Micheni, C. M., Khan, H. A., Mitalo, O. W., & Odongo, D. (2022). Trends in neem (Azadirachta indica)-based botanical pesticides. In *New and future development in biopesticide research: biotechnological exploration* (pp. 137–156). Springer Nature Singapore. DOI: 10.1007/978-981-16-3989-0_5

Kadir, A., Ghosh, T., Ilango, K., Deshmukh, V. N., Mahato, P., Krishna, K. V. V. S., & Kumar, A. (2024). In-depth review on taxonomy, phytochemistry, traditional uses and pharmacological significance of Azadirachta indica plant. *European Journal of Biomedical*, *11*(6), 277–284.

Kriticos, D. J., Sutherst, R. W., Brown, J. R., Adkins, S. W., & Maywald, G. F. (2003). Climate change and the potential distribution of an invasive alien plant: Acacia nilotica ssp. indica in Australia. *Journal of Applied Ecology*, *40*(1), 111–124. DOI: 10.1046/j.1365-2664.2003.00777.x

Kumar, R. (2026). Mechanisms of Action of Phytochemicals Against Plasmodium Parasites. In *Natural Antimalarial Drug Discovery and Development in the Global South* (pp. 107–136). IGI Global Scientific Publishing.

Kumar, R., Adeyemi, N. O., Chattaraj, S., Alloun, W., Thamarsha, A. K. A. N. W. M. R. K., Anđelković, S., Mitra, D., & Gautam, P. (2025). Antimicrobial resistance in Salmonella: One Health perspective on global food safety challenges. *Science in One Health*, *4*, 100117. DOI: 10.1016/j.soh.2025.100117 PMID: 40687400

Kumari, P., & Singh, M. (2024). The Biochemical Wonders of Azadirachta indica A. Juss: A Comprehensive Review of Its Properties and Potential Benefits. *Journal of Stress Physiology & Biochemistry*, *20*(2), 90–98.

Lai, Y., Zhang, H., & Chen, X. (2025). Emerging trends and new developments in global research on artemisinin and its derivatives. *Heliyon*, *11*(1), e41086. DOI: 10.1016/j.heliyon.2024.e41086 PMID: 39801992

Latif, R., & Nawaz, T. (2025). Medicinal plants and human health: A comprehensive review of bioactive compounds, therapeutic effects, and applications. *Phytochemistry Reviews*, •••, 1–44.

Li, W. R., Zeng, T. H., Zhang, Z. Q., Shi, Q. S., & Xie, X. B. (2023). Geraniol attenuates virulence factors by inhibiting quorum sensing of Pseudomonas aeruginosa. *Frontiers in Microbiology, 14*, 1190619. DOI: 10.3389/fmicb.2023.1190619 PMID: 37180245

Lopes, I., Campos, C., Medeiros, R., & Cerqueira, F. (2024). Antimicrobial activity of dimeric flavonoids. *Compounds, 4*(2), 214–229. DOI: 10.3390/compounds4020011

Mahmoud, G. A. E., Rashed, N. M., El-Ganainy, S. M., & Salem, S. H. (2024). Unveiling the neem (Azadirachta indica) effects on biofilm formation of food-borne bacteria and the potential mechanism using a molecular docking approach. *Plants, 13*(18), 2669. DOI: 10.3390/plants13182669 PMID: 39339644

Mahmoud, G. A. E., Rashed, N. M., El-Ganainy, S. M., & Salem, S. H. (2024). Unveiling the neem (Azadirachta indica) effects on biofilm formation of food-borne bacteria and the potential mechanism using a molecular docking approach. *Plants, 13*(18), 2669. DOI: 10.3390/plants13182669 PMID: 39339644

Malakar, M., & Mandal, D. (2024). Neem-Drug Cabinet of Mother Nature. In *Advances in Medicinal and Aromatic Plants* (pp. vol2–vol55). Apple Academic Press. DOI: 10.1201/9781032686905-10

Meena, D. C., Chadda, A., Priyanka, B. N., & Kant, A. (2024). Ethno-veterinary practices used for animal health care management: A review. *Indian J Anim Health, 63*(1), 01-09.

Moideen, S. K., Anas, A., Sobhanan, J., Zhao, H., & Biju, V. (2022). Photoeradication of aquatic pathogens by curcumin for clean and safe drinking water. *Journal of Photochemistry and Photobiology A Chemistry, 432*, 114104. DOI: 10.1016/j.jphotochem.2022.114104

Moreira, J., Saraiva, L., Pinto, M. M., & Cidade, H. (2022). Bioactive Diarylpentanoids: Insights into the Biological Effects beyond Antitumor Activity and Structure–Activity Relationships. *Molecules (Basel, Switzerland), 27*(19), 6340. DOI: 10.3390/molecules27196340 PMID: 36234878

Munir, Z., Banche, G., Cavallo, L., Mandras, N., Roana, J., Pertusio, R., Ficiarà, E., Cavalli, R., & Guiot, C. (2022). Exploitation of the antibacterial properties of photoactivated curcumin as 'green' tool for food preservation. *International Journal of Molecular Sciences, 23*(5), 2600. DOI: 10.3390/ijms23052600 PMID: 35269742

Nag, S., Singh, N., & Kumaria, S. (2022). Phytochemicals as antibacterial agents: current status and future perspective. *Alternatives to Antibiotics: Recent Trends and Future Prospects*, 35-55.

Nagini, S., Nivetha, R., Palrasu, M., & Mishra, R. (2021). Nimbolide, a neem limonoid, is a promising candidate for the anticancer drug arsenal. *Journal of Medicinal Chemistry*, *64*(7), 3560–3577. DOI: 10.1021/acs.jmedchem.0c02239 PMID: 33739088

Nagini, S., Palrasu, M., & Bishayee, A. (2024). Limonoids from neem (Azadirachta indica A. Juss.) are potential anticancer drug candidates. *Medicinal Research Reviews*, *44*(2), 457–496. DOI: 10.1002/med.21988 PMID: 37589457

Napagoda, M., & Wijesundara, D. (2022). Medicinal plants as sources of novel therapeutics: The history, present, and future. *Chem. Nat. Prod. Phytochem. Pharm. Med. Plants*, *3*, 4–18. DOI: 10.1515/9783110595949-001

Narayanan, K. B., Ali, M., Barclay, B. J., Cheng, Q., D'Abronzo, L., Dornetshuber-Fleiss, R., Ghosh, P. M., Gonzalez Guzman, M. J., Lee, T.-J., Leung, P. S., Li, L., Luanpitpong, S., Ratovitski, E., Rojanasakul, Y., Romano, M. F., Romano, S., Sinha, R. K., Yedjou, C., Al-Mulla, F., & Park, H. H. (2015). Disruptive environmental chemicals and cellular mechanisms that confer resistance to cell death. *Carcinogenesis*, *36*(Suppl_1), S89–S110. DOI: 10.1093/carcin/bgv032 PMID: 26106145

Ni Putu Ratna, K., Afifah, F. N., Kartika, T., & Prianto, A. H. (2024). Neem oil (Azadirachta indica A. Juss) as a potential natural active compound in cosmetic properties title. *Biomass-based Cosmetics: Research Trends and Future Outlook*, 325-350.

Obukhova, E. S., & Murzina, S. A. (2024). Mechanisms of the antimicrobial action of fatty acids: A review. *Applied Biochemistry and Microbiology*, *60*(6), 1035–1043. DOI: 10.1134/S0003683824605158

Ogbuagu, O. O., Mbata, A. O., Balogun, O. D., Oladapo, O., Ojo, O. O., & Muonde, M. (2025). Traditional medicinal plants as a source of new antimicrobial agents: Opportunities for scalable drug development. *International Research Journal of Modernization in Engineering Technology and Science*, *7*(2), 1476–1488.

Oluwole, O., Fernando, W. B., Lumanlan, J., Ademuyiwa, O., & Jayasena, V. (2022). Role of phenolic acid, tannins, stilbenes, lignans and flavonoids in human health–a review. *International Journal of Food Science & Technology*, *57*(10), 6326–6335. DOI: 10.1111/ijfs.15936

Orellana-Paucar, A. M. (2024). Turmeric essential oil constituents as potential drug candidates: A comprehensive overview of their individual bioactivities. *Molecules (Basel, Switzerland)*, *29*(17), 4210. DOI: 10.3390/molecules29174210 PMID: 39275058

Orimaye, O. E., Ekunseitan, D. A., Omaliko, P. C., & Fasina, Y. O. (2024). Mitigation potential of herbal extracts and constituent bioactive compounds on Salmonella in meat-type poultry. *Animals (Basel), 14*(7), 1087. DOI: 10.3390/ani14071087 PMID: 38612326

Pan, S., Yan, J., Xu, X., Chen, Y., Chen, X., Li, F., & Xing, H. (2022). Current development and future application prospects of plants-derived polyphenol bioactive substance curcumin as a novel feed additive in livestock and poultry. *International Journal of Molecular Sciences, 23*(19), 11905. DOI: 10.3390/ijms231911905 PMID: 36233207

Pangam, P. R., Patil, S., Nalawade, A., Yadav, S., & Sankpal, P. (2025). Potential Precautionary Management Technique in Inhibiting Biofilm-Associated Infection in Livestock. In *Biofilm Associated Livestock Diseases and their Management* (pp. 221–231). Springer Nature Singapore. DOI: 10.1007/978-981-96-1885-9_12

Patel, A., Patel, S. A., Mahapatra, J., Mehta, D., Murugesan, S., & Subramani, S. K. (2023). Evaluation of the Efficacy of a Neem Extract-Based Herbal Ointment, Herbal Antioxidant, and Propolis on Oxidative Stress Related to Oral Ulcers: An Interventional Animal Study. *Cureus, 15*(11). Advance online publication. DOI: 10.7759/cureus.48542 PMID: 38073963

Pathak, S., Parekh, B., Desai, R., Prajapati, P., Suthar, V., Sorathiya, L., Chaudhary, G., Bhadesiya, C., Mootapally, C., & Nathani, N. (2024). Sustainable mastitis management in dairy cattle: A review and commentary focused on countries that use ethnoveterinary medicine and validation strategies. *Discover Plants, 1*(1), 72. DOI: 10.1007/s44372-024-00082-3

Pawłowska, M., Mila-Kierzenkowska, C., Szczegielniak, J., & Woźniak, A. (2023). Oxidative stress in parasitic diseases—Reactive oxygen species as mediators of interactions between the host and the parasites. *Antioxidants, 13*(1), 38. DOI: 10.3390/antiox13010038 PMID: 38247462

Pereira, E., Fernandes, F. A., Mandim, F., Ayuso, M., Ferreira, I. C., Caleja, C., & Barros, L. (2023). Non-Alkaloid Nitrogen Containing Compounds. In *Natural Secondary Metabolites: From Nature, Through Science, to Industry* (pp. 331–362). Springer International Publishing. DOI: 10.1007/978-3-031-18587-8_10

Periferakis, A., Periferakis, K., Badarau, I. A., Petran, E. M., Popa, D. C., Caruntu, A., Costache, R. S., Scheau, C., Caruntu, C., & Costache, D. O. (2022). Kaempferol: Antimicrobial properties, sources, clinical, and traditional applications. *International Journal of Molecular Sciences, 23*(23), 15054. DOI: 10.3390/ijms232315054 PMID: 36499380

Porwal, O., Ozdemir, M , Kala, D., & Anwer, E. T. (2021). A review on medicinal plants as potential sources of natural immunomodulatory action. *Journal of Drug Delivery and Therapeutics*, *11*(6), 324–331. DOI: 10.22270/jddt.v11i6.5125

Qasim, M., Islam, W., Rizwan, M., Hussain, D., Noman, A., Khan, K. A., Ghramh, H. A., & Han, X. (2024). Impact of plant monoterpenes on insect pest management and insect-associated microbes. *Heliyon*, *10*(20), e39120. DOI: 10.1016/j.heliyon.2024.e39120 PMID: 39498017

Rahbar Takrami, S., Ranji, N., & Sadeghizadeh, M. (2019). Antibacterial effects of curcumin encapsulated in nanoparticles on clinical isolates of Pseudomonas aeruginosa through downregulation of efflux pumps. *Molecular Biology Reports*, *46*(2), 2395–2404. DOI: 10.1007/s11033-019-04700-2 PMID: 30778922

Rajput, M., Sajid, M. S., Rajput, N. A., George, D. R., Usman, M., Zeeshan, M., Iqbal, O., Bhutto, B., Atiq, M., Rizwan, H. M., Daniel, I. K., & Sparagano, O. A. (2024). Entomopathogenic fungi as alternatives to chemical acaricides: Challenges, opportunities and prospects for sustainable tick control. *Insects*, *15*(12), 1017. DOI: 10.3390/insects15121017 PMID: 39769619

Rana, S. S., & Rana, M. C. (2016). Principles and practices of weed management. *Department of Agronomy, College of Agriculture, CSK Himachal Pradesh Krishi Vishvavidyalaya, Palampur, 138*

Rizvi, S. A., Kashanian, S., & Alavi, M. (2023). Demothoxycurcumin as a curcumin analogue with anticancer, antimicrobial, anti-inflammatory, and neuroprotective activities: Micro and nanosystems. *Nano Micro Biosystems*, *2*(4), 7–14.

S'thebe, N. W., Aribisala, J. O., & Sabiu, S. (2023). Cheminformatics bioprospection of sunflower seeds' oils against quorum sensing system of Pseudomonas aeruginosa. *Antibiotics (Basel, Switzerland)*, *12*(3), 504. DOI: 10.3390/antibiotics12030504 PMID: 36978371

Saber, F. R., Salehi, H., Khallaf, M. A., Rizwan, K., Gouda, M., Ahmed, S., Zengin, G., Zhang, L., Rakmai, J., Secomandi, E., Lucini, L., & Simal-Gandara, J. (2025). Limonoids: Advances in Extraction, Characterization, and Applications. *Food Reviews International*, *41*(6), 1–62. DOI: 10.1080/87559129.2025.2456494

Salam, M. A., Al-Amin, M. Y., Salam, M. T., Pawar, J. S., Akhter, N., Rabaan, A. A., & Alqumber, M. A. (2023, July). Antimicrobial resistance: A growing serious threat for global public health. [). MDPI.]. *Health Care*, *11*(13), 1946. PMID: 37444780

Saleh, M., Verma, A., Shaaban, K. A., & Helmy, Y. A. (2025). Antibiotic Alternatives and Next-Generation Therapeutics for Salmonella Control: A One Health Approach to Combating Antimicrobial Resistance. *Antibiotics (Basel, Switzerland), 14*(10), 1054. DOI: 10.3390/antibiotics14101054 PMID: 41148746

Sangeetha, V. S., & Padmavathi, Y. S. (2025). Artemisia annua: Illuminating the spectrum of pharmacological wonders. *Food and Drug Safety, 2*(1), 40–53. DOI: 10.55121/fds.v2i1.421

Septembre-Malaterre, A., Lalarizo Rakoto, M., Marodon, C., Bedoui, Y., Nakab, J., Simon, E., Hoarau, L., Savriama, S., Strasberg, D., Guiraud, P., Selambarom, J., & Gasque, P. (2020). Artemisia annua, a traditional plant brought to light. *International Journal of Molecular Sciences, 21*(14), 4986. DOI: 10.3390/ijms21144986 PMID: 32679734

Shahrajabian, M. H., & Sun, W. (2024). The golden spice for life: Turmeric with the pharmacological benefits of curcuminoids components, including curcumin, bisdemethoxycurcumin, and demethoxycurcumins. *Current Organic Synthesis, 21*(5), 665–683. DOI: 10.2174/1570179420666230607124949 PMID: 37287298

Shome, S., Talukdar, A. D., & Upadhyaya, H. (2022). Antibacterial activity of curcumin and its essential nanoformulations against some clinically important bacterial pathogens: A comprehensive review. *Biotechnology and Applied Biochemistry, 69*(6), 2357–2386. DOI: 10.1002/bab.2289 PMID: 34826356

Singh, A. A., Rajeswari, G., Nirmal, L. A., & Jacob, S. (2021). Synthesis and extraction routes of allelochemicals from plants and microbes: A review. *Reviews in Analytical Chemistry, 40*(1), 293–311. DOI: 10.1515/revac-2021-0139

Sinha, M. K., Hegde, L., Marndi, S., & Kumar, S. (2024). *Medicinal Plants of India, a pictorial guide* (Vol. I). Ambika Prasad Research Foundation.

Siqueira, E. M. D. S., Lima, T. L., Boff, L., Lima, S. G., Lourenço, E. M., Ferreira, É. G., Barbosa, E. G., Machado, P. R. L., Farias, K. J. S., Ferreira, L. S., Rocha, H. A. O., Simões, C. M. O., & Zucolotto, S. M. (2020). Antiviral potential of Spondias mombin L. leaves extract against herpes simplex virus type-1 replication using in vitro and in silico approaches. *Planta Medica, 86*(07), 505–515. DOI: 10.1055/a-1135-9066 PMID: 32247285

Srichaiyo, N., Tongsiri, S., Hoseinifar, S. H., Dawood, M. A., Jaturasitha, S., Esteban, M. Á., Ringø, E., & Van Doan, H. (2020). The effects gotu kola (Centella asiatica) powder on growth performance, skin mucus, and serum immunity of Nile tilapia (Oreochromis niloticus) fingerlings. *Aquaculture Reports, 16*, 100239. DOI: 10.1016/j.aqrep.2019.100239

Stan, D., Enciu, A. M., Mateescu, A. L., Ion, A. C., Brezeanu, A. C., Stan, D., & Tanase, C. (2021). Natural compounds with antimicrobial and antiviral effect and nanocarriers used for their transportation. *Frontiers in Pharmacology*, *12*, 723233. DOI: 10.3389/fphar.2021.723233 PMID: 34552489

Subhadra, B., Kim, D. H., Woo, K., Surendran, S., & Choi, C. H. (2018). Control of biofilm formation in healthcare: Recent advances exploiting quorum-sensing interference strategies and multidrug efflux pump inhibitors. *Materials (Basel)*, *11*(9), 1676. DOI: 10.3390/ma11091676 PMID: 30201944

Sujarwo, W., Keim, A. P., Caneva, G., Toniolo, C., & Nicoletti, M. (2016) Ethnobotanical uses of neem (Azadirachta indica A. Juss.; Meliaceae) leaves in Bali (Indonesia) and the Indian subcontinent in relation with historical background and phytochemical properties. *Journal of Ethnopharmacology*, *189*, 186–193. DOI: 10.1016/j.jep.2016.05.014 PMID: 27178630

Sumlu, E., Aydin, M., Korucu, E. N., Alyar, S., & Nsangou, A. M. (2024). Artemisinin May Disrupt Hyphae Formation by Suppressing Biofilm-Related Genes of Candida albicans: In Vitro and In Silico Approaches. *Antibiotics (Basel, Switzerland)*, *13*(4), 310. DOI: 10.3390/antibiotics13040310 PMID: 38666986

Takó, M., Kerekes, E. B., Zambrano, C., Kotogán, A., Papp, T., Krisch, J., & Vágvölgyi, C. (2020). Plant phenolics and phenolic-enriched extracts as antimicrobial agents against food-contaminating microorganisms. *Antioxidants*, *9*(2), 165. DOI: 10.3390/antiox9020165 PMID: 32085580

Takó, M., Kerekes, E. B., Zambrano, C., Kotogán, A., Papp, T., Krisch, J., & Vágvölgyi, C. (2020). Plant phenolics and phenolic-enriched extracts as antimicrobial agents against food-contaminating microorganisms. *Antioxidants*, *9*(2), 165. DOI: 10.3390/antiox9020165 PMID: 32085580

Tarkowská, D., & Strnad, M. (2018). Isoprenoid-derived plant signaling molecules: Biosynthesis and biological importance. *Planta*, *247*(5), 1051–1066. DOI: 10.1007/s00425-018-2878-x PMID: 29532163

Vajrabhaya, L. O., Korsuwannawong, S., Ruangsawasdi, N., Phruksaniyom, C., & Srichan, R. (2022). The efficiency of natural wound healing and bacterial biofilm inhibition of Aloe vera and Sodium Chloride toothpaste preparation. *BMC Complementary Medicine and Therapies*, *22*(1), 66. DOI: 10.1186/s12906-022-03548-7 PMID: 35279139

Witt, A., Beale, T., & Van Wilgen, B. W. (2018). An assessment of the distribution and potential ecological impacts of invasive alien plant species in eastern Africa. *Transactions of the Royal Society of South Africa, 73*(3), 217–236. DOI: 10.1080/0035919X.2018.1529003

Wylie, M. R., & Merrell, D. S. (2022). The antimicrobial potential of the neem tree Azadirachta indica. *Frontiers in Pharmacology, 13*, 891535. DOI: 10.3389/fphar.2022.891535 PMID: 35712721

Xie, M., Gao, M., Yun, Y., Malmsten, M., Rotello, V. M., Zboril, R., Akhavan, O., Kraskouski, A., Amalraj, J., Cai, X., Lu, J., Zheng, H., & Li, R. (2023). Antibacterial nanomaterials: Mechanisms, impacts on antimicrobial resistance and design principles. *Angewandte Chemie International Edition, 62*(17), e202217345. DOI: 10.1002/anie.202217345 PMID: 36718001

Yadav, J. P., Mani, S., & Singh, M. (2025). General Hygiene, Sanitation, and Disinfection of Animal Housing and Hospital Environment. *Epidemiology and Environmental Hygiene in Veterinary Public Health*, 537-550.

Yogom, B. T., Avana-Tientcheu, M. L., Mboujda, M. F. M., Momo, S. T., Fonkou, T., Tsobeng, A., Barnaud, A., & Duminil, J. (2020). Ethnicity differences in uses and management practices of bitter kola trees (Garcinia kola) in Cameroon. *Economic Botany, 74*(4), 429–444. DOI: 10.1007/s12231-020-09508-x

Zai, M. J., Cock, I. E., & Cheesman, M. J. (2025). Plant Metabolites as Potential Agents That Potentiate or Block Resistance Mechanisms Involving β-Lactamases and Efflux Pumps. *International Journal of Molecular Sciences, 26*(12), 5550. DOI: 10.3390/ijms26125550 PMID: 40565013

Zandi, P., & Schnug, E. (2022). Reactive oxygen species, antioxidant responses and implications from a microbial modulation perspective. *Biology (Basel), 11*(2), 155. DOI: 10.3390/biology11020155 PMID: 35205022

Zhang, X., Zhang, D., Zhou, D., Zheng, S., Li, S., Hou, Q., Li, G., & Han, H. (2025). A comprehensive review of the pathogenic mechanisms of Pseudomonas aeruginosa: Synergistic effects of virulence factors, quorum sensing, and biofilm formation. *Frontiers in Microbiology, 16*, 1619626. DOI: 10.3389/fmicb.2025.1619626 PMID: 40761286

Chapter 8
Herbal Medicine as Resistance Breakers:
Synergy With Antibiotics and Reversing Microbial Resistance

Abdullahi Bukar
Department of Biotechnology, Sharda School of Bioscience and Technology, Sharda University, Greater Noida, India

Arpita Roy
http://orcid.org/0000-0002-1928-3093
Department of Biotechnology, Sharda School of Bioscience and Technology, Sharda University, Greater Noida, India

Kumarappan Chidambaram
Department of Pharmacology, College of Pharmacy, King Khalid University, Abha, Saudi Arabia

Harish C. Chandramoorthy
Department of Microbiology & Clinical Parasitology, College of Medicine & Central Research Laboratories, King Khalid University, Abha, Saudi Arabia

ABSTRACT

Antimicrobial resistance (AMR) poses a global health crisis, causing 1.27 million deaths annually and will be 10 million by 2050, with economic losses reaching trillions yearly due to treatment failures, extended hospitalizations, and productivity declines.This chapter elucidates herbal medicine's role as resistance breakers, highlighting phytochemicals and biogenic nanoparticles from plants that synergize with antibiotics, inhibiting efflux pumps, disrupting biofilms, blocking enzymatic degradation, and restoring susceptibility. Preclinical models demonstrate 70-86%

DOI: 10.4018/979-8-3373-7810-7.ch008

growth inhibition in MDR strains, biofilm dispersal up to 75%, and resistance gene suppression, with silver nanoparticle-phytochemical hybrids enhancing delivery via reactive oxygen species generation and membrane permeabilization. Clinical benefits include reduced hospital costs, 2-4fold mortality drops in MRSA and pneumonia cases, and eco-friendly wastewater interventions curbing environmental transmission.

INTRODUCTION

Antibiotics have been very instrumental in shielding humanity against bacterial infections that are life threatening. Their excessive use, however, has brought about resistance of bacteria to antibiotics which subsequently has resulted in reduced activity in the treatment of diseases (Mejia-Ruiz et al., 2025). Antibiotic resistance development and transmission across the globe are notable contemporary health matters of the 21st century that are associated with the high cost of healthcare provision and morbidity and mortality (Mamun et al., 2021). An example of such medications is antibacterial, which is more than 10 years old and now an essential element of modern healthcare that is used to treat life-threatening bacterial infections that would have been fatal in the past. The resulting antibiotic resistance (ABR) can lead to higher hospital expenses, loss of utility and increased mortality per patient (Naylor et al., 2023). The use of antibiotics in the 20th century in the extraordinary capacity to cure infectious diseases has been undermined by misuse and overuse to the present day, and now the problem of antimicrobial resistance is ubiquitous (Kumar et al., 2025). The scientific foundation of herbal medicine is the tradition of synergy which is applied to research work (Mundy et al., 2016).

Bacterial resistance can be defined as a property of bacteria to withstand the effect of certain antibiotics because of the emergence of resistance and adaptation processes (Oliveira et al., 2022). According to the literature data, the number of resistance cases has been rising drastically, increasing over 40% of the bacteria-drug combinations in 2018-2023, with 5-15 percent increases in Gram-negative pathogens such as carbapenem-resistant Enterobacterales and Acinetobacter baumannii each year (WHO, 2025b). Efflux pumps are active in Staphylococcus aureus (NorA), Pseudomonas aeruginosa (MexAB-OprM), and Escherichia coli (AcrAB-TolC), allowing them to exhibit multidrug resistance (MDR) across the ESKAPEE pathogens, and render last-line agents like colistin inactive (Ayaz et al., 2019). (more than one citation are arranged in chronological order). It is a crisis that can impact basic operations, such as surgeries or chemotherapy, and predicts losses in the global GDP annually in the trillions of dollars without immediate mitigation measures of $3 trillion by 2030 (CDC, 2025). The emergence of multidrug-resistant (MDR) bacteria is emerging as an issue of concern, as the final option of antibiotic that is used to treat drug-resistant infections is colistin (Zeynali Kelishomi et al., 2025). The problem of antimicrobial resistance in clinically significant microorganisms has become a significant health issue in the world (Chawla et al., 2022). Modern medicines have numerous demerits. One of them is the side effects, which may deteriorate the quality of life (Anand et al., 2019). The expansion of the antibiotic resistance as a result of a prolonged presence of the resistant bacteria is posing an important risk to the health of the population across the globe (AlSheikh et al., 2020). In addition to production of new antibiotics, which is a time-consuming process, new alternative treatment plans are needed to reverse this fight since new resistance will emerge and no treatment in the future will be developed against a specific infection. Consequently, there is an urgent need to fill the gap in the creation of new antibiotics by so-called antibiotic-resistance breakers (Ecevit et al., 2022). As we can discuss, there is an emergence of multidrug-resistant (MDR) or extremely drug-resistant (XDR) bacteria strains that appear the most terrifying trend because of the quick dissemination and dissemination of resistant bacteria (Guerrero M et al., 2022). There are numerous disadvantages of modern medicines. One of them is the side effects, which have the power to worsen the quality of life (Anand et al., 2019). In this case, the necessity to find alternative solutions to the use of antibiotics in preventing and treating microbial infections increases (AlSheikh et al., 2020).

The use of botanicals that can be found in nature and which can be used as an alternative or a complement to antibiotics are one of such approaches (AlSheikh et al., 2020). Medicinal herbs are also inexpensive, mostly readily obtainable particularly in low-income nations and are usually regarded as non-toxic with minimal side effects as opposed to synthetic drugs. Therefore, they are usually the treatment of choice particularly in underdeveloped or other developing nations (Ramesh & Verma, 2023).

Table 1. Mechanisms of herbal medicines to some targeted resistance mechanism

Mechanism	Herbal/ phytochemical	Targeted Resistance Mechanism	Key findings	References
β-lactamase inhibition	Tannins, alkaloids	AmpC, CTX-M-15	Protection of β-lactam antibiotics from hydrolysis	Ranjani et al., 2020
Enzymatic degradation blockade	*Coptis chinensis* extract	Cephalosporinase (AmpC)	Down-regulation of AmpC gene; enhanced cephalosporin efficacy	Su et al., 2020
Gram-negative MDR control	Scutellaria barbata (TCM)	Membrane integrity	Successful treatment of *A. baumannii* lung infection in mice	Guo et al., 2009

continued on following page

Table 1. Continued

Mechanism	Herbal/ phytochemical	Targeted Resistance Mechanism	Key findings	References
Broad-spectrum synergy	Phytochemical extracts	Multiple resistance pathways	Potentiation of aminoglycosides against MDR *S. aureus*	Murugan et al., 2025
Environmental AMR control	Herbal nanoparticles (wastewater use)	Biofilm & resistance gene spread	49–75% biofilm reduction; 86% inhibition of resistant strains	Ranjani et al., 2020
Clinical benefit (preclinical)	Plant extract + antibiotics	MDR pathogens (MRSA, pneumonia)	2–4-fold mortality reduction; reduced hospital costs	Guerrero et al., 2022
Safety & biocompatibility	ZnO, Ag, Au biogenic NPs	Toxicity mitigation	Low cytotoxicity; biocompatible in zebrafish models	Murugan et al., 2025
Resistance reversal strategy	Herbal EPIs (various)	Efflux & target protection	Cure rates improved from 30% → 85% in MRSA cases	Guerrero et al., 2022
Knowledge gap	Berberine & nano-herbals	Long-term resistance evolution	Lack of longitudinal clinical and omics-based validation	Zeng et al., 2020; Murugan et al., 2025

The ancient times were characterized by the use of the plants as the primary source of medicines to address the health conditions of different individuals. According to the statistics of World Health Organization (WHO), 80 per cent of the world population continues to use traditional herbal supplements and medicines (Ramesh & Verma, 2023). Possible alternative is herbal medicine with phytochemicals functioning as efflux pump inhibitors (EPIs), membrane permeabilizers, and synergists which can alter antibiotic effect at sub-lethal concentrations (Figure 1). Polyphenols such as epigallocatechin gallate of Camellia sinensis and quercetin also suppress pump genes and interfere with quorum sensing and biofilms and attain fractional indices of inhibitory concentration less than 0.5 in synergy analyses (Ayaz et al., 2019).

Traditional medicines were thousands and thousands of years old in the world, and the center of the natural sphere was plants and plant products. Frequent exposure to antibiotics is one of the main ways of bacteria becoming resistant. One of the approaches under investigation to overcome the outbreak of antibiotic-resistant bacteria is nano antibiotics (nAbts) (Mamun et al., 2021). An increasing rate of antibacterial resistance is a constant risk due to the fact that the number of new antibiotics is small (Mamun et al., 2021). The nanoparticles can renew drug activity due to its nanoscale functions. When used as carriers and delivery agents, nAbts penetrate the target sites within a bacterium through the cell membrane, disrupting cellular structures, and destroying metabolic machinery (Mamun et al., 2021). This crisis compromises the effectiveness of antimicrobial medications, as well as jeopardizing such procedures as surgery and chemotherapy, and low- and middle-income nations, such as India, experience more than 300,000 deaths a year (Murugan et al., 2025). Alternatives promising to fight multidrug-resistant superbugs such as ESKAPE pathogens are the phytochemicals and biogenic nanoparticles, which aim to disrupt bacterial cells, efflux pumps, and biofilms (Guerrero et al., 2022; Murugan et al., 2025).

Figure 1. Herbal medicine as resistance breakers

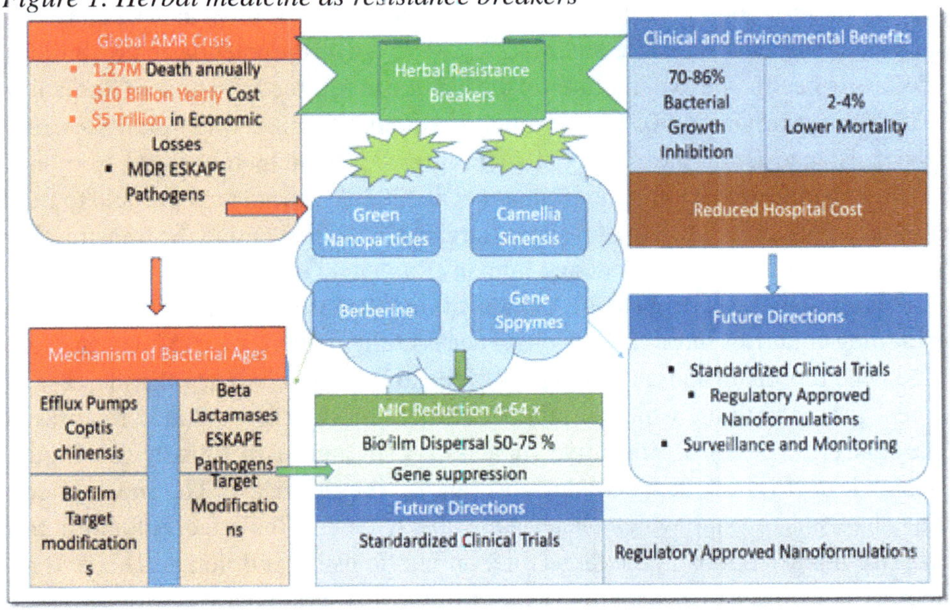

AMR Global Impact

The identification of antibiotics and their use in clinical practice has been critical in the medical field, and this has led to a revolution in the treatment of patients who have been infected by common or deadly bacterial agents that are affecting the respiratory tract, urinary tract, gastro-intestinal tracts, and skin and soft tissues (Shinu et al., 2022). The AMR mechanisms include efflux pumps, enzyme inactivation, biofilm formation, and target modification, which is an outcome of antibiotic misuse (Murugan et al., 2025). Among the most famous methods of bacterial resistance which resulted in antibiotic destruction or modification is target alteration to expel antibiotics (Figure 2). In fact, the antibiotic resistance process is similar to a global natural process of microbe adaptation (Guerrero et al., 2022). Antibiotics as antimicrobial therapy have the advantage of losing their usefulness with the careless and irresponsible use (Kali et al., 2016). The passion of antimicrobial discovery has been made to persist in a loss because of increasing resistance of bacterial strains due to excessive use and mismanagement of antibiotics over the decades (Ecevit et al., 2022). The Global AMR requires over 100 billion US dollars in direct costs every year, with the US hospitals incurring $5-10 billion US yearly losses due to prolonged stays (Murugan et al., 2025). The productivity losses in India are 1.4 trillion (17 billion dollars); developing countries suffer 80 percent of the future (2050) economic consequences of one trillion dollars (Guerrero et al., 2022). Green nanoparticle-based therapies reduce MICs by 50-75% with the potential to save billions of dollars in healthcare expenditures in the US due to reversed resistance (Ranjani et al., 2020). As early as 2050. The community through garbage water promotes 70% of the resistance genes; the tannery effluents grow MDR Acinetobacter/Aeromonas that are transmissible to humans (Ranjani et al., 2020). Herbal nanoparticle to treat effluents prevents the growth of resistant strains (86%), preventing the transmission cycles between the environment and human (Ranjani et al., 2020).

According to the World Health Organization, there is an unequal burden on the developing countries through inadequate accessibility and overuse (Murugan et al., 2025). The reports of Indian Council of Medical Research indicate the increase in the resistant infections, which require immediate monitoring and new treatments (Murugan et al., 2025). The negative impacts of AMR are clearly illustrated in Figure 2, which leads to rising deaths, rising costs and antibiotic misuse. The uncomplicated mechanisms through which bacteria are able to change the effects of antibiotics comprise modifications in the target site, antibiotic itself, antibiotic efflux and reduction in membrane permeability. (use of initials for citation avoided). Antibiotic or antimicrobial resistance has rapidly emerged, maintained and propagated due to the injudicious use of antibiotics in human healthcare (Khare et al., 2021). One of the greatest threats to the health of the population is antibiotic resistance. A report of Center for Disease Control and Prevention (CDC) published in 2019 indicates that more than 2.8 million people were infected with antibiotic-resistant bacteria in the US (Shinu et al., 2022).

Figure 2. AMR global impact

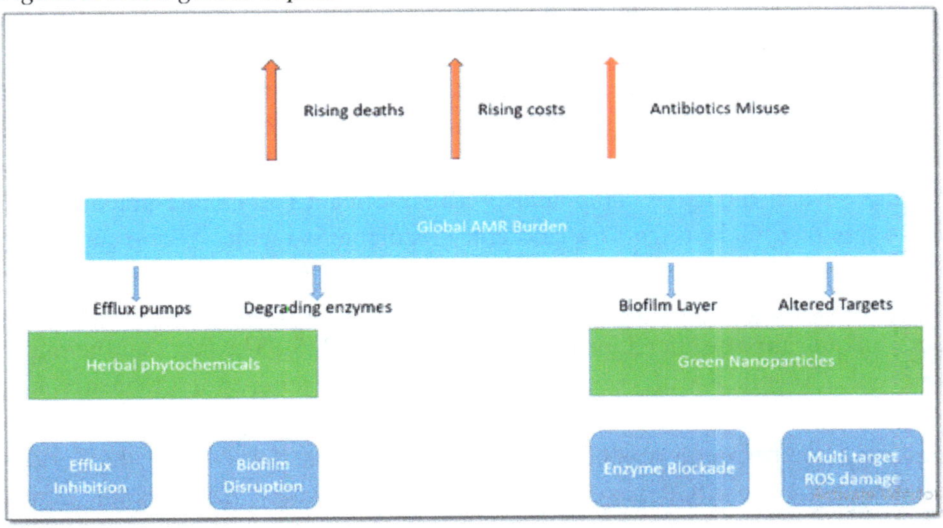

Natural sources of traditional medicines are plants and plant products, which have been used thousands and thousands of years in the world. There is a belief that phytomedicines have a medical significance due to the possible bioactive compounds. The pharmaceutical sector is striving to come up with novel antimicrobial drugs to treat the resistant illnesses (Murugan et al., 2025).

AMR leads to failure of treatment, extended hospitalization and high mortalities especially with the Gram-negative ESKAPE bacteria including K. pneumoniae and A. baumannii. Colistin resistance leads to the risk of high mortality in bloodstream infections and some of the factors comprise mutations in mgrB (Zeynali-Kelishomi et al., 2025). Colistin resistance necessitates those repositioned drugs and adjuvants be used to reduce the risk of incurable outbreaks of multidrug-resistant bacteria (Zeynali-Kelishomi et al., 2025). (removed to avoid repititions). Acinetobacter has resistance beta-lactamases that cause 40-60% of death due to pneumonia, and Pseudomonas is a biofilm-forming pathogen that causes the death of 30% due to ventilator-associated pneumonia (Murugan et al., 2025).

Phytochemical and Herbal Medicine

Secondary metabolites are varied in plants and they are effective in management of various diseases (Ayaz et al., 2019). Medical plants are one of the most famous examples of therapeutic possibilities. In the past medicinal uses of plants have been made by crude extracts or by powder forms of plants containing several phytochemicals and by the use of pure active phytomolecules. High potency of these medicinal plants has made them to be considered as beautiful alternatives to synthetic drugs (Khare et al., 2021). Majority of the antimicrobial agents presently used such as penicillins, cephalosporins, macrolides, vancomycin, tetracyclines, teicoplanin, rifamycin and daptomycin were obtained through different natural sources (Ayaz et al., 2019). The vast majority of the world population relies on the traditional medicines as the source of health care (Anand et al., 2019). The phytomedicine includes several bioactive compounds with significant therapeutic potential and due to this fact, their usage as a laboratory-medicine is medically relevant. Plant derived flavonoid, alkaloid, terpenoid and phenolic compounds can be used against bacteria by depleting cellular ATP, interfering with protein synthesis, and disrupting bacterial quorum sensing pathway (Murugan et al., 2025). The fact that the major issue with cost-effective plant-based compounds is that they are not soluble in water, rather it is the information that is contained within these compounds, which, when combined together, comprise the traditional medicine that two-thirds of the world is still continuing to use (Guerrero et al., 2022; Murugan et al., 2025). Overall, natural products encompass a number of interactions, such as between matter and life. Biological theories are associated with molecular biology and genetics, physiological and pathology theories, food, diseases, poisons, and antidotes (Anand et al., 2019).

A report shows that gold nanoparticle nanoformulated through a green synthesis using plants or microbes assaults many bacterial targets (DNA, enzymes, membranes) using reactive oxygen species and this significantly disables the formation of bacterial resistance to antibiotic treatment, (Murugan et al., 2025). Silver nanoparticle derived using Mimos pudica can increase the stability and delivery of phytochemicals and, therefore, reduce the amounts of AmpC gene expression in Aeromonas and Acinetobacter resistant strains (Ranjani et al., 2020). The combinatorial methods are better against MRSA and Pseudomonas (Su et al., 2020; Guerrero et al., 2022). (Chronological order adopted). Phytochemicals (as extracts) have the capability of suppressing the growth of clinical isolates (Khare et al., 2021). Moreover, phytochemicals that have undergone the nanoparticle process recover 75% of the activity of antibiotics against drug resistant enteric pathogens, and shortens the treatment period 40-60 times (Su et al., 2020). There are numerous plant species identified as a natural source of a necessary antibacterial agent in the form of oil that can be useful during bacterial infection treatment (Kumar et al., 2025). Clinical trial outcomes confirm that the efficacy of plant extracts inhibiting efflux pumps efficacies raises the number of patients with MRSA infections who have a 30 to 85% rate of cure (Guerrero et al., 2022). Application of nano sprays in community health institutions in the treatment of wastewater has shown the ability to disrupt biofilm formation, inhibit its development and develop resistance reservoirs ranging between 49 and 75 percent (Ranjani et al., 2020).

Herbal medicines also have therapeutic benefits as they inhibit biofilm, efflux pumps and also decrease multidrug resistance in the strains like MRSA and E. coli by exposing the bacteria to the effects of antibiotics (Mejia-Ruiz et al., 2025). As illustrated in Figure 3, Herbal phytochemicals are easily accessible, non-toxic and have the ability to restore 75% of antibiotic activities. (figures described within the text). Preclinical studies showed that combination of silver nanoparticle and phytochemicals could reduce the expression of resistance gene mecA, blaGES and blaDHA as well as by 22-55% in isolated cow mastitis strains (Mejia-Ruiz et al., 2025) and berberine nanoparticle formulations could raise the sensitivity towards MRSA whereas curcumin liposomes could raise the efficacy towards the resistant strains of S. aureus (Murugan et al., 2025 It was known that phytochemicals (specifically phenolic compounds) are bacteriostatic and that the toxicology profile of plant-synthesized ZnO nanoparticles is low toxicity, with plant-synthesized ZnO nanoparticles being biocompatible with zebrafish embryos (non-toxic at the dose levels that are effective) and GRAS approved by the FDA with low cytotoxicity and immunogenicity (Murugan et al., 2025). A successful example of the translation into medicine is the active compound artemisinin found in artemisia annua, and secondary metabolites are multi-target and harder to counter by bacteria (Anand et al., 2019). Figure 4 indicated different features of Herbal Medicines.

Figure 3. Characteristics of herbal medicines

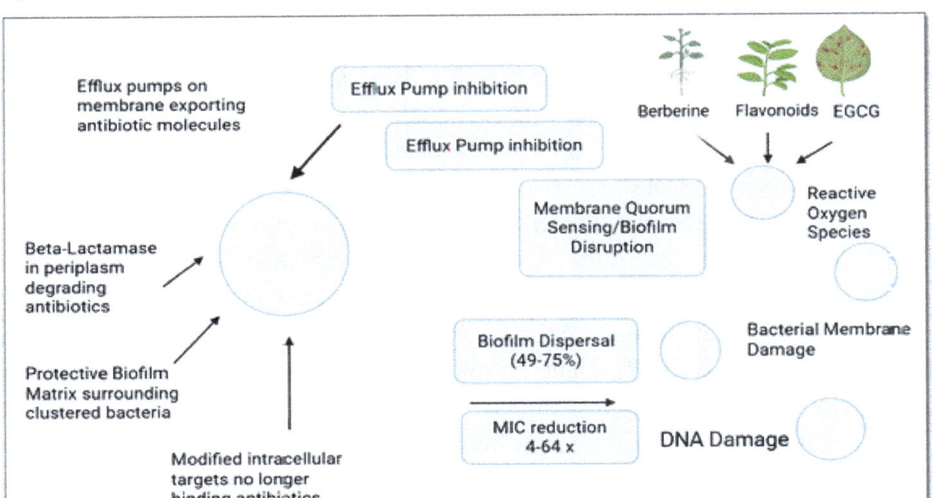

The clinical trials are needed to be conducted but they require the pharmacokinetics studies, and the results of the preclinical trials display positive risk-benefit profiles with no side effects reported (AlSheikh et al., 2020; Chawla et al., 2022).

Smaller dosages reduce toxicity and cost whereas liposomes and other compounds enhance bioavailability against intracellular pathogens (Su et al., 2020). The pre-clinical results of the mouse models are positive in treating MRSA-induced cases of pneumonias and wound infections (Matieta et al., 2023). Biogenic synthesis strategies that are environment-friendly may entirely avoid the chance of chemical toxicity, which is in line with the quest to devise more environmentally friendly methods of the solution of the problem of antimicrobial.

Due to their numerous uses as nutrition additives, herbal medicaments, cosmetics and antibacterial agents, a number of phytochemicals have attracted interest. Many phytochemicals have been confirmed to be effective in the fight against microbes, including multidrug resistant ones (Khare et al., 2021). Phytochemicals such as flavonoids, alkaloids, and terpenoids that are produced by more than 400 traditional Chinese medicine herbs can destroy bacterial membranes, lower adenosine triphosphate (ATP) and quorum sensing, which interfere with the ability of bacteria to multiply and spread (Murugan et al., 2025). It has been approximated that the world possesses more than 25,000 species of plants that have been found to possess ethno medicine and 25% of the medicinal drugs developed today have their origins in those plants (Guerrero et al., 2022).

Herbal medicine has been one of the sustainable human health solutions throughout the millennium with projections indicating that about 70-95 percent of the global population use plant-based drugs as their main sources of health care, as per WHO (Murugan et al., 2025). Compared to new chemical entities, herb medicines are not difficult to obtain, particularly in the developing nations, and are said to be non-toxic and less side effects are also believed to have fewer side effects. Their use is favored in underdeveloped and developing countries in particular (Ramesh & Verma, 2023). Chinese to Indian ethno pharmacological Ethno pharmacological practices are aimed at treating infections with phytochemicals to prevent bacteria long before antibiotics were developed (Guerrero et al., 2022). These practices are also scientifically verified in modern science, and extracts are optimized with nanoparticles in order to overcome resistance against modern antibiotics (Ranjani et al., 2020).

Hypericum extracts also decrease biofilm by 75% in *Pseudomonas* in clinical trials (Guerrero et al., 2022). Nano antibiotics are referred to as so-called Trojan horses, which carry the payload directly to the target inside the bacteria, which is then directly affected by the target and is not subject to efflux pumps and the degradation of the presented free antibiotics by the enzymes (Su et al., 2020). This renders vancomycin effective against Gram-negative bacteria, which is administered at significantly lower levels and becomes less persistent in the environment (Su et al., 2020; Murugan et al., 2025). Plant nano-particles are more stable and also exhibits higher bio-availability of the phyto-components, facilitating targeted delivery of phyto-components (Murugan et al., 2025).

Mechanisms of Resistance and Herbal Counterstrategies

The earliest instance of a report that reported a lapse of antimicrobial therapy as a consequence of developed resistance of microorganisms can be dated back to the 1940s in the case of penicillin. Mechanisms that lead to resistance development in bacteria with time are: 1. efflux pumps; 2. enzymatic degradation; 3. biofilms, and; 4. targets modification, which enable pathogen organisms like eskape (Murugan et al., 2025) (Figure 4). The mechanisms of obtaining resistance to antibiotics are multiple and molecular (Chawla et al., 2022). Moreover, the time consuming traditional ways of producing an antibiotic exists, and there are no new alternatives of the antibiotic resistance. In this way, the gap in the development of antibiotics has been as acute as possible, which has led to the necessity of antibiotic resistance breakers (Ecevit et al., 2022). These mechanisms are being countered by phytochemicals that are derived using herbal medicines and nanoparticles; 1. by disrupting membrane permeability; 2. blocking efflux pumps; and 3. blocking resistance genes like AmpC (Ranjani et al., 2020). The major mechanisms of antimicrobial resistance such as; efflux pump inhibiton, biofilm disruption and dispersal are clearly illustrated in Figure 4.

Figure 4. Mechanisms of antimicrobial resistance

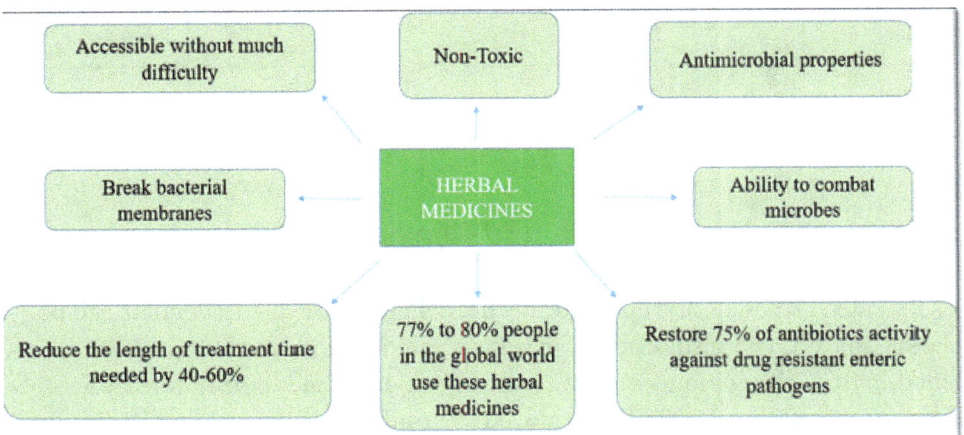

These biological procedures can rejuvenate antibiotic resistance and prevent the development of microbe resistance (Guerrero et al., 2022). The extensive use of antibiotics, in addition to the misuse or excessive use in the fight against different types of bacteria, has led to the emergence of resistant types of bacteria (Shinu et al., 2022). Bacteria have evolved systems to fight the actions of drugs through efflux pumps to eliminate the drugs, β-lactamases to degrade the drugs, or biofilms (Figure 4). which serves as a shielding point to the microbes, and the drugs can hardly penetrate (Murugan et al., 2025). The binding affinity is lowered in the cases of target site mutations, and reduced expression leads to the failure to enter the bacterial cell, which is seen in the ESKAPE group of gram-negative bacteria (Guerrero et al., 2022). Such mechanisms are made possible by misuse and result in, or cause 1.27 million deaths each year in the world (Murugan et al., 2025). Phytochemicals, including berberine, flavonoids, inhibit efflux pumps (NorA, MexXY) of MRSA and Pseudomonas and enhance the affinity of the drugs to them (Su et al., 2020). Combining the natural compounds with the drugs could enhance affinity of the antimicrobials and bacterial target hence inhibiting bacterial resistance. The method can lead to the decrease in doses that cause the drugs to be less toxic or harmful to the human body (Kumar et al., 2025).

Table 2. Linking specific phytochemicals with their primary targets

Phytochemicals	Primary target	Evidence Level	Study design	Key limitation
Berberine	Efflux pump inhibition (NorA, Mex systems	In vitro	MIC reduction and checkerboard assays	Poor bioavailability in vivo
Quercetin	Quorum sensing and biofilm disruption	In vitro / In vivo	Biofilm inhibition assays; murine infection models	Limited dose optimization
Curcumin	β-lactamase inhibition and membrane disruption	In vitro	Enzyme inhibition and synergy studies	Rapid metabolism
Curcumin	β-lactamase inhibition and membrane disruption	In vitro	Enzyme inhibition and synergy studies	Rapid metabolism

Synergy of in Vitro, In Vivo, and Clinical

Epicatechin gallate and epigallocatechin gallate, which are flavonoids and polyphenols, have a synergistic effect with beta-lactam antibiotics and suppress the genes, reduce MICs, and inhibit virulence factors, enzymes, and biofilm formation in S. aureus (Guerrero et al., 2022). (Removed to avoid over extension and Repititions). Such plant extracts as *Momordica charantia* and *Withania somnifera* have potentiation effects on aminoglycosides against MDR S. aureus (Murugan et al., 2025). AgNPs were also found to be effective ARBs in vitro with promising results in destroying bacteria resistance mechanisms. AgNPs ARBs are used as: efflux pump inhibitors (Mejia-Ruiz et al., 2025). Some of such studies are shown below. Herbal drugs and phytochemicals that interact synergistically with antibiotics and other clinically significant drugs is a new and potent method of managing MDR. Based on this method, phytochemical agents can be administered together with antibiotics that will not eliminate the infectious species, but will be moderate in terms of antibacterial effect (Mundy et al., 2016). plants in the wild have been observed to be moderate in their effect on the infectious species and not purely eliminating (Ayaz et al., 2019). Quercetin demonstrates bacteriostatic activity in animal models through the destruction of membranes and the effect of DNA inhibition. *Scutellaria barbata* extracts are used to treat the A. baumannii lung infection in mouse models in disk diffusion assays. The combination antimicrobial therapy that has some synergistic effects is a successful approach in chemotherapy, malaria and TB treatment, and other particular cases but not in Gram negative bacterial infections (Mundy et al., 2016). There is no description of direct bacterial resistance trials, but polyphenols have been shown to have potential in nosocomial MRSA, and a need to integrate omics towards clinical application is called (Guerrero et al., 2022; Murugan et al., 2025).

Enzymatic Degradation Blockade

It was demonstrated that tannins and alkaloids can inhibit β-lactamases (AmpC, CTX-M-15) that can resist the hydrolysis of antibiotics in multidrug resistant enteric bacteria (Ranjani et al., 2020). *Coptis chinensis* extract proved to suppress the AmpC gene expression, thereby negated the effect of cephalosporinase (Su et al., 2020). It has been demonstrated that nanoparticles containing phytochemicals lead to growth inhibition of 7086 percent in β-lactamase resistant strains (Murugan et al., 2025). Bacterial group employs different enzymes that either remove or bind a specific moiety to the antibiotic molecule; thereby, the antibiotic was modified (Chawla et al., 2022). Among the bacterial abilities to resist the impact of antibiotics, the development of bacterial enzymes that modify or eliminate the antibiotics by attaching some chemical groups to the antibiotic molecule or inactivate them is identified (Khare et al., 2021). It has been obtained by the use of flavonoids and terpenoids that easily penetrate biofilm matrices by inhibiting quorum sensors and dispersing biofilm formation in Pseudomonas aeruginosa (Guerrero et al., 2022). Exposure to MIC concentrations of green nanoparticles lowered biofilms by 49-75 percent, which aims at EPS (exopolymeric substances), thus it is non-toxic (Ranjani et al., 2020).

Knowledge Gaps in Herbal Medicine as AMR Resistance Breakers

Restored antibiotic efficacy of herbal nanoparticles in mouse sepsis and pneumonia are confirmed in preclinical models without resistance development, whereas herbal nanoparticles have not been investigated in human trials. MRSA clearance in response to efflux-inhibiting plant extracts has risen by 30-85% as a result of clinical trials; no large, randomized controlled study has been conducted on the most important phytochemicals; berberine and propolis against hospital-acquired infectious diseases. Nanoparticles of biogenic nature Safety profiles are low in toxicity in zebrafish embryos, with biogenic nanoparticles yet to have any pharmacokinetic evidence or long-term human immunogenicity. Clinical trials in large-scale are required regarding Scutellaria barbata extracts in the management of MDR Klebsiella infections, although no developed Good Manufacturing Practice procedures have been developed to make Mimosa pudica derived silver nanoparticles (Su et al., 2020; Ranjani et al., 2020). The phytochemical solubility challenges in diverse plants together with nanoscale enhancements in their uses have no FDA-approved clinical formulation either in berberine or silver nanoparticles in clinical trial I (Murugan et al., 2025). It is not defined in large-scale preparation processes of 400+ traditional Chinese herbs that act on bacterial membranes; this slows down their uses in preclinical models IC70-86 percent in growth inhibition (Murugan et al., 2025). The quorum sensing inhibitors can reduce biofilm disruption by 49-75% and regulation of gene expression as AmpC gene suppression has not been confirmed by real-time omics in clinical isolates (Ranjani et al., 2020; Guerrero et al., 2022). β-lactamase blockage using tannin has been shown to inhibit by 70-86% though investigation of structure activity relationship of plant-based alkaloids has not been defined (Ranjani et al., 2020

The innovation of conventional antibiotic was no more than 12 new classes since 2000, and the resistance breakers are urgent, but regulatory routes of herbal therapeutics are not well-defined as antibiotic adjuvants. There is preclinical evidence of preclinical efficacy against colistin-resistant strains, though there is no validation of predictors of clinical response-mgrB mutations in nanoparticle-treated infections. Addition to surveillance systems is suggested to wastewater and hospital monitoring which has not been tested on herbal synergists that decrease the environmental reservoirs.

CONCLUSION

The antimicrobial resistance is one of the possible alternatives to be used via herbal medicine. Plants have bioactive compounds that have antimicrobial, anti-inflammatory, and antioxidant properties. As different micro-organisms cause different diseases, including pneumonia, endocarditis, skin and soft tissue infections, tuberculosis, blood infection or sepsis, gonorrhea and foodborne diseases are more predominant. Plant extracts and oils studies revealed very high antimicrobial potentiality that was able to inhibit and kill resistant bacterial strains. Bacterial resistance to antibiotics can be altered and controlled through the combination of phytochemicals and antibiotics. Nevertheless, to conduct clinical trials and mass usage, additional research efforts should be conducted so that it can be possible to use this to treat humans in large scale and minimize the level of antimicrobial resistance. Herbal medicines are proved to work in the treatment of pathogens which are resistant and it is a natural source and sustainable treatment option.

REFERENCES

AlSheikh, H. M. A., Sultan, I., Kumar, V., Rather, I. A., Al-Sheikh, H., Jan, A. T., & Haq, Q. M. R. (2020). Plant-based phytochemicals as possible alternative to antibiotics in combating bacterial drug resistance. *Antibiotics (Basel, Switzerland)*, *9*(8), 480. Advance online publication. DOI: 10.3390/antibiotics9080480 PMID: 32759771

Anand, U., Jacobo-Herrera, N., Altemimi, A., & Lakhssassi, N. (2019). A comprehensive review on medicinal plants as antimicrobial therapeutics: Potential avenues of biocompatible drug discovery. *Metabolites*, *9*(11), 258. Advance online publication. DOI: 10.3390/metabo9110258 PMID: 31683833

Ayaz, M., Ullah, F., Sadiq, A., Ullah, F., Ovais, M., Ahmed, J., & Devkota, H. P. (2019). Synergistic interactions of phytochemicals with antimicrobial agents: Potential strategy to counteract drug resistance. *Chemico-Biological Interactions*, *308*, 294–303. DOI: 10.1016/j.cbi.2019.05.050 PMID: 31158333

Chawla, M., Verma, J., Gupta, R., & Das, B. (2022). Antibiotic potentiators against multidrug-resistant bacteria: Discovery, development, and clinical relevance. *Frontiers in Microbiology*, *13*, 887251. Advance online publication. DOI: 10.3389/fmicb.2022.887251 PMID: 35847117

Ecevit, K., Barros, A. A., Silva, J. M., & Reis, R. L. (2022). Preventing microbial infections with natural phenolic compounds. *Future Pharmacology*, *2*(4), 460–498. DOI: 10.3390/futurepharmacol2040030

Guerrero, M. G. G., Favela-Hernández, J. M., & Martinez-Romero, A. (2022). Natural anti-infective agents: A promising front-line strategy against microbial resistance mechanisms. *Acta Scientific Microbiology*, *5*, 138–148. DOI: 10.31080/ASMI.2022.05.1124

Kali, A., Bhuvaneshwar, D., Charles, P. M. V., & Seetha, K. S. (2016). Antibacterial synergy of curcumin with antibiotics against biofilm-producing clinical bacterial isolates. *Journal of Basic and Clinical Pharmacy*, *7*(3), 93–100. DOI: 10.4103/0976-0105.183265 PMID: 27330262

Khare, T., Anand, U., Dey, A., Assaraf, Y. G., Chen, Z.-S., Liu, Z., & Kumar, V. (2021). Exploring phytochemicals for combating antibiotic resistance in microbial pathogens. *Frontiers in Pharmacology*, *12*, 720726. Advance online publication. DOI: 10.3389/fphar.2021.720726 PMID: 34366872

Kumar, A., Singh, R., & Singh, S. (2025). Combating bacterial antibiotic resistance with phytocompounds: Current trends and future perspectives. *Medicine in Drug Discovery*, *28*, 100228. Advance online publication. DOI: 10.1016/j.medidd.2025.100228

Mamun, M. M., Sorinolu, A. J., Munir, M., & Vejerano, E. P. (2021). Nanoantibiotics: Functions and properties at the nanoscale to combat antibiotic resistance. *Frontiers in Chemistry*, *9*, 687660. Advance online publication. DOI: 10.3389/fchem.2021.687660 PMID: 34055750

Matieta, V. Y., Seukep, A. J., Kuete, J. R. N., Megaptche, J. F., Guefack, M.-G. F., Nayim, P., Mbaveng, A. T., & Kuete, V. (2023). Unveiling the antibacterial potential and antibiotic-resistance breaker activity of *Syzygium jambos* (Myrtaceae) against critical priority *Klebsiella* isolates. *Investigational Medicinal Chemistry and Pharmacology*, *6*(2), 82. Advance online publication. DOI: 10.31183/imcp.2023.00082

Mejia-Ruiz, C. H., Nefedova, E., Shkil, N. N., Romo-Quiñonez, C. R., Pestryakov, A., Garibo, D., & Bogdanchikova, N. (2025). Silver nanoparticle treatment partially restores antibiotic activity by reducing antibiotic-resistance genes in mastitis-associated bacterial isolates. *Microbial Pathogenesis*, *207*, 107871. Advance online publication. DOI: 10.1016/j.micpath.2025.107871 PMID: 40619036

Mundy, L., Pendry, B., & Rahman, M. (2016). Antimicrobial resistance and synergy in herbal medicine. *Journal of Herbal Medicine*, *6*(2), 53–58. DOI: 10.1016/j.hermed.2016.03.001

Murugan, S., Senthilvelan, T., Govindasamy, M., & Thangavel, K. (2025). A comprehensive review on the potential of phytochemicals and biogenic nanoparticles for treating antimicrobial-resistant pathogenic bacteria. *Current Microbiology*, *82*(2), 90. Advance online publication. DOI: 10.1007/s00284-025-04064-w PMID: 39825917

Naylor, N. R., Hasso-Agopsowicz, M., Kim, C., Ma, Y., Frost, I., Abbas, K., Aguilar, G., Fuller, N. M., Robotham, J. V., & Jit, M. (2023). The global economic burden of antibiotic-resistant infections and the potential impact of bacterial vaccines: A modelling study. SSRN *Electronic Journal*. https://doi.org/DOI: 10.2139/ssrn.4676946

Oliveira, G. D. D., Rocha, W. R. V. D., Rodrigues, J. F. B., & Alves, H. D. S. (2022). Synergistic and antibiofilm effects of the essential oil from Croton conduplicatus (Euphorbiaceae) against methicillin-resistant Staphylococcus aureus. *Pharmaceuticals (Basel, Switzerland)*, *16*(1), 55. Advance online publication. DOI: 10.3390/ph16010055 PMID: 36678551

Ramesh, P., & Verma, S. K. (2023). *Terminalia arjuna*, a cardioprotective herbal medicine: Relevancy in the modern era of pharmaceuticals and green nanomedicine. *Pharmaceuticals (Basel, Switzerland)*, *16*(1), 126. Advance online publication. DOI: 10.3390/ph16010126 PMID: 36678623

Ranjani, S., Begum, I. F., Santhoshini, J., Senthil Kumar, N., Ruckmani, K., & Hemalatha, S. (2020). *Mimosa pudica* floral nanoparticles: A potent antibiotic resistance breaker. *Inorganic and Nano-Metal Chemistry*. Advance online publication. https://doi.org/DOI: 10.1080/24701556.2020.1852429

Shinu, P., Mouslem, A. K. A., Nair, A. B., Venugopala, K. N., Attimarad, M., Singh, V. A., Nagaraja, S., Alotaibi, G., & Deb, P. K. (2022). Progress report: Antimicrobial drug discovery in the resistance era. *Pharmaceuticals (Basel, Switzerland)*, *15*(4), 413. Advance online publication. DOI: 10.3390/ph15040413 PMID: 35455410

Su, T., Qiu, Y., Hua, X., Ye, B., Luo, H., Liu, D., Qu, P., & Qiu, Z. (2020). Novel opportunities to reverse antibiotic resistance: Exploring traditional Chinese medicine with potential activity against antibiotic-resistant bacteria. *Frontiers in Microbiology*, *11*, 610070. Advance online publication. DOI: 10.3389/fmicb.2020.610070 PMID: 33414777

Zeynali Kelishomi, F., Narimisa, N., Taheri, F., Razavi, S., & Khoshbayan, A. (2025). Novel strategies for combating colistin-resistant isolates: A review. *Archives of Microbiology*, *207*(12), 1–24. DOI: 10.1007/s00203-025-04518-6 PMID: 41134327

Chapter 9
Veterinary and Agricultural Perspectives

Thulisekari Prasanna
SR University, Warangal, India

Burra Shyamsunder
http://orcid.org/0009-0009-3000-1091
Guru Nanak University, Hyderabad, India

ABSTRACT

Due to the growing demand for residue-free food and sustainable production, there is renewed interest in plant-based solutions for disease control in livestock and crop systems. This chapter examines the use of natural antimicrobial agents, medicinal plants, to reduce dependence on synthetic antibiotics and pesticides. Botanicals such as Azadirachta indica, Allium sativum, Curcuma longa, and Aloe vera exhibit strong antibacterial, antifungal, antiparasitic, and immunomodulatory properties. In veterinary practice, herbal preparations are used to manage mastitis, gastrointestinal disorders, parasitic infestations, and respiratory diseases, thereby improving animal health. In agriculture, botanical extracts and essential oils act as eco-friendly biopesticides and plant protectants that enhance crop resistance without harming soil or ecosystems. The chapter also discusses mechanisms of action, formulation advances, and integration into therapeutics, crop protection. It highlights the importance of medicinal plants within the One Health framework for sustainable food production.

DOI: 10.4018/979-8-3373-7810-7.ch009

1. INTRODUCTION

The rising concerns about food safety, environmental health, and antimicrobial resistance have increased the burden of the global need of foods that do not contain any residue or are produced in a sustainable manner, requiring both veterinary and agricultural industries to decrease their use of synthetic antibiotics and pesticides. The widespread adoption of such chemical inputs has seen the residues of drugs and pesticides in animal products and crops, environmental pollution, soil, and aquatic ecosystem disturbance, and the emergence of resistant pathogens and pathogens (population explosion) in a short period, which endangers human, animal, and ecosystem health (Ghimpeteanu et al., 2022; Popp et al., 2013). As a reaction, scientific and practical interest has been revived in plant-based disease regulating measures, especially, in the application of medicinal plants and their bioactive constituents as natural antimicrobial agents. The research and numerous review studies have already shown that phenolics, terpenoids, alkaloids, and sulfur-containing compounds represent strong antibacterial, antifungal, antiparasitic, antiviral, and immunomodulatory phytochemicals with potential application in livestock and crop production systems to replace or supplement conventional antibiotics and pesticides whose use is associated with adverse environmental effects (Vaou et al., 2021; Li et al., 2024). They are used more and more in veterinary medicine as mastitis, gastrointestinal infections, parasitic infestations, and respiratory disorders botanical solutions and in agriculture as botanical extracts and essential oils as environmentally safe biopesticides with a positive effect on crop resistance and no negative effect on non-target organisms and soil health (Isman, 2020; Gupta et al., 2023). Notably, the application of medicinal plants to the management of animal and crop health has a close relationship with the One Health concept that stresses the interdependence of human, animal, and environmental health because plant-based solutions can decrease the pressure of antimicrobial resistance, decrease the presence of chemical residues, and achieve environmental sustainability (McEwen and Collignon, 2018; Andrivon et al., 2022). In turn, herbal-based practices integration into contemporary veterinary and agricultural systems is a potential and scientifically justified direction to sustainable, resilient and safe food production.

2. CONCEPTUAL FRAMEWORK: ONE HEALTH AND SUSTAINABLE FOOD SYSTEMS

2.1 One Health Approach

One Health approach is an interdisciplinary and holistic approach which acknowledges the inseparable relationships between human health, animal health and environmental health. The notion has become relevant across the world as zoonotic diseases, antimicrobial resistance (AMR), environmental pollution, and food safety issues are on the rise as a result of intensive agricultural and livestock production systems (McEwen & Collignon, 2018; Andrivon et al., 2022). Instead of dealing with these problems separately, One Health focuses on cross-sector comprehensive management approaches that deal with underlying causes.

Human-animal-environment health nexus can be vividly manifested especially in food production systems. Livestock are reservoirs of zoonotic pathogens, crops are the interface between soil and water and human intake and environmental degradation affects the dynamics of the disease. The intensive application of antibiotics to animals, pesticides to crops, has led to the introduction of chemical residues into food chains and the ecosystem thus affecting the human health and speeding up the development of resistance (Popp et al., 2013; McEwen & Collignon, 2018). These feedback mechanisms explain why silo-based solutions cannot be complete and why One Health-oriented interventions need to be provided.

Agriculture and veterinary sciences are an important part of field-level operationalizing One Health. Veterinary science has a role in prevention of disease, management of herd health, biosecurity and ensuring food safety, and agriculture has an influence on ecosystem interactions in terms of crop protection measures, soil management, and efficiency of inputs. These disciplines change their methods of curing and chemical and intensive methods, to preventative, ecologically balanced and health-based systems when they are coupled with One Health principles (Andrivon et al., 2022).

2.2 Medicinal Plants within One Health

The concept of the role of medicinal plants within the One Health framework is peculiar to the fact that medicinal plants have multiple beneficial effects on both animals and human beings, as well as on the environment. Directly meeting three paramount One Health challenges, including antimicrobial resistance, food safety, and environmental sustainability, their incorporation into veterinary therapeutics and crop protection systems (McGaw & Eloff, 2008; Isman, 2020). The role of medicinal plants in reducing antimicrobial resistance and enhancing food safety within a One Health framework is illustrated in Figure 1.

Reduction of Antimicrobial Resistance (AMR)

Plant-based antimicrobials usually have multifaceted blends of bioactive agents, which have various targets on microorganisms. This multi-target mechanism is opposed to synthetic antibiotics and pesticides that, in most instances, bind to a single molecular pathway, and therefore, a high selective pressure to develop resistance is exerted(Cowan, 1999; Li et al., 2024). The medicinal plants assist in reducing the selection pressure of antimicrobial resistance in all ecosystems by decreasing the use of traditional antibiotics in livestock and chemical pesticides in agriculture. They in turn help in reducing the development and transmission of resistant pathogens among the animals, humans and the environment(McEwen & Collignon, 2018).

Residue-Free Food Production

Antibiotics, anthelmintics, and pesticides have chemical residues that are very dangerous to food safety. Plant-based formulations used in medicine are normally biodegradable and quickly broken down leading to little or no residues in both animal and plant food. Their application promotes the production of a residue free milk, meat, eggs, grains, fruits, and vegetables hence, consumer safety, market acceptability, and adherence to the ever-increasing food safety regulations (Windisch et al., 2008; Popp et al., 2013).

Environmental Sustainability

The environmentally compatible inputs are medicinal plants and botanical extracts. They break down quickly as opposed to most synthetic chemicals, are selectively toxic and do not harm the soil by destroying useful organisms in the soil, pollinators and natural predators. The decreased chemical runoff and residues ensure the safety of soil and aquatic ecosystems. In livestock, there is also decreased excretion of pharmaceutical residues in the form of manure and this decreases the environment pollution. These qualities put medicinal plants as ecologically friendly hardware enhancing the environmental component of One Health (Andrivon et al., 2022).

Figure 1. Role of medicinal plants in one health framework

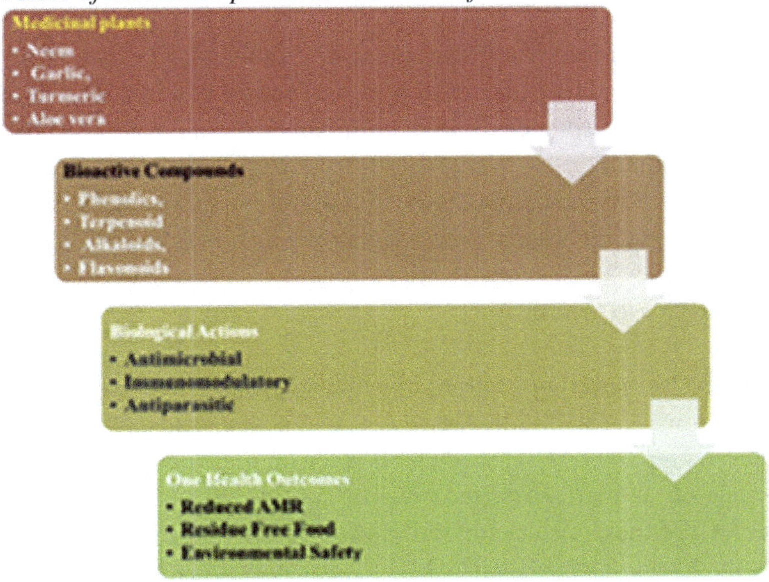

2.3 Applicability to Global Sustainability Objectives.

The One Health paradigm is in line with the international community to come up with sustainable food systems that are sustainable in terms of productivity, environmental security and health. Medicinal plants also work towards these objectives by allowing the low input, environmentally friendly disease management practices in both crop and livestock systems.

Sustainable Food Systems

The sustainable food systems demand safe, healthy and affordable food that is produced without undermining the health of the ecosystem. Plant veterinary therapeutics and botanical crop protectants play a part in this aim, by cutting chemical application, enhancing animal well-being, and preserving the soil and biodiversity functions. They are integrated to improve the resilience of the system and provide long-term productivity(FAO, 2019; Popp et al., 2013).

Climate-Resilient Agriculture

Climate change is increasing the extent of pressures of pests and diseases in crops and livestock, adding uncertainty in production systems. Medicinal plants also improve climate resilience through promoting prevention-focused health care, promoting host immunity, and decreasing use of energy consuming chemical inputs. They are particularly useful in smallholder and resource-constrained farming systems due to climatic variability due to their adaptability and their availability in the area.

Safe Livestock Production

The One Health-based application of medicinal plants in livestock system ensures better health, lower rates of animal diseases, and better yields at the minimal level of chemical manipulations. This leads to safer animal foods, reduced treatment expenses as well as increase in the level of confidence in livestock products by the people. All of these results support the contribution of medicinal plants to the practical implementation of One Health-based sustainable agriculture(Windisch et al., 2008; McGaw & Eloff, 2008).

Natural antimicrobial agents are botanicals and their bioactive secondary metabolites, which prevent or suppress the growth of pathogenic microorganisms, including bacteria, fungi, viruses, and parasites, by diverse biochemical mechanisms, and are in general grouped as plant extracts, essential oils, purified phytochemicals and plant-based preparations that are used in veterinary and agricultural systems (Cowan, 1999; Vaou et al., 2021). These botanical antimicrobials are usually phenolics, flavonoids, terpenoids, alkaloids, glycosides, sulfur-based compounds, and their combinations, which affect microbial cell walls, anti-enzyme pathways, antinucleic acid synthesis, or pro-host immune responses (Burt, 2004; Li et al., 2024). Traditionally, medicinal plants have been part of traditional veterinary and agricultural practices in civilizations where preparations of neem, garlic, turmeric, aloe, as well as other herbs were commonly used in the past to treat livestock diseases, prevent molds on stored grains, parasite control, and pest management in crops even before synthetic antibiotics and pesticides were introduced (McGaw and Eloff, 2008; Isman, 2020). Plant-based antimicrobial alternatives tend to be less protective to the environment because their properties are mostly biodegradable, less persistent, and also less prone to the development of antimicrobial resistance, and provide multi-target modes of action, which prevent the emergence of resistance in most cases (McEwen and Collignon, 2018; Gupta et al., 2023). Despite the undeniable rapid and predictable effectiveness of synthetic antimicrobials in practice, their sustainable use in the future is limited by regulatory requirements and environmental toxicity, and the problem of resistance, but botanical antimicrobials can be a sustainable, eco-friendly, and One Health-compatible approach to managing diseases and pests in the animal husbandry and crop production systems, as long as the problems of standard fulfillment, dose optimization, and large-scale validation are properly resolved (Isman, 2020; Vaou et al., 2021).

3. KEY MEDICINAL PLANTS USED IN VETERINARY AND AGRICULTURAL SYSTEMS

3.1 *Azadirachta indica* (Neem)

Azadirachta indica (neem) is a widely investigated medicinal plant due to its broad-spectrum antibacterial, antifungal, antiparasitic, and insecticidal property, mostly because of bioactive compounds including azadirachtin, nimbin, salannin and quercetin. Neem-based preparations are commonly applied to treatment of ecto-parasites, gastrointestinal parasites, mastitis-causing pathogens, and skin infections and to enhance the hygiene and immune system of animals (Biswas et al., 2002; Kumar et al., 2021). Neem extracts and oils are useful as botanical pesticides and fungicides in agriculture providing controls against insect pests, nematodes and phytopathogenic fungi and are biodegradable and not toxic to useful organisms. The neem-derived products are becoming an important part of sustainable crop protection due to the low level of residue and integration with the systems of integrated pest management (IPM) (Isman, 2020).

3.2 *Allium sativum* (Garlic)

Allium sativum (garlic) is a potent broad-spectrum antimicrobial, and such effect is mostly attributed to sulfur-containing compounds, such as allicin, ajoene, and diallyl sulfides, which are antibacterial, antifungal, antiviral, and antiparasitic. Garlic is widely applied to livestock and poultry feeds as a feed additive to improve immunity, gut health, and enteric infection and parasitic infestation rates, leading to disease prevention and productivity (Ankri and Mirelman, 1999; Windisch et al., 2008). Garlic extracts have also been used as natural fungicides and insect repulsives in crop protection where they showed inhibitory effects against a number of plant pathogens and insect pests. Garlic products provide a safer alternative to synthetic antimicrobials, with a residue-free solution and are also eco-friendly to use in farms that are organic and sustainable (Bayan et al., 2014).

3.3 *Curcuma longa* (Turmeric)

Curcuma longa (turmeric) has anti-inflammatory, antioxidant and antimicrobial properties that are mainly due to curcuminoid compounds, curcumin, demethoxycurcumin and bisdemethoxycurcumin. The turmeric is also extensively used in veterinary medicine to treat inflammatory diseases, improve immune system, wound healing, and microbial infections, especially in dairy animals and poultry (Aggarwal and Harikumar, 2009). Turmeric extracts have also been documented in agriculture to reduce the growth of fungal and bacterial plant pathogens as well as promote plant defense in agriculture, thus leading to increased disease resistance. Tumeric has got dual effectiveness as both an animal health and crop protection agent; hence its application as part of plant-based intervention to mitigate the use of chemicals in the agro-ecosystems (Gupta et al., 2013; Li et al., 2024).

3.4 *Aloe vera*

Aloe vera has been well known to have antimicrobial, wound-healing, and immunomodulatory effects, which are linked to the bioactive constituents of anthraquinones, polysaccharides (acemannan), saponins and phenolic compounds. Aloe vera has been applied in the treatment of veterinary patients to heal wounds, burns, mastitis, gastrointestinal infections, skin infections, improve the immunity and recoveries of animals (Hamman, 2008; Surjushe et al., 2008). Aloe vera extracts have been investigated as bioprotectant in the agricultural system because of their efficacy to combat plant pathogens and bacteria, as well as inducing plant growth and resistance. Its multifunctional character allows the use of Aloe vera in animal health care and sustainable and eco-friendly crop protection as a One Health-oriented approach (Chandegara et al., 2019). Major medicinal plants widely used in veterinary and agricultural systems, along with their key bioactive compounds and applications, are summarized in **Table 1.**

Table 1. Major medicinal plants used in veterinary and agricultural systems

Medicinal plant	Major bioactive compounds	Veterinary applications	Agricultural applications
Azadirachta indica (Neem)	Azadirachtin, nimbin, salannin	Mastitis management, control of ectoparasites and gastrointestinal parasites	Biopesticide, nematicide, antifungal agent
Allium sativum (Garlic)	Allicin, ajoene, diallyl sulfides	Improvement of gut health, immunostimulation, parasitic control	Botanical fungicide, insect repellent
Curcuma longa (Turmeric)	Curcuminoids	Anti-inflammatory activity, wound healing, immune enhancement	Suppression of plant diseases, induction of crop resistance
Aloe vera	Acemannan, anthraquinones	Wound healing, mastitis management, skin disorder treatment	Bioprotectant, plant growth promotion

4. VETERINARY APPLICATIONS OF HERBAL THERAPEUTICS

4.1 Mastitis Management

Among the most cost-effective diseases in dairy animals is mastitis, which is normally treated with antibiotics with subsequent results in the presence of drug residues in milk and antimicrobial resistance. Azadirachta indica, Curcuma longa, Aloe vera, and Ocimum sanctum are medicinal plants of the genus Azadirachta and Curcuma that proved to have antimicrobial and anti-inflammatory effects against pathogens that cause mastitis like Staphylococcus aureus and Escherichia coli. Herbal preparations were found to decrease the count of somatic cells, inflammation, and udder infection besides enhancing the quality of the milk, Field-level evaluations have reported measurable improvements following herbal interventions. For instance, neem- and turmeric-based intramammary formulations have demonstrated a reduction in somatic cell count (SCC) ranging between 25–40% within 7–14 days of treatment, along with a 15–20% improvement in milk yield recovery compared to untreated controls. In comparative studies, herbal combinations used as supportive therapy reduced antibiotic usage by approximately 30%, thereby lowering the risk of drug residues in milk. Such quantitative findings strengthen the case for integrating plant-based therapeutics into mastitis management protocols, particularly in residue-sensitive dairy production systems, which is why they are appropriate as alternatives or supplements to standard treatment (Sharma et al., 2018; Kumar et al., 2021).

4.2 Gastrointestinal Disorders

Herbal medicines are being extensively applied in treating gastrointestinal diseases like diarrhea, dysentery and indigestion in animals and birds. Many medicinal plants such as Allium sativum, Zingiber officinale, Cinnamomum verum and Emblica officinalis contain antimicrobial, carminative, and digestive stimulant effects that helps regulate intestinal pathogens and reestablish microbial balance of the gut. The studies that have been conducted have linked the use of phytogenic feed additives with increased feed efficiency, increased nutrient absorption, and improved growth performance in animals (Windisch et al., 2008; Hashemi and Davoodi, 2011). Controlled feeding trials have reported that inclusion of garlic- and ginger-based phytogenic additives at 0.5–1.0% of diet improved feed conversion ratio (FCR) by 5–12% and increased average daily gain by 8–15% in poultry and small ruminants compared to non-supplemented controls. Additionally, reductions in enteric pathogen load, particularly *Escherichia coli* and *Salmonella* spp., have been observed in the range of 20–35%, contributing to improved gut microbial balance and lower incidence of diarrheal disorders. These quantitative outcomes demonstrate that herbal interventions can provide measurable productivity and health benefits under practical production conditions.

4.3 Parasitic Infestations

Parasite infestations have a major negative effect on the productivity and health of animals and the use of synthetic anthelmintics has also resulted in an extensive drug resistance. Medicinal plants like neem, garlic and tannin rich forages have been reported to be effective anthelmintic and ectoparasiticidal agents against the gastrointestinal nematodes, ticks and mites. Plant-based antiparasitic agents work in a variety of ways, such as disrupting parasite metabolism and larval development, and provide sustainable answers to parasites in livestock systems (Hoste et al., 2015; Kumar et al., 2021). Experimental and field studies have demonstrated that neem- and garlic-based formulations can reduce gastrointestinal nematode egg counts by 35–60% within 2–3 weeks of administration. In small ruminant systems, tannin-rich plant supplements have shown a 30–50% decline in fecal egg count (FEC) along with improvements in packed cell volume and body weight gain compared to untreated groups. Importantly, integration of botanical anthelmintics in rotational parasite management programs has been associated with delayed development of anthelmintic resistance, highlighting their role as resistance-mitigating agents in sustainable livestock production.

4.4 Respiratory Diseases

The most common symptoms of respiratory diseases in animals are inflammation, microbiological infection, and weakening of immunity. Herbal extracts of Curcuma longa, Ocimum sanctum, Zingiber officinale, and Glycyrrhiza glabra are commonly employed in the treatment of respiratory disease in aiding their anti-inflammatory, bronchodilatory, expectorant and immunomodulatory effects. These herbal remedies assist in alleviating breathing problems, promoting immune system, and speeding up recovery especially as supportive care in veterinary preventive care (Gautam et al., 2010; Vaou et al., 2021).

4.5 Impact on Animal Health, Immunity, and Productivity

It has been demonstrated that the use of herbal therapeutics in the veterinary practice has a beneficial effect on the overall wellness of the animals by increasing the immune competency, decreasing the occurrence of diseases, and increasing the physiological resistance to stress and infections. Plants containing bioactive molecules, including polyphenols, flavonoids, terpenoids, and polysaccharides, have been shown to induce innate and adaptive immune response by enhancing phagocytic, antibody as well as antioxidant defense responses. It has been observed that in livestock fed with herbal formulations, several studies have recorded improvements in growth rate, feed conversion efficiency, milk yield and reproductive performance, mainly because of improved gut quality, low inflammatory load, and better metabolic efficacy (Hashemi and Davoodi, 2011; Windisch et al., 2008). Also, decreasing reliance on synthetic substances decreases the risk of adverse side effects and residue buildup, which leads to improved animal welfare and marketability of animal food (Kumar et al., 2021). Beyond therapeutic applications, the routine inclusion of standardized herbal formulations in preventive herd health programs can reduce disease incidence over time, thereby strengthening long-term productivity and minimizing reliance on synthetic antimicrobials.

4.6 Integration of Herbal Formulations in Preventive Veterinary Care

It has also been established that use of herbal therapeutics in the veterinary practice positively impacts the general wellness of the animals through boosting the immune competency, reduction of disease cases and enhancing the physiological resistance to stress and infections. Plants that contain bioactive molecules such as polyphenols, flavonoids, terpenoids and polysaccharides are able to activate both innate and adaptive immune response through strengthening of phagocytic, antibody and antioxidant protective responses. It is noted that in livestock that has been given herbal formulations, some studies have reported growth rate enhancement, increase in feed ratio, milk production and reproduction, primarily due to the improved quality of the gut, reduction of inflammatory burden, as well as enhancement of metabolic effectiveness (Hashemi and Davoodi, 2011; Windisch et al., 2008). In addition, reducing the use of synthetic materials will reduce the chances of side effects and accumulated residues, which translates to better animal welfare and commercialization of animal food (Kumar et al., 2021). The major veterinary applications of medicinal plants in the management of mastitis, gastrointestinal disorders, parasitic infestations, and respiratory diseases are summarized in **Table 2.**

Table 2. Veterinary applications of medicinal plants in disease management

Disease / Condition	Medicinal plants used	Mode of action	Key benefits
Mastitis	Neem, turmeric, *Aloe vera*	Antibacterial, anti-inflammatory	Reduced somatic cell count (SCC), residue-free milk
Gastrointestinal disorders	Garlic, ginger, cinnamon	Antimicrobial activity, gut microbiota modulation	Improved digestion and feed efficiency
Parasitic infestations	Neem, garlic, tannin-rich plants	Anthelmintic and ectoparasiticidal action	Reduced development of drug resistance
Respiratory diseases	Tulsi, turmeric, licorice	Anti-inflammatory and immunomodulatory effects	Faster recovery and improved respiratory health

5. AGRICULTURAL APPLICATIONS OF BOTANICAL EXTRACTS

5.1 Botanical Extracts and Essential Oils as Biopesticides

Increasing use of botanical extracts and essential oils of medicinal plants like Azadirachta indica, Allium sativum, Eucalyptus spp. and Ocimum spp. as biopesticides is because of their insecticidal, antifeedant, ovicidal and growth-regulating activities. These plant-based materials include various bioactive substances, such as terpenoids, phenolics and alkaloids, which work on various physiological targets on pests, thus lowering the occurrence of resistance development. In comparison to synthetic pesticides, botanical biopesticides are biodegradable and have low persistence in the environment, which makes them the appropriate part of an integrated pest management (IPM) system (Isman, 2020; Gupta et al., 2023). Field-based evaluations have reported that neem seed kernel extract (NSKE) at 5% concentration can reduce brown planthopper and aphid populations by 40–65% within 5–7 days of application, with yield improvements ranging between 8–18% compared to untreated controls. Similarly, essential oil–based formulations have demonstrated 50–70% suppression of early-stage larval infestations in vegetable crops. Although synthetic pesticides may show faster knockdown effects, botanical products have shown comparable season-long pest regulation when applied in integrated pest management (IPM) programs, with significantly lower residue levels in harvested produce. Recent global assessments indicate that botanical biopesticides represent one of the fastest-growing segments in sustainable crop protection markets, driven by regulatory restrictions on synthetic pesticides and increasing consumer demand for residue-free produce (Marrone, 2022).

5.2 Use of Botanical Extracts as Fungicides

Various phytopathogenic fungi such as Fusarium, Alternaria, Rhizoctonia, and Phytophthora species have been highly antagonized using plant extracts and essential oils. Certain compounds like thymol, eugenol, curcumin and allicin prevent germination of spores, induce disruption of cell membranes of fungi and also prevent enzymes. Research has shown that botanical fungicides have the potential to minimise disease occurrences in crops and reduce chemical residues and negative impact on non-target organisms in addition to making disease management measures safer and more sustainable (Burt, 2004; Li et al., 2024). Experimental trials have demonstrated that garlic- and clove oil–based formulations can reduce fungal disease severity index (DSI) by 35–60% in crops affected by *Fusarium* and *Alternaria* species under field conditions. In rice systems, neem-based botanical fungicides have shown up to 45% reduction in sheath blight incidence compared to untreated controls, with yield increases ranging from 6–15%. Although chemical fungicides may provide rapid curative effects, botanical fungicides have shown consistent protective action when applied preventively, contributing to lower cumulative residue levels and improved soil microbial activity over time.

5.3 Botanical Insect Repellents

A number of medicinal plants have been traditionally and scientifically known to have insect repellent properties especially with reference to agricultural pests and disease vectors. Neem, citronella, eucalyptus and garlic essential oils are repulsive because they disrupt the insects sense of smell and appetite. The botanical repellents are particularly useful in organic and low-input farms: they do not impair the ecological balance in the agro-ecosystems since they deter insects without inflicting damage on useful insects like pollinators and natural enemies (Regnault-Roger et al., 2012; Isman, 2020).

5.4 Enhancement of Crop Resistance and Yield Stability

In addition to direct effects on the suppression of pests and pathogens, botanical extracts have the potential to improve plant defense, enabling systemic resistance, and inducing the expression of enzymes and antioxidants that can be related to defense. Plant-based formulations have also been found to enhance the crop vigor, stress resistance, and yield stability in biotic stress environments. Increased crop productivity and yield sustainability Botanical products reduce pest pressure and disease severity, which are indirectly related to crop productivity and yield in agricultural systems susceptible to climate, especially smallholder agriculture (Popp et al., 2013; Gupta et al., 2023).

5.5 Benefits to Soil Health, Biodiversity, and Agro-Ecosystems

Application of botanical extracts in farming has a great environmental advantage since it leads to better soil health, microbial diversity, and minimal land and water pollution. As opposed to most of the synthetic agrochemicals, botanical products break down quickly and are not accumulated in the soil, thus maintaining the beneficial soil microorganisms and ecosystem processes. Further, lower chemical loading sustains biodiversity by keeping pollinators, natural enemies, and other non-targets safe, and these issues will help in sustaining resilient and sustainable agro-ecosystems in compliance with climate-smart and One Health-oriented agricultural systems (Isman, 2020; Li et al., 2024). A comparative assessment of synthetic pesticides and botanical extracts with respect to resistance development, environmental persistence, residue concerns, and compatibility with sustainable farming systems is presented in **Table 3.**

Table 3. Comparison of synthetic agrochemicals and botanical extracts in crop protection

Aspect	Synthetic Pesticides	Botanical Extracts
Source	Chemically synthesized	Plant-derived (extracts, essential oils)
Mode of action	Usually single-target	Multi-target (complex phytochemicals)
Resistance development	High risk over repeated use	Lower risk due to multiple modes
Environmental persistence	Often persistent	Rapidly biodegradable
Impact on non-target organisms	Frequently harmful	Generally safer
Residue concerns	High	Minimal to negligible

continued on following page

Table 3. Continued

Aspect	Synthetic Pesticides	Botanical Extracts
Soil health impact	May disrupt microbial activity	Supports soil microbial diversity
Suitability for IPM/organic farming	Limited	Highly suitable

5.6 Crop-Specific Applications of Botanical Extracts

Rice and Cereals

The extracted botanical resources that are commonly applied in rice-based systems include the neem seed kernel extract (NSKE), extract of garlic, extract of eucalyptus and citronella essential oils to control insect pests (brown planthopper, stem borer) and fungal diseases (Rhizoctonia solani, Magnaporthe oryzae). Field experiments show that products based on neem are effective to suppress the population of pests and leave the natural enemies intact and reduce the level of pesticides in grain, thus ensuring sustainable rice cultivation (Isman, 2020; Gupta et al., 2023).

Vegetable Crops

There is also high response of vegetable crops to botanical fungicides and insect repellent, which are created using neem, garlic, turmeric and clove oil. These botanicals have demonstrated considerable efficacy in aphids, whiteflies, thrips and fungal pathogens (Alternaria, Fusarium, Phytophthora) and have also been observed to enhance quality and marketability of crops because of lowered chemical residues (Burt, 2004; Regnault-Roger et al., 2012).

Oilseeds and Pulses

The botanical extracts are utilized in oil seed and pulse crops in both the field area and as after harvest protection. Plant-based repellents, neem oil and a solution of garlic extracts have shown to be effective against pod borers, aphids and have helped in maintaining the yield and minimizing post-harvest losses. The practices apply specifically to smallholder and organic systems of farming where synthetic inputs are not readily available (Popp et al., 2013; Isman, 2020).

Contribution to Sustainable Agro-Ecosystems

By increasing biodiversity, diversity of soil micro-organisms and reducing the amount of chemicals used in agro-eco systems, the addition of botanical extracts to crop protection systems enhances the ecological factors within the agro-eco systems. Botanical products produce better crop resistance and yield stability during biotic stress, therefore, supporting climate-resilient and sustainable agriculture. The fact that they can be used with integrated pest management (IPM), organic farming, and One Health-oriented agricultural strategies contributes to their increased importance to modern crop production systems (Isman, 2020; Li et al., 2024). The major agricultural applications of botanical extracts, including their use as biopesticides, fungicides, insect repellents, and resistance enhancers, are summarized in **Table 4**

Table 4. Agricultural applications of botanical extracts

Application	Botanical source	Target	Advantage over chemicals
Biopesticides	Neem, eucalyptus	Insect pests	Low resistance risk
Fungicides	Garlic, turmeric, clove oil	Plant pathogens	Minimal residues
Insect repellents	Citronella, neem	Field pests	Safe to pollinators
Resistance enhancement	Multiple botanicals	Crops	Yield stability

6. MECHANISMS OF ACTION OF PLANT-BASED ANTIMICROBIALS

6.1 Disruption of Microbial Cell Membranes

Dramatic destabilization of the microbial cell membrane is one of the main mechanisms through which antimicrobials derived out of plants act. Phytochemicals (e.g., terpenoids and phenolics and essential oils components e.g., thymol, carvacrol, eugenol and allicin) react with lipid bilayers making the membrane more permeable, and leading to intracellular content leakage. This causes loss of membrane integrity, proton motive force collapsing and subsequent cell death in bacteria and fungi. This is so because such membrane-targeted activity facilitates the elimination of a wide range of pathogen and the likelihood of resistance is minimal since this mode of action is non-specific (Burt, 2004; Cowan, 1999).

6.2 Inhibition of Enzymatic and Metabolic Pathways

Antimicrobials obtained by use of plants are likely to prevent some crucial enzymes and metabolic pathways that support the growth and survival of microbes. There are flavonoids, alkaloids, and phenolic acids that disrupt enzymatic reaction of energy metabolism, protein synthesis, and nucleic acid replication in the body. Taking the case of allicin present in garlic, it blocks sulphydryl-containing enzymes whereas curcumin interferes with various signalling and metabolic processes in microbial cells. Such antimicrobial impacts lead to a decrease in the growth and virulence of the microorganisms, which are used in veterinary practice and crop protection (Ankri & Mirelman, 1999; Li et al., 2024).

6.3 Antioxidant and Immunomodulatory Mechanisms

In addition to the direct effects on antimicrobials, various medicinal plants exhibit antioxidant and immunomodulatory effects, which increase host defense. Polyphenols, flavonoids, and curcuminos scavenge reactive oxygen species (ROS), alleviate oxidative stress, and prevent the damage of host tissues by inflammation. These compounds in animals activate the immune responses by improving macrophage action, antibody secretions, and control of cytokines, and hence, enhance immunity to infections. These immunomodulatory properties can help prevent and cure diseases, and plant-based antimicrobials can be the important ingredients of preventive veterinary treatment (Hashemi and Davoodi, 2011; Aggarwal and Harikumar, 2009).

6.4 Synergistic Effects of Phytochemicals

The main benefit of using plant-based antimicrobials is the synergism effect that or a combination of various phytochemicals found in botanical extracts. These agents have a combined effect on various microbial targets, leading to increased antimicrobial effects and decreased chances of emergence of resistance. Combination of plant-based substances with traditional antimicrobials may also lead to synergism, which is associated with the reduction of doses and improved therapeutic results. This type of multi-target and synergistic activity sets botanical antimicrobials in contrast to single-compound synthetic drugs and forms the basis of their success in a sustainable veterinary and agricultural disease management system (Vaou et al., 2021; Isman, 2020).

7. FORMULATION AND DELIVERY STRATEGIES

7.1 Herbal Extracts, Essential Oils, and Nanoparticles

The sources of plant-based antimicrobials most commonly are herbal extracts and essential oils, but the most common limitations to their practical use are volatility, lack of solubility in water, and degradability under field or gastrointestinal environments. The recent progress in the field of formulation science resulted in the creation of nano-based delivery systems, namely, the nanoemulsions, nano-based polymeric nanoparticles, and a lipid-based nanocarrier that increase the bioavailability, stability, and antimicrobial effect of phytochemicals. Delivery of nanoparticles enhances specific activity against pathogens in livestock and in plants and enables the use of lower dosages and less environmental loss which results in the overall efficacy of botanical products (Prasad et al., 2017; Li et al., 2024).

7.2 Feed Additives and Veterinary Formulations

Plant-based antimicrobials are also being used more often in veterinary practice in the form of feed additives, boluses, drenches, and topical preparations to aid in prevention and treatment of health. Phyto-genic feed supplements of garlic, turmeric, neem and essential oils have been found to improve the gut health, immune response and growth performance of livestock and chicken. These formulations would help decrease the occurrence of diseases and decreased dependence on synthetic antibiotics especially when it happens in a critical production phase. Accuracy and reproducibility of dosage and other therapeutic effects is also enhanced by standardized herbal formulations (Windisch et al., 2008; Hashemi and Davoodi, 2011).

7.3 Encapsulation and Controlled-Release Technologies

Encapsulation technologies such as microencapsulation, liposomal delivery and biodegradable polymer coating are increasingly becoming more popular methods to protect sensitive phytochemicals against oxidative, light and enzymatic degradation. CdRS provides added persistence and effectiveness of active compounds to animal and crop systems due to the gradual and sustained release of active compounds. In farming, controlled-release botanical preparations cut back on the number of applications and decrease losses to leaching and volatilization and in veterinary practice they help to prolong antimicrobial action and enhance adherence. They are critical in transferring laboratory efficacy of botanicals to the field-level success (Burt, 2004; Vaou et al., 2021).

7.4 Shelf Life, Stability, and Efficacy Improvement

Achieving shelf life and stability of efficacy as a result of differences in phytochemical composition and susceptibility to degradation is one of the biggest complications in commercializing plant-based antimicrobials. More sophisticated formulation methods like incorporation of antioxidants, pH adjustment, emulsification and encapsulation can greatly help in increasing the stability and storage of the product. Enhancement of shelf life is not only helpful in maintaining uniform antimicrobial activities but also in the production of large scale products, their regulatory acceptance, and adoption by the farmers. In turn, the optimization of the formulation is the key to the successful introduction of botanical antimicrobials into standard veterinary and agricultural procedures (Isman, 2020; Li et al., 2024).

8. CHALLENGES AND CONSTRAINTS

8.1 Variability and Instability of Phytochemical Composition

Among the most significant limitations of the use of medicinal plants as the antimicrobial agents, one should mention the fact that the phytochemical composition is subject to various changes based on the plant genotype, agro-climatic factors, soil type, harvesting time, and post-harvest treatments. This inconsistency usually results in unreliable antimicrobial actions and restricts reproducibility between research and real-world practice. Moreover, most of the bioactive compounds found in plant extracts and essential oils are chemically unstable and degrade under light, heat, oxygen and microbial activity thereby negatively affecting shelf life and efficacy (Cowan, 1999; Vaou et al., 2021).

8.2 Standardization and Quality Control Issues

The unification of extraction, formulation and dosage protocols is a major impediment to large scale uptake of plant-based antimicrobials. Differences in the solvents used in extraction, methodology used in the process, and storing environment may significantly influence the content and activity of the active constituents. Furthermore, the product consistency and approval of the product by the regulatory bodies are complicated by the issues of quality control such as contamination, adulteration, and batch-to-batch variability. Lack of generally accepted standards of references and validated biomarkers further reduces the comparability of the research results and reliability of commercial products (Isman, 2020; Li et al., 2024).

8.3 Regulatory and Safety Concerns

The regulatory systems relating to the acceptance and application of botanical antimicrobials in veterinary and agriculture systems are piecemeal and in different countries. Although plant-based products can be viewed as generally safe, lack of toxicological data, residues profile, and possible reactions with conventional drugs can be considered safe issues. In the veterinary use, absence of clear guidelines to withdrawal periods and maximum residue limits causes some challenges to commercialization and acceptance by the farmers. To close such regulatory gaps, effective safety testing, harmonized policies, and evidence-based risk testing are needed (McEwen and Collignon, 2018; Vaou et al., 2021).

8.4 Limited Large-Scale Field Validation

Even though several laboratory and small-scale field tests prove the antimicrobial activity of medicinal plants, there is a dearth of a large, multi-location field studies to substantiate their abilities within varied production frameworks. There is a lack of validation of the field, which inhibits assurance within the stakeholders and inhibits both policy endorsement and business investment. To scale up botanical interventions, the field testing, which is region-specific, needs to be conducted over the long term as well as assess the efficacy, cost-effectiveness, environmental effects, and compatibility against the current veterinary and agricultural practices (Isman, 2020; Gupta et al., 2023). In addition to biological efficacy, large-scale validation must include cost–benefit analysis, comparison with conventional antibiotic and pesticide programs, and assessment of long-term productivity impacts. Multi-location trials across different agro-climatic zones are essential to establish reproducibility and farmer-level feasibility. Public–private partnerships involving research institutions, veterinary universities, agri-startups, and farmer producer organizations (FPOs) can facilitate pilot commercialization models. Such collaborative validation frameworks will help bridge the gap between experimental research and field-scale adoption.

9. REGULATORY FRAMEWORK AND ACCEPTANCE

9.1 Current Regulatory Status of Herbal Products

Regulatory status of veterinary and agricultural herbal products are widely diverse among countries and regions and are associated with differences in legal classification, risk and regulatory priorities. Botanical products are licensed in most countries as a feed additive, veterinary herbal medication, biopesticide or as a traditional remedy, and in many cases, with less rigorous systems than synthetic drugs or pesticides. Herbal veterinary products in India are regulated by the Drugs and Cosmetics Act (1940) of traditional medicine and the Central Insecticides Board and Registration Committee (CIBRC) of botanical pesticides, but food and feed safety issues are dealt with by the Food Safety and Standards Authority of India (FSSAI). Nonetheless, when there is no international regulation and strict division of products, it tends to create uncertainty in the approval channels, delaying commercialization and broader adoption (Isman, 2020; Li et al., 2024).

9.2 Guidelines for Veterinary and Agricultural Botanicals

A set of national and international procedures have been put out to provide guidelines that can help in the safe use of botanical products in veterinary and agricultural systems. These are generally found to focus on botanical identification, standardization of active ingredients, safety analysis and efficacy validation. In the case of veterinary botanicals, some organisations, including the World Organisation for Animal Health (WOAH, now OIE), and national drug registration agencies, suggest that the evidence-based assessment of herbal preparations, including toxicity tests and residue analysis, should be carried out. Bio-based pesticides are also finding more applications in agriculture, both through the adoption of botanical pesticides as part of Integrated Pest Management (IPM), and the encouragement of regulatory authorities to adopt reduced-risk biopesticides. These attempts still fail to ensure stable regulatory acceptance in the regions due to the absence of a standardized testing procedure and reference standards (McEwen and Collignon, 2018; Gupta et al., 2023). To improve regulatory clarity and commercialization efficiency, a structured pathway is required that includes: (i) botanical authentication and phytochemical standardization, (ii) toxicity and residue profiling under laboratory and field conditions, (iii) multi-location efficacy trials, and (iv) dossier submission aligned with national regulatory authorities. Harmonization of standards between veterinary drug regulators, biopesticide registration bodies, and food safety authorities would reduce approval delays. Establishing dedicated regulatory categories for scientifically validated botanical products could further streamline market entry and encourage responsible private-sector investment. International policy discussions increasingly emphasize harmonized regulatory frameworks for botanicals to accelerate sustainable input adoption while ensuring safety and quality standards (OECD, 2023)

9.3 Market Acceptance and Farmer Adoption Barriers

Several socio-economic and technical factors affect the market acceptance and adoption of herbal based veterinary and agricultural products by farmers. The low recognition of botanical substitutes, uneven performance across the fields, increased initial expenses and delay of action as compared to synthetic chemicals often discourage usage. Moreover, the lack of information about the quality of the product, shelf life, and compliance with the regulations decreases the confidence of farmers. Supporting extensions, pilot projects, and labeling are needed to enhance the trust and adoption. It has been shown that when botanical products are in government recommendations, ICAR/SAU field validation, and part of the official package of practices, the likelihood of farmer acceptance rises, and the implementation of policy-based promotion and capacity-building efforts is necessary (Popp et al., 2013; Isman, 2020).

10. ROLE OF MEDICINAL PLANTS IN THE ONE HEALTH APPROACH

10.1 Interconnection Between Human, Animal, and Environmental Health

One Health approach acknowledges the inherent interconnections between human health, animal health, and environmental sustainability and focuses on the fact that any intervention in any area is bound to impact other areas. The medicinal plants and plant-based antimicrobials can also be easily incorporated into this framework, because they are currently used together in human traditional medicine, veterinary therapeutics, and agricultural crop protection. Through decreasing the use of synthetic antibiotics and pesticides, medicinal plants facilitate the restriction of the chemical residues in foodstuffs, lesser environmental degradation, and safer ecosystems, thus, aiding the comprehensive health outcome of human, animal, and environmental interfaces (McEwen and Collignon, 2018; Andrivon et al., 2022).

10.2 Reduction of Antimicrobial Resistance (AMR)

The increased and improper use of antibiotics in human medicine, livestock production, and agriculture is a one health challenge that has complicated antimicrobial resistance globally. Plant-based antimicrobials can also help solve AMR by providing an alternative disease control approach with multi-target modes of action, which decrease selective pressure against resistant strains. It has been linked to a decrease in the use of antibiotics, hence lessening the possibility of the transfer of resistant pathogens between animals and humans via food chains and the environment due to the use of herbal therapeutics in the preventive care and supportive treatment of livestock (McEwen and Collignon, 2018; Vaou et al., 2021). Also, botanical biopesticides in agriculture have reduced the distribution of resistant plant pathogen and also depleted the antimicrobial remnants into soil and water structures. Recent systematic analyses further confirm that phytogenic antimicrobials exhibit significantly lower resistance induction rates compared to conventional antibiotics in livestock-associated pathogens (Tang et al., 2022; Roth et al., 2023).

10.3 Environmental Safety and Ecosystem Preservation

Plant-based products that are used as medicines tend to be biodegradable, have low environmental persistence, and low adverse effects on non-target organisms; hence, they are safer and greener alternatives to synthetic chemicals. Their application in agriculture has aided in the maintenance of the soil microbial diversity, safeguarding of beneficial insects including the pollinators, natural enemies, and mitigating water body contamination. In the veterinary systems, the decrease in excretion of synthetic drug remains by the animal waste also protects the soil as well as the water ecosystem. As a result, medicinal plant adoption helps preserve and sustain ecosystems and strengthen the sustainability and climate-intelligent agricultural methods in accordance with the principles of One Health (Isman, 2020; Li et al., 2024).

10.4 Case Studies and Real-Life Examples

There are a number of examples in real life of the effective use of medicinal plants in a One Health approach. In India, biopesticides and herbal veterinary preparations designed by ICAR using neem have also decreased the use of chemical pesticides and reliance on antibiotics in mixed crop-livestock systems thereby enhancing food security and environmental well-being. Smallholder dairy systems which employ ethnoveterinary solutions such as turmeric, neem and aloe vera to treat mastitis and wounds have led to reduced cost of treatment and reduced milk wastage by treatment withdrawal periods. Globally, the use of plant-based feeds as a substitute to antibiotic growth promoters in swine and poultry feeding systems in some regions of Europe has helped not only to lower the incidence of antibiotic growth promoters but also in line with regulatory antibiotic bans and health objectives. These illustrations indicate the practical applicability of medicinal plants to integrate and operationalize the concept of One Health using integrated and sustainable solutions (Isman, 2020, Kumar et al., 2021).

11. SOCIO-ECONOMIC DIMENSIONS OF MEDICINAL PLANT–BASED VETERINARY AND AGRICULTURAL SYSTEMS

11.1 Economic Viability and Cost–Benefit Considerations

In addition to biological efficacy, economic feasibility and risk perception by farmers are strongly contributing factors to the adoption of medicinal plants in the veterinary and agricultural systems. Synthetic antibiotics, anthelmintics, and chemical pesticide procedures tend to be costly to keep up, both in terms of purchase prices, frequent application, and losses during withdrawal in livestock products, and in expenses incurred to remediate the environment in the long term, which are typically not factored into the conventional analysis of the economic system of the farm level (Popp et al., 2013; McEwen and Collignon, 2018).

Another economic model that is based on low input costs, local source, and less reliance on external inputs is through medicinal plant-based interventions in particular in smallholder and marginal farming systems. The use of ethnoveterinary preparations made using neem (Azadirachta indica), turmeric (Curcuma longa), garlic (Allium sativum), and aloe vera (Aloe vera) is based on the presence of widely found plants, which reduces the amount of money spent on pharmaceutical and pesticide purchases by a significant margin (McGaw and Eloff, 2008; Kumar et al., 2021).

Medicinal plants promote a decrease in the occurrence and frequency of diseases when incorporated in preventive animal health and crop protection programs, which leads to the reduction of the labor, veterinary, and chemical expenses. The use of phytogenic feed additives as a preventive measure has proven effective in boosting immunity, gut health and productivity of livestock, thus increasing net farm income and decreasing economic susceptibility (Windisch et al., 2008; Hashemi and Davoodi, 2011). When used at the early stages in crop systems, botanical extracts help in suppressing outbreaks of pests and diseases, which minimizes the use of recurrent sprays of chemicals and the expenses incurred (Isman, 2020).

In a long-term view, agro-based systems based on medicinal plants increase the economic performance and resilience by protecting farmers against the prices fluctuations in agrochemicals markets and decreasing the dependency on imported inputs. Moreover, animal and vegetable products that are residue-free are typically priced highly in organic and niche markets, which once again increases the profitability of farms (Popp et al., 2013).

11.2 Smallholder Livelihoods and Indigenous Knowledge Systems

Livestock and crop health Traditional knowledge systems have long maintained the health of livestock and crop using medicinal plants, especially in areas where there is poor access to modern veterinary and farming services. Ethnoveterinary and ethno-agricultural practices are the examples of locally customized, low-cost, and culturally rooted solutions that have been developed over the generations of experience learning (McGaw & Eloff, 2008; Kumar et al., 2021).

Medicinal plants are also essential livelihood resources to smallholder farmers as they can independently manage their ailments and also cut on the reliance of external service providers. The application of herbal treatment of mastitis, parasitism, wounds, and gastrointestinal complications is prevalent in mixed farming, which reduces the treatment expenses and enhances the livestock survival and performance (Hoste et al., 2015; Kumar et al., 2021). In the same way, botanical crop protectants facilitate the management of pests and diseases in agricultural systems that have low input levels, especially when synthetic pesticides are unavailable and unaffordable (Isman, 2020).

The accolade and institutionalisation of native knowledge systems are scientifically validated and credibility is increased, which makes them easier to integrate into the modern framework of veterinary and agricultural systems. Participatory research, farmer field school, community-based extension initiatives are especially important in the process of reconciling the traditional knowledge with formal science, so that the locally applicable innovations could be sustainable by preservation, improvement, and scaling (FAO, 2019; Andrivon et al., 2022).

In addition to economic value, maintenance of indigenous knowledge helps in social sustainability in that it enhances cultural identity, community cohesion, and intergenerational transfer of knowledge. Therefore, medicinal plants are helpful to the rural populations as sources of bio-inputs besides as socio-cultural resources.

11.3 Gender Roles and Rural Entrepreneurship

Medicinal systems of plants of veterinary and agricultural systems revolve around gender aspects. Women are mostly main care givers of livestock in most rural societies and hold knowledge of traditional information as far as herbal medicine, seeds and household nutrition are concerned. Their participation makes them the major players in the adaptation and spreading of medicinal plant activities (FAO, 2011).

There is a strong potential of the rural entrepreneurship of women in the medicinal plant cultivation, processing, and formulation. Nursery development, drying and processing of plant materials, preparation of herbal formulations, and local marketing are some of the activities which allow diversification of income with minimum capital investment. Such businesses especially are open to women who have a low level of land ownership or finances (FAO, 2019).

Women increase their economic contribution through community-based units of herbal production, self-help groups, and cooperatives, and augment the strength of local value chains. More household food security, access to health services, and educational achievements are also produced by increased income due to medicinal plant enterprises, bringing more extended social benefits (World Bank, 2020).

In developmental terms, the idea of women empowerment on the basis of medicinal plants based livelihoods will conform to the inclusive growth and sustainable development goals. With training, extension, and market connections, the women-oriented programs may grow beyond subsistence activities to successful rural businesses that strengthen sustainable veterinary and agricultural systems (FAO, 2011; World Bank, 2020).

12. CLIMATE CHANGE, RESILIENCE, AND BOTANICAL INTERVENTIONS

Climate change has become one of the most important forces that are changing the dynamics of diseases, the population of pests and the stability of the ecosystem in agricultural and livestock production systems. The increase in average temperature, changes in precipitation patterns, growth in the frequency of droughts and floods, and extreme weather events are augmenting biotic stress and undermining the success of traditional disease and pest management methods. The changes are making crops and livestock more susceptible to infections, infestations, and loss of productivity, especially in tropical and subtropical areas where smallholder systems of farming are the predominant types (FAO, 2018; IPCC, 2022).

In livestock systems, the effects of climate change include changes in the survival of pathogens, distribution of vectors, and host susceptibility. Heat stress reduces animal immunity, metabolic loads, and is associated with the predisposition of animals to gastrointestinal, respiratory, and vector-carried diseases. Likewise, in crop-based systems, high temperatures and moisture encourage the growth of fungi and bacteria, whereas alterations in precipitation change the life cycles of insects, as well as their population boom. In these circumstances, the application of only chemical antibiotics and pesticides turns out to be more unsustainable as the degree of resistance, the loss of effectiveness, and the increasing risk of harm to the environment increase (McEwen and Collignon, 2018; Andrivon et al., 2022).

The adaptive and resilient responses by plants to climate-related problems due to the biochemical diversity and the secondary metabolites that respond to stress render medicinal plants the best solution. The natural production of phenolics, terpenoids, alkaloids, and flavonoids are defense mechanisms against biotic and abiotic stresses (heat, drought, ultraviolet radiation, and pathogen attack, etc.) in plants. The same compounds have antimicrobial, immunomodulatory and pest-repellent effects which are exploited in veterinary and agricultural purposes. Notably, the production of numerous bioactive phytochemicals is intensified in response to stress which is why medicinal plants are particularly useful in agro-ecosystems that are stressed by climatic conditions (Li et al., 2024). Climate change–induced pest and disease dynamics and their cascading effects on agro-veterinary systems are illustrated in Figure 6.

Figure 2. Climate change–driven pathways leading to vulnerability in agro-veterinary systems

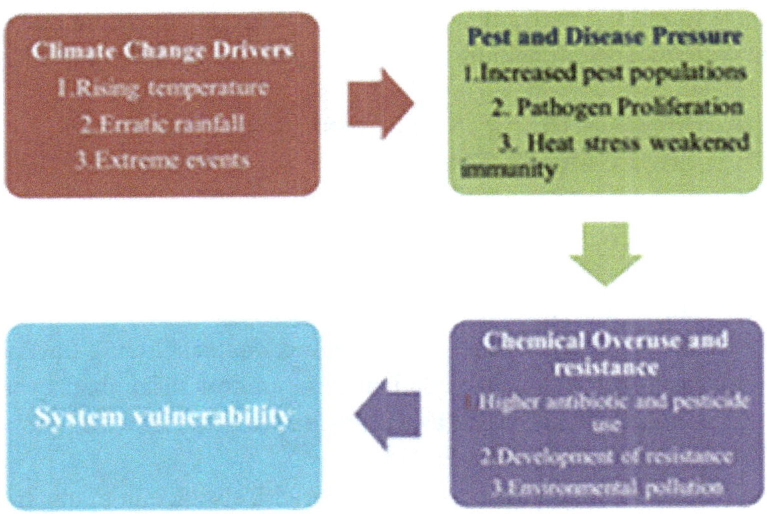

12.1 Botanical Antimicrobials and Preventive Disease Management

The benefit of botanical antimicrobials is that they help in the prevention of diseases, which is the foundation of climate-resilient food. Plant-based formulations help to suppress pathogen activities at an early phase and enhance host defense mechanisms in place of responding to disease outbreaks by intensive chemical interventions. Phytogenic feed additives and herbal therapeutics have been shown to enhance gut health, antioxidant capability, and immune responsiveness in livestock and makes them less susceptible to climate stress inducing diseases (Hashemi and Davoodi, 2011; Windisch et al., 2008). Controlled studies under heat-stress conditions have reported that phytogenic supplementation can reduce oxidative stress biomarkers (such as malondialdehyde levels) by 15–25% and improve immune indices, including antibody titers, by 10–20% compared to non-supplemented animals. Additionally, incidence of heat stress–associated diarrheal and respiratory disorders has been observed to decline by approximately 18–30% in herds receiving herbal immune-support formulations. These findings indicate that botanical interventions contribute measurable resilience benefits in climate-vulnerable production systems.

Botanical extracts and essential oils can also be used in crop production as biopesticides and biofungicides to prevent the growth of pathogens and pests. Their multi-target mechanisms of action can be used in changing climatic conditions unlike most synthetic chemicals whose activity is affected by changes in temperature or the change in rainfall. Preventive botanical applications therefore make disease management strategies in a changing climate more stable and reliable (Isman, 2020).

12.2 Strengthening Plant and Animal Immunity under Climate Stress

The impact of climate stress on both plants and animals is a reduction in the innate and adaptive immune responses. Medicinal plants are very essential in immunomodulation, which allows the organisms to adapt to physiological stresses caused by heat, drought, and food imbalances. Plant-derived antioxidants including flavonoids and curcuminoids in animals alleviate oxidative stress, control inflammatory pathways, and promote immune cell activities, as well as lead to better disease resistance and recovery (Aggarwal & Harikumar, 2009).

Plants Botanical formulations causing systemic resistance in plants trigger the activation of defense-related signaling pathways, antioxidants, and enzymes. This increased defense ability increases resistance to pathogens and pests in changing climatic conditions. Medicinal plants promote long-term resistance and lessen reliance on recurrent chemotherapy by enhancing host immunity instead of focusing on pathogens..

12.3 Reducing Dependence on Fossil Fuel–Based Agrochemicals

Synthetic antibiotics and pesticides are energy-consuming substances to produce, formulate, and transport and have a heavy reliance on fossil fuels. Climate smart agriculture focuses on the minimization of greenhouse gas (GHG) emission without compromising on productivity. The use of medicinal plant-based interventions can help reduce the use of synthetic agrochemicals and consequent carbon emission through mitigation.

The products obtained in botanical sources are usually locally produced, low-energy-processing, and lower-environmental application. Their introduction to agricultural systems minimizes the carbon footprint of disease and pest management and promotes the low-input and decentralized model of agriculture. In cattle feed, a decrease in the use of antibiotics also reduces the emissions related to the production of pharmaceuticals and waste disposal (FAO, 2018; Popp et al., 2013).

12.4 Medicinal Plants in Climate-Smart Agriculture and Livestock Systems

Climate-smart agriculture (CSA) has three interdependent goals, namely, increased productivity, better adaptation, and reduction of climate change effects. The medicinal plants are in tandem with all three CSA pillars. They facilitate productivity by decreasing the occurrence of diseases and enhancing host resilience. Their preventive effect and adaptive biochemical characteristics increase their adaptation to climatic stress on system level. At the same time, less use of chemical inputs is part of mitigation due to the reduced emission and impact on the environment.

Medicinal plants provide local and affordable means of climate resiliency, locally adjustable by smallholder and vulnerable agricultural communities. Their incorporation into the systems of mixed crops and livestock increases general farm sustainability, climate risks, and food security. Therefore, botanical interventions are a feasible operationalization of climate-smart and One Health-oriented agriculture, which connects the ecological stewardship with the health of animals, safeguarding crops, and rural livelihoods.

13. FUTURE PROSPECTS AND RESEARCH NEEDS

13.1 Integration into Mainstream Veterinary and Agricultural Practices

To become successful in the context of the successful introduction of medicinal plant-based products into mainstream veterinary and agricultural systems, it is essential that these products should undergo a shift in terms of going beyond the status of a complementary or alternative input into an evidence-based, routinely recommended intervention. Further to this, the application of validated herbal therapeutics, as well as botanical crop protectants, should be included in official packages of practices, herd health programs, and integrated pest and disease management (IPM/IDM) approaches. More effort needs to be done in collaborating with the researchers, veterinarians, agronomists, and extension agencies in order to demonstrate field-level consistency and generate confidence among the farmers and the practitioners. The adoption will be stimulated further with the help of capacity building, training farmers, and demonstration trials that emphasize economic and ecological gains (Isman, 2020; Kumar et al., 2021).

13.2 Advances in Formulation and Standardization

Development of formulation technologies is a very important research area that is of critical research priority to enhance reliability and commercial viability of botanical antimicrobials. The future studies will focus on standardization of active phytochemicals, establishment of good quality control measures, and optimization of delivery systems including nanoformulations, encapsulation and controlled-release products. These innovations may be used to increase stability, bioavailability and specificity of targets with minimum doses variability. It will be necessary to create phytochemical markers and harmonized analysis methods that would guarantee batch-to-batch reproducibility and regulatory standards to get ready to bridge the gap between lab work and commercial deployment (Li et al., 2024; Vaou et al., 2021).

13.3 Policy Support and Interdisciplinary Research

Development of formulation technologies is a very important research area that is of critical research priority to enhance reliability and commercial viability of botanical antimicrobials. The future studies will focus on standardization of active phytochemicals, establishment of good quality control measures, and optimization of delivery systems including nanoformulations, encapsulation and controlled-release products. These innovations may be used to increase stability, bioavailability and specificity of targets with minimum doses variability. It will be necessary to create phytochemical markers and harmonized analysis methods that would guarantee batch-to-batch reproducibility and regulatory standards to get ready to bridge the gap between lab work and commercial deployment (Li et al., 2024; Vaou et al., 2021).

13.4 Scaling-Up for Sustainable Food Production

To address the increasing demand of sustainable and residue-free food, there is the need to scale-up the application of medicinal plants in veterinary and agricultural systems. This will necessitate multi-location and multi-location field experimentation lasting over time to determine efficacy, costs to benefit ratios and environmental effects in different agro-ecological settings. Establishing production systems that are scalable to produce high-quality plant raw materials, enhancing supply chains, and making them affordable to the smallholder farmers are some of the major challenges to be overcome. Plant-based interventions, when properly scaled, would allow considerable reduction in chemical inputs, resiliency in food production systems, and sustainable intensification in accordance with climate-smart agriculture and One Health objectives (Popp et al., 2013; Isman, 2020). A viable scaling strategy should include organized cultivation of medicinal plants under contract farming or cooperative models to ensure consistent raw material quality and supply. Establishment of decentralized primary processing units for drying, extraction, and formulation at the district or cluster level can reduce post-harvest losses and improve value addition. Integration with Farmer Producer Organizations (FPOs), veterinary extension services, and agri-input distribution networks can facilitate last-mile delivery. Furthermore, digital traceability systems and quality certification mechanisms will enhance market credibility, particularly for residue-free and export-oriented products. Such structured supply-chain models are essential to transition botanical interventions from niche applications to mainstream agro-veterinary practice.

14. CONCLUSION

This chapter has exhaustively discussed the importance of medicinal plants in sustainable veterinary and agricultural system in the wider One Health paradigm and the importance of this concept in the face of modern challenges of antimicrobial resistance, food safety, environmental degradation, climate change, and socio-economic resilience. Based on the evidence synthesis provided in the entire chapter, medicinal plants are not such a traditional or alternative input, but a scientifically supported and multifunctional resource that can change the way food production works in the modern world.

Medicinal plants like Azadirachta indica, Allium sativum, Curcuma longa and Aloe vera have been also indicated to have a broad range of antimicrobial, antiparasitic, anti-inflammatory and immunomodulatory effects that have a direct implication to livestock disease management and crop protection. Their application in animal care- especially in treatment of mastitis, alimentary ailments, parasitism and respiratory issues help to improve animal well-being, their immunity, disease rates and improved efficiency, but at the same time reduce dependence on synthetic antibiotics and resultant residues. Botanical extracts and essential oils are good biopesticides, fungicides, insect repellers and resistance enhancers in agriculture, providing alternatives to harmful chemicals that do not damage soil health, harm non-target organisms and enhance the biodiversity.

The convergence that is offered by the integration of medicinal plants is a rare intersection between human health, animal health, and environmental sustainability, and viewed through the lens of One Health. Plant-based interventions are effective in reducing the spread of antimicrobial resistance and protecting the health of the population by diminishing the antimicrobial selection pressure, limiting chemical residues in food and ecosystems, as well as, enhancing biodegradable and low-persistence inputs. The attributes have been very important especially in the aspect of the intensification of the agricultural systems and the increasing consumer demand of safe and residue free food.

The chapter also demonstrates the significance of medicinal plants in climate-smart agriculture, in which their preventive, resilience-promoting, and low-carbon properties are quite consistent with the objectives of productivity, adaptation, and mitigation. The botanical interventions would boost the immunity of plants and animals by sustaining them in climate stress conditions, decrease reliance on fossil fuel-based agrochemicals, and improve resilience of the system at the system level within susceptible agro-ecosystems, particularly those in smallholder and resource-constrained environments.

Medicinally, plant-based systems which are socio-economically beneficial because they reduce the cost of inputs, decrease reliance on external inputs and provide livelihood opportunities through cultivating, processing and value addition. The acknowledgment and scientific credibility of the indigenous knowledge systems does not only maintain the cultural heritage but also promotes the inclusive innovation process, and women participation in medicinal plant-based business enhances gender equity and rural entrepreneurship.

Although they hold serious potential, the chapter admits the persistent issues associated with the variability of phytochemicals, their standardization, formulation stability, regulatory acceptance and lack of large-scale field validation. It will be necessary to mainstream medicinal plants into regular veterinary and agricultural practice by dealing with these constraints using interdisciplinary research, improvements in formulation and delivery technology, harmonized regulatory frameworks, and strong policy support.

To sum up, medicinal plants are a strategic, evidence-based approach towards attaining sustainable, resilient, and health-focused food systems. Their application in veterinary care, agricultural protection, and climate-sensitive agriculture offers a viable way of realizing the One Health concept in the field level. Plant-based interventions can be critical in providing a sustainable and safe food production, protecting human health and integrity of the environment, with further scientific validation, policy coordination and involvement of the stakeholders.

ACKNOWLEDGMENT

Burra Shyamsunder served as the corresponding author for this chapter.

REFERENCES

Aggarwal, B. B., & Harikumar, K. B. (2009). Potential therapeutic effects of curcumin. *The International Journal of Biochemistry & Cell Biology*, *41*(1), 40–59. DOI: 10.1016/j.biocel.2008.06.010 PMID: 18662800

Andrivon, D., Montarry, J., & Corbière, R. (2022). Plant health in a One Health world: Missing links and hidden connections. *Plant Pathology*, *71*(2), 244–255. DOI: 10.1111/ppa.13463

Ankri, S., & Mirelman, D. (1999). Antimicrobial properties of allicin from garlic. *Microbes and Infection*, *1*(2), 125–129. DOI: 10.1016/S1286-4579(99)80003-3 PMID: 10594976

Bayan, L., Koulivand, P. H., & Gorji, A. (2014). Garlic: A review of potential therapeutic effects. *Avicenna Journal of Phytomedicine*, *4*(1), 1–14. PMID: 25050296

Biswas, K., Chattopadhyay, I., Banerjee, R. K., & Bandyopadhyay, U. (2002). Biological activities and medicinal properties of neem (*Azadirachta indica*). *Current Science*, *82*(11), 1336–1345.

Burt, S. (2004). Essential oils: Their antibacterial properties and potential applications. *International Journal of Food Microbiology*, *94*(3), 223–253. DOI: 10.1016/j.ijfoodmicro.2004.03.022 PMID: 15246235

Chandegara, V. K., Varshney, A., & Patel, R. M. (2019). Antimicrobial and plant growth-promoting potential of *Aloe vera* extracts. *Journal of Applied and Natural Science*, *11*(2), 355–361.

Cowan, M. M. (1999). Plant products as antimicrobial agents. *Clinical Microbiology Reviews*, *12*(4), 564–582. DOI: 10.1128/CMR.12.4.564 PMID: 10515903

FAO. (2011). *The role of women in agriculture*. Food and Agriculture Organization of the United Nations.

FAO. (2018). *Climate-smart agriculture sourcebook*. Food and Agriculture Organization of the United Nations.

FAO. (2019). *Agroecology and indigenous knowledge systems*. Food and Agriculture Organization of the United Nations.

Gautam, M., Goel, S., Ghatule, R. R., & Singh, A. (2010). Immunomodulatory activity of *Ocimum sanctum* (Tulsi). *Journal of Ethnopharmacology*, *130*(2), 329–334. PMID: 20488234

Ghimpe eanu, O. M., Pogurschi, E. N., Popa, D. C., Dragomir, N., Drăgotoiu, T., Mihai, O. D., & Petcu, C. D. (2022). Antibiotic use in livestock and residues in food—A public health concern. *Foods*, *11*(10), 1430. DOI: 10.3390/foods11101430 PMID: 35627000

Gupta, I., Kumar, A., & Singh, P. (2023). Plant essential oils as biopesticides: Mechanisms, formulation, and applications in sustainable agriculture. *Frontiers in Sustainable Food Systems*, *7*, 1189456.

Hamman, J. H. (2008). Composition and applications of *Aloe vera* leaf gel. *Molecules (Basel, Switzerland)*, *13*(8), 1599–1616. DOI: 10.3390/molecules13081599 PMID: 18794775

Hashemi, S. R., & Davoodi, H. (2011). Herbal plants as immuno-stimulators in animal nutrition. *Journal of Medicinal Plants Research*, *5*(19), 4870–4878.

Hoste, H., Torres-Acosta, J. F. J., Sandoval-Castro, C. A., Mueller-Harvey, I., Sotiraki, S., Louvandini, H., & Terrill, T. H. (2015). Alternative strategies to control gastrointestinal nematodes in livestock. *Veterinary Parasitology*, *212*(3–4), 1–10.

IPCC. (2022). [*Impacts, adaptation and vulnerability*. Intergovernmental Panel on Climate Change.]. *Climatic Change*, 2022.

Isman, M. B. (2020). Botanical insecticides in the twenty-first century—Fulfilling their promise? *Annual Review of Entomology*, *65*(1), 233–249. DOI: 10.1146/annurev-ento-011019-025010 PMID: 31594414

Kumar, S., Kumar, R., Khan, A., & Sharma, P. (2021). Ethnoveterinary medicinal plants and their role in animal health management. *Journal of Ethnopharmacology*, *268*, 113576.

Li, S., Chen, G., Zhang, C., Wu, M., Wu, S., & Liu, Q. (2024). Natural antimicrobial compounds from plants: Advances, applications, and future prospects. *Trends in Food Science & Technology*, *144*, 104339.

Marrone, P. G. (2022). Pesticidal natural products—Status and future potential. *Pest Management Science*, *78*(9), 3554–3566. PMID: 30941861

. McEwen, S. A., & Collignon, P. J. (2018). Antimicrobial resistance: A One Health perspective. *Microbiology Spectrum, 6*(2), ARBA-0009-2017.

McGaw, L. J., & Eloff, J. N. (2008). Ethnoveterinary use of medicinal plants in animal health. *Journal of Ethnopharmacology, 119*(3), 559–574. DOI: 10.1016/j.jep.2008.06.013 PMID: 18620038

OECD. (2023). *Regulatory approaches for biopesticides and botanical products in sustainable agriculture*. Organisation for Economic Co-operation and Development Publishing.

Popp, J., Pető, K., & Nagy, J. (2013). Pesticide productivity and food security: A review. *Agronomy for Sustainable Development, 33*(1), 243–255. DOI: 10.1007/s13593-012-0105-x

Prasad, R., Kumar, V., & Prasad, K. S. (2017). Nanotechnology in sustainable agriculture: Role of nanomaterials. *Journal of Nanobiotechnology, 15*, 39.

Regnault-Roger, C., Vincent, C., & Arnason, J. T. (2012). Essential oils in insect control: Low-risk products in a high-stakes world. *Annual Review of Entomology, 57*(1), 405–424. DOI: 10.1146/annurev-ento-120710-100554 PMID: 21942843

Roth, N., Käsbohrer, A., Mayrhofer, S., Zitz, U., Hofacre, C., & Domig, K. J. (2023). The application of antibiotics in broiler production and the resulting antimicrobial resistance in *Escherichia coli*: A systematic review. *Poultry Science, 102*(2), 102345. PMID: 36571873

Sharma, N., Maiti, S. K., & Kundu, A. (2018). Herbal approaches for mastitis management in dairy animals. *Veterinary World, 11*(7), 937–945.

Surjushe, A., Vasani, R., & Saple, D. G. (2008). *Aloe vera*: A short review. *Indian Journal of Dermatology, 53*(4), 163–166. DOI: 10.4103/0019-5154.44785 PMID: 19882025

Tang, K. L., Caffrey, N. P., Nóbrega, D. B., Cork, S. C., Ronksley, P. E., Barkema, H. W., & Polachek, A. J. (2022). Restricting antibiotic use in food-producing animals and its associations with antimicrobial resistance in animals and humans: A systematic review and meta-analysis. *The Lancet. Planetary Health, 6*(1), e47–e60.

Vaou, N., Stavropoulou, E., Voidarou, C., Tsigalou, C., & Bezirtzoglou, E. (2021). Advances in medicinal plant antimicrobial activity: Challenges and future perspectives. *Microorganisms, 9*(10), 2041. DOI: 10.3390/microorganisms9102041 PMID: 34683362

Windisch, W., Schedle, K., Plitzner, C., & Kroismayr, A. (2008). Use of phytogenic products as feed additives in animal nutrition. *Journal of Animal Science*, *86*(14, suppl_14), E140–E148. DOI: 10.2527/jas.2007-0459 PMID: 18073277

World Bank. (2020). *Harvesting prosperity: Technology and productivity growth in agriculture*. World Bank Publications.

Chapter 10
Plant-Derived Bioactive Compounds for Mitigating Environmental Antimicrobial Resistance Reservoirs:
Focus on Groundwater

Asma Outagant
http://orcid.org/0009-0000-3534-4602
Faculté de Médecine et de Pharmacie, Université Cadi Ayyad, Marrakech, Morocco

Souad Loqman
Faculté de Médecine et de Pharmacie, Université Cadi Ayyad, Morocco & Laboratoire de Microbiologie, CHU Mohammed VI, Marrakech, Morocco

ABSTRACT

Water is essential for sustaining life; nevertheless, more than 780 million people worldwide still lack access to safe drinking water. The growing prevalence of multidrug-resistant bacteria highlights the urgent need for sustainable antimicrobial strategies. In this context, plant-derived bioactive compounds have gained considerable attention as promising natural antibacterial agents. Medicinal plants are rich sources of secondary metabolites, including alkaloids, phenolics, flavonoids, which have demonstrated broad-spectrum antimicrobial activity. Although numerous studies have reported the antibacterial potential of these compounds, limited attention has been given to their purification and application in water treatment, particularly in groundwater systems. Groundwater is a major source of drinking

DOI: 10.4018/979-8-3373-7810-7.ch010

water and may serve as an important environmental reservoir of AMR. This book chapter provides a comprehensive overview of plant-derived bioactive compounds as potential tools for mitigating environmental AMR reservoirs, with a specific focus on groundwater and drinking water systems.

INTRODUCTION

The occurrence and dissemination of antimicrobial resistance (AMR) in the environment, become a worldwide concern for both human health and ecosystems. The over- and misuse of clinically-relevant antimicrobials on human and animals, have accelerated the antimicrobial resistance theses past decades (de Kraker et al., 2016). According to the World Health Organization (WHO), approximately, 5 million deaths per year are linked to AMR, with projections reaching 10 million deaths annually by 2050 (WHO, 2024). Polluted environments are enriched with antibiotic resistant bacteria (ARBs) and ARGs which can be transferred into humans and animals through so called mobile genetic elements.

Several studies have reported AMR contamination in aquatic environments, including seawater, surface water, and increasingly, groundwater (Abuzerr et al., 2024; Adelowo et al., 2018; Adiba et al., 2024; Banjo et al., 2020; Tuts et al., 2025), often due to untreated hospital waste, wastewater effluents, landfill leachate, and human and animal excreta (Koudoum et al., 2025; Matloko et al., 2021) (Solaiman et al., 2022), (Foyle et al., 2023; Fu et al., 2022; Kipré et al., 2025). Groundwater is a primary drinking water source for over 2.2 billion people worldwide and is critical to agriculture and food production, particularly in arid and semi-arid regions. However, it also serves as a reservoir for ARBs and ARGs, contributing to infectious disease risk (Abuzerr et al., 2024; De Lambert et al., 2021; Sidhu et al., 2013).

With limited treatment options for drug-resistant microbes, the WHO prioritizes the discovery of new and effective antimicrobial agents (WHO, 2024). Key Gram positive and Gram negative bacteria such as *E. coli*, *P. aeruginosa*, *S. aureus*, *E. spp.*, *K. pneumoniae*, *A. baumannii*, *Enterobacter spp*, becomes a cause of life-threatening infections in healthcare settings worldwide (De Oliveira et al., 2020). In this regard, natural sources of antibacterial compounds, particularly plants, have gained significant attention. Medicinal plants produce a variety of derived-bioactive compounds, including phenolics, alkaloids, flavonoids, coumarins, terpenoids, tannins, and polypeptides, which exhibit diverse antimicrobial activities (Bagher-zadehsurbagh et al., 2025; Benhalima et al., 2025; Kakkar et al., 2025; Mane et al., 2020; M.E. Tsie et al., 2025).

Although numerous researches have explored the antibacterial potential of Phyto-compounds, their application in water treatment, especially groundwater disinfection, remains limited. Some plant extracts, such as *Moringa oleifera*, peppermint, *Azadirachta indica*, and *Curcuma longa*, have demonstrated activity against multidrug-resistant pathogens and show promise for use in water purification (Abd El-Lateef et al., 2025; Gouda et al., 2025; Pritchard et al., 2009). Plant metabolites act via multiple mechanisms, including disrupting microbial membranes, interfering with metabolic pathways, and modulating gene expression (Adeeyo & Odelade, 2018; Angelini, 2024; Huang et al., 2022; Jubair et al., 2021). Therefore, plant and their natural products may have a significant role in mitigating the environmental antimicrobial resistance reservoirs, especially groundwater.

This chapter aims to provide a comprehensive overview of plant-derived natural products as potential tools for mitigating environmental AMR reservoirs, with a focus on groundwater and drinking water systems, and to discuss their mechanisms of action and potential applications in water disinfection.

1. LIMITATIONS OF THERAPEUTIC DRUGS ON CONTROLLING AMR BACTERIA

Elevated rates of antibiotic resistance and bacterial infections represent a major global health concern (WHO, 2024). Several factors have enabled pathogens to develop new resistance mechanisms against currently available antibiotics, including inappropriate use and over-prescription, which often lead to treatment failure. Such practices prolong the duration of illness, increase treatment costs, and elevate the risk of mortality (Tanwar et al., 2014). Common infections affected by these practices include sepsis, diarrhea, urinary tract infections, and sexually transmitted infections. Consequently, many existing antibiotics must be administered in combination to achieve effective treatment.

With the growing prevalence of antimicrobial resistance, combination therapy has emerged as a preferred approach (Coates et al., 2011). This involves using two or more antimicrobial agents simultaneously to inhibit targets in the same or different pathways, thereby reducing the development of resistance and overcoming the limitations of monotherapy (Sheard et al., 2019; Worthington & Melander, 2013). However, the use of combination therapies is limited by several factors. Pharmacokinetic and pharmacodynamic communications can appear, whereby drugs are influenced each other. These interactions may be synergistic, additive, or antagonistic, such as when bacteriostatic and bactericidal drugs are combined (Ocampo et al., 2014). Additionally, alterations in drug absorption, distribution, metabolism, or elimination can change effective drug concentrations in tissues and blood, potentially accelerating resistance due to sub-inhibitory exposure (Coates et al., 2020). Another significant limitation is toxicity; for example, piperacillin-tazobactam and vancomycin combination has been associated with acute kidney injury (Al Yami, 2017).

2. PLANT AND NATURAL PRODUCTS POTENTIAL FOR AMR

Plants have been traditionally used to address various health problems over time. With the increasing failure of synthetic medications against antimicrobial-resistant (AMR) pathogens and their potential adverse effects, plant-derived therapeutic agents have gained attention as alternative or complementary treatments. These agents are recognized for their minimal undesirable effects and reduced toxicity, promoting their adoption in healthcare systems (Valli et al., 2012). However, rigorous scientific validation is required for their use in modern medicine, including assessments of safety, standardization, and dosage optimization (Bagherzadehsurbagh et al., 2025; Izah et al., 2024). Different parts of plant seeds, roots, bark, leaves and flowers contain metabolites essential for survival and environmental interactions.

Plant bioactive compounds, including phenolics, flavonoids, alkaloids, terpenoids and essential oil have recently attracted attention as potential tools to combat antimicrobial resistance and as leads for antibiotic development (Angelini, 2024). For example, Mwalongo et al., determined the antimicrobial efficacy of ethanol-based extracts under in vitro conditions from Six Tanzanian medicinal plants; *Plectranthus barbatus, Tridax procumbens, Sida acuta, Ocimum gratissimum, Euphorbia hirta* and *Lantana camara* against clinical strains, such as *E. coli, K. pneumoniae, S. aureus, Methicillin-resistant S aureus* (MRSA) *C albicans*, *E faecalis*. Showing that medicinal plant *Lantana camara Plectranthus barbatus,* and *Euphorbia hirta,* exhibiting the potential antimicrobial activity (Mwalongo et al., 2025).

In other review, *Polygonum plebejum, P. persicaria, Rumex hastatus, and R. dentatus,* crude extracts, exhibit antibacterial properties, and inhibiting bacteria such as *C. frundii, E. coli* and *S. aureus* (Angelini, 2024; Sultana, 2018). The leaf extract and aqueous leaf extracts of *T. peruviana, E. suaveolens,* and *E. hirta* exhibit antibacterial activities against extended-spectrum beta-lactamase (ESBL) producing bacteria, such as *E. coli, K. pneumonia,* and *Pseudomonas* in addition to MRSA (Angelini, 2024). Moreover several studies have demonstrated the antibacterial potential of Phyto-therapeutic plants, for instance *P. barbatus, L. camara, E. hirta O. gratissimum, S. acuta, and T. procumbens,* for the uropathogens therapy (Andriana et al., 2019; Cordeiro et al., 2022; Dubey & Padhy, 2013; Ferdosh, 2023; Hassan et al., 2022; Kin et al., 2018; Mwalongo et al., 2025).

In vitro research confirm the efficacy of these plant extracts: *P. barbatus* is effective against MRSA, ESBL and carbapenemase producing *K. pneumoniae* (Cordeiro et al., 2022). *L. camara acts against MDR ESBL-producing bacteria (C. freundii,*

A. baumannii, Proteus spp., Pseudomonas aeruginosa), MRSA, and vancomycin-resistant Enterococcus faecalis (Dubey & Padhy, 2013). And *T. procumbens* targeting S. aureus clinically resistant (Adjapong et al., 2017). Furthermore, combined plant extracts often exhibit synergistic or additive antimicrobial effects, making herb-herb combinations a promising alternative against AMR(Bagherzadehsurbagh et al., 2025; Mwalongo et al., 2025).

Therefore, preventing bacterial resistance is crucial to maintain the one health safety and environmental balance integrity, which may result from developing effective therapies by establishing new antimicrobial agents as reported by the WHO, using plant-derived bioactive compounds as an alternative for AMR solution, which implies the effective comprehension of their modes of action.

3. ANTIMICROBIAL MECHANISMS OF PLANT-DERIVED BIOACTIVE COMPOUNDS AGAINST RESISTANT BACTERIA

Plant-derived active compounds, also called secondary metabolites, act via multiple mechanisms, including structure and function membrane disruption, blocking of DNA/RNA synthesis, metabolism perturbation, induction of cytoplasmic coagulation, and disruption of cell-to-cell communication (Radulovic et al., 2013). These compounds are classified into polyphenols and flavonoids, alkaloids, terpenes, and others based on their biosynthetic origin and phytochemical properties (Bagherzadehsurbagh et al., 2025; Jubair et al., 2021). According to literature, these metabolites were the most studied plant derived compounds for antimicrobial resistance activity. With Polyphenols, flavonoids and alkaloids exhibit the most antimicrobial resistance efficacy. Fig 1 represents the most Plant-derived active compounds mechanisms against resistant bacteria.

a. Polyphenols

Polyphenols are abundant bioactive compounds in plants with significant antimicrobial activity. They include flavonoids, stilbenes, lignans, and phenolic acids. Their mechanisms include altering membrane permeability, changing environmental pH, inhibiting biofilm formation, degrading cell walls via peptidoglycan binding, inhibiting extracellular enzymes, disrupting microbial processes, and chelating essential minerals (Bagherzadehsurbagh et al., 2025; Daglia, 2012; Yang & Zhang, 2019).

A variety of mechanisms puts polyphenols in the position of one of the most potent agents against many types of MDR bacteria, for instance, polyphenols interfere with amino acids of enzymes, by hydrogen bonds effecting protein synthesis and increasing membrane permeability (Jubair et al., 2021). Improving the alterations of enzyme-based antimicrobial resistance strategies. Various phenolic compounds identified from *Oxalis corniculata* extract such as Phenylpropanoids has proven strong antibacterial effect versus *K pneumoniae, E coli*, *Citrobacter spp.* and *S typhi multi-drug resistant* (Bakal et al., 2017). Similarly, extracts from *Myrtus communis* and *Cinnamomun zeylanicum* with polyphenols the most found compounds have demonstrated significant potential in vitro inhibition of various pathogenic bacteria. That, inhibition zones diameter ranged from 23 to 28 mm, and Myrtle extract exhibit more powerful antibacterial activity than penicillin at minimum inhibitory concentration of 30 mg/mL (Masoumian & Zandi, 2016). polyphenols shown an enhanced synergistic interaction with tetracycline and β-lactams, providing enzymatic inactivation by bacteria (Bagherzadehsurbagh et al., 2025). In vitro assays of (Mourabiti et al., 2025) shows a ranges of MIC and MBC of 12.5 to 25 mg/L and of 25–50 mg/L (Mourabiti et al., 2025) due to polyphenols. In addition, the combination of quercetin and cefotaxime improved bacterial susceptibility and suppressed β-lactam hydrolysis, ensuring total inhibition at 200 mg/L. However, the intrinsic variability of polyphenol composition across different plant sources, is impacted by several variables such as growth conditions, geographic position, and methods of extraction. Thus, complicating of the standardization and reliability of observed antimicrobial potential (Bagherzadehsurbagh et al., 2025).

b. Flavonoids

Flavonoids are the most common plant polyphenols, exhibiting antimicrobial activity (Abbas et al., 2017). The B chain of certain flavonols, for example myricetin, robinetin and flavanols as epigallocatechin gallate, act with synergistic modes of action, such as disrupting DNA biosynthesis, bonds of hydrogen between B chain and DNA bases, (Bagherzadehsurbagh et al., 2025). Flavonols Quercetin, is widely present in apples, onions, tea and kale has shown large antibacterial properties. And characterize by its interaction with hydroxyl groups of bacterial cell membranes, due to cell membrane different composites and permeability, especially versus Gram positive bacteria (Nguyen & Bhattacharya, 2022; Wang et al., 2018). Moreover, quercetin confirmed enhanced antibacterial effect targeting MRSA (Angelini, 2024). There is a differences in MIC of flavonols according to bacterial species, for example, quercetin shows MIC of 20 mg/mL for *P aeruginosa and S aureus* (Jaisinghani, 2017), and *Lactobacillus and Streptococcus* at ranges of 1-8 mg/mL of MICs. Flavonoids connect with beta-ketoacyl acyl carrier protein synthase (KAS) II and III enzymes of cell membrane precursors, including Flavanones such as taxifolin, eriodictyol and naringenin inhibit KAS III, which exhibit modest antibacterial efficacy (Angelini, 2024). Despite the antibacterial effects of these compounds demonstrated in vitro, the main limitation lies in their relevance in a clinical context that demand meticulous assessment of attainable concentrations and dosages in vivo with cytotoxicity consideration (Bagherzadehsurbagh et al., 2025).

c. Alkaloids

Alkaloids, nitrogen containing basic compounds, are found in leaves, roots, bark, flowers, and seeds of many plants, including Fabaceae, Solanaceae, Amaryllidaceae, Asteraceae, Papaveraceae, Rutaceae, Apocynaceae, and Rubiaceae (Y. Liu et al., 2020).

Alkaloids are classed into isoquinoline alkaloids and indole alkaloids (Bagherzadehsurbagh et al., 2025; Faisal et al., 2023).) and act via efflux pump inhibition, respiratory inhibition, QS disruption, nucleic acid synthesis inhibition, interference with cell proliferation, enzyme inhibition, and modulation of virulence determinants (Othmani et al., 2020; Thawabteh et al., 2019). All these strategies provide an important role together decreasing the efficacy of antimicrobial resistance of bacteria. For instance, in a study reported by (Kashyap et al., 2022) alkaloids extracts of *Curcuma amada* were demonstrated for their inactivation efficacy of different pathogenic bacteria including *V cholerae, C albicans* and *S aureus.* Alkaloids can inhibit bacterial biofilms, and proteins, such as pyruvate kinase and filamentous temperature-sensitive protein Z (Bagherzadehsurbagh et al., 2025; Jubair et al., 2021; Y. Liu et al., 2020). Berberine, an isoquinoline alkaloid extracted from *Cortex phellodendri* and *Rhizoma coptidis,* blocks the FtsZ protein-mediated cell division process. Moreover, berberine also disrupt DNA, targets RNA polymerase, gyrase and topoisomerase IV (Angelini, 2024; Bagherzadehsurbagh et al., 2025). In addition, Piperidine extract of *Piper longum or Piper nigrum,* show a significant decrease of minimum inhibition concentration of mutant *S. aureus* growth, in combination with ciprofloxacin, (Angelini, 2024).

d. Terpenes

Terpenes, or isoprenoids, are a diverse group of organic molecules present in almost all forms of life (Lei et al., 2019). They are built from five isoprene carbon units and categorized to monoterpenes such as D-limonene, geraniol, carvone, and to diterpenes, for example retinoic acid, retinol, and into triterpenes as oleanolic acid and betulinic acid. and tetraterpenes as α-carotene, β-carotene, and lutein (Jubair et al., 2021; Nawrot et al., 2014), Monoterpenes, which constitute the main volatile components of essential oils, are particularly biologically active.

Terpenes act against multidrug-resistant (MDR) bacteria through various mechanisms, including membrane disruption, antibiofilm activity, inhibition of quorum sensing (QS), and suppression of ATP production, targeting the formation of biofilm and QS is especially relevant for long-term infections in which conventional antibiotics are frequently ineffective (Angelini, 2024; Huang et al., 2022; Salikin et al., 2024). Monoterpenes such as carvacrol, eugenol, thymol, and menthol inhibit bacterial growth, efflux pumps, and disrupt membranes (Zacchino et al., 2017). Thymol exhibit potent efficacy versus *S. aureus* and *E. coli*, while geraniol inhibits efflux pumps in resistant Gram-negative *Enterobacter* species (Bergman et al., 2019).

e. Peptides

In addition to these metabolites, peptides also present an antimicrobial effect against MDR bacteria. Antimicrobial peptides (AMPs) have a reduced size (2 to 10 kDa), exhibit amphiphilic properties, and a positive charge at neutral pH making them more interaction with microbial membrane (Dos Santos-Silva et al., 2020; Slavokhotova & Rogozhin, 2020). Peptides extracted from plant contain a compacted and stabilized spatial structure due to disulfide intramolecular bonds, providing a stability against temperature variation, synthetic compounds and enzymatic function, in addition to hydrolases and protein synthesis inhibition (Angelini, 2024).which serving as water disinfection. Studies reported that, *Moringa oleifera*, medicinal plant, disrupt the membrane of *P. aeruginosa*, *E. coli, S typhimurium* and MRSA, due to the rapid membrane targeting mechanisms of AMPs (Bagherzadehsurbagh et al., 2025; Shin et al., 2018).

Despite the highly specific interactions protein-based compounds with target bacteria, even MDR pathogens (Srivastava et al., 2025). Several challenges were found related to their therapeutic development in vivo, such as stability, oral absorbance efficiency, and proteolytic degradation vulnerability and in large-scale purification. Table 1 summarized the bioactive compounds and the target pathogens of some plants.

Figure 1. Plant-derived active compounds mechanisms against resistant bacteria

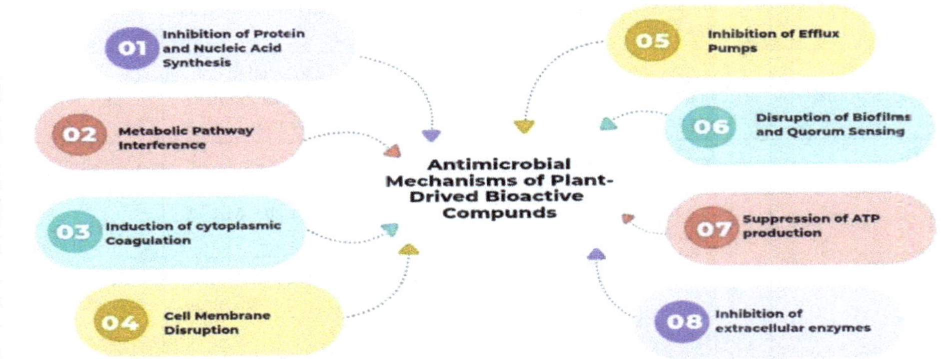

Figure 1: Plant-derived active compounds mechanisms against resistant bacteria.

Table 1. Bioactive compounds of different plant, extraction method and target pathogens of some plants

Plant name	Plant part use	Plant active products	Target pathogens	Reference
L. camara Verbenaceae	Leaves (Soxhlet extraction)	Tannins, saponins Phenolics, Coumarins, steroids, terpenoids,	Escherichia coli K. pneumoniae S. aureus, MRSA	(Mansoori et al., 2020)
Arundo donax	Whole powder biomass	Phenolics and flavonoids	S. aureus, P. aeruginosa, E. coli,	(Benhalima et al., 2025)
Curcuma amada	Rhizome	Phenols, Alkaloids, Tannins, Flavonoids, Terpenoids,	E coli, S typhi, S enterica, S aureus, and others	(Kashyap et al., 2022)
Moringa oleifera	Leaves	Coumarin, Tannin, Alkaloids, Polyphenols, Terpenes, Sterol Flavonoids, Anthraquinones, Saponin,	E coli, Klebsiella spp. P. aeruginosa,	(Reiter et al., 2017)
T. procumbens Asteraceae	Leaves (Ethanolic extracts)	Flavonoids, puerarin, esculetin, oleanolic acid, betulinic acid, centaurein,	E. coli, S. aureus, MRSA K. pneumoniae	(Mwalongo et al., 2025)
Myrtus communis and *Cinnamomun zeylanicum*	Seeds Hydro-Alcoholic Extraction	Polyphenols	Escherichia coli Staphylococcus aureus and others	(Masoumian & Zandi, 2016)

continued on following page

Table 1. Continued

Plant name	Plant part use	Plant active products	Target pathogens	Reference
E. hirta Euphorbiaceae	Foliage (Ethanolic extracts)	Quercetin and kaempferol, caffeic acid, hydroquinone, Ocoumaric acid	Escherichia coli K. pneumoniae S. aureus, MRSA	(Mwalongo et al., 2025)
S. acuta Malvaceae	Leaves and flowers (Ethanolic extracts)	Indoloq uinoline alkaloids	Escherichia coli K. pneumoniae S. aureus, MRSA	(Mwalongo et al., 2025)
Matricaria recutita	Flowers part	Polyphenols, Alkaloids, Flavonoids, Saponin, Coumarin, Sterol, Tannin, Terpenes,	E coli, Klebsiella spp., P aeruginosa,	(Reiter et al., 2017)
P. barbatus Lamiaceae	Leaves (Ethanolic extracts)	Flavonoids, alkaloids, cinnamic derivatives, ellagic acid, Triterpenes	Escherichia coli K. pneumoniae S. aureus, MRSA	(Mwalongo et al., 2025)
Oxalis corniculata	Not mentioned	Alkaloids, Terpenoids, Phenylpropanoids, Polyketide,	E. coli, K. pneumoniae, S. typhi MDR, C koseri	(Bakal et al., 2017; Subramani et al., 2017)
Thymus vulgaris	Commercial compounds	Furnesol and nerolidol	S aureus	(Togashi et al., 2010)
Nigella sativa	Crude extract and oil	Not mentioned	Salmonella spp.	(Bakal et al., 2017)
Berberis vulgaris	Extract	Alkaloids	Staphylococcus spp. Bacillus, Listeria,	(Vaou et al., 2021)
Carex baccans	Fruit extract	Curcusinol	A. baumannii, P. aeruginosa,	(T. Liu et al., 2024)
Chelidonium majas	Root Extract	Chelyrethrine	P. aeruginosa	(Pacyga et al., 2024)
Bauhinia thonningii Schumach	Leaves extract	Quercetin, 3,4′-dimethyl ether	MRSA	(Nouemsi et al., 2025)
Origanum heracleoticum	-	Organic Volatile compounds	K pneumoniae	(Semenzato et al., 2024)
Curcuma longa	-	Turmerone and others	MRSA, P. aeruginosa A. baumannii	(Kirmani et al., 2024)
Polygonum acuminatum	Leaves, stem, and root extract	Pantanal macrophytes	E coli, E. faecium, S. aureus,	(Takahashi et al., 2024)

Consequently, Plant-Derived Bioactive Compounds with theirs diverse mechanisms in different bioactivities, and natural properties as reported by data, may ensure the water disinfection from pathogenic microorganisms with a low environmental impact, through inactivation, destruction and their entire removal, which avoid the contamination of water in supply systems and maintain the safest quality of water (Bagherzadehsurbagh et al., 2025). Therefore, plant and their natural products may have a significant role in mitigating the environmental antimicrobial resistance reservoirs.

4. ENVIRONMENTAL APPLICATION OF PLANTS ACTIVE PRODUCTS

Water contamination, particularly by pathogenic microorganisms, has recently become a significant concern for environmental scientists and government authorities because of the risks it presents to humans and fauna (Y. su et al., 2024). Several studies reported the antimicrobial resistance contamination of aquatic environment such as surface water, and recently groundwater, serving as a reservoir for antibiotic resistance bacteria and genes, leading to several infectious diseases. These pathogens coming from several pathways, leaching groundwater through infiltration. However, some populations in developing countries continue to rely heavily on the use of these open water sources, especially groundwater supplies used as primary drinking water sources by 2.2 billion people worldwide. According to the World Health Organization (WHO), lack of safe drinking water is implicated in approximately one million deaths annually in developing countries (WHO, 2014).

Over time, various physic-chemical treatment approaches are employed for the disinfection of water, which generally require extended treatment durations, restricted capacity and elevated cost (Collivignarelli et al., 2018). Chlorine (Cl_2), sodium hypochlorite (NaOCl), chloramines (NH_2Cl), and ozone (O_3), are some of the most synthetic chemicals using as disinfectants of water, in addition to UV irradiation. However, they present many limitations. For instance, the use of chlorine, sodium hypochlorite and ozone risk for the formation of disinfection by-products (DBPs) such as haloacetic acids, trihalomethanes, and bromate, having carcinogenic proprieties and health effects. Chloramines, has a lower disinfection efficiency, in addition to the odour and taste problems (Collivignarelli et al., 2018). The purpose of using UV irradiation is to destroying bacteria by breaking down DNA bonds and the formation of thymine dimers. However, some bacteria, have the photo reactivation repair mechanism (Gogate, 2007). For example, thyme oils were more effective in reducing E. coli O157:H7 compared to ozonation and chlorine dioxide as reported by the study of (Singh et al., 2002). In addition, these conventional disinfectants were strongly influenced by conditions such as pH, temperature, which affect the entire removal of microorganisms. Therefore, the emergence of multidrug resistant pathogens in water and undesirable effects of conventional antimicrobials call for alternative approaches of water treatment of plant origin (Benhalima et al., 2025; Yahya et al., 2014). (Benhalima et al., 2025)

Plants as previously mentioned, have numerous natural metabolites such as Terpenes, flavonoids and phenolic, which aid in the defense against antimicrobial resistant agents and water disinfection (Abd El-Lateef et al., 2025; Hammerschmidt, 2014). However, there is a limited literature on the practical applying of plants as disinfectants in water treatment. Particularly, research on groundwater purification using plant extracts is narrow in scope. For instance, a study reported by (Abatneh et al., 2014) highlighted that extracts of *Moringa oleifera*, *Jatropha curcas*, and *guar gum* showed a turbidity reduction with 90% (50 NTU initial turbidity) and coliforms and *E.coli* reduction, which improve the well water quality. Another study of (Pritchard et al., 2009) in Malawi, using extracts of the same plants, demonstrated a percentage of 90% turbidity reduction (49 NTU) and 80% of Coliforms reduction in shallow well water with *M. oleifera* showing the best efficiency.

M. oleifera is the most studied plant as demonstrated by several studies, regarding their green coagulant and disinfection potential. (Aney et al., 2009) highlighted some studies determining that *Moringa* seeds carried Pterygospermin antibiotic which is accountable for pathogens degradation. Moreover, a review conducted by (Shah et al., 2026) showed that *M. oleifera* extracts using 0.5 to 1.5 M of saline medium (NaCl) ensure a 1 to 3 log of *E. coli* and coliforms elimination at ranges between 50 and 300 mg/L in drinking water. Another study of (Alam et al., 2020) using *Moringa oleifera* seeds and leaves hexane extracts for groundwater disinfection showing the purification efficacy of combined seeds and leaves in comparison to the individual extracts. However, several limitation related to their potential toxicity were reported(Shah et al., 2024).

Luffa cylindrica have been reported in (Shaheed et al., 2009) for its use in water treatment, 86% of coliform inactivation was achieved with preference of total coliforms than faecal coliforms. Also seed extracts were more performant in coliform inactivation than fruit extracts. In addition, the use of chloroform and crude ethyl acetate fraction extracts of *G. latifolium,* and *Z. zanthoxyloides,* at concentrations between 25 and 500 mg/ml, showed an antimicrobial activity versus *E. coli, Klebsiella sp, P. aeruginosa,* and others, with inhibition diameter ranging from 12 to 38 mm (Adeeyo et al., 2020). The results of (Adeeyo & Odelade, 2018) demonstrated the effectiveness of *Terminalia glaucescens in* water treatment, thus the inhibition diameter against *S. pneumoniae* is 20.5 mm using Ethanol, ethyl acetate and chloroform extracts in different concentrations ranging from 25 mg/ml to 500 mg/ml with ethanolic extract more effective than ethyl acetate and chloroform extracts. Moreover, commercial neem oil of *Azadirachta indica*, using for water treatment, showed 90% of *E.coli* inactivation with dose 2,13 g/l, and lower than 70% inactivation of total coliforms (Matthews et al., 2025). Another study using thyme oils recorded an inhibitory and bactericidal activity versus *E. coli* isolates, with minimum inhibitory concentration and minimum bactericidal concentration values of 512 μg/mL (Brito-Junior et al., 2025).

Regarding the co-working derived products, (Winward et al., 2008) reported the 100% effectiveness antimicrobial activity of Origanum oil and Thyme oil plant for disinfection of coliform in grey water, at 468 mg/L concentration for 30 min, with a limited efficacy in the presence of particles and organics. In addition, several mixtures of these plant extracts, showed more optimization of bacterial inactivation. Additionally, plant extracts and oils from ginger, turmeric, mango ginger, lavender, and Tulsi have been used in combination with acoustic cavitation to enhance disinfection efficiency. These studies demonstrated rapid microbial inactivation, achieving up to 96% disinfection in 15 minutes with ginger oil and 88% with mango ginger extract, relative to 44% with cavitation alone (Mane et al., 2020).

Advanced research from 2020 to 2025 regarding water disinfection, highlighted several approaches, for instance, studies reported the combination of plant extracts with biopolymers or nanomaterials to enhance antimicrobial efficacy. For example, (Benhalima et al., 2025) developed a bio-nanocomposite using xanthan gum and carboxymethyl cellulose strengthened with the pouder of *A. donax* whole biomass encapsulating ZnO nanoparticles, which effectively neutralized bacteria in water, with an antibacterial effect versus *E. coli, S. aureus*, and *P. aeruginosa* after 12h (initial concentration 10^5 CFU mL^{-1}) using just 1g of bio-nanocomposite. In addition, (Gouda et al., 2025) investigated poly-vinyl alcohol aerogels incorporated with garlic peel extract, demonstrating strong antibacterial activity and biofilm inhibition against *K. pneumoniae*, *S. aureus*, *S. enterica*, and *Streptococcus* species, Inhibition zones were 28 mm for *S. enterica* and 25 mm for K. pneumoniae, at concentration of 100 μg/ml. and complete bacterial inactivation for disinfection essays within 150 to 240 min.

Moreover, a study reported by (M.E. Tsie et al., 2025) investigated the performance of composite based on *Moringa oleifera* (CFA-MO) serving for coagulation and bacterial disinfection in raw water treatment, thus, a percentage of 96% turbidity reduction and 100% disinfection bacterial removal was showed using 0.4 g/L of total mass. Another study evaluated sponges made of PVA/sodium alginate incorporated with thyme extract (Thymol and Carvacrol) Raw wastewater, tap water, and groundwater. Pathogens inactivation within 4 to 6 hours for disinfecting groundwater, tap water, and wastewater achieving significant bacterial reductions within 4 to 6 hours, with differences observed between Gram-positive and Gram-negative bacteria (Abd El-Lateef et al., 2025). Some Medicinal plants, their extraction method in addition to theirs studies results were showing in Table 2.

Table 2. Medicinal plants, target antimicrobial action and studies results

Plant name	Plant part use and extraction method	Target antimicrobial action	Results	Reference
Thymus vulgaris	Thyme extract (Thymol and Carvacrol)	Raw wastewater, tap water, and groundwater.	Pathogens inactivation within 4 to 6 hours	(Abd El-Lateef et al., 2025)
M. oleifera	Moringa seed extract (Simple aqueouse extraction)	Groundwater treatment	90% turbidity reduction (50 NTU)	(Abatneh et al., 2014)
M. oleifera	Extract	Groundwater treatment	90% turbidity reduction (49 NTU) 80% Coliforms reduction	(Pritchard et al., 2009)
Guar gum	Extract	Groundwater treatment	90% turbidity reduction (49 NTU) 80% Coliforms reduction	(Pritchard et al., 2009)
Arundo donax	Pouder of whole plant	Water treatment	Antibacterial effect after 12h (with initial concentration 10^5 CFU mL^{-1})	(Benhalima et al., 2025)
Jatropha curcas	Extract	Groundwater treatment	90% turbidity reduction (49 NTU) 80% Coliforms reduction	(Pritchard et al., 2009)
Acacia nilotica	Whole plant (maceration extraction method)	Groundwater and surface water treatment	Reduction of bacterial colonies with 99.86%	(Kirui et al., 2015)
Acacia anthelmintica	Whole plant (maceration extraction method)	Groundwater and surface water	Reduction of bacterial colonies with 9.47%	(Kirui et al., 2015)
Jatropha carcus	Seeds	Groundwater treatment	90% turbidity reduction (50 NTU)	(Abatneh et al., 2014)
Allium sativum L.	Garlic peel extract	Water disinfection	Antibacterial effect Inhibition zone of *K. pneumoniae* (25 mm), *S. enterica* (28 mm) with a EPA concentration of 100 μg/ml. Complete bacterial inactivation for disinfection essays within 150 to 240 min.	(Gouda et al., 2025).
Azadirachta indica	Commercial neem oil	Water treatment	90% Inactivation of *E.coli* with dose 2,13 $g{\cdot}L^{-1}$, inactivation of total coliforms ≤ 70%	(Matthews et al., 2025).

continued on following page

Table 2. Continued

Plant name	Plant part use and extraction method	Target antimicrobial action	Results	Reference
Terminalia glaucescens,	Extraction with chloroform and Ethanol, ethyl acetate. Different concentrations from 25 mg/ml to 500 mg/ml,	Water treatment	Ethanolic extract efficacy than chloroform and ethyl acetate extracts, inhibition zone of 20.5 mm against S. pneumoniae	(Adeeyo & Odelade, 2018)
Zanthoxylum zanthoxyloides,	Crude ethyl acetate and chloroform fractions. Concentrations between 25 and 500 mg/ml	Water treatment	Antimicrobial activity of extract against E. coli, P. aeruginosa, Klebsiella sp, S. pneumoniae and others, With inhibition diameter ranging from 12 to 38 mm	(Adeeyo et al., 2020)
Gongronema latifolium	Crude chloroform and ethyl acetate fractions. Concentrations between 25 and 500 mg/ml	Water treatment	Antimicrobial activity of extract versus E. coli, Klebsiella sp, P. aeruginosa and others, with inhibition diameter ranging from 12 to 38 mm	(Adeeyo et al., 2020)
O. basilicum var. thyrsiflora, O. basilicum var. purpureum, O. citriodorum Vis.	(oils obtained by steam distillation)	Water treatment	O. citriodorum Oil showed the strongest antimicrobial activity.	(Avetisyan et al., 2017)
Moringa oleifera	Seeds and leaves (hexane extract)	Groundwater	Removal efficacy of combined Seeds and leaves relative to the individual extracts,	(Alam et al., 2020)
Moringa oleifera	Composite based on Moringa oleifera with 0.4 g/L mass	Raw water	100% bacterial disinfection and 96% Turbidity reduction	(M.E. Tsie et al., 2025)
Thymus vulgaris	Thyme oils	Water treatment	inhibitory and bactericidal activity, versus Escherichia coli isolates, with MIC and MBC values of 512 μg/mL	(Brito-Junior et al., 2025)
Acacia etbaica	Whole plant (maceration extraction method)	Groundwater and surface water treatment	Reduction of bacterial colonies with 9.47%	(Kirui et al., 2015)

continued on following page

Table 2. Continued

Plant name	Plant part use and extraction method	Target antimicrobial action	Results	Reference
Moringa oleifera	Leaves hexane extract	Groundwater	Improving water quality Escherichia coli count from 315 to 41	(Pandey et al., 2020)
Moringa, Mango Cactus and Okra	Cactus pads, Moringa seeds, Mango kernels and Okra seed pods	Freshwater		(Othmani et al., 2020)
Moringa oleifera	Saline medium Extraction (0.5 to 1.5 M)	Drinking water	1 to 3 log of E. coli and coliforms destruction at 50-300 mg/L	(Shah et al., 2026)

Reported data demonstrated that the plant extracts offer non-toxic and natural effect compared to conventional disinfectants for treating water, which improve the water quality and maintain the one health safety. However, several limitation and boundaries were found for plant and derived compounds using for water disinfection.

5. CHALLENGES AND LIMITATION OF USING PLANT AND THEIR DERIVED COMPOUNDS IN WATER DISINFECTION

The adoption of plant and their derived compounds as alternative to water disinfection encounters several challenges. plant may increase the organic load of water and consequently contribute to water recontamination and promote microbial growth (Shah et al., 2024), which compromising the water sustainability. Regarding toxicity, The interactions of organic matter with other chemical products may produce disinfection byproduct (DBP) precursor (Cheshire & Mitch, 2025; Shah et al., 2026). In addition, some part plant extract using for water purification, may generate a toxicity or mutagenicity due to instability or organic matter fixation, as previously tested on bacteria, fish and unicellular eukaryote what can generate a potential effect on both human health and environment balance (WOK GRABOw et al., 1985), whereby undermine the water disinfection efficacity. This, involves further research on their toxicity and secondary effects to address uncertainties related to its long-term safety.

Furthermore, several variabilities undermine the standardization and disinfection consistency of plant use in water purification including, plant extract composition depending on species, soil nature, season and climate, plant part used and extraction method. In addition, The time demanding applicability of plant extracted for water disinfection and coagulation, which may reduce its industrial implementation

(Pritchard et al., 2010; Shah et al., 2024). In addition to challenges attributed to universal doses and stable concentration of plant and derived active compounds, in addition to regulatory validation in water disinfection which restricted the large-scale applicability, and treatment plants integration.

CONCLUSION

This book chapter highlighted the emerging applications of medicinal plants and their active products in water treatment, as well as the associated challenges in combating antimicrobial pathogens during water disinfection. Antimicrobial resistance represents a major threat to public health due to the dissemination of multidrug-resistant pathogens. Plant-derived active products have been recognized as environmentally friendly, generally less toxic, and possessing broad medicinal applications and demonstrated considerable efficacy against resistant microorganisms. Their extracts have been successfully applied in water treatment to reduce microbial loads and have shown synergistic effects when combined with conventional antibiotics, enhancing overall antimicrobial efficacy. However, additional studies are required to elucidate the mechanisms of action underlying the antimicrobial activities of plant-derived products and enhancing their effectiveness for environmental applications. Research on other effective plants and conduct studies on the mechanisms of derivatives on water disinfection. In addition, in the field research were needed preventing in large-scale purification, ensuring in vivo stability. Studies on their toxicity, immunogenicity, and long-term effects were necessary. Also, mitigating standardization conditions, attributed to doses and concentrations in water disinfections, and developing effective delivery systems providing a clear direction for drug resistance, which maintain water security, environment protection and human health safety.

REFERENCES

Abatneh, Y., Sahu, O., & Yimer, S. (2014). Purification of drinking water by low cost method in Ethiopia. *Applied Water Science, 4*(4), 357–362. DOI: 10.1007/s13201-013-0151-9

Abbas, M., Saeed, F., Anjum, F. M., Afzaal, M., Tufail, T., Bashir, M. S., Ishtiaq, A., Hussain, S., & Suleria, H. A. R. (2017). Natural polyphenols: An overview. *International Journal of Food Properties, 20*(8), 1689–1699. DOI: 10.1080/10942912.2016.1220393

Abd El-Lateef, H. M., Khalaf, M. M., Abou Taleb, M. F., Alali, I., & Gouda, M.Abd El-Lateef. (2025). Ecologically benign synthesis of thyme polyphenolic bioactive compounds loaded Poly(vinyl Alcohol)/Sodium Alginate sponges for sustainable point-of-use applications: Characterization and inactivation of waterborne pathogens in water disinfection. *Journal of Environmental Chemical Engineering, 13*(2), 115704. DOI: 10.1016/j.jece.2025.115704

Abuzerr, S., Hadi, M., Zinszer, K., Nasseri, S., Yunesian, M., Mahvi, A. H., Nabizadeh, R., & Mohammed, S. H. (2024). Quantitative microbial risk assessment for Escherichia Coli O157: H7 via drinking water in the Gaza Strip, Palestine. *SAGE Open Medicine, 12*, 20503121241258071. DOI: 10.1177/20503121241258071 PMID: 38846513

Adeeyo, A. O., Odelade, K. A., Msagati, T. A. M., & Odiyo, J. O. (2020). Antimicrobial potencies of selected native African herbs against water microbes. *Journal of King Saud University. Science, 32*(4), 2349–2357. DOI: 10.1016/j.jksus.2020.03.013

Adeeyo, J. O., & Odelade, K. (2018). Chemical Profiling and Antimicrobial Properties of Phyto-Active Extracts from Terminalia glaucescens Stem Against Water Microbial Contaminants. *The Open Biotechnology Journal, 12*(1), 1–15. DOI: 10.2174/1874070701812010001

Adelowo, O. O., Caucci, S., Banjo, O. A., Nnanna, O. C., Awotipe, E. O., Peters, F. B., Fagade, O. E., & Berendonk, T. U. (2018). Extended Spectrum Beta-Lactamase (ESBL)-producing bacteria isolated from hospital wastewaters, rivers and aquaculture sources in Nigeria. *Environmental Science and Pollution Research International, 25*(3), 2744–2755. DOI: 10.1007/s11356-017-0686-7 PMID: 29139076

Adiba, F., Naufal Prakosa, B. R., Awaliyah, A. R., Heruputri, N. V., Kasuga, I., Priadi, C. R., & Rahmatika, I. (2024). Occurrence of extended-spectrum beta-lactamase producing escherichia coli in groundwater. *E3S Web of Conferences, 485*, 04007. DOI: 10.1051/e3sconf/202448504007

Adjapong, G., Mills-Robertson, F. C., Appenteng, M. A., Ocloo, A., & Garrill, A. (2017). Antimicrobial Activities of Six Selected Plants against Multi-drug Resistant Staphylococcus aureus. *European Journal of Medicinal Plants*, *18*(1), 1–12. DOI: 10.9734/EJMP/2017/30464

Al Yami, M. S. (2017). Comparison of the incidence of acute kidney injury during treatment with vancomycin in combination with piperacillin–tazobactam or with meropenem. *Journal of Infection and Public Health*, *10*(6), 770–773. DOI: 10.1016/j.jiph.2016.11.007 PMID: 28209320

Alam, M. W., Pandey, P., Khan, F., Souayeh, B., & Farhan, M. (2020). Study to Investigate the Potential of Combined Extract of Leaves and Seeds of Moringa oleifera in Groundwater Purification. *International Journal of Environmental Research and Public Health*, *17*(20), 7468. DOI: 10.3390/ijerph17207468 PMID: 33066518

Andriana, Y., Xuan, T. D., Quy, T. N., Minh, T. N., Van, T. M., & Viet, T. D. (2019). Antihyperuricemia, Antioxidant, and Antibacterial Activities of Tridax procumbens L. *Foods*, *8*(1), 21. DOI: 10.3390/foods8010021 PMID: 30634624

Aney, T. Kulkarni, & Kiran Bhise. (2009). Pharmacological And Pharmaceutical Potential Of Moringa oleifera: A Review. | Request. *ResearchGate*. https://www.researchgate.net/publication/40886200_Pharmacological_And_Pharmaceutical_Potential_Of_Moringa_oleifera_A_Review

Angelini, P. (2024). Plant-Derived Antimicrobials and Their Crucial Role in Combating Antimicrobial Resistance. *Antibiotics (Basel, Switzerland)*, *13*(8), 746. DOI: 10.3390/antibiotics13080746 PMID: 39200046

Avetisyan, A., Markosian, A., Petrosyan, M., Sahakyan, N., Babayan, A., Aloyan, S., & Trchounian, A. (2017). Chemical composition and some biological activities of the essential oils from basil Ocimum different cultivars. *BMC Complementary and Alternative Medicine*, *17*(1), 60. DOI: 10.1186/s12906-017-1587-5 PMID: 28103929

Bagherzadehsurbagh, E., Cankiran, G., Ceylan, N., & Azizoglu, R. O. (2025). Plant and fungal antimicrobials targeting antibacterial resistance: A review. *Journal of Herbal Medicine*, *54*, 101071. DOI: 10.1016/j.hermed.2025.101071

Bakal, S. N., Bereswill, S., & Heimesaat, M. M. (2017). Finding Novel Antibiotic Substances from Medicinal Plants—Antimicrobial Properties of Nigella Sativa Directed against Multidrug-resistant Bacteria. *European Journal of Microbiology & Immunology*, *7*(1), 92–98. DOI: 10.1556/1886.2017.00001 PMID: 28386474

Banjo, O. A., Adekanmbi, A. O., & Oyelade, A. A. (2020). Occurrence of CTX-M, SHV and TEM β-lactamase genes in Extended Spectrum Beta-Lactamase (ESBL)-producing bacteria recovered from wastewater of a privately-owned hospital in Nigeria and a hand-dug well within its vicinity. *Gene Reports*, *21*, 100970. DOI: 10.1016/j.genrep.2020.100970

Benhalima, T., Doufene, N., Ferfera-Harrar, H., & Sadi, A. (2025). Bionanocomposite hydrogel based on carboxymethyl cellulose/xanthan gum/ZnO reinforced with Arundo donax biomass: Production, characterization, swelling study and bacterial disinfection properties in water. *International Journal of Biological Macromolecules*, *320*, 145957. DOI: 10.1016/j.ijbiomac.2025.145957 PMID: 40653230

Bergman, M. E., Davis, B., & Phillips, M. A. (2019). Medically Useful Plant Terpenoids: Biosynthesis, Occurrence, and Mechanism of Action. *Molecules (Basel, Switzerland)*, *24*(21), 3961. DOI: 10.3390/molecules24213961 PMID: 31683764

Brito-Junior, L., Brito, H. C., Simões, M. M., Farias, J. H. A., Santos, B., Marques, F. M. C., Medeiros, M. A., Alves, M. S., Pereira, C. T., Diniz, A. F., Vilela, V. L. R., & Oliveira-Filho, A. (2025). Antibacterial potential of Thymus vulgaris (Thyme) oil against Escherichia coli strains isolated from food. *Brazilian Journal of Biology = Revista Brasleira De Biologia, 85*, e292165. https://doi.org/DOI: 10.1590/1519-6984.292165

Cheshire, M. M., & Mitch, W. A. (2025). Algae-derived organic matter in drinking water sources and the formation of disinfection byproducts: A critical review. *Current Opinion in Environmental Science & Health*, *48*, 100675. DOI: 10.1016/j.coesh.2025.100675

Coates, A. R. M., Halls, G., & Hu, Y. (2011). Novel classes of antibiotics or more of the same? *British Journal of Pharmacology*, *163*(1), 184–194. DOI: 10.1111/j.1476-5381.2011.01250.x PMID: 21323894

Coates, A. R. M., Hu, Y., Holt, J., & Yeh, P. (2020). Antibiotic combination therapy against resistant bacterial infections: Synergy, rejuvenation and resistance reduction. *Expert Review of Anti-Infective Therapy*, *18*(1), 5–15. DOI: 10.1080/14787210.2020.1705155 PMID: 31847614

Collivignarelli, M. C., Abbà, A., Benigna, I., Sorlini, S., & Torretta, V. (2018). Overview of the Main Disinfection Processes for Wastewater and Drinking Water Treatment Plants. *Sustainability (Basel)*, *10*(1), 86. DOI: 10.3390/su10010086

Cordeiro, M. F., Nunes, T. R. S., Bezerra, F. G., Damasco, P. K. M., Silva, W. V., Ferreira, M. R. A., Magalhães, C. M. C., Soares, L. L., Cavalcanti, I. M. F., Pitta, M. G. R., & Rêgo, M. J. B. M. (2022). Phytochemical characterization and biological activities of *Plectranthus barbatus* Andrews. *Brazilian Journal of Biology, 82*, e236297. DOI: 10.1590/1519-6984.236297 PMID: 33787716

Daglia, M. (2012). Polyphenols as antimicrobial agents. *Current Opinion in Biotechnology, 23*(2), 174–181. DOI 10.1016/j.copbio.2011.08.007 PMID: 21925860

de Kraker, M. E. A., Stewardson, A. J., & Harbarth, S. (2016). Will 10 Million People Die a Year due to Antimicrobial Resistance by 2050? *PLoS Medicine, 13*(11), e1002184. DOI: 10.1371/journal.pmed.1002184 PMID: 27898664

De Lambert, J. R., Walsh, J. F., Scher, D. P., Firnstahl, A. D., & Borchardt, M. A. (2021). Microbial pathogens and contaminants of emerging concern in groundwater at an urban subsurface stormwater infiltration site. *The Science of the Total Environment, 775*, 145738. DOI: 10.1016/j.scitotenv.2021.145738 PMID: 33631564

De Oliveira, D. M. P., Forde, B. M., Kidd, T. J., Harris, P. N. A., Schembri, M. A., Beatson, S. A., Paterson, D. L., & Walker, M. J. (2020). Antimicrobial Resistance in ESKAPE Pathogens. *Clinical Microbiology Reviews, 33*(3), . https://doi.org/DOI: 10.1128/CMR.00181-19

Dos Santos-Silva, C. A., Zupin, L., Oliveira-Lima, M., Vilela, L. M. B., Bezerra-Neto, J. P., Ferreira-Neto, J. R., Ferreira, J. D. C., de Oliveira-Silva, R. L., Pires, C. de J., Aburjaile, F. F., de Oliveira, M. F., Kido, E. A., Crovella, S., & Benko-Iseppon, A. M. (2020). Plant Antimicrobial Peptides: State of the Art, In Silico Prediction and Perspectives in the Omics Era. *Bioinformatics and Biology Insights, 14*, 1177932220952739. DOI: 10.1177/1177932220952739 PMID: 32952397

Dubey, D., & Padhy, R. N. (2013). Antibacterial activity of Lantana camara L. against multidrug resistant pathogens from ICU patients of a teaching hospital. *Journal of Herbal Medicine, 3*(2), 65–75. DOI: 10.1016/j.hermed.2012.12.002

Faisal, S., Badshah, S. L., Kubra, B., Emwas, A.-H., & Jaremko, M. (2023). Alkaloids as potential antivirals. A comprehensive review. *Natural Products and Bioprospecting, 13*(1), 4. DOI: 10.1007/s13659-022-00366-9 PMID: 36598588

Ferdosh, S. (2023). Ethnobotanical Review of Selected Medicinal Plants in Guam for the Treatment of Urinary Tract Ailments and Their Pharmacological Properties. *Scientia Pharmaceutica, 91*(3), 43. DOI: 10.3390/scipharm91030043

Foyle, L., Burnett, M., Creaser, A., Hens, R., Keough, J., Madin, L., Price, R., Smith, H., Stone, S., & Kinobe, R. T. (2023). Prevalence and distribution of antimicrobial resistance in effluent wastewater from animal slaughter facilities: A systematic review. *Environmental Pollution*, *318*, 120848. DOI: 10.1016/j.envpol.2022.120848 PMID: 36563990

Fu, C., Xu, B., Chen, H., Zhao, X., Li, G., Zheng, Y., Qiu, W., Zheng, C., Duan, L., & Wang, W. (2022). Occurrence and distribution of antibiotics in groundwater, surface water, and sediment in Xiong'an New Area, China, and their relationship with antibiotic resistance genes. *The Science of the Total Environment*, *807*, 151011. DOI: 10.1016/j.scitotenv.2021.151011 PMID: 34715223

Gogate, P. R. (2007). Application of cavitational reactors for water disinfection: Current status and path forward. *Journal of Environmental Management*, *85*(4), 801–815. DOI: 10.1016/j.jenvman.2007.07.001 PMID: 17714855

Gouda, M., Abou Taleb, M. F., Albassami, N. A., & Alqahtani, N. K. (2025). Manufacturing of poly(vinyl alcohol) aerogel enriched with waste garlic peel extract as an alternative eco-friendly solution for inactivation of waterborne pathogens during water disinfection. *Journal of Environmental Chemical Engineering*, *13*(1), 115053. DOI: 10.1016/j.jece.2024.115053

Hammerschmidt, R. (2014). Chlorogenic acid: A versatile defense compound. *Physiological and Molecular Plant Pathology*, *88*, iii–iv. DOI: 10.1016/j.pmpp.2014.11.002

Hassan, M., Musa, F. M., Alíyu, F., & Adamu, A. (2022). Antibacterial Activity of Ethanol Leaf Extract of Sida acuta Against Some Clinical Bacterial Isolates. *International Journal of Life Sciences and Biotechnology*, *5*(3), 572–580. DOI: 10.38001/ijlsb.1115771

Huang, W., Wang, Y., Tian, W., Cui, X., Tu, P., Li, J., Shi, S., & Liu, X. (2022). Biosynthesis Investigations of Terpenoid, Alkaloid, and Flavonoid Antimicrobial Agents Derived from Medicinal Plants. *Antibiotics (Basel, Switzerland)*, *11*(10), 1380. DOI: 10.3390/antibiotics11101380 PMID: 36290037

Izah, S. C., Joshua, M. T., Torru, K. E., Ngun, C. T., Ogwu, M. C., & Hait, M. (2024). Antimicrobial Resistance and the Role of Herbal Medicine: Challenges, Opportunities, and Future Prospects. In *Herbal Medicine Phytochemistry* (pp. 1–26). Springer., DOI: 10.1007/978-3-031-21973-3_71-1

Jaisinghani, R. N. (2017). Antibacterial Properties of Quercetin. *Microbiology Research*, *8*(1), 6877. DOI: 10.4081/mr.2017.6877

Jubair, N., Rajagopal, M., Chinnappan, S., Abdullah, N. B., & Fatima, A. (2021). Review on the Antibacterial Mechanism of Plant-Derived Compounds against Multidrug-Resistant Bacteria (MDR). *Evidence-Based Complementary and Alternative Medicine : eCAM, 3663315*, 1–30. Advance online publication. DOI: 10.1155/2021/3663315 PMID: 34447454

Kakkar, S., Dharavat, N., & Pammi, S. V. N. (2025). A Moringa oleifera, a natural coagulant, as a potential future approach for sustainable water purification: A patent based study. *Results in Engineering, 27*, 106630. DOI: 10.1016/j.rineng.2025.106630

Kashyap, J. D., Bhardwaj, M. H., & Pal, D. (2022). *(PDF) In-vitro Antibacterial and Antifungal Activity of Curcuma Amada Roxb. Against Human Pathogens*. ResearchGate., DOI: 10.48048/tis.2022.4839

Kin, A., Yaki, L. M., Abubakar, I., Olusola, L. F., & Zubairu, R. (2018). Antibacterial activity of Ocimum gratissimum (scent leaf) on some pathogenic gastrointestinal bacteria. *African Journal of Microbiological Research, 12*(40), 923–929. DOI: 10.5897/AJMR2018.8847

Kipré, Ainyakou-Sanga, Gblossi Goualie, Sylla, Kouamé Kra, Kakou-NGazoa, & Alfred Karou. (2025). Antimicrobial Resistance in Salmonella spp. Isolated from Poultry Waste in Abidjan District, Côte d'ivoire. *International Journal of Agriculture and Biosciences*. Advance online publication. DOI: 10.47278/journal.ijab/2025.036

Kirmani, F., Saddiqe, Z., Saleem, S., Ali, F., & Haq, F. (2024). Phytochemical investigation and antibacterial activity of Curcuma longa against multi-drug resistant bacteria. *South African Journal of Botany, 164*, 137–145. DOI: 10.1016/j.sajb.2023.11.042

Kirui, J. K., Kotut, K., & Okemo, P. O. (2015). Efficacy of aqueous plant extract in disinfecting water of different physicochemical properties. *Journal of Water and Health, 13*(3), 848–852. DOI: 10.2166/wh.2015.002 PMID: 26322770

Koudoum, P. L., Founou, R. C., Founou, L. L., Foueyem, M. D., Guemkam, G. W., Deuguen, R., Nke, G. A., Gonsu, H., & Choukem, S. P. (2025). Plasmid-mediated quinolone resistance among extended-spectrum β-lactamase-producing Escherichia coli and Klebsiella pneumoniae isolated from hospitalized patients, hospital environment and wastewaters in Cameroon. *Scientific Reports, 15*(1), 36526. DOI: 10.1038/s41598-025-15465-z PMID: 41120393

Lei, J., Sun, L. C., Huang, S., Zhu, C., Li, P., He, J., Mackey, V., Coy, D. H., & He, Q. Y. (2019). The antimicrobial peptides and their potential clinical applications. *American Journal of Translational Research, 11*(7), 3919–3931. PMID: 31396309

Liu, T., Wang, Z.-J., Shi, Y.-Z., Tao, R., Huang, H., Zhao, Y.-L., & Luo, X.-D. (2024). Curcusinol from the fruit of Carex baccans with antibacterial activity against multidrug-resistant strains. *Journal of Ethnopharmacology, 318*(Pt A), 116892. DOI: 10.1016/j.jep.2023.116892

Liu, Y., Cui, Y., Lu, L., Gong, Y., Han, W., & Piao, G. (2020). Natural indole-containing alkaloids and their antibacterial activities. *Archiv der Pharmazie, 353*(10), 2000120. DOI: 10.1002/ardp.202000120 PMID: 32557757

Mane, M. B., Bhandari, V. M., Balapure, K., & Ranade, V. V. (2020). Destroying antimicrobial resistant bacteria (AMR) and difficult, opportunistic pathogen using cavitation and natural oils/plant extract. *Ultrasonics Sonochemistry, 69*, 105272. DOI: 10.1016/j.ultsonch.2020.105272 PMID: 32739732

Mansoori, A., Singh, N., Dubey, S. K., Thakur, T. K., Alkan, N., Das, S. N., & Kumar, A. (2020). Phytochemical Characterization and Assessment of Crude Extracts From Lantana camara L. for Antioxidant and Antimicrobial Activity. *Frontiers in Agronomy, 2*, 582268. Advance online publication. DOI: 10.3389/fagro.2020.582268

Masoumian, M., & Zandi, M. (2016). *Antimicrobial Activity of Some Medicinal Plant Extracts against Multidrug Resistant Bacteria* (Research Article 19; Issue 11, p. e10080). Zahedan Journal of Research in Medical Sciences. https://brieflands.com/journals/zjrms/articles/10080#abstract

Matloko, K., Fri, J., Ateba, T. P., Molale-Tom, L. G., & Ateba, C. N. (2021). Evidence of potentially unrelated AmpC beta-lactamase producing Enterobacteriaceae from cattle, cattle products and hospital environments commonly harboring the blaACC resistance determinant. *PLoS One, 16*(7), e0253647. DOI: 10.1371/journal.pone.0253647 PMID: 34324493

Matthews, Templeton, Tripathi, & Bhattarai. (2025). (PDF) Disinfection of Waterborne Coliform Bacteria by Neem Oil. *ResearchGate*. DOI: 10.1089/ees.2009.0058

Mourabiti, F., Jouga, F., Calvo, L. G., Villarino, R.-A., Zouheir, Y., Soukri, A., de Miguel, T., & El Khalfi, B. (2025). Polyphenols as Potential β-Lactamase Inhibitors: An Integrated Computational and Experimental Study. *Molecules (Basel, Switzerland), 30*(22), 4416. DOI: 10.3390/molecules30224416 PMID: 41302475

Mwalongo, O., Kisula, L., Credo, D., & Sauli, E. (2025). Comparative evaluation of antimicrobial activity and synergistic effects of selected Tanzanian medicinal plants against urinary tract infection pathogens. *South African Journal of Botany, 185*, 130–143. DOI: 10.1016/j.sajb.2025.07.057

Nawrot, barylski, & niwicki. (2014). (PDF) Plant antimicrobial peptides. *ResearchGate*. DOI: 10.1007/s12223-013-0280-4

Nguyen, T. L. A., & Bhattacharya, D. (2022). Antimicrobial Activity of Quercetin: An Approach to Its Mechanistic Principle. *Molecules (Basel, Switzerland)*, *27*(8), 2494. DOI: 10.3390/molecules27082494 PMID: 35458691

Nouemsi, B. J., Peron, L., & Valaire, Y. (2025). (PDF) A new flavonol derivative and other compounds from the leaves of Bauhinia thonningii Schum with activity against multidrug-resistant bacteria. *ResearchGate*. https://www.researchgate.net/publication/364031581_A_new_flavonol_derivative_and_other_compounds_from_the_leaves_of_Bauhinia_thonningii_Schum_with_activity_against_multidrug-resistant_bacteria

Ocampo, P. S., Lázár, V., Papp, B., Arnoldini, M., Abel zur Wiesch, P., Busa-Fekete, R., Fekete, G., Pál, C., Ackermann, M., & Bonhoeffer, S. (2014). Antagonism between bacteriostatic and bactericidal antibiotics is prevalent. *Antimicrobial Agents and Chemotherapy*, *58*(8), 4573–4582. DOI: 10.1128/AAC.02463-14 PMID: 24867991

Othmani, B., Rasteiro, M. G., & Khadhraoui, M. (2020). Toward green technology: A review on some efficient model plant-based coagulants/flocculants for freshwater and wastewater remediation. *Clean Technologies and Environmental Policy*, *22*(5), 1025–1040. DOI: 10.1007/s10098-020-01858-3

Pacyga, K., Pacyga, P., Topola, E., Viscardi, S., & Duda-Madej, A. (2024). Bioactive Compounds from Plant Origin as Natural Antimicrobial Agents for the Treatment of Wound Infections. *International Journal of Molecular Sciences*, *25*(4), 2100. DOI: 10.3390/ijms25042100 PMID: 38396777

Pandey, P., Khan, F., Ahmad, V., Singh, A., Shamshad, T., & Mishra, R. (2020). Combined efficacy of Azadirachta indica and Moringa oleifera leaves extract as a potential coagulant in ground water treatment. *SN Applied Sciences*, *2*(7), 1300. DOI: 10.1007/s42452-020-3124-2

Pritchard, M., Craven, T., Mkandawire, T., Edmondson, A. S., & O'Neill, J. G. (2010). A study of the parameters affecting the effectiveness of *Moringa oleifera* in drinking water purification. *Physics and Chemistry of the Earth Parts A/B/C*, *35*(13), 791–797. DOI: 10.1016/j.pce.2010.07.020

Pritchard, M., Mkandawire, T., Edmondson, A., O'Neill, J. G., & Kululanga, G. (2009). Potential of using plant extracts for purification of shallow well water in Malawi. *Physics and Chemistry of the Earth Parts A/B/C*, *34*(13), 799–805. DOI: 10.1016/j.pce.2009.07.001

Radulovic, N. S., Blagojevic, P. D., Stojanovic-Radic, Z. Z., & Stojanovic, N. M. (2013). Antimicrobial Plant Metabolites: Structural Diversity and Mechanism of Action. *Current Medicinal Chemistry*, *20*(7), 932–952. DOI: 10.2174/092986713805219136 PMID: 23210781

Reiter, J., Levina, N., van der Linden, M., Gruhlke, M., Martin, C., & Slusarenko, A. J. (2017). Diallylthiosulfinate (Allicin), a Volatile Antimicrobial from Garlic (Allium sativum), Kills Human Lung Pathogenic Bacteria, Including MDR Strains, as a Vapor. *Molecules (Basel, Switzerland)*, *22*(10), 1711. DOI: 10.3390/molecules22101711 PMID: 29023413

Salikin, N. H., Keong, L. C., Azemin, W.-A., Philip, N., Yusuf, N., Daud, S. A., & Rashid, S. A. (2024). Combating multidrug-resistant (MDR) Staphylococcus aureus infection using terpene and its derivative. *World Journal of Microbiology & Biotechnology*, *40*(12), 402. DOI: 10.1007/s11274-024-04190-w PMID: 39627623

Semenzato, G., Bernacchi, A., Amata, S., Bechini, A., Berti, F., Calonico, C., Catania, V., Esposito, A., Puglia, A. M., Piccionello, A. P., Emiliani, G., Biffi, S., & Fani, R. (2024). Antibacterial Properties of Bacterial Endophytes Isolated from the Medicinal Plant Origanum heracleoticum L. *Frontiers in Bioscience (Landmark Edition)*, *29*(3), 111. DOI: 10.31083/j.fbl2903111 PMID: 38538270

Shah, A., Arjunan, A., Batool, M., Chike, O., Dhir, A., Arafat, A., Manning, G., Zakharova, J., Abbas, N., Mehran, M. T., Rubab, R., & Husnain, M. (2026). Green yet constrained: Reassessing *Moringa oleifera's* role in drinking water and wastewater treatment. *Cleaner Water*, *5*, 100202. DOI: 10.1016/j.clwat.2025.100202

Shah, A., Arjunan, A., Manning, G., Zakharova, J., Andraulaki, I., & Batool, M. (2024). The effect of dose, settling time, shelf life, storage temperature and extractant on *Moringa oleifera* Lam. Protein coagulation efficiency. *Environmental Nanotechnology, Monitoring & Management*, *21*, 100919. DOI: 10.1016/j.enmm.2024.100919

Shaheed, A., Templeton, M. R., Matthews, R. L., Tripathi, S. K., & Bhattarai, K. (2009). Disinfection of waterborne coliform bacteria using Luffa cylindrica fruit and seed extracts. *Environmental Technology*, *30*(13), 1435–1440. DOI: 10.1080/09593330903193485 PMID: 20088208

Sheard, O'Brien-Simpson, D. Wade, & Separovic. (2019). (PDF) Combating bacterial resistance by combination of antibiotics with antimicrobial peptides. *ResearchGate*. DOI: 10.1515/pac-2018-0707

Shin, J., Prabhakaran, V.-S., & Kim, K.-S. (2018). The multi-faceted potential of plant-derived metabolites as antimicrobial agents against multidrug-resistant pathogens. *Microbial Pathogenesis, 116*, 209–214. DOI: 10.1016/j.micpath.2018.01.043 PMID: 29407230

Sidhu, J. P. S., Ahmed, W., Gernjak, W., Aryal, R., McCarthy, D., Palmer, A., Kolotelo, P., & Toze, S. (2013). Sewage pollution in urban stormwater runoff as evident from the widespread presence of multiple microbial and chemical source tracking markers. *The Science of the Total Environment, 463–464*, 488–496. DOI: 10.1016/j.scitotenv.2013.06.020 PMID: 23831795

Singh, Rakesh Singh, Bhunia, & L. Stroshine. (2002). Efficacy of Chlorine Dioxide, Ozone, and Thyme Essential Oil or a Sequential Washing in Killing Escherichia coli O157:H7 on Lettuce and Baby Carrots | Request PDF. *ResearchGate*. DOI: 10.1006/fstl.2002.0933

Slavokhotova, A. A., & Rogozhin, E. A. (2020). Defense Peptides From the α-Hairpinin Family Are Components of Plant Innate Immunity. *Frontiers in Plant Science, 11*, 465. DOI: 10.3389/fpls.2020.00465 PMID: 32391035

Solaiman, S., Handy, E., Brinks, T., Goon, K., Bollinger, C., Sapkota, A. R., Sharma, M., & Micallef, S. A. (2022). Extended Spectrum β-Lactamase Activity and Cephalosporin Resistance in Escherichia coli from U.S. Mid-Atlantic Surface and Reclaimed Water. *Applied and Environmental Microbiology, 88*(15), e00837–e22. DOI: 10.1128/aem.00837-22 PMID: 35862684

Srivastava, S., Kumar, A., Gupta, R., & Malik, A. (2025). Geospatial mapping and cluster analysis of antibiotic-resistant Escherichia coli in drinking water of semi-arid areas. *Environmental Monitoring and Assessment, 197*(5), 582. DOI: 10.1007/s10661-025-13982-y PMID: 40272580

Subramani, R., Narayanasamy, M., & Feussner, K.-D. (2017). Plant-derived antimicrobials to fight against multi-drug-resistant human pathogens. *3 Biotech, 7*(3), 172. DOI: 10.1007/s13205-017-0848-9

Sultana, S. (2018). *Medicinal plants combating against human pathogens: A review.* https://www.academia.edu/120040005/Medicinal_plants_combating_against_human_pathogens_A_review

Takahashi, N., Nakasato, J. A., de Jesus, G. S., Micheletti, A. C., Pott, A., Yoshida, N. C., & Paulo, P. L. (2024). Antimicrobial activity of Pantanal macrophytes against multidrug resistant bacteria shows potential for improving nature-based solutions. *Plant Stress, 11*, 100304. DOI: 10.1016/j.stress.2023.100304

Tanwar, J., Das, S., Fatima, Z., & Hameed, S. (2014). Multidrug Resistance: An Emerging Crisis. *Interdisciplinary Perspectives on Infectious Diseases, 2014*(1), 541340. DOI: 10.1155/2014/541340 PMID: 25140175

Thawabteh, A., Juma, S., Bader, M., Karaman, D., Scrano, L., Bufo, S. A., & Karaman, R. (2019). The Biological Activity of Natural Alkaloids against Herbivores, Cancerous Cells and Pathogens. *Toxins, 11*(11), 656. DOI: 10.3390/toxins11110656 PMID: 31717922

Togashi, N., Hamashima, H., Shiraishi, A., Inoue, Y., & Takano, A. (2010). Antibacterial Activities Against Staphylococcus aureus of Terpene Alcohols With Aliphatic Carbon Chain. *The Journal of Essential Oil Research, 22*(3), 263–269. DOI: 10.1080/10412905.2010.9700321

Tsie, M. E., Ntuli, F., Lekgoba, T., & Kandjou, V. (2025). Hybridization of fly ash – Moringa based composite for coagulation and disinfection of raw water. *Next Materials, 8*, 100722. DOI: 10.1016/j.nxmate.2025.100722

Tuts, L., Heyndrickx, M., Becue, I., Boon, N., De Maesschalck, P., & Rasschaert, G. (2025). Tracking antibiotics and antibiotic-resistant E. coli in the aquatic environment linked to agriculture. *Environmental Pollution, 374*, 126265. DOI: 10.1016/j.envpol.2025.126265 PMID: 40252751

Valli, Pivatto, & Danuello. (2012). (PDF) Tropical biodiversity: Has it been a potential source of secondary metabolites useful for medicinal chemistry? *ResearchGate*. https://doi.org/DOI: 10.1590/S0100-40422012001100036

Vaou, N., Stavropoulou, E., Voidarou, C., Tsigalou, C., & Bezirtzoglou, E. (2021). Towards Advances in Medicinal Plant Antimicrobial Activity: A Review Study on Challenges and Future Perspectives. *Microorganisms, 9*(10), 2041. DOI: 10.3390/microorganisms9102041 PMID: 34683362

Wang, S., Yao, J., Zhou, B., Yang, J., Chaudry, M. T., Wang, M., Xiao, F., Li, Y., & Yin, W. (2018). Bacteriostatic Effect of Quercetin as an Antibiotic Alternative In Vivo and Its Antibacterial Mechanism In Vitro. *Journal of Food Protection, 81*(1), 68–78. DOI: 10.4315/0362-028X.JFP-17-214 PMID: 29271686

WHO. (2014). *Preventing diarrhoea through better water, sanitation and hygiene Exposures and impacts in low- and middle-income countries*. https://www.who.int/publications/i/item/9789241564823

WHO. (2024). *WHO bacterial priority pathogens list, 2024: Bacterial pathogens of public health importance to guide research, development and strategies to prevent and control antimicrobial resistance*. https://www.who.int/publications/i/item/9789240093461

Winward, G. P., Avery, L. M., Stephenson, T., & Jefferson, B. (2008). Essential oils for the disinfection of grey water. *Water Research*, *42*(8), 2260–2268. DOI: 10.1016/j.watres.2007.12.004 PMID: 18221769

WOK GRABOw, J. L., MORGAN, W., & Jahn, S. A. A. (1985). Toxicity and mutagenicity evaluation of water coagulated with Moringa oleifera seed preparations using fish, protozoan, bacterial, coliphage, enzyme and Ames Salmonella assays. *Water S.A.*, *11*(1), 9.

Worthington, R. J., & Melander, C. (2013). Combination approaches to combat multidrug-resistant bacteria. *Trends in Biotechnology*, *31*(3), 177–184. DOI: 10.1016/j.tibtech.2012.12.006 PMID: 23333434

Y. su, Rui Gao, Fang Huang, & Liang. (2024). Occurrence, transmission and risks assessment of pathogens in aquatic environments accessible to humans. *Journal of Environmental Management, 354*, 120331. DOI: 10.1016/j.jenvman.2024.120331

Yahya, Ibrahim, Zawawi, & Abd Hamid. (2014). (PDF) Biofilm Killing Effects of Chromolaena odorata Extracts against Pseudomonas aeruginosa. *ResearchGate*. DOI: 10.3923/rjphyto.2014.64.73

Yang, Y., & Zhang, T. (2019). Antimicrobial Activities of Tea Polyphenol on Phytopathogens: A Review. *Molecules (Basel, Switzerland)*, *24*(4), 816. DOI: 10.3390/molecules24040816 PMID: 30823535

Zacchino, S. A., Butassi, E., Cordisco, E., & Svetaz, L. A. (2017). Hybrid combinations containing natural products and antimicrobial drugs that interfere with bacterial and fungal biofilms. *Phytomedicine : International Journal of Phytotherapy and Phytopharmacology*, *37*, 14–26. DOI: 10.1016/j.phymed.2017.10.021 PMID: 29174600

Chapter 11

Natural Defense Mechanisms and Bioactive Compounds in Ipomoea biloba Leaves:

Antimicrobial Profiling of the Ethanol Extract

R. Vallipriya
http://orcid.org/0009-0008-6692-603X
SRM Campus, Pavendar Bharathidasan College of Arts and Science, Bharathidasan University, India

Shobana Devi Paulraj
http://orcid.org/0000-0002-8843-4721
Bharathidasan University, India

S. Srinivasan
http://orcid.org/0009-0002-0179-9849
Graphic Era University, India

ABSTRACT

Ipomoea biloba (Ipomoea pes-caprae), a coastal medicinal plant of the family Convolvulaceae, is traditionally used to treat infections, inflammation, and gastro-intestinal disorders. The present study investigates the natural defense mechanisms and antimicrobial potential of the ethanol leaf extract of I. biloba, with emphasis on its bioactive compounds. Phytochemical screening revealed the presence of alkaloids, flavonoids, phenolics, tannins, saponins, terpenoids, steroids, carbohydrates, pro-

DOI: 10.4018/979-8-3373-7810-7.ch011

teins, and amino acids, indicating a rich secondary metabolite profile. Antimicrobial activity assessed using the disc diffusion method demonstrated effective inhibition against selected Gram-positive and Gram-negative bacterial pathogens, with measurable zones of inhibition and defined minimum inhibitory concentrations. The observed antimicrobial efficacy is attributed to the synergistic action of phenolics and flavonoids, which also contribute to antioxidant and plant defense functions.

INTRODUCTION

Most of the vigorously threatening illnesses in public health are caused by pathogens such as fungi, bacteria, parasites, and viruses, despite many advances in medical science. Many medicinal plants play an essential role in controlling numerous diseases and exhibit antimicrobial activity. Herbal and medicinally important plants are used as antimicrobial drugs in many countries; they are the main source of potentially active and most powerful drugs in the modern era, and plant parts are directly extracted and used as drugs (Abdallah et al., 2023). They also have different properties, with medicinally active compounds that act as antimicrobial agents. Moreover, these plants have been used as indigenous medicine and are also widely and easily available in villages and coastal areas, although they are present in smaller amounts compared to modern medicine. The large genus *Ipomoea* in the Convolvulaceae family has more than 500 species (Meira et al., 2012). The medicinal plant *Ipomoea biloba* is a rich source of antimicrobial activity. *Ipomoea biloba* is also known as *Ipomoea pes-caprae*, and its common name is beach morning glory. This species mainly grows in coastal environments with salt tolerance and remarkable bioactive compounds. This plant thrives in tropical and subtropical regions of Asia (India). Ipomoea biloba is found worldwide. It occurs in coastal areas of tropical South America, North America, Africa, Asia, and Australia (Akinniyi et al., 2022). According to the Global Register, this species is found in Anguilla, Spain, South Africa, and the Marshall Islands.

Taxonomic Ipomoea biloba Classifications

Figure 1. Scientific classification

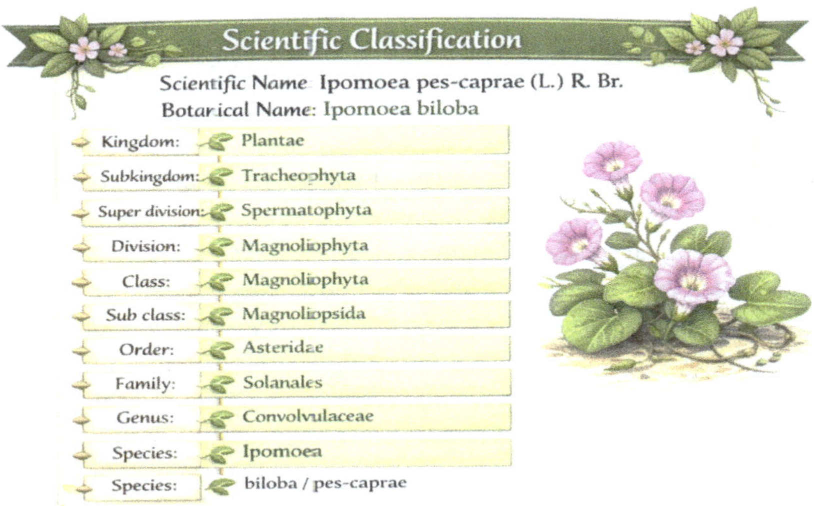

Figure 2. Ipomoea biloba original plant leaf and Herbarium plant leaf

The name *biloba* is derived from its two lobes (Figure 2), which are the main morphological structures The leaves are ovate (egg-shaped) or circular with notched tips. This is the main feature that differentiates it from other *Ipomoea* species (Brown & Frank, 2020). The phylogenetic differentiation of this plant helps scientists understand its evolutionary relationship with other species that have medicinal and edible properties. This species is mainly a stabilizer in salty sand and is colonized on beaches and in subtropical and tropical regions worldwide (Muñoz-Rodríguez et al., 2023). It has deep taproots and sprawls, which helps prevent erosion in coastal sand. This plant is a prostrate, perennial, hairless (glabrous) vine that produces a showy, funnel-shaped pink and purple flower that opens in the morning. The leaves of these plants are very simple, arranged alternately, dark green, leather-like, and hairless (glabrous). The bases of the leaves are truncate and cordate (heart-shaped), and the apex of the leaf is usually deeply cleft. These leaves are reddish in color when young, and when they mature, they appear yellowish green (Brown & Frank, 2020).

Indigenous plant-use studies have shown that *Ipomoea biloba* is used to treat analgesic conditions and inflammation, with the leaves and whole plant commonly utilized (Yeshi et al., 2022). The antinociceptive characteristics were observed in mice. In vitro, the crude extract of this plant inhibited prostaglandin synthesis. Leaf extracts have high antioxidant and cytotoxic activities (Vallipriya & Shabana Begum, 2021; Parekh et al., 2012). It possesses insulinogenic, hypoglycemic, anti-hemolytic, antispasmodic, anti-inflammatory, and antihistaminic properties

Ipomoea biloba is used for various medicinal purposes based on traditional practices. Some realistic examples are as follows: Australian Aborigines used heated leaves for wounds, skin infections, sore inflammations, and the stings of poisonous fishes, manta rays, and insects (Yeshi et al., 2022). In Brazil, this plant is used to treat inflammation, gastrointestinal disorders, and pain. This plant is used in India for protection against evil spirits and during ritual baths. Other recommendations include its use to treat kidney problems and hypertension. This plant is sometimes used to treat disorders of the digestive system, pain in internal and external parts, inflammation, strains, fatigue, joint inflammation, intestinal infections, diarrhea with blood, and rheumatic diseases (Akinniyi et al., 2022).

This plant has multiple activities that are antioxidant activity, anti-inflammatory activity, antipyretic activity, anti-diabetic activity, anti-ulcer effect, cytotoxic activity, antifungal activity, cardioprotective activity, and anti-tumor activity. This plant has better activity against many diseases, so researchers chose this plant for the study Natural Defense Mechanisms and Bioactive Compounds in *Ipomoea biloba* Leaves: Antimicrobial Profiling of the Ethanol Extract. In the future, this plant can be used in further pharmacological studies.

BACKGROUND OF THE STUDY

Medicinal plants are important sources of therapeutic agents due to their diverse bioactive compounds and natural defense mechanisms against microbial pathogens. The rising challenge of antimicrobial resistance has renewed interest in plant-based antimicrobials that are effective, affordable, and environmentally sustainable. Consequently, the study of plant defense systems and secondary metabolites has gained increasing scientific relevance.

Ipomoea biloba (family: Convolvulaceae) is a coastal and tropical plant traditionally used to treat infections, inflammation, and gastrointestinal disorders. Its survival in harsh environments suggests the presence of strong defense mechanisms mediated by secondary metabolites such as phenolics, flavonoids, alkaloids, terpenoids, and glycosides. These compounds protect the plant from microbial attack and environmental stress while offering potential pharmacological benefits. Leaves are key sites for the synthesis of defensive phytochemicals, and ethanol is an effective solvent for extracting a wide range of antimicrobial bioactive compounds. Therefore, the *EEIB* leaves represent a promising source for antimicrobial investigation. Despite its traditional use, scientific evidence on the antimicrobial potential of *Ipomoea biloba* leaf ethanol extract remains limited. This study aims to evaluate its antimicrobial efficacy, thereby supporting traditional knowledge and contributing to the search for novel plant-derived antimicrobial agents to address antimicrobial resistance.

Significance of the study

1. To examine the antimicrobial activity of *EEIB* in the in-vitro studies.
2. To determine the phytochemical analysis of the *EEIB* leaves.
3. This study suggests that it may support the use of plant-derived antimicrobials in the pharmaceutical, food preservation, and healthcare industries.
4. Scientific understanding of basic medicinal knowledge and provides laboratory-based validation of ethnobotanical practices.
5. This study contributes to the growing field of natural product research and encourages further isolation, purification, and characterization of specific antimicrobial compounds.

Objectives

1. The *EEIB* was determined for its yield of extraction.
2. The *EEIB* was analyzed for its phytochemical constituents.
3. The *EEIB* was assessed for its antimicrobial activity against the selected organisms.
4. To characterize the major bioactive compounds.

5. To contribute scientific insights that may support future herbal medicine and drug development applications.

Research Questions

1. What phytochemical compounds are present in the *EEIB* leaves?
2. Does the *EEIB* leave significant antimicrobial effect?
3. Is there is any relationship between *EEIB* leaves and antimicrobial effect?

CRITICAL REVIEW OF EXISTING LITERATURE

In the last few decades, pharmacological industries have developed many new antibiotics, but microorganisms have become resistant to them. The WHO released the Priority Pathogen List (PPL) in 2016 to guide researchers in developing new antibiotic guidelines. Most of the old and conventional medicines are used based on these guidelines and are nowadays practiced against a variety of infections. In developing nations, above 80 percent of the world's population uses traditional medicines solely for their medical needs (Bhatia et al., 2021). *Ipomoea biloba* plants contribute to environmental sustainability in coastal areas and help control shoreline erosion. (Martinez et al., 2016; Das ,2017).

MOLECULAR AND BIOCHEMICAL DEFENSE STRATEGIES IN PLANTS

Plants have an inbuilt variety of natural defense mechanisms that protect them against pathogens, stress, and pests. Defense mechanisms may be structural or biochemical in nature. Structural defenses may include physical barriers, such as cell walls, thorns, cuticles, and trichomes, which help prevent the entry of toxic organisms (Demis, 2024). Biochemical defenses may involve the production of compounds, such as phytoalexins, enzymes, and secondary metabolites that prevent pathogen growth and neutralize toxic compounds (Figure 3). Indirect defenses may involve the production of compounds that attract predators and endophytic fungi, which protect the host. Natural defense mechanisms that help plants survive, maintain, and adapt to diverse and challenging environments (Li et al., 2019; Roychowdhury et al., 2025).

Figure 3. Natural defense mechanisms

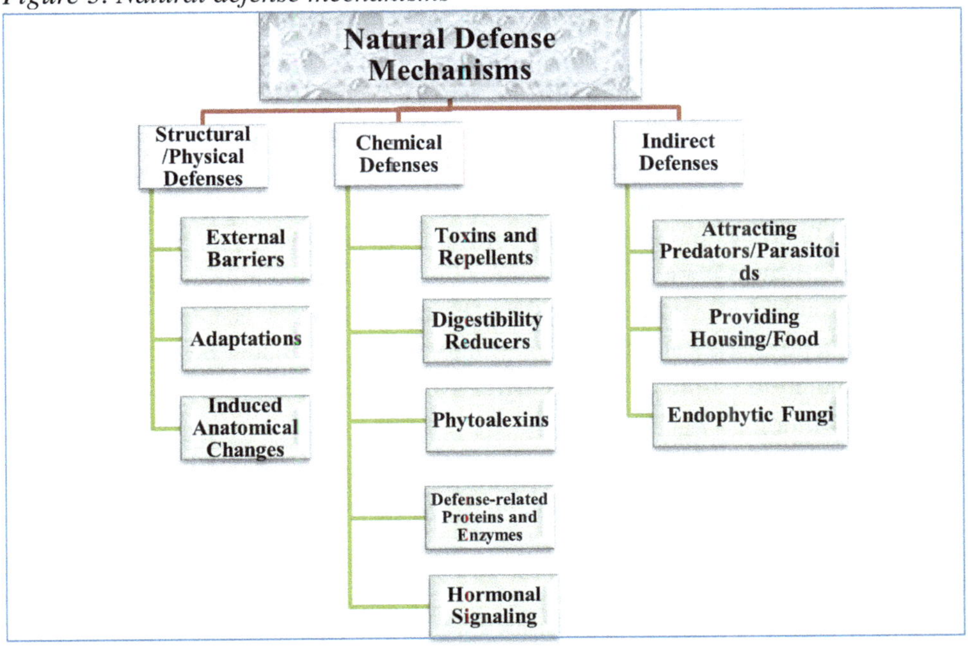

Ipomoea biloba plant, having a well-developed immune system, helps in multiple ecosystems. Both biotic and abiotic factors activate the immune system, MTI and ETI. It produces a long-term hypersensitive response against pathogens (Muthamilarasan & Prasad, 2013). Finally, *Ipomoea biloba* not only supports ecological adaptability but also enhances pharmacological value, providing a basis for future research. (Figure 4).

Figure 4. Signal-mediated induced immune responses in plants

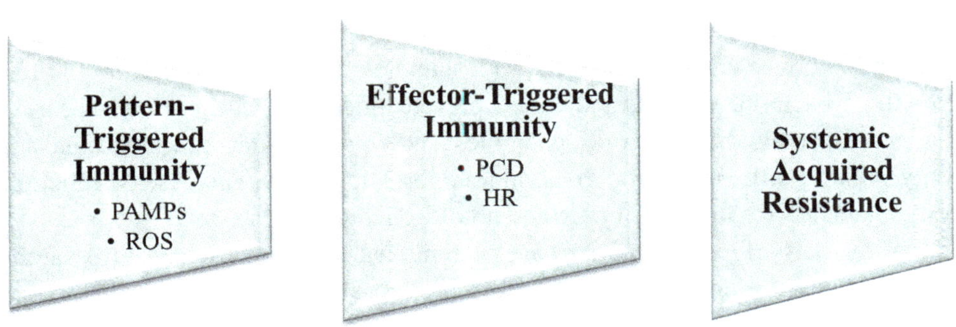

Plant growth mostly depends on defense activity compounds that are mainly saponins, phenolic compounds, flavonoids, terpenoids, and alkaloids. These secondary metabolites actively participate in survival and the immune system. Bioactive compounds are not only used in direct primary physiological processes but also have sustainability in pharmacolog (Bhatla & Lal, 2023). The role of plant defense activity is carried out by secondary metabolites, and they act as protective agents and control various biotic and abiotic stresses. The main defense-rich compounds—phenolics, alkaloids, and terpenoids—prevent the growth of pathogens, bacteria, fungi, and viruses. Other compounds, such as saponins, tannins, and cyanogenic glycosides, protect plants from insects and herbivorous animals. Other compounds, mainly carotenoids, flavonoids, and other antioxidants, help plants protect themselves from UV radiation, salinity, drought, oxidative stress, and temperature fluctuations (Bassman, 2004).

Figure 5. Defense activities and protective roles of Ipomoea biloba

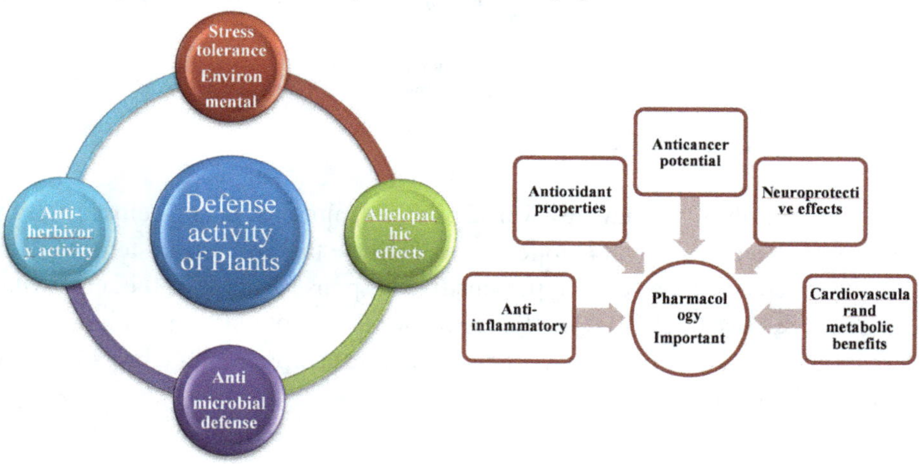

Plants produce chemicals that act as a shield against UV radiation, inhibit the growth of bacteria, viruses, and fungi, prevent infections, and control diseases. Flavonoids, alkaloids, terpenoids, phenolics, and saponins exhibit antimicrobial activity, provide antioxidant properties, and support allelopathy. This defense function comprehensively withstands biotic and abiotic challenges (Figure 5).

Various parts of *Ipomoea biloba* have been studied by different scientific researchers and are used in folk medicine. According to previous study, the evaluation of the root extract for in vitro antimalarial activity in *Ipomoea pes-caprae* shows that methanolic extracts have very good antimalarial activity (Pothula & Kanikaram, 2015).

The diverse bioactive and natural defense activities of this plant will be scientifically researched. *Ipomoea biloba* has effective antimicrobial and antifungal properties; it is used against both bacteria and fungi. The biological natural defense properties of the medicinal plant *Ipomoea biloba* in antimicrobial activity against Gram-positive bacteria (*Bacillus species and Coccus*) and Gram-negative *Escherichia coli, Pseudomonas aeruginosa, and Klebsia*). This study focused collectively on the antioxidant activity, phytochemical composition, and antibacterial efficacy of *Ipomoea biloba* extracts collected from the leaves.

EXTRACTION PROCESS AND APPROPRIATE SOLVENT SELECTION

In this basic pilot study, different solvents were used with *Ipomoea biloba* leaf powder. The important solvents are:

1. Water / Aqueous
2. Hexane
3. Petroleum
4. Methanol
5. Ethanol

Preliminary phytochemical studies were conducted using all the above extractions, and finally, ethanol was chosen for further analysis (Figure 6). Because the ethanol extract has better extraction of the plant bioactive compounds, it is used in pharmaceutical phytochemical analysis and has also given positive results for many compounds. When compared to water/aqueous and methanol, it also gives more positive and richer phytochemical compounds.

The following are the reasons for better extraction.

Figure 6. Preparation of ethanol extract

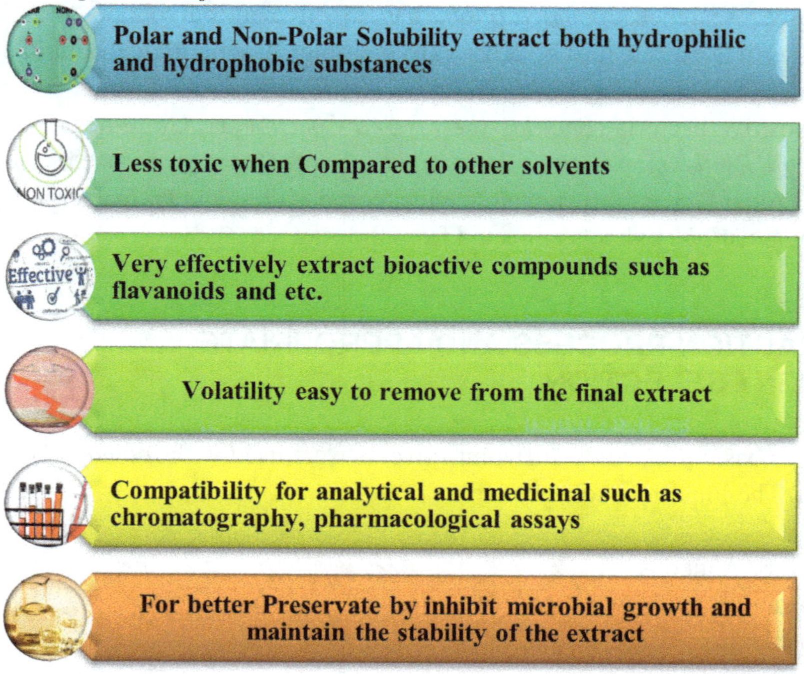

This is the reason ethanol is used to extract a broad spectrum of compounds. In previous studies, the flower extract of *Ipomoea pes-caprae* was analyzed. Phytochemical compounds were examined in five different solvents: ethanol, methanol, n-butanol, petroleum ether, and chloroform, among which ethanol were found to be the best (Sheeba et al., 2021).

Previous studies examined different phytochemical compounds of the *Ipomoea pes-caprae* stem, and the compounds were analyzed using chloroform, hexane, ethyl acetate, methanol, acetone, and aqueous solvents, among which ethyl acetate gave the best results (Venugopal et al., 2015). The phytochemical screening of the plant *Ipomoea pes-caprae* was conducted using different extracts from the leaf. Six different extracts—hexane extract, petroleum ether extract, chloroform extract, methanolic extract, ethanolic extract, and aqueous extract—were analyzed, among which ethanol was found to be the best (Anitha, 2014).According to the previous study carried out three different extractions of n-hexane, methanol, and ethyl acetate. The results did not show a better extraction solvent (Nuskiya et al., 2023).

These findings demonstrate that *Ipomoea biloba* leaf extract is pharmacologically potent and contains multiple bioactive compounds. Ethanol is an effective solvent for pharmacological extraction. It extracts a wide range of phytochemicals, mainly alkaloids, phenolics, and flavonoids. These extracts are mainly very powerful anti-inflammatory, antimicrobial, and antioxidant agents, and many therapeutically active compounds are present in them. Previous studies related to this *Ipomoea* species show that it contains important bioactive compounds, so it is suggested that *Ipomoea biloba* may also possess similar or unique bioactive compounds. There is very limited literature on *Ipomoea biloba*; this gap may be due to its phytochemical and biological activities. Further pharmacological identification may encourage its use in herbal medicine and drug development.

ECOLOGICAL, PHARMACOLOGICAL, AND MEDICINAL SIGNIFICANCE OF *IPOMOEA BILOBA*

Ipomoea biloba plays a vital role in human health through its secondary metabolites that are at high levels of phenolics and flavonoids, which help avoid oxidative stress, caused by environmental factors and pathogens, and are neutralized by reactive oxygen species. These bioactive molecules strengthen plants against biotic and abiotic stresses (Ashraf et al., 2024).

The medicinal roles of *Ipomoea biloba* are mediated by secondary metabolites; mainly, they act as antioxidant agents, prevent oxidative stress in living organisms, and also reduce chronic diseases such as inflammation, microbial action, and hepatodamage. New drugs are developed through the phytochemical effects of *Ipomoea biloba* natural health products with biological activities. In conclusion, the secondary metabolites of *Ipomoea biloba* are important for the ecological survival of the plant and provide valuable pharmacological benefits, making the species significant for both botanical defense studies and natural product research.

SECONDARY METABOLITES AND BIOACTIVE COMPOUNDS IN *IPOMOEA BILOBA*: THERAPEUTIC INSIGHTS

The bioactive compounds present in this plant extract are important phytochemicals, including (Figure 7)quercetin 3-O-β-D-glucofuranoside, (Figure 8)β amyrin acetate, (Figure 9)α amyrin acetate, (Figure 10)betulinic acid, and (Figure 11)glochidone (Srivastava & Rauniyar, 2020). Simultaneously, terpenoids, tannins, saponins, alkaloids, steroids,phenolic compounds, glycosides, and flavonoids are rich in bioactive profiles present in the *Ipomoea biloba* leaf extract (Islamy et al., 2024).

Figure 7. Quercetin 3-O-β-D-glucofuranoside

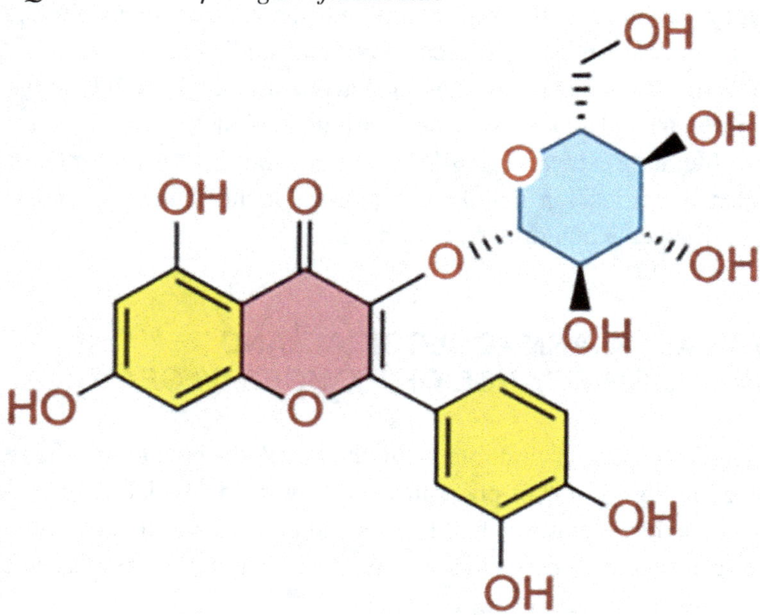

Figure 8. β amyrin acetate

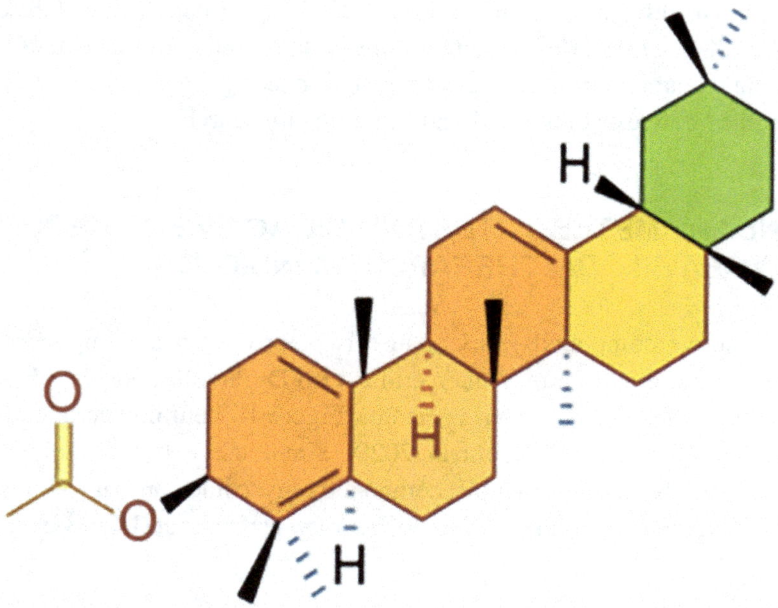

Figure 9. α amyrin acetate

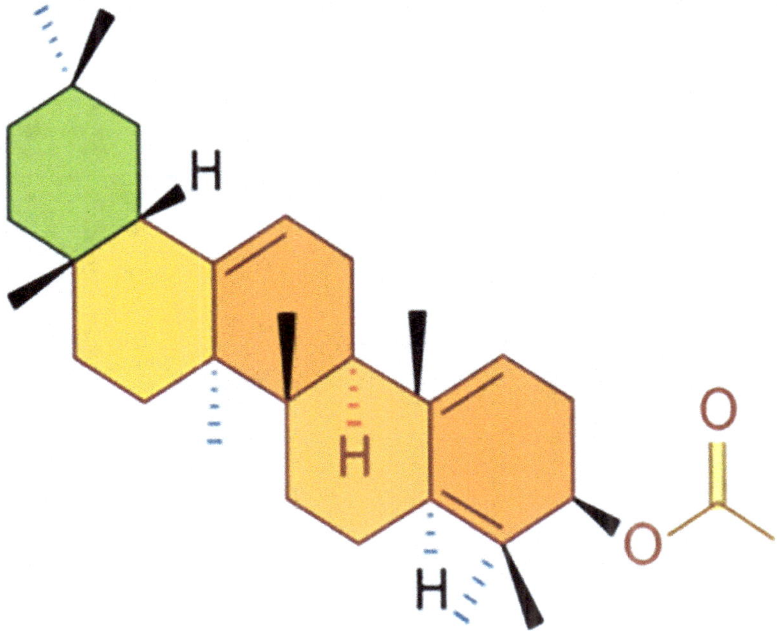

Phytochemical analysis indicated the presence of secondary metabolites, namely triterpenes, alkaloids, steroids, coumarins, carbohydrates, tannins, saponins, phenols, phlobatannins, and flavonoids in *Ipomoea pes-caprae* extracts. Secondary metabolites and different chemicals have been extracted from *Ipomoea pes-caprae*, which have a broad range of bioactive components. Based on these studies, the natural products isolated from *Ipomoea pes-caprae* were classified and categorized according to their bioactive components reported in previous studies.

BIOACTIVITY OF EXTRACTS FROM DIFFERENT PLANT PARTS

Therapeutic and Pharmacological Potential of *Ipomoea biloba*

These plant defense properties mainly activate pharmacological roles and prevent oxidative stress–related disorders, such as cancer, aging, inhibition of cancer cell proliferation, and cardiovascular diseases. Tissue repair mechanisms, such as reducing inflammation, are involved. Antibiotic, antiviral, antidiabetic, hepatoprotective, and neuroprotective activities are activated by pharmacological roles. All these activities combine and contribute to the synthesis of new drugs.

Table 1. Reported pharmacological activities of Ipomoea biloba

Pharmacological Activity	Author Name	Plant Part	Extract Type	Experimental Model	Key Findings
Antioxidant	Vallipriya & Begum, (2021)	Leaf	Ethanol/ Methanol	In vitro assays	High phenolic & flavonoid content
Antifungal	Ramanathan et al., (2012)	Whole plant / unspecified	Ethanol extract	In vitro (agar well diffusion method)	Demonstrated antifungal activity against tested fungi
Antimicrobial	Ramanathan et al., (2012)	Leaf	Ethanol	Disc diffusion	Effective against Gram ± bacteria
Antihyperglycemic / Antidiabetic	Suhasini et al., (2019)	Whole plant / leaf	Ethanol	In vivo (STZ-induced diabetic rats) 300 mg/ kg b.w	Reduced oxidative stress in the liver; restored carbohydrate-metabolizing enzymes; improved glucose utilization and inhibited hepatic glucose synthesis
Anti-aging / Collagen-promoting	Panichakul et al., (2022)	Leaves / fibroblastic cells	70% and 95% ethanol extracts	In vitro (CCD-986sk fibroblasts, scratch assay)	Inhibited collagenase, promoted cell proliferation, enhanced collagen production, exhibited antioxidant and wound-healing activity
Antipyretic	Singh et al., (2011)	Whole plant / leaf	Aqueous	In vivo (rats)	Exhibited significant antipyretic activity comparable to paracetamol; lowered yeast-induced body temperature
Cardioprotective	Vallipriya & Begum (2021)	Leaves	Ethanolic leaf extract	In vivo (albino rats, isoproterenol-induced cardiotoxicity)	Reduced myocardial damage enzymes (ALT, AST, LDH, CK), lipids, cholesterol, triglycerides; improved antioxidant status
Gastroprotective / Anti-ulcer	Venkataramana et al., (2003)	Whole plant / leaf	Ethanol extract	In vivo (Wistar rats, aspirin- and pylorus-induced ulcers) (200 & 400 mg/kg)	Exhibited anti-ulcer activity; protected gastric mucosa
Anticancer	Meira et al., (2012)	Aerial parts	Methanol	Cancer cell lines	Mild cytotoxic activity
Diuretic properties	Lorenzi & Abreu Matos, 2002; Pereda-Miranda et al., (2005)	Root	Methanol / Ethanol	Phytochemical screening, chromatographic analysis (TLC, GC–MS, NMR)	Identified alkaloids, flavonoids, tannins, saponins, terpenoids, steroids, and phenolic compounds. Isolation of resin glycosides **and** nor-isoprenoids, contributing to anti-inflammatory, antioxidant, and antimicrobial properties.
Antitumor / Chemopreventive	Manigauha et al., (2015)	Whole plant	Methanol, aqueous, petroleum ether extracts, swaras	In vivo (melanoma B16F10 in mice)	Methanol and aqueous extracts showed high anticancer activity; effects attributed to alkaloids, tannins, flavonoids, glycosides, and terpenoids
Antioxidant, antibacterial, antifungal, and anticancer activities	Sheeba et al., (2021)	Flower	Aqueous / Methanolic	In vitro assays	Exhibited strong antioxidant activity; showed broad-spectrum antibacterial activity against Gram-positive (*Staphylococcus aureus, Bacillus subtilis, Streptococcus mutans*) and Gram-negative bacteria (*Proteus vulgaris, Klebsiella pneumoniae, Escherichia coli*); demonstrated antifungal activity against *Aspergillus flavus, Aspergillus niger,* and *Penicillium* sp.; displayed anticancer activity against A549 lung cancer cells.

continued on following page

Table 1. Continued

Pharmacological Activity	Author Name	Plant Part	Extract Type	Experimental Model	Key Findings
Antioxidant, antimicrobial, anticancer, nanoparticle-mediated activity	Veeramani et al., (2018)	Stem	Aqueous stem extract; stem extract for GC–MS and HPLC analysis	**In vitro** (free radical scavenging assays, GC–MS, HPLC profiling, antibacterial assays, MTT cytotoxicity assay)	Stem extracts showed strong free-radical scavenging activity and revealed diverse bioactive compounds through GC–MS and HPLC analyses. Identified compounds included phenols, steroids, terpenoids, and glycosides. Major terpenoids detected were β-caryophyllene, α-humulene, germacrene D, limonene, α-terpineol, (E)-nerolidol, guaiol, and α-cadinol, along with 8-cedren-13-ol, hexadecanoic acid, fatty acids, and phytol. HPLC analysis identified and quantified phenolic acids (caffeic, gallic, chlorogenic, salicylic acids) and **flavonoids** (quercetin-3-O-glucoside, isorhamnetin derivatives, kaempferol). The aqueous stem extract was also used for green synthesis of silver nanoparticles (AgNPs), which were small, stable, eco-friendly, and exhibited antibacterial activity against *Escherichia coli, Bacillus* sp., and *Pseudomonas aeruginosa*, as well as anticancer activity confirmed by the MTT assay.
Cytotoxic / Antiproliferative	Meira et al., (2012)	Aerial parts / flowers / leaves	Lipophilic glycosides, hydro-ethanol extract	In vitro (human cancer cell lines: nasopharyngeal, colon, cervical, ovarian; S. aureus resistance strains)	Mild cytotoxicity (ED_{50} = 5–20 μg/mL); resistance-modifying activity against *S. aureus*; isoquercitrin detected

Genomic Insights and Therapeutic Potential of *Ipomoea biloba*

In recent studies, *Ipomoea biloba* was found to possess a chromosomal-level genome comprising 15 chromosomes with approximately 34,000 protein-coding genes, which are perfectly coded for high salt tolerance, wherein pre-existing genes express stability under saline conditions (Cheng et al., 2023). In RNA sequence analysis, root as well as leaf tissues show HKT genes and dehydrins related to salt stress, which regulate Na^+ and K^+ balancing (Guo et al., 2025). The important secondary metabolite compounds are mainly phenolic compounds; they are involved in the ROS process, which is associated with human therapeutic processes, including diabetes and other human disorders. Pro-inflammatory compounds, cytokines such as IL-1β and IL-6, are present in *Ipomoea biloba*, which is used to treat chronic inflammation (Majid et al., 2018). Phytoestrogenic properties are present in this plant, which leads to interaction with cellular aging (Panichakul et al., 2022). This study shows anti-arthritic, anti-inflammatory, and potential anticancer activities, particularly against MCF-7 (HBC) and A549 cell lines (Chanda et al., 2020).

Resin glycosides were identified as inhibitors of multidrug resistance (MDR), which was revealed through molecular docking analysis. This is not directly involved in bacterial activity, but it suppresses efflux mechanisms associated with antibiotic resistance in pathogens and increases intracellular antibiotic accumulation (Corona-Castañeda et al., 2016: Li et al., 2025). This compound acts as a potential antibiotic, similar to tetracycline, and significantly reduces the MIC. In the modern era, *Ipomoea biloba* leaf extract is used as an adjuvant therapeutic compound for humans and animals because many synthetic antibiotics are losing resistance activity.

This phytochemical analysis and antimicrobial activity; further studies will explain the bioactive compounds, namely quercetin-3-O-galactoside and caffeoylquinic acids (CQAs), with molecular activity centered on the Nrf2–Keap1 signaling pathway. These particular compounds are used in UV radiation applications, especially in cancer therapy and also in neurodegenerative diseases.

In the molecular defense mechanism, salt tolerance and UV radiation are present in the plant *Ipomoea biloba*. This plant well balances salinity stress. Su et al. (2023) identified four NAC genes that are involved in salt stress. Sixty-five WRKY family members were identified from transcriptome data under salt stress, and 17 genes were upregulated (Wu et al., 2024). The UV activity of this plant was also shown to produce positive results and control human skin abnormalities such as edema, urticaria, and irritation (Situmorang et al., 2024). In natural plants, defense mechanisms prevent their tissues from UV radiation through the phenylpropanoid pathway. Stabilizing cellular structure will be carried out through increased flavonoids, anthocyanins, and resin glycosides; they will also act as UV filters and scavenge ROS (Mahdavian 2024).

Phytochemical compounds are isolated using GC–MS and NMR spectroscopy, which include CQA (chlorogenic acid) derivatives and resin glycosides (Mutha et al., 2025). These compounds are activated in the plant defense mechanism; they also act as controllers of environmental stressors. (salinity and ultraviolet (UV) radiation) (Kumar et al., 2020; Chowdhary et al., 2021). The main secondary metabolites are involved in significant human health benefits; mainly, these are mediated by oxidative stress and the Nrf2 pathway, and by modulation of cell survival inhibition through AKT1 signaling (Shiau et al., 2022; Goyal et al., 2024).

Ipomoea biloba has a dual action: one is anti-inflammatory, which is controlled by polyphenols that can manage inflammation in chronic diseases, and MDR reversal (multidrug resistance reversal). It is expressed through terpenoids, alkaloids, or flavonoids, which can resist pathogenic organisms (Karthikeyan et al., 2020). Future research is mainly focused on next-generation sequencing (RNA-Seq), in which stress-related genes are activated in response to environmental stress (e.g., salinity, oxidative stress, light, heat, or drought). This will create a bioactive fingerprint. This will ensure reproducibility, quality control, and clinical reliability of *Ipomoea biloba* –derived interventions (Liu et al., 2020).

GC–MS and HPLC-Based Phytochemical Profiling

The CP3800 Saturn 2200 GC–MS analysis was carried out on *Ipomoea pes-caprae*. The temperature was maintained from 80 to 350 °C at 3 °C/min for 55 min. The acetone extract was analyzed using mass spectrometry. In this analysis, 29 compounds were identified. The compounds include 2-furancarboxaldehyde, clindamycin, and 5-(hydroxymethyl), which are some of the identified compounds.

Collagenase inhibitors and antioxidants were isolated from the *in vitro* analysis of ethanolic extracts of *Ipomoea pes-caprae*. In this, chlorogenic acid, ferulic acid, quercetin (Q), and 3, 5-di-caffeoylquinic acid was identified from the HPLC analysis of IPC70 and IPC95 extracts. These compounds enhance collagen production with non-cytotoxic concentrations, were found to promote cell proliferation, and support wound healing. These active compounds mainly exhibit anti-aging activity (Tasanee et al., 2022).

Chromatographic techniques were used to isolate compounds from *Ipomoea pes-caprae*, family Convolvulaceae. In HPLC-UV analysis, the aerial parts of the plant were used for extraction with different ethanol concentrations. In this study, different chromatographic analyses clearly showed a high level of ethyl caffeate in 50°GL, which is a lower strength of ethanol (Da Silva Barth et al., 2019). This shows the anti-inflammatory effect on paw oedema induced by trypsin. Previous study explained that HPLC analysis identified chlorogenic acid, gallic acid, catechin, and quercetin, along with other phenolic compounds such as syringic acid, caffeic acid, ferulic acid, coumarin, naringenin, kaempferol, and quercetin (Qasima et al., 2017).

Bioactive Compounds in *Ipomoea* Species

Many *Ipomoea* species have better active compounds; some of the reviews are given here. *Ipomoea batatas* is the common name for sweet potato; its leaves contain bioactive compounds, which are mainly anthocyanins and polyphenolic compounds, and are analysed using the HPLC analysis system and NMR spectroscopic methods. In this analysis, there are fifteen different anthocyanins and six different polyphenolic compounds. This is mainly in radical scavenging activity and biologically active compounds (Islam *et al.*, 2002). The main phytochemical compounds present are *Ipomoea batatas* are phenolic compounds, flavonoids, and carotenoids, it has many nutrients such as protein, vitamin C, carbohydrates, fiber, β-carotene, and some mineral-rich compounds. It will have antioxidant, anticancer, aphrodisiac, and anti-inflammatory activity (Suhendy et al., 2023).

In previous study their systematic review, mainly focus on the genus *Ipomoea* (Convolvulaceae), are highlighting its bioactive and ethnobotanical importance in treating ailments such as cancer, diabetes, dermatological problems, and joint problems, wound healing, malaria, sexually transmitted infections (STDs), and gastrointestinal problems. *Ipomoea* species also have some bioactive compounds that help control antibacterial, anti-inflammatory, antioxidant, antiviral, cytotoxic, analgesic, hallucinogenic, and anticancer activities (Kobe Borlongan et al., 2024).

Another important Convolvulaceae family member is *Ipomoea pes-tigridis*. It exhibits analgesic, anti-inflammatory, antimicrobial, thrombocytic activities and cytotoxic effectsIt has rich bioactive and phytochemical compounds, including flavonoids, saponins, alkaloids, and tannins, which are present in the leaf and root extracts (Selvam & Acharya, 2015). Ipomoea Alba of the Convolvulaceae family is also a tropical plant used in America.The whole plant is used in snakebite (McDonald, 1991). It is rich in flavonoids, alkaloids, anthraquinones, simple phenolics, saponin, anthracene glycosides, polyoses, and cardenolides, including leucoanthocyanin. Albinoside XI with strong potential for vinblastine susceptibility used in human breast carcinoma cells to prevent multidrug resistance activity (Ambika & Nair, 2019).

Ipomoea aquatica Forssk (Water morning glory), also from the Convolvulaceae family, has calystegines, N-transferuloyltyramine, and N-cis-feruloyltyramine observed from the roots of *Ipomoea aquatica*, which will inhibit prostaglandin synthesis. *Ipomoea aquatica* aqueous extract acts as an oral hypoglycaemic drug. Isochlorogenic acids A, B, and C of *Ipomoea aquatica* are observed in this plant, which help in the treatment of diabetes and obesity, and also exhibit antifungal activity. Ipomoea aquatica contains, in its crude extract, a particular compound, 7-O-β-D-glucopiranosil-dihidroquercetina-3-O-β-D-glucopiranoside, which exhibits activity against cancer. Ipomoea aquatica, in the investigation, shows diuretic activity.

The plants in the *Ipomoea* genus used for medicine and pathological conditions, such as algesic and inflammatory conditions, kidney ailments, obstipation, and digestive problems. Some of the species have analgesic, antimicrobial, hypotensive, and anticancer properties(Narayanan & Suseem, 2024; Bhaumik et al., 2025). Some of the pharmacological studies have better evidence in this species, such as the antinociceptive action in *Ipomoea pes-caprae*. The root of Ipomoea stans contains an active substance that has a vasorelaxant effect and acts as an antispasmodic agent.

The plant *Ipomoea carnea*, also from the Convolvulaceae family, has several applications, mostly for medicinal purposes. It has been shown that different extracts of the *Ipomoea carnea* plant have antifungal, antibacterial, antioxidant, anticancer, antimicrobial, immunomodulatory, hepatoprotective, anticonvulsant, anti-inflammatory, anxiolytic, antidiabetic, wound-healing, sedative, and cardiovascular activities (Srivastava & Shukla 2015).

Another plant from the *Ipomoea* species is *Ipomoea* cairica Sweet, commonly known as Railroad creeper. The aqueous extract of *Ipomoea* cairica has a specific preventive effect against respiratory syncytial virus. The ethanolic extract has an antinociceptive effect (Ferreira et al., 2006) as well as anti-inflammatory and antioxidant activities. The special effect is the prevention of HIV. The important phytochemical compounds are Trachelogenin, Scopoletin, Lignans, Umbelliferone, Matairesinol, Arctigenin, and Coumarins (Fig 12).

Figure 12. Phytochemical compounds

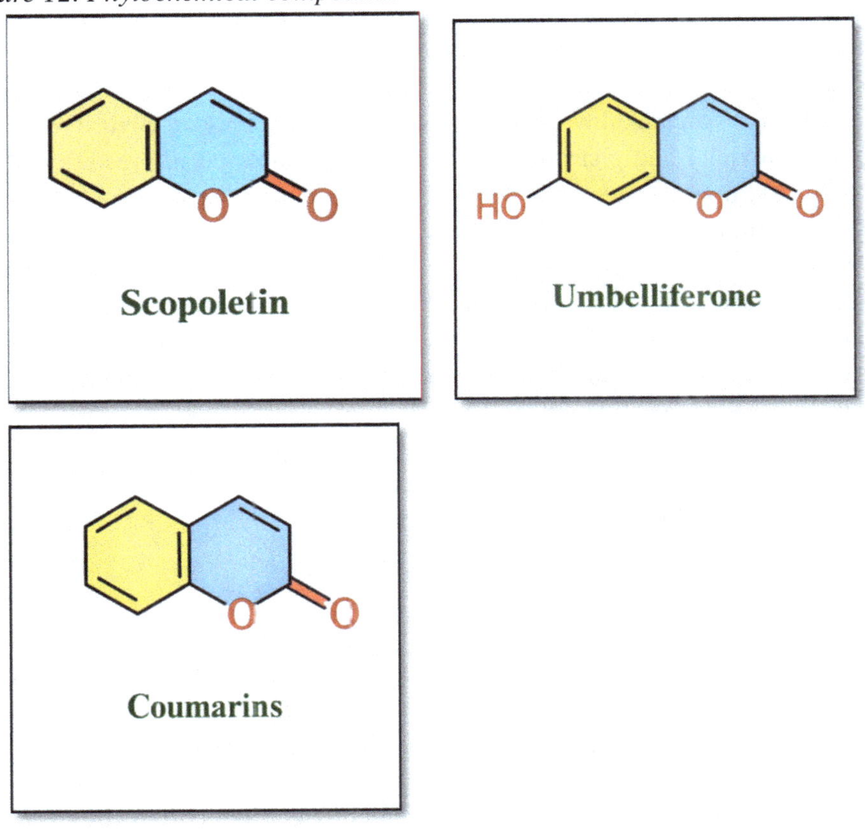

In the review, the focus is mainly on the Convolvulaceae family, including *Ipomoea pes-caprae*. It can help to prevent gastrointestinal issues and inflammation. It is also used for colic, rheumatism, whitlow, piles, and oedema. The leaves of this plant are also used for strain, fatigue, rheumatism, arthritis, anti-inflammatory purposes, menorrhagia, anti-hemolytic activity, anticancer effects, skin diseases, and as an antispasmodic. The same leaves also inhibit diarrhea, platelet aggregation, piles, and vomiting.

Antimicrobial Screening of Plant Extracts Common methods and their relevance.

Gaps in existing research related to *Ipomoea biloba*

The huge heterogeneity of plant extracts sources, mainly species, location, and season, requires reliable cross-sectional studies for comparative analysis of antimicrobial activity. In the present studies, the focus is mainly on crude extracts without phytochemical characterization or isolation of bioactive compounds, and their mechanisms are unknown. Most studies are focused only on in vitro or in vivo tests, while toxicity and pharmacokinetics are not addressed in preclinical diagnostics. In addition, the interaction of plant extracts with conventional antibiotics and their potential to target antibiotic-resistant pathogens has not been evaluated. Finally, the main issues of reproducibility, quality control, and source variation are sometimes inadequately addressed. In this paper, these gaps hinder the identification, optimization, and safe use of plant-derived antimicrobials for clinical applications, underscoring the need for standardized methods, phytochemical focus, mechanistic investigations, in vivo validation, and sustainability assessments.

MATERIALS AND METHODS

Processing of Medicinal Plants

The fresh and properly cleaned mature leaves (Fig 12) of *Ipomoea biloba* were collected in Khalli Hills, Namakkal District. All the leaves used for the study were collected, washed, shaded, and dried for the plants. The dried leaves were finely powdered using a mechanical grinder, and the powder was utilized for the extracts. 10 g of the powder was weighed and used with nearly 100 ml of ethanol for cold percolation purposes. The extract provides the final result; the filtrate may be stored at 4 °C for further use.

Phytochemical Analysis

Phytochemical analysis of the ethanol leaf extract of *Ipomoea biloba* was individually performed using different quantitative tests for alkaloids, flavonoids, saponins, carbohydrates, tannins, steroids, terpenoids, phenolic compounds, catechins, proteins, and amino acids.

Extraction of Phytochemicals

The phytochemicals were extracted in ethanol and stored in air-tight containers at 4°C.

Test for Alkaloids

Take a test tube and add a small amount of extract, then add a few drops of Mayer's reagent. A precipitate indicates the presence of alkaloids.

Test for Flavonoids

Take a test tube and add 1 ml of extract, then add a few drops of dilute ferric chloride solution. The color change to pale green or red-brown indicates the presence of flavonoids.

Test for Saponins

Take a test tube and add 1 ml of extract, then add 1 ml of alcohol, dilute 20 ml of distilled water, and shake well for 15 minutes. The formation of a 1 cm of foam indicates presence of saponins.

Test for Tannins

Take a test tube add 5 ml of extract solution, then add 1 ml of lead acetate solution. The appearance of a flocculent brown precipitate indicates the presence of tannins.

Test for Steroid

Take a test tube and add a small amount of aqueous extract, then add a few crystals of sodium nitrate and heat gently for a minute. Next, allow it to cool and add 0.5 ml of concentrated H_2SO_4.

Test for Terpenoids

Take a test tube and add a small amount of the extract, then filter it. Next, gently warm this filtrate with tin and thionyl chloride. The appearance of a pink-colored solution indicates the presence of terpenoids.

Test Cultures

The bacterial cultures were obtained from the Microbiology Lab of GVN Hospital Private Limited, Trichy, and subcultured.

ANTIMICROBIAL ASSAY

Effect of Antibacterial Activity

The antibacterial effects of the extract against bacterial strains have been studied. The antibacterial effects of the *EEIB* were examined against different strains.

Antibacterial Activity

The antibacterial activities against selected bacteria were carried out using the standard disc diffusion method.

Strains

Bacterial strains were collected from urinary tract–infected patients at G.V.N. Medical College, Trichy. These strains were inoculated into agar and incubated at 37 °C for 24 hours.

Selected Strains (Gram Negative & Gram Positive)

1. E.coli
2. Salmonella Para ‘ A’
3. Staphylococcus areause
4. Salmonella Typhi
5. Entercoccoi
6. Pseudomonas Aeruginosa
7. Klebsiella pneumonia
8. Streptococcus

9. Proteus Species
10. Enterobacter agglumins

Antimicrobial Agents

Antimicrobial plants (*Ipomoea biloba*) mature leaves were collected and shade-dried, and the leaves were powdered and mixed with the solvent (ethanol). Ten grams of powder was taken and mixed with 100 ml of solvent. After overnight incubation, the extract was separated. Then, fresh solvent was add (this separation was carried out three times), extract was kept in /incubator or oven for drying.

ANTIMICROBIAL ACTIVITY

Disc Diffusion Method

The disinfectant was tested against the activity of strains using the disc diffusion method. Bacterial cultures were maintained in nutrient broth, and the diluted bacterial culture (0.2 ml) was spread on sterile Mueller–Hinton agar plates. The desired volume of the disinfectant was applied to sterile Whatman No. 1 filter paper discs, 6 mm in diameter. These discs were allowed to dry before being placed on top of the agar plates. This plate, containing three paper discs with the desired volume of disinfectants, was tested in triplicate. These plates were incubated at 37 °C for 24 hours, after which inhibition zones were recorded. The antibacterial activities of the plant extract (10 µg/100 µL) and ampicillin (10 µg), which served as the standard antibiotic, were indicated by clear zones of inhibition.

Preparation of Inoculums

Colonies of pure culture were touched with a sterile loop from the subculture plates and transferred to tubes containing nutrient broth. This tube is incubated at 37 °C for 4–6 hours.

Inoculation Pour Plate Method

An inoculum was taken from the mother slant and mixed with Mueller–Hinton agar medium (spread with the bacterial strains). It was then poured into the Petri dish. All the plates were prepared in this manner with various pathogenic bacteria on each plate. The lids of the plates were kept at room temperature for 5–10 minutes to reduce the moisture content.

Application of Antimicrobial Discs

The filter paper discs soaked in the working standard solution of the plants with solvent at a concentration of 10 μg/100 μl were removed from the respective vials with the help of flamed and cooled forceps. It was carefully ensured that each disc was at least 15 mm away from another to avoid overlapping of the zone of inhibition.

Incubation After exact placement of these discs, the Petri plates was incubated aerobically at 37 °C for 16–18 hours.

Minimal inhibitory Concentration of Starins

From the serial dilution method, a 10–60 mg/mL concentration of the EEIB was taken for MIC. Then, the inhibitory zone was measured using a scale, and the minimal inhibitory concentration was also determined.

RESULTS

Phytochemical Constituents & Qualitative Phytochemical Analysis of Secondary Metabolites

Figure 14. Qualitative phytochemical analysis of Ipomoea Biloba leaf extract

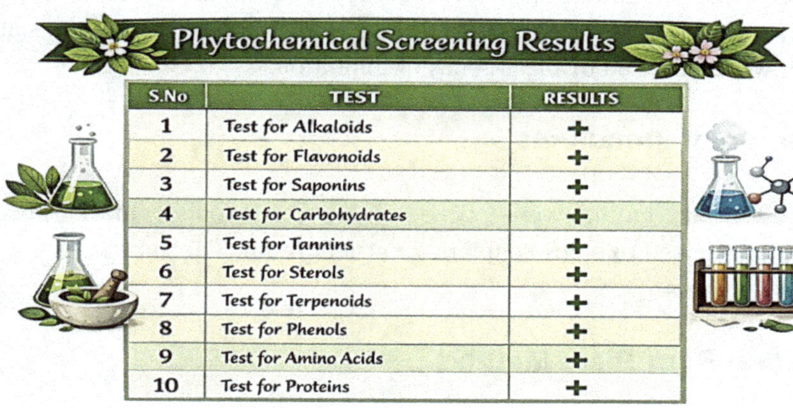

Phytochemical Screening Results

S.No	TEST	RESULTS
1	Test for Alkaloids	+
2	Test for Flavonoids	+
3	Test for Saponins	+
4	Test for Carbohydrates	+
5	Test for Tannins	+
6	Test for Sterols	+
7	Test for Terpenoids	+
8	Test for Phenols	+
9	Test for Amino Acids	+
10	Test for Proteins	+

Figure shows the (+) shows the presence of the phytochemical compounds (-) shows shows the absence of the phytochemical compounds in the *EEIB*.

In Vitro Antimicrobial Activity

Figure 15. Antimicrobial activity expressed as zone of inhibition (mm in diameter) of ethanol extract (0.5 g/2 mL)

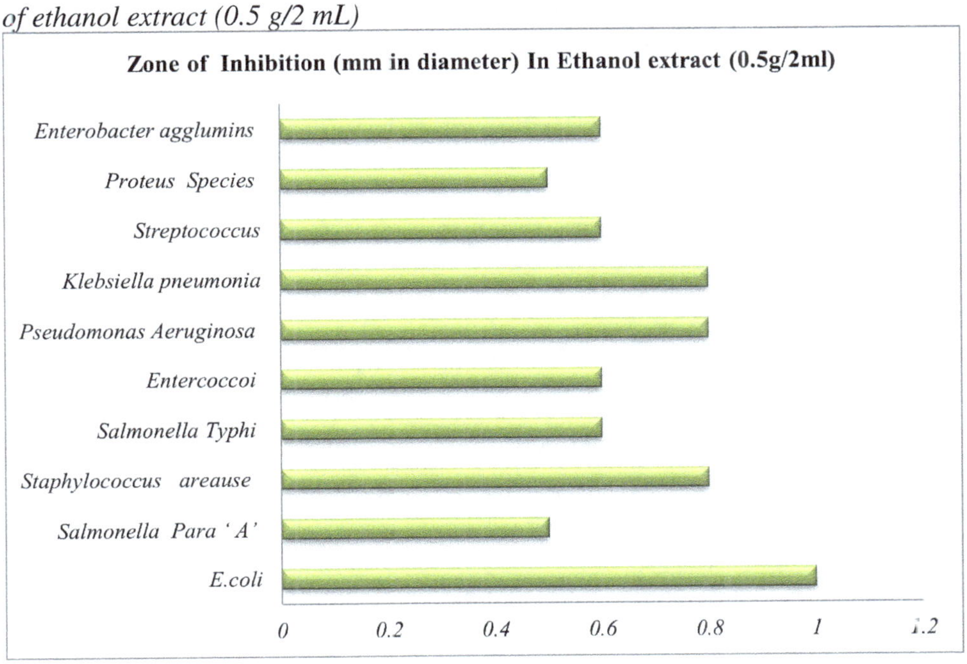

Figure 16. Antimicrobial activity

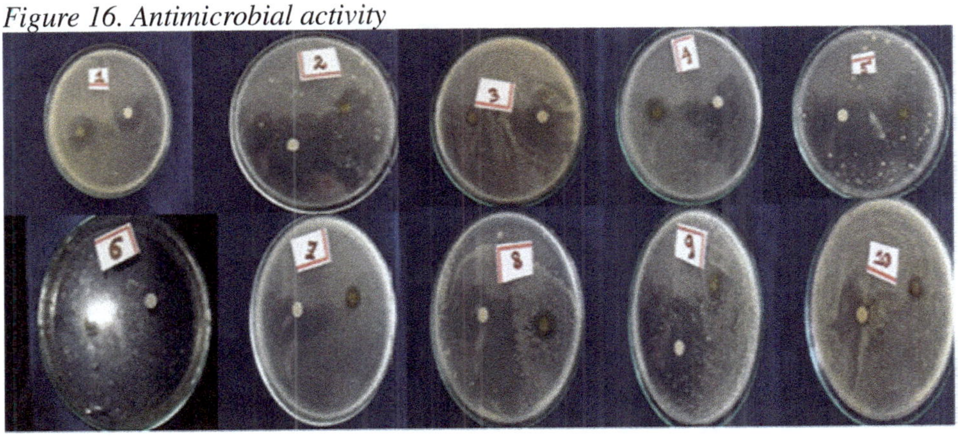

It was explained that the antimicrobial activity of the selected species was determined by the zone of inhibition against the following: *E. coli*, *Salmonella* Para 'A', *Staphylococcus aureus*, *Salmonella Typhi*, *Enterococcus*, *Pseudomonas aeruginosa*, *Klebsiella pneumoniae*, *Streptococcus*, *Proteus* species, and *Enterobacter agglomerans*, against ampicillin, represented as an arrow.

Minimum Inhibitory Concentration (MIC) of *EEIB* Against Test Strains (mg/mL)

Figure 17. Minimum inhibitory concentration (MIC) values of tested strains (mg/mL)

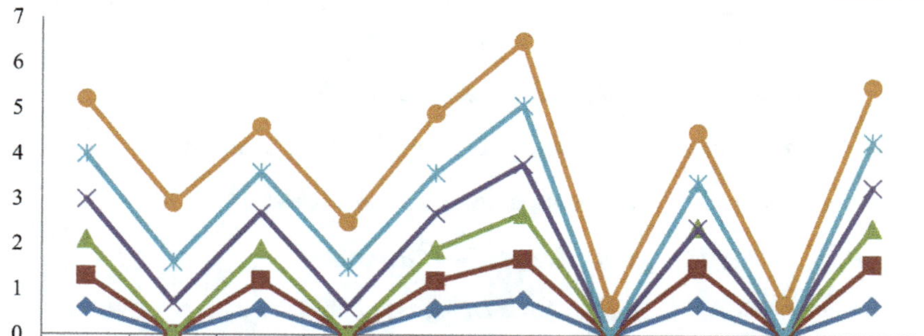

	E.coli	Salmonella Para 'A'	Staphylococcus areause	Salmonella Typhi	Entercocci	Pseudomonas Aeruginosa	Klebsiella pneumonia	Streptococcus	Proteus Species	Enterobacter agglumins
60	1.2	1.3	1	1	1.3	1.4	0.7	1.1	0.7	1.2
50	1	0.9	0.9	0.9	0.9	1.3		1		1
40	0.9	0.7	0.8	0.6	0.8	1.1				0.9
30	0.8		0.7		0.7	1		0.9		0.8
20	0.7		0.6		0.6	0.9		0.8		0.9
10	0.6		0.6		0.6	0.8		0.7		0.7

The MIC of the strains was determined at 10 mg/ml to 60 mg/ml for *E. coli, Salmonella Para 'A', Staphylococcus aureus, Salmonella Typhi, Enterococcus, Pseudomonas aeruginosa, Klebsiella pneumoniae, Streptococcus, Proteus species, and Enterobacter agglomerans.*

PRINCIPAL FINDINGS

Results of Qualitative Phytochemical Screening

Figure 14 expresses that phytochemical analysis was done on the EEIB and indicated the presence of phytochemical compounds, namely alkaloids, saponins, flavonoids, carbohydrates, tannins, sterols, terpenoids, phenols, amino acids, and proteins. These similar results were indicated in the ethanolic leaf extract reported by (Anitha, 2014). Phytochemical analysis is most important in the evaluation of new and possible medicines derived from plants and also in determining the active principles responsible for known biological importance of plant compounds. In further investigations, isolation of compounds plays an important role. This type of solvent shows the extraction of phytochemical compounds.

According to previous reports, medicinal plants are active in the treatment of different diseases and have potential pharmacologically active compounds. The biologically active and potent compounds include alkaloids, phenolics, flavonoids, tannins, saponins, steroids, glycosides, terpenes, etc., are the very important phytochemicals that have biological activities. The bioactive compounds that present in the plant which can be identify the phytochemical compounds (Shaikh & Patil, 2020).

The phytochemical analysis of proteins, tannins, carbohydrates, steroids, amino acids, saponins, glycosides, and flavonoids found in the flower extract indicates their use as antimicrobial, anticancer, and antioxidant agents. That Ipomoea biloba, an aquatic plant, is used as an herbal remedy for various diseases.

The *Ipomoea biloba* plants were collected and extracted for biochemical compounds such as enzymes, antioxidants, amino acids, calcium, iron, chlorophyll content, and total phenols from the plant using different methods. According to the previous study evaluating the antimicrobial, phytochemical components, antioxidant, and toxic properties of six coastal plants, namely *Catharanthus roseus, Ipomoea biloba, Vitex negundo, Cymbopogon citratus, Pandanus tectorius, and Thespesia populnea*. The study analyzed the phytochemical components in these six coastal species (Prakash et al., 2016).

The *Ipomoea pes-caprae* (*Ipomoea* biloba) and Merremia umbellata belong to different genera and contain different compounds. In this study, both medicinal plants were compared for their phytochemical compounds as well as their cytotoxic and antioxidant activities. Phytochemical tests revealed that tannins are present in *Ipomoea pes-caprae (Ganjir et al., 2013).*

The study on *Ipomoea biloba* reveals several important findings related to its phytochemical composition, antioxidant activity, and potential role in plant defense. The leaf extract showed a high concentration of total phenolics and flavonoids, indicating a rich presence of secondary metabolites known for their protective and therapeutic properties. Various extracts analyzed, the methanol/ethanol extract shows the strongest antioxidant activity, suggesting its effectiveness in scavenging free radicals and reducing oxidative stress.

Phenolics, flavonoids, and other phytoconstituents suggest that *Ipomoea biloba* employs a robust chemical defense system to protect itself against pathogens and environmental stressors. Compounds not only contribute to the plant's innate defense but also highlight potential applications in natural medicine. The antioxidant and possible antimicrobial properties of the extract further support its relevance in pharmacological research.

The study confirms that *Ipomoea biloba* is a valuable source of bioactive compounds with strong antioxidant potential. These findings provide a foundation for future research on isolating individual compounds, exploring their mechanisms of action, and assessing their therapeutic or agricultural applications.

Antimicrobial Activity of *EEIB*

Fig 15 & 16 According to the previous study, flower extracts were tested against three Gram +ve bacteria *(Bacillus subtilis, Streptococcus mutans, and Staphylococcus aureus),* three Gram −ve bacteria *(Klebsiella pneumoniae, Escherichia coli, and Proteus vulgaris),* and fungi *(Aspergillus flavus, Penicillium sp., and Aspergillus niger)* using the disc diffusion method. The ethanolic flower extract showed antimicrobial activity against all the pathogens, and the zone of inhibition was recorded for S*taphylococcus species (Sheeba et al., 2021).* The plant *Ipomoea pes-caprae* contains bioactive compounds such as alkaloids, phenolics, saponins, and flavonoids (Nuralyza & Hanifa 2025).

Ipomoea pes-caprae (Convolvulaceae) is a rich source of bioactive compounds. The current study shows that the ethyl acetate extract has potential antibacterial activity against *S. aureus*. The ethyl acetate extract contains bioactive compounds such as flavonoids and alkaloids, while the aqueous extract is a rich source of phenolic substances. The methanol extract contains alkaloids, saponins, flavonoids, and tannins, making it a potential novel drug for treatment against microbes (Gazali et al., 2024).

According to the previous study, the antibacterial activity against UTI pathogens was analyzed using silver nitrate ions from *Ipomoea biloba* extract. These silver nanoparticles showed potential antibacterial effects against UTI microbes such as P. aeruginosa, E. coli, K. pneumoniae, S. aureus, and *Enterobacter sp*. These silver nanoparticles have antibiotic properties, are low cost, and exhibit no toxic effects (Satyavani et al., 2013). According to previous study,*Ipomoea biloba* showed antibacterial activity, which was evaluated using the paper disc method against *B. cereus, E. coli, S. typhi, M. luteus, C. tropicalis, and C. albicans (*Nuskiya et al., 2023).

Relevance of the Results to Plant Defense and Bioactive Compound Research

The findings of this study hold significant value for advancing the understanding of plant defense mechanisms and the role of bioactive compounds in protecting plants against biotic and abiotic stress. Identifying key bioactive constituents—such as phenolics, flavonoids, alkaloids, and terpenoids—highlights their critical function in antioxidant activity, antimicrobial properties, and response to pathogen attacks. These results support the concept that plants synthesize diverse defensive metabolites as part of their innate immune system, contributing to disease resistance and stress tolerance.

The observed antioxidant potential of the plant extract further strengthens its relevance, as oxidative stress is a major component of pathogen-induced damage and cell injury. When free radicals, the compounds help maintain cellular integrity, enhancing the plant's ability to withstand infections. Moreover, the strong antimicrobial activity demonstrated by the extract suggests potential applications in natural pesticide development, reducing reliance on synthetic chemicals.

The study contributes valuable evidence to natural product research, emphasizing the importance of plant-derived molecules in ecological defense and their possible use in pharmaceuticals, agriculture, and biotechnology. These results can guide future investigations into isolating specific active compounds, understanding their mechanisms, and exploring their commercial potential.

The results of this study on *Ipomoea biloba* provide valuable insights into the plant's natural defense mechanisms and its rich profile of bioactive compounds. The high levels of phenolics, flavonoids, and other secondary metabolites identified in the ethanol/methanol extracts highlight the plant's strong antioxidant potential, which plays a crucial role in protecting plant tissues from oxidative stress caused by pathogens and environmental pressures. The petroleum ether, ethyl acetate, chloroform, and methanol extracts of leaves, and the minimum inhibitory concentrations were determined accordingly (Christhu Uthayam & Vijayarengan, 2019).

The demonstrated antioxidant and antimicrobial activities suggest that *Ipomoea biloba* utilizes these bioactive compounds as part of its intrinsic defense strategy, helping to inhibit pathogen growth and enhance resilience against biotic stress. These findings support its traditional medicinal use and confirm that the plant possesses significant therapeutic potential.

Furthermore, the study contributes to natural product research by identifying *Ipomoea biloba* as a valuable source of biologically active compounds that can be explored for pharmaceutical, nutraceutical, and agricultural applications. The results also establish a foundation for future work focused on isolating specific active components, understanding their molecular mechanisms, and evaluating their potential in developing eco-friendly antimicrobial agents or plant protectants.

LIMITATIONS

Few bacterial strains were used. In general, there is a broad range of pathogens. This study focused on identifying antimicrobial activity using crude extracts. Since there are many bioactive compounds in this plant extract. These experiments were conducted only under laboratory conditions and not in living organisms, so the complexity may differ in in vivo systems. Therefore, their effectiveness in living organisms may vary. In this study, it is not clarified how the extract inhibits microbial growth, limiting its potential for drug development.

CONCLUSION AND RECOMMENDATIONS

In this study, the significant antimicrobial and phytochemical potential of plant extract was reported. The main screening of phytochemical tests confirmed the presence of major pharmacologically active compounds, including flavonoids, alkaloids, tannins, saponins, and phenols, which are responsible for various therapeutic properties. The extract mainly, contributes to its ability to inhibit the growth of pathogenic microorganisms. This study mainly focused on the antimicrobial assay of that plant extract to determine its suitability for inhibitory activity. Suggests that this plant extract is a natural bioactive compound with strong antimicrobial properties. Future studies should mainly cover compound isolation, mechanism of action, evaluation of plant toxicity, and necessary in vivo validation for therapeutic and pharmaceutical applications. This study provides a foundation and supports future research aimed at developing plant-based antimicrobial formulations.

REFERENCES

Abdallah, E. M., Alhatlani, B. Y., de Paula Menezes, R., & Martins, C. H. G. (2023). Back to nature: Medicinal plants as promising sources for antibacterial drugs in the post-antibiotic era. *Plants*, *12*(17), 3077. DOI: 10.3390/plants12173077 PMID: 37687324

Akinniyi, G., Lee, J., Kim, H., Lee, J. G., & Yang, I. (2022). A medicinal halophyte Ipomoea pes-caprae (Linn.) R. Br.: A review of its botany, traditional uses, phytochemistry, and bioactivity. *Marine Drugs*, *20*(5), 329. DOI: 10.3390/md20050329 PMID: 35621980

Ambika, A. P., & Nair, S. N. (2019). Wound healing activity of plants from the convolvulaceae family. *Advances in Wound Care (New Rochelle, N.Y.)*, *8*(1), 28–37. DOI: 10.1089/wound.2017.0781 PMID: 30705787

Anitha, P. (2014). *Pharmacognostic, Phytochemical and Pharmacological Evaluation of the Leaves of Ipomoea Pes-Caprae Linn. R. Br. Convolvulaceae* (Doctoral dissertation, College of Pharmacy, Madurai Medical College, Madurai).

Ashraf, M. V., Khan, S., Misri, S., Gaira, K. S., Rawat, S., Rawat, B., Khan, M. A. H., Shah, A. A., Asgher, M., & Ahmad, S. (2024). High-altitude medicinal plants as promising source of phytochemical antioxidants to combat lifestyle-associated oxidative stress-induced disorders. *Pharmaceuticals (Basel, Switzerland)*, *17*(8), 975. DOI: 10.3390/ph17080975 PMID: 39204080

Bassman, J. H. (2004). Ecosystem Consequences of Enhanced Solar Ultraviolet Radiation: Secondary Plant Metabolites as Mediators of Multiple Trophic Interactions in Terrestrial Plant Communities. *Photochemistry and Photobiology*, *79*(5), 382–398. DOI: 10.1111/j.1751-1097.2004.tb00025.x PMID: 15191046

Bhatia, P., Sharma, A., George, A. J., Anvitha, D., Kumar, P., Dwivedi, V. P., & Chandra, N. S. (2021). Antibacterial activity of medicinal plants against ESKAPE: An update. *Heliyon*, *7*(2), e06310. Advance online publication. DOI: 10.1016/j.heliyon.2021.e06310 PMID: 33718642

Bhatla, S. C., & Lal, M. A. (2023). Secondary metabolites. In *Plant physiology, development and metabolism* (pp. 765–808). Springer Nature Singapore., DOI: 10.1007/978-981-99-5736-1_33

Bhaumik, A., Saha, S., Baishya, U., Kashyap, G., Kuli, B., Rahman, F., & Sarkar, N. (2025). Screening of Bioactive Molecules and Phytopharmacological Potential of Ipomoea Cairica: A Comprehensive Review. *American Journal of Psychiatric Rehabilitation*, *28*(1), 210–215. DOI: 10.69980/ajpr.v28i1.80

Kobe Borlongan, Mark Anthony Jariol Torres, Cesar Demayo, Orven Ebarle Llantos (2024). A Systematic Review on Ethnobotany and Bioactive Compounds of the Genus Ipomoea (Convolvulaceae) Himalayan Journal of Health Sciences DOI:DOI: 10.22270/hjhs.v9i2.191

Brown, S. H., & Frank, M. S. (2020). Railroad Vine (Ipomoea pes-caprae): Identification and Uses: ENH1323/EP587, 10/2020. *Edis, 2020*(5) https://doi.org/DOI: 10.32473/edis-ep587-2020

Chanda, S., Donga, S., & Pande, J. (2020). Green synthesized zinc oxide nanoparticles: A review of antimicrobial, antioxidant, cytotoxic and photocatalytic properties. *Zinc Oxide: Production, Properties and Applications*, 1-45.

Cheng, Y., Wang, Y., Sun, J., Liao, Z., Ye, K., Hu, B., Dong, C., Li, Z., Deng, F., Wang, L., Cao, S., Pan, C., Zheng, P., Cai, L., Cao, L., Wang, S., Aslam, M., Hong Wang, H., & Qin, Y. (2023). Unveiling the genomic blueprint of salt stress: Insights from *Ipomoea pes-caprae* L. *Seed Biology*, *2*(1), 0. Advance online publication. DOI: 10.48130/SeedBio-2023-0021

Chowdhary, V., Alooparampil, S., Pandya, R. V., & Tank, J. G. (2021). Physiological function of phenolic compounds in plant defense system. In *Phenolic compounds-chemistry, synthesis, diversity, non-conventional industrial, pharmaceutical and therapeutic applications*. IntechOpen., DOI: 10.5772/intechopen.101131

Christhu Uthayam, M., & Vijayarengan, P. (2019). Antimicrobial Activity of Ipomoea pes-caprae L. against Selected Microbial Species. DOI: https://doi.org/DOI: 10.32628/IJSRST1962134

Corona-Castañeda, B., Rosas-Ramírez, D., Castañeda-Gómez, J., Aparicio-Cuevas, M. A., Fragoso-Serrano, M., Figueroa-González, G., & Pereda-Miranda, R. (2016). Resin glycosides from Ipomoea wolcottiana as modulators of the multidrug resistance phenotype in vitro. *Phytochemistry*, *123*, 48–57. DOI: 10.1016/j.phytochem.2016.01.004 PMID: 26774597

Da Silva Barth, C., de Souza, H. G. T., Rocha, L. W., de Souza Madeira, C. R., Assis, C., Bonomini, T., & Bresolin, T. M. B. (2019). RP-HPLC and LC–MS–MS determination of a bioactive artefact from Ipomoea pes-caprae extract. *Revista Brasileira de Farmacognosia*, *29*(5), 570–577. DOI: 10.1016/j.bjp.2019.05.005

Das, R. A bio-geographical study on major existed and endangered phyto-resources over fragile and sensitive coastal tract of midnapore in west bengal download-libre.pdf.

Demis, E. (2024). Mechanism of plant resistance to insects, weeds and pathogens. *Mid. East Res. J. Agr. Food Sci*, *4*(2), 76–85. DOI: 10.36348/merjafs.2024.v04i02.005

Ferreira, A. A., Amaral, F. A., Duarte, I. D. G., Oliveira, P. M., Alves, R. B., Silveira, D., Azevedo, A. O., Raslan, D. S., & Castro, M. S. A. (2006). Antinociceptive effect from Ipomoea cairica extract. *Journal of Ethnopharmacology, 105*(1-2), 148–153. DOI: 10.1016/j.jep.2005.10.012 PMID: 16307856

Ganjir, M., Behera, D. R., & Bhatnagar, S. (2013). Phytochemical analysis, cytotoxic and antioxidant potential of Ipomoea pes caprae (L) R. Br and Merremia umbellata (L.) H. Hallier. *International Journal of Scientific & Technology Research, 2*(5), 80–83.

Gazali, M., Umar, W., Nufus, H., Fadly, S. A. W., & Syafitri, R. (2024). The Investigation of the Ipomea pes-caprae Leaf Extract as Antimicrobial of Staphylococcus aureus. In *IOP Conference Series: Earth and Environmental Science* (Vol. 1410, No. 1, p. 012011). IOP Publishing. DOI: DOI: 10.1088/1755-1315/1410/1/012011

Goyal, A., Sikarwar, O., Verma, A., Solanki, K., Agrawal, N., Dubey, N., & Yadav, H. N. (2024). Unveiling myricetin's pharmacological potency: A comprehensive exploration of the molecular pathways with special focus on PI3K/AKT and Nrf2 signaling. *Journal of Biochemical and Molecular Toxicology, 38*(6), e23739. DOI: 10.1002/jbt.23739 PMID: 38769721

Guo, Z., Sun, J., Chen, X., Li, H., Liang, S., Liu, F., Qu, T., Wang, H., Li, X., Ou, Z., Feng, H., Ma, J., Wang, S., Wang, L., Tang, B., Wang, G., Qin, Y., & Cheng, Y. (2025). Comparative analysis of HKT genes in *Ipomoea pes-caprae* unveils conserved Na+/K+ symporter functions within the gene family. *Frontiers in Plant Science, 16*, 1538669. DOI: 10.3389/fpls.2025.1538669 PMID: 40247947

Islam, S., Yoshimoto, M., Ishiguro, K., & Yamakawa, O. (2002). Bioactive compounds in Ipomoea batatas leaves. In *XXVI International Horticultural Congress: Issues and Advances in Postharvest Horticulture 628* (pp. 693-699). DOI: DOI: 10.17660/ActaHortic.2003.628.88

Islamy, R. A., Hasan, V., Mamat, N. B., Kilawati, Y., & Maimunah, Y. (2024). Various solvent extracts of Ipomoea pes-caprae: A promising source of natural bioactive compounds compare with vitamin C. *Iraqi Journal of Agricultural Sciences, 55*(5), 1602–1611. DOI: 10.36103/5vd4j587

Karthikeyan, G., Swamy, M. K., Viknesh, M. R., Shurya, R., & Sudhakar, N. (2020). Bioactive Phytocompounds to fight against antimicrobial resistance. In *Plant-derived Bioactives: Production, Properties and Therapeutic Applications* (pp. 335-381). Singapore: Springer Singapore https://doi.org/DOI: 10.1007/978-981-15-1761-7_14

Kumar, S., Abedin, M. M., Singh, A. K., & Das, S. (2020). Role of phenolic compounds in plant-defensive mechanisms. In *Plant phenolics in sustainable agriculture* (Vol. 1, pp. 517–532). Springer Singapore., DOI: 10.1007/978-981-15-4890-1_22

Li, C. Y., Yan, M. K., Wang, L. X., Shen, J. Q., Xiong, J., Wang, W. Q., & Xuan, L. J. (2025). UPLC-TOF-MS/MDF oriented isolation of calonyctins KR: Resin glycosides from Ipomoea muricata with cytotoxic and multidrug resistance reversal activities. *Phytochemistry*, *235*, 114454. DOI: 10.1016/j.phytochem.2025.114454 PMID: 39986411

Li, N., Han, X., Feng, D., Yuan, D., & Huang, L. J. (2019). Signaling crosstalk between salicylic acid and ethylene/jasmonate in plant defense: Do we understand what they are whispering? *International Journal of Molecular Sciences*, *20*(3), 671. DOI: 10.3390/ijms20030671 PMID: 30720746

Liu, Y., Dai, X. B., Zhao, L. K., Huo, K. S., Jin, P. F., Zhao, D. L., Zhou, Z., Tang, J., Xiao, S., & Cao, Q. H. (2020). RNA-seq reveals the salt tolerance of Ipomoea pes-caprae, a wild relative of sweet potato. *Journal of Plant Physiology*, *255*, 153276. DOI: 10.1016/j.jplph.2020.153276 PMID: 33059125

Lorenzi, H., & Abreu Matos, F. J. (2002). *Plantas medicinales no Brasil. Nativas e exóticas*. Instituto Plantarum de Estudos da Flora Ltda.

Mahdavian, K. (2024). Effects of ultraviolet radiation on plants and their protective mechanisms. *Russian Journal of Plant Physiology: a Comprehensive Russian Journal on Modern Phytophysiology*, *71*(6), 184. DOI: 10.1134/S1021443724607481

Majid, M., Nasir, B., Zahra, S. S., Khan, M. R., Mirza, B., & Haq, I. U. (2018). Ipomoea batatas L. Lam. ameliorates acute and chronic inflammations by suppressing inflammatory mediators, a comprehensive exploration using in vitro and in vivo models. *BMC Complementary and Alternative Medicine*, *18*(1), 216. DOI: 10.1186/s12906-018-2279-5 PMID: 30005651

Manigauha, A., Kharya, M. D., & Ganesh, N. (2015). In vivo antitumor potential of Ipomoea pes-caprae on melanoma cancer. *Pharmacognosy Magazine*, *11*(42), 426. DOI: 10.4103/0973-1296.153099 PMID: 25829785

Martinez, M. L., Silva, R., Mendoza, E., Odériz, I., & Pérez-Maqueo, O. (2016). Coastal dunes and plants: An ecosystem-based alternative to reduce dune face erosion. *Journal of Coastal Research*, *75*(sp1), 303–307. DOI: 10.2112/SI75-061.1

McDonald, A. N. D. R. E. W. (1991). Origin and diversity of Mexican Convolvulaceae. *Anales del Instituto de Biología. Serie Botánica*, *62*(1), 65–82.

Meira, M., Silva, E. P. D., David, J. M., & David, J. P. (2012). Review of the genus *Ipomoea*: Traditional uses, chemistry and biological activities. *Revista Brasileira de Farmacognosia*, *22*(3), 682–713. DOI: 10.1590/S0102-695X2012005000025

Muñoz-Rodríguez, P., Wood, J. R., Wells, T., Carruthers, T., Sumadijaya, A., & Scotland, R. W. (2023). The challenges of classifying big genera such as Ipomoea. *Taxon*, *72*(6), 1201–1215. DOI: 10.1002/tax.12887 PMID: 40687846

Mutha, R. E., Kalaskar, M., & Khan, Z. G. (2025). Modern Analytical Techniques for Quality Control and Chemical Identification of Phytochemicals. *Pharmacognosy and Phytochemistry: Principles, Techniques, and Clinical Applications*, 167-188 https://doi.org/DOI: 10.1002/9781394203680.ch09

Muthamilarasan, M., & Prasad, M. (2013). Plant innate immunity: An updated insight into defense mechanism. *Journal of Biosciences*, *38*(2), 433–449. DOI: 10.1007/s12038-013-9302-2 PMID: 23660678

Narayanan, L., & Suseem, S. R. (2024). Pharmacological insights into Ipomoea staphylina: Therapeutic activities and the isolated bioactive metabolic compounds. *Mongolian Journal of Chemistry*, *25*(52), 1–9. Advance online publication. DOI: 10.5564/mjc.v25i52.3195

Nuralyza, I., & Hanifa, N. I. (2025). Potential of Ipomoea pes-caprae Extract as an Antibacterial: A Literature Review. *Jurnal Biologi Tropis*, *25*(2), 1573–1582. DOI: 10.29303/jbt.v25i2.8839

Nuskiya, A., Sibero, M. T., Setyati, W. A., Andriani, C., & Hendryanti, D. N. (2023). Bioprospecting of katang-katang leaves (*Ipomoea pes-caprae*) from Sumba Island, East Nusa Tenggara: antimicrobial, antioxidant and secondary metabolites content. In *IOP Conference Series: Earth and Environmental Science* (Vol. 1260, No. 1, p. 012054). IOP Publishing. DOI: 10.1088/1755-1315/1260/1/012054

Panichakul, T., Ponnikorn, S., Tupchiangmai, W., Haritakun, W., & Srisanga, K. (2022). Skin anti-aging potential of *Ipomoea pes-caprae* ethanolic extracts on promoting cell proliferation and collagen production in human fibroblasts (CCD-986sk Cells). *Pharmaceuticals (Basel, Switzerland)*, *15*(8), 969. DOI: 10.3390/ph15080969 PMID: 36015117

Parekh, K. K., Patel, A. M., Modi, A. J., & Chandrashekhar, H. R. (2012). Antioxidant and cytotoxic activities of few selected Ipomoea species http://www.pharmacologia.co.uk

Pereda-Miranda, R., Escalante-Sánchez, E., & Escobedo-Martínez, C. (2005). Characterization of Lipophilic Pentasaccharides from Beach Morning Glory (*Ipomoea pes-caprae*). *Journal of Natural Products*, *68*(2), 226–230. DOI: 10.1021/np0496340 PMID: 15730248

Pothula, V. V. S., & Kanikaram, S. (2015). In vitro antiplasmodial efficacy of mangrove plant, *Ipomoea pes-caprae* against Plasmodium falciparum (3D7 strain). *Asian Pacific Journal of Tropical Disease*, *5*(12), 947–956. DOI: 10.1016/S2222-1808(15)60963-X

Prakash, S., Ramasubburayan, R., Ramkumar, V. S., Kannapiran, E., Palavesam, A., & Immanuel, G. (2016). In vitro—Scientific evaluation on antimicrobial, antioxidant, cytotoxic properties and phytochemical constituents of traditional coastal medicinal plants. *Biomedicine & Pharmacotherapy = Biomédecine & Pharmacothérapie*, *83*, 648–657. DOI: 10.1016/j.biopha.2016.07.019 PMID: 27459122

Qasim, M., Abideen, Z., Adnan, M. Y., Gulzar, S., Gul, B., Rasheed, M., & Khan, M. A. (2017). Antioxidant properties, phenolic composition, bioactive compounds and nutritive value of medicinal halophytes commonly used as herbal teas. *South African Journal of Botany*, *110*, 240–250. DOI: 10.1016/j.sajb.2016.10.005

Ramanathan, T., Shanmugapriya, R., Thirunavukkarasu, P., & Renugadevi, G. (2012). Invitro antifungal activity from flower extracts of *Ipomoea pes-caprae*. *International Journal of Advances in Pharmaceutical Research*, *3*(1), 732–735.

Roychowdhury, R., Hada, A., Biswas, S., Mishra, S., Prusty, M. R., Das, S. P., Ray, S., Kumar, A., & Sarker, U. (2025). Jasmonic acid (JA) in plant immune response: Unravelling complex molecular mechanisms and networking of defence signalling against pathogens. *Journal of Plant Growth Regulation*, *44*(1), 89–114. DOI: 10.1007/s00344-024-11264-4

Satyavani, K., Gurudeeban, S., Ramanathan, T., & Balasubramanian, T. (2013). *Ipomoea pes-caprae* Mediated Silver Nanoparticles and their. *Science international*, *1*(5), 155-159. DOI: p 155-159.DOI: 10.5567/sciintl.2013.155.159

Selvam, N. T., & Acharya, M. V. (2015). Review of *Ipomoea pes-tigridis* L.: Traditional uses, botanical characteristics, chemistry and biological activities. *International Journal of Pharmaceutical Sciences and Research*, *6*, 1443–1448.

Shaikh, J. R., & Patil, M. (2020). Qualitative tests for preliminary phytochemical screening: An overview. *International Journal of Chemical Studies*, *8*(2), 603–608. DOI: 10.22271/chemi.2020.v8.i2i.8834

Sheeba, S. N., Ariharan, V. N., Mary, J. V. J., Bai, S. M. M., & Paul, J. V. (2021). Phytochemical and Biological Screening of Organic Solvent Extracts of Ipomoea Pes-Caprae flower. *Annals of the Romanian Society for Cell Biology*, *25*(3), 7800–7821.

Shiau, J. P., Chuang, Y. T., Tang, J. Y., Yang, K. H., Chang, F. R., Hou, M. F., Yen, C.-Y., & Chang, H. W. (2022). The impact of oxidative stress and AKT pathway on cancer cell functions and its application to natural products. *Antioxidants*, *11*(9), 1845. DOI: 10.3390/antiox11091845 PMID: 36139919

Singh, A., Agrawal, A., Jain, A., Jain, S., & Darwhekar, G. N. (2011). Antipyretic Activity of Aqueous Leaf Extract of Ipomoea biloba. *Inventi Rapid: Ethnopharmacology*.

Situmorang, E. C., Bara, R., Warouw, V., Wagey, B., Mantiri, D., Sondak, C., & Posangi, J. (2024). Testing of UV protection activity, SPF determination, and irritation assessment of coastal vegetation *Ipomoea pes-caprae. International Journal of Health & Medical Research, 3*(11), 03 DOI: .DOI: 10.58806/ijhmr.2024.v3i11n03

Srivastava, D., & Rauniyar, N. (2020). *Medicinal plants of genus Ipomoea: A glimpse of potential bioactive compounds of genus Ipomoea and its detail.* Lap Lambert Academic Publishing.

Srivastava, D., & Shukla, K. (2015). Pharmaceutical efficacy of *Ipomoea* carnea. In *Biological Forum* (Vol. 7, No. 1, p. 225). Research Trend.URL: https://researchtrend .net/bf12/39%20DEEPA%20SRIVASTAVA.pdf

Su, Y., Liu, Y., Xiao, S., Wang, Y., Deng, Y., Zhao, L., Wang, Y., Zhao, D., Dai, X., Zhou, Z., & Cao, Q. (2023). Isolation, characterization, and functional verification of salt stress response genes of NAC transcription factors in *Ipomoea pes-caprae. Frontiers in Plant Science*, *14*, 1119282. DOI: 10.3389/fpls.2023.1119282 PMID: 36818867

Suhasini, S., Elanchezhiyan, C., & Chandirasegaran, G. (2019). Ameliorative effect of Ipomoea pes-caprae ethanolic leaf extract on carbohydrate metabolizing enzymes and oxidative status in Streptozotocin-induced diabetic Wistar rats. *Journal of Drug Delivery and Therapeutics*, *9*(4-A), 167–175. DOI: 10.22270/jddt.v9i4-A.3448

Suhendy, H., Fidrianny, I., & Insanu, M. (2023). Phytochemical compounds and pharmacological activities of Ipomoea batatas L.: An updated review. *Farmatsiia (Sofia)*, *70*(4), 1283–1294. DOI: 10.3897/pharmacia.70.e108922

Vallipriya, R., & Begum, M. S. (2021). Effect of *Ipomoea biloba* leaves on cardioprotective activity against the isoproterenol induced myocardial infarction in rats. *Research Journal of Agricultural Science, 12*(1), 281–285. http://rjas.org/Article/DownloadFile/3565

Vallipriya, R., & Begum, M. S. (2021). Protective Effect of Ipomoea biloba on Myocardial Antioxidant Status in Isoproterenol Induced Myocardial Infarction Male Albino Rats. *Asian Journal of Research in Biochemistry, 9*(3), 36–45. DOI: 10.9734/ajrb/2021/v9i330204

Veeramani, S., Ravindran, E., Ramadoss, P., Joseph, C., Shanmugam, K., & Renganathan, S. (2018). Silver nanoparticles green synthesis with Aq. extract of stems *Ipomoea pes-caprae*, characterization, antimicrobial and anti-cancer potential. *International Journal of Medical Nano Research, 5*(024), 10–23937. DOI: 10.23937/2378-3664.1410024

Venkataramana, N. D., Atleeb, W. C., Prabhuc, T. P., & Kannana, R. (2013). Anti-inflammatory potential of ethanolic extracts from aerial parts of *Ipomoea pes-caprae* (L.) R. Br using cotton pellet induced granuloma model. *Journal of Applied Pharmaceutical Science, 3*(7), 061-063. DOI:DOI: 10.7324/JAPS.2013.3711

Wu, J., Su, Y., Pan, Z., Wang, Y., Zhang, Y., Li, L., Jiang, J., & Cao, X. (2024). Identification of WRKY transcription factors in *Ipomoea pes-caprae* and functional role of IpWRKY16 in sweet potato salt stress response. *BMC Plant Biology, 24*(1), 1190. DOI: 10.1186/s12870-024-05928-2 PMID: 39702019

Yeshi, K., Turpin, G., Jamtsho, T., & Wangchuk, P. (2022). Indigenous uses, phytochemical analysis, and anti-inflammatory properties of Australian tropical medicinal plants. *Molecules (Basel, Switzerland), 27*(12), 3849. DOI: 10.3390/molecules27123849 PMID: 35744969

APPENDIX

Abbreviations

µg: Microgram
µL: Microliter
AST: Antimicrobial Susceptibility Testing
CFU: Colony Forming Unit
CLSI: Clinical and Laboratory Standards Institute
DMSO: Dimethyl Sulfoxide
***EEIB*:** Ethanol extract of *Ipomoea biloba*
ETI: Effector-Triggered Immunity
EtOH: Ethanol
GC-MS: Gas Chromatography–Mass Spectrometry
GC–MS: Gas Chromatography–Mass Spectrometry
GC–MS: Gas Chromatography–Mass Spectrometry
GI: Gastrointestinal
HBC: Human breast cancer
HPLC: High-Performance Liquid Chromatography
HR: Programmed Cell Death
IPC70: *Ipomoea pes-caprae* 70 extracts
IPC95: *Ipomoea pes-caprae* 85 extracts
MAMPs: Microbe-associated molecular patterns
MBC: Minimum Bactericidal Concentration
MDR Reversal: multidrug resistance reversal
mg: Milligram
MIC: Minimum Inhibitory Concentration
mL: Milliliter
mm: Millimeter
MTI–MAMP: Triggered Immunity
NMR: Nuclear Magnetic Resonance
PAMPs: Pathogen-Associated Molecular Patterns
PBS: Phosphate Buffered Saline
PCD: Programmed Cell Death
ROS: Reactive Oxygen Species
RT: Room Temperature
SD: Standard Deviation
SEM: Scanning Electron Microscopy
STZ: Streptozotocin

UV: Ultraviolet
WHO: World Health Organization
ZDI: Zone of Inhibition

Chapter 12
Herbal Medicine and Antimicrobial Stewardship:
Bridging One Health and Traditional Knowledge

Parthiban Pandian
http://orcid.org/0000-0002-3582-3416
Saveetha Institute of Medical and Technical Sciences (SIMATS), India

Sivaranjani Varadharajan
http://orcid.org/0009-0002-5795-4377
Periyar Maniammai Institute of Science and Technology, India

Gomathi Rengaraj
http://orcid.org/0000-0002-7367-0094
Holy Cross College, Trichy, India

E. Mohanraj
http://orcid.org/0009-0008-6520-5273
Sri Rangapoopathi College of Pharmacy, India

ABSTRACT

This chapter per the authors examines the potential of herbal medicine as a complementary strategy for mitigating antimicrobial resistance within a One Health framework. It synthesizes evidence on the antimicrobial, antivirulence, and immunomodulatory properties of plant-derived compounds and evaluates their application across human healthcare, veterinary medicine, aquaculture, and environmental management. The chapter highlights mechanistic pathways through which phyto-

DOI: 10.4018/979-8-3373-7810-7.ch012

chemicals reduce resistance pressure, including biofilm inhibition, quorum sensing disruption, and synergistic enhancement of conventional antibiotics. Real-world case studies demonstrate reductions in antibiotic use without compromising health outcomes. Structural, regulatory, and scientific barriers to adoption are critically assessed, alongside emerging research priorities and policy needs. Overall, the chapter underscores the role of validated, standardized herbal interventions in supporting antimicrobial stewardship and advancing sustainable One Health solutions.

1. GLOBAL ANTIMICROBIAL RESISTANCE: EPIDEMIOLOGICAL AND SYSTEMIC DIMENSIONS

1.1 Conceptual Foundations and Clinical Implications of AMR

Antimicrobial resistance (AMR) poses a major threat to modern medicine, reversing decades of progress against infectious diseases. According to the WHO (2023), multidrug-resistant bacterial infections kill 4.95 million people annually, mostly in low- and middle-income countries. Globally, antimicrobial resistance kills more people in areas with fewer healthcare resources. Sub-Saharan Africa has five times the drug-resistant infection death rate of wealthier areas. This gap exists because not everyone has access to tests, treatments, and infection prevention methods. Less effective antibiotics mean fewer treatment options, longer recovery times, more deaths, and higher healthcare costs (Aslam et al., 2018; Antimicrobial Resistance Collaborators, 2022).

Resistance to antimicrobials can develop due to evolutionary selection. Resistant strains spread faster due to overprescribing, misdosing, short treatment periods, and farming with drugs (O'Neill, 2016). Murray et al. (2022) contend that indiscriminate antibiotic use for minor infections or those that would resolve spontaneously imposes stress on microbial ecosystems. Adaptive resistance can influence both treatment and therapy. This puts at risk procedures such as transplants, cancer treatments, surgeries, and other procedures that require effective infection control.

1.2 Global Burden and Epidemiology

Recent epidemiological assessments have a tendency to emphasize the expanse of AMR as well as the geographic diversity of its types. Antimicrobial resistance directly caused 1.27 million deaths and contributed to 4.95 million more by 2019, according to the Global Burden of Disease study. Murray et al. (2022) found that South Asia and sub-Saharan Africa were the regions with the most problems in their study. These numbers show how hard it is to treat microorganisms, how many different healthcare systems there are, how well they find infections, and how to give antibiotics the right way.

The rise of bacteria like ESBL-producing Enterobacteriaceae, carbapenem-resistant strains, and MRSA in hospitals and communities highlights the antimicrobial resistance crisis (World Health Organization, 2023; Marino et al., 2025). Infections from multidrug-resistant Gram-negative bacteria such E. coli, Klebsiella pneumoniae, and Acinetobacter baumannii often cause high mortality. In the absence of adequate control and surveillance, pathogens may continue to exist. Uncontrolled outpatient prescribing and inadequate primary care diagnostics spread ESBL-producing E. coli. Hospitals without good infection control and real-time surveillance have developed carbapenem-resistant pathogens.

1.3 Drivers of Resistance: Human, Animal, and Environmental Factors

Environmental factors, human health, and animal agriculture all play a role in the development of antimicrobial resistance. Antibiotics are becoming less effective due to underuse or overuse, according to Kasse et al. (2024). Their performance has declined. Patient requests, testing errors, and precise prescribing worsen the situation.

Farmers feed animals' penicillin to grow and stay healthy. Direct contact or the food chain can spread antibiotic-resistant bacteria (Van Boeckel et al., 2019). This activity promotes that bacteria. Drug waste, farm runoff, and untreated sewage release antibiotics and resistance genes. This causes bacteria to share genes (Kraemer et al., 2019).

1.4 Effects on Socioeconomic Conditions and Health Care Systems

Not only will resistance to antimicrobials hurt health, but it will also hurt the economy and society as a whole. These repercussions will be significant. More diagnostic tests, longer hospital stays, more expensive second-line treatments, and more treatment failures make healthcare systems work harder. Sihombing et al. (2023) estimate that AMR could cost over $100 trillion by 2050. This is because people believe no good solutions will exist.

It is likely that these estimates do not take into account additional costs, such as decreased productivity, long-term disability, food insecurity, and increased poverty in areas that are vulnerable. Not only is it difficult to obtain medical care in regions with limited resources, but it is also difficult to monitor the spread of diseases in these regions. The fact that there is a limited capacity to respond and high infection rates demonstrates how important it is to find solutions that are both affordable and adaptable.

1.5 The rationale behind the integration of herbal medicine and its scope

In light of the fact that the production of new antibiotics has slowed down, as well as the rapid emergence of resistance, it is necessary for us to investigate additional and supportive strategies. According to Jamshidi-Kia et al. (2018) and Anand et al. (2019), traditional knowledge systems that have been around for a significant amount of time serve as the foundation for herbal medicines. These drugs may contain many bioactive chemicals. These can fight germs, reduce inflammation, alter the immune system, and prevent biofilm.

One of the most important things that we need to do is monitor the use of herbal medicine and demonstrate how effective it is in combating antibiotic resistance. Instead of using antibiotics more often, we should focus on proper use and antimicrobial stewardship programs to improve their efficacy.

2. ANTIMICROBIAL RESISTANCE AS A ONE HEALTH CHALLENGE

2.1 Conceptualizing One Health in AMR

Antimicrobial resistance is a complex problem that impacts numerous domains. The One Health idea says that we need to look at the big picture when fighting resistance because people, animals, and the environment are all connected (Destoumieux-Garzón et al., 2018). By altering the composition of microbial communities, antimicrobials confer an increased level of resistance on these communities. Various components of this intricate system interact with one another and have an effect on one another.

Antibiotics treat many diseases, including some in humans and animals. They influence decisions that affect more than disease treatment. Antibiotics and antibiotic-resistant bacteria can enter the environment in small amounts, spreading resistance genes and sustaining them (Wu et al., 2025). Utilizing, drinking, or touching them can all be effective ways for people and animals to strengthen their immune systems. Collaborating with one another is of utmost importance in the fight against antimicrobial resistance in every domain.

2.2 Human Health Drivers of AMR

In human health, AMR is strongly connected to the misuse of antibiotics. Even with guidelines, many prescriptions for respiratory, urinary, and digestive infections still remain unnecessary or are not given correctly (Dyar et al., 2018). Because testing is often limited, doctors sometimes have to guess the cause of illness, and patient demands, or failure to follow instructions, can make things worse.

As a result of these issues, the number of bacteria that are resistant to antibiotics has increased in hospitals as well as in the general civilian population. The prevalence of methicillin-resistant Staphylococcus aureus (MRSA), vancomycin-resistant enterococci, and specific strains of Escherichia coli has been on the rise outside of large hospitals, which poses a significant threat to the majority of the population (Antimicrobial Resistance Collaborators, 2022).

2.3 Animal Health and Livestock Practices

In certain regions, livestock accounts for more than 70 percent of the total antibiotic usage (Van Boeckel et al., 2019). Additionally, antibiotic administration in animals that are used for food production constitutes a significant portion of the global antibiotic consumption. For a long time, antibiotics have been used for more than just treating illnesses. For instance, they have been utilized to promote growth and prevent disease. This practice has resulted in persistent selection pressure within animal production systems.

Ethnoveterinary treatments and plant-based feed additives (Gadde et al., 2017; Umair et al., 2024) can help animals stay healthy and productive without the use of antibiotics. To make sure these parts always work together, they need to be carefully managed and watched.

2.4 The Effects of AMR on the Environment

Environmental systems not only serve as enzymes but also as storage locations for antimicrobial resistance, which helps to spread the disease. According to Kraemer et al. (2019), antimicrobials found in wastewater treatment plants, pharmaceutical waste, and agricultural runoff facilitate the exchange of genes between bacteria. Through the maintenance of genes that confer resistance in individuals, these conditions contribute to the process of evolution.

In aquaculture, water, and land soils, resistance genes are rapidly spreading throughout the environment. According to Li et al.'s 2025 research, pathogenic microorganisms that possess genetic characteristics that make them more resistant to their environment can live for a longer period of time. Therefore, it is necessary for each nation's plans to combat antimicrobial resistance to incorporate measures to safeguard the environment. This should include more stringent regulations regarding the disposal of waste, improved methods for treating wastewater, and reduced levels of pollution on the part of agricultural practices.

2.5 Cross-Sector Transmission Pathways

The transmission of the disease can take place through the consumption of contaminated food products, the sharing of water supplies, the exposure of workers in workplaces that involve the handling of animals, and through global trade and travel systems (Destoumieux-Garzón et al., 2018; Van Boeckel et al., 2019). The fact that these routes are interconnected demonstrates that containment measures that are effective in just one sector in isolation are insufficient.

In order to put a stop to transmission, it is necessary for all of the nodes that are part of the network to work together at the same time. For quick problem detection and prevention across sectors, integrated surveillance, coordinated regulation, and shared data systems are essential.

2.6 The Integration of Stewardship Across Different Sectors

The prescribing practices of hospitals have been the primary focus of antimicrobial stewardship programs according to traditional standards. A One Health-informed approach goes beyond clinical settings and incorporates environmental regulations as well as the systems that are involved in the production of animals. When it comes to human healthcare, stewardship attributes a significant amount of importance to the management of infections, the provision of assistance with diagnoses, and the prudent prescription of medications. It exerts a significant amount of pressure on vaccination strategies, alternative treatments that have been sanctioned, and the restriction of preventative measures in veterinary settings (Pulcini et al., 2017).

Working together to manage these areas allows them to make a greater difference by simultaneously lowering the selective pressure in multiple reservoirs. This allows them to make a more significant impact. As a result of this integration, antimicrobial resistance mitigation is transformed from a reactive strategy into a proactive, comprehensive approach that encompasses the entire system.

3. TRADITIONAL HERBAL SYSTEMS AND ANTIMICROBIAL FOUNDATIONS

3.1 Overview of Traditional Medicine Systems

In order to treat infections and inflammation, people have relied on traditional medical systems for a significant amount of time (WHO, 2019). Some examples of these systems include Ayurveda, Traditional Chinese Medicine (TCM), Unani, and a variety of indigenous remedies. Sure, their theories and ways of diagnosing things don't always line up, but they all focus on bringing the whole body back into balance, not just attacking the germs themselves. Most of their treatments come from plants—decoctions, powders, tinctures, oils, even fermented mixtures are all staples.

Around 80% of people worldwide relies on traditional medicine for aspects of basic healthcare, particularly in regions, with low- and middle-income, where access to modern drugs are not always easily available (Bari et al., 2021). Many plants that have been used traditionally for infectious conditions have shown measurable antimicrobial properties in laboratory studies. Phytochemicals such as flavonoids, alkaloids, tannins, and also essential oils exhibit activity against fungi, bacteria and viruses thereby supporting the pharmacological effectiveness of the traditional practices (Mehdizadeh & Moghaddam, 2018).

3.2 Historical Perspectives on Herbal Antimicrobials

Classical medical texts from the diverse cultures provide plant based strategies for the cure of infectious diseases. The Ayurvedic Charaka Samhita and the Unani Canon of Medicine describe plant formulations for infections, notably wound care, dysentery, respiratory infections, and febrile illnesses. According to Jamshidi-Kia et al. (2018), two medicinal plants that have been known to have healing properties for a long time are Azadirachta indica (also known as neem) and Curcuma longa (also known as turmeric). These include the ability to stop infections, lower inflammation, and help wounds heal.

Traditional Chinese Medicine (TCM) also uses plants like Scutellaria baicalensis and Isatis tinctoria to treat fevers and infections. Long-term use of these treatments is a strong foundation for modern drug research. However, historical use is not the same thing as clinical evidence. Over the course of the past few years, research on antimicrobials has been concentrating more and more on traditional formulations in an effort to discover bioactive compounds and investigate the possibility of synergistic mechanisms.

3.3 Phytochemicals with Antimicrobial Activity

Medical plants kill microbes because their secondary metabolites have different effects on living things. Khameneh et al. (2019) discovered that alkaloids such as berberine and sanguinarine cause damage to the cell walls of microorganisms and prevent them from producing nucleic acids. Flavonoids like quercetin and kaempferol inhibit quorum sensing and regulate bacterial enzymes. This leads to less harmful bacteria and biofilm formation (Martínez-Rodríguez et al., 2025).

Tannins and phenolic compounds can attach to proteins and break down microbial membranes, affecting microbes differently, according to Angelini (2024). Essential oils contain terpenes and terpenoids, which have the ability to alter membranes and inhibit efflux pumps, thereby making bacteria more susceptible to antibiotic attacks. (Tariq et al., 2019). These plant chemicals affect many molecular pathways, so bacteria may not become resistant quickly.

3.4 Practices of Ethnoveterinary Medicine

Herbal medicine benefits humans and farm animals like chickens and cows. Azadirachta indica, Allium sativum (garlic), and Ocimum sanctum help stomach and breathing issues, boost the immune system, and keep animals healthy in traditional veterinary medicine (Arya et al., 2024).

Plant-based feed helps animals grow faster and get sick less often, reducing their antibiotic use (Harbarth et al. 2015; Gadde et al. 2017). The use of these techniques can assist in reducing the number of harmful microbes that are present in agricultural areas and can also slow the growth of resistant strains that are associated with animal farming systems.

3.5 Validation of Contemporary Practices and Data from Clinical Studies

Over the course of the past decade, there has been an increase in the number of experimental and clinical studies concerned with traditional plant remedies. Curcuma longa extracts have been shown to inhibit the growth of Staphylococcus aureus and Escherichia coli, whereas the leaf extracts of Azadirachta indica have been found to be effective against Salmonella enterica and Pseudomonas aeruginosa. Both of these strains of bacteria are important in the treatment of bacterial infections. Garlic contains an extract called allicin, which has been shown to be effective against gram-positive and gram-negative bacteria, according to a study that was published not too long ago (Anand et al., 2019; Sun et al., 2025).

Despite the fact that many of the most recent discoveries are still in the preliminary stages of development or are dependent on laboratory models, they do provide scientific support for the conventional applications of antimicrobials. It is essential to have well-planned studies, standardized preparations, and comprehensive evaluations of both the safety of medications and the ways in which herbal remedies and medications interact with one another in order to guarantee that clinical practice is effective. On the other hand, there is an increasing body of evidence suggesting that herbal remedies might be useful in larger-scale plans to exercise control over antibiotic use.

4. PHYTOCHEMICAL MECHANISMS OF ACTION

4.1 Mechanistic Rationale for Phytochemicals in Antimicrobial Therapy

In the realm of antimicrobial agents, phytochemicals are a distinct category that can exert their influence on various components of microbial systems. In contrast to conventional antibiotics, which typically concentrate on a single enzymatic or structural target, bioactive compounds derived from plants have the ability to disrupt multiple pathways simultaneously. These pathways include disrupting membrane integrity, affecting enzymatic activity, hindering nucleic acid replication, altering quorum sensing mechanisms, and influencing virulence expression (Martínez-Rodríguez et al., 2025).

According to the data presented in Table 1, certain medicinal plants possess antimicrobial properties and have the ability to modify resistance by inhibiting efflux pumps, destabilizing membranes, disrupting metabolic processes, and compromising biofilms. Recent mechanistic syntheses (Zhang et al., 2025) have shown that many phytochemicals have regulatory effects at sub-inhibitory concentrations. Antimicrobial stewardship aims to reduce selective pressure while maintaining therapeutic efficacy, which matches this trait.

Table 1. Major herbal sources with antimicrobial and resistance-modifying potential

Medicinal Plant	Bioactive Compounds	Targeted Microorganisms	Mechanism of Action	Key References
Curcuma longa	Curcumin	*S. aureus, E. coli*	Cell membrane disruption, efflux pump inhibition	Pacyga et al., 2024; Zhang et al., 2025
Azadirachta indica	Nimbin, azadirachtin	Gram-positive bacteria	cell wall and membrane disruption	Alzohairy, 2016
Allium sativum	Allicin	MDR bacteria	Enzyme inactivation, quorum sensing disruption	Martínez-Rodríguez et al., 2025
Ocimum sanctum	Eugenol, rosmarinic acid	Respiratory pathogens	Anti-biofilm, immunomodulation	Ogal et al., 2021
Camellia sinensis	Catechins	*P. aeruginosa*	Synergistic activity with antibiotics	Khameneh et al., 2019

The following table provides a list of significant medicinal plants, the active compounds that they contain, the microbes that they go after, and the ways in which they contribute to a reduction in antimicrobial resistance.

4.2 Alkaloids: Genomic and Transport System Disruption

Nitrogen-rich heterocyclic alkaloids inhibit enzymes and bacterial reproduction. According to Li et al.'s research from 2025, berberine disrupts the replication process in Gram-positive and Gram-negative bacteria by acting as an inhibitor of DNA topoisomerase and by intercalating into nucleic acid structures. It is clear that genetic and structural cells are of critical importance because sanguinarine has the ability to reduce the potential of cell walls and membranes. According to Chowdhury et al.'s 2019 research, alkaloids inhibit multidrug efflux systems such as NorA and AcrAB-TolC, which play a significant role in the development of multidrug resistance.

According to Chowdhury et al.'s 2019 research, alkaloids have the ability to prevent transporters such as NorA and AcrAB-TolC from moving drugs out of cells, which thus enables certain bacteria to resist multiple drugs. This indicates that alkaloids have the ability to kill germs immediately. Through the use of molecular modeling, Angelini (2024) demonstrated that alkaloids have the ability to alter resistance by adhering to the surfaces of transporters.

4.3 Flavonoids: Quorum Sensing Inhibition and Virulence Modulation

Flavonoids and polyphenols eliminate the ability of germs to communicate with one another. Both kaempferol and quercetin disrupt the quorum sensing pathways, which are responsible for the formation of biofilms by bacteria and the subsequent illness of humans (Angelini, 2024). In addition to protecting cells from damage, flavonoids also play a role in the fight against germs. There is a possibility that fewer decisions will make it simpler to give in.

The anti-virulence mechanism is crucial because biofilms prevent antibiotics from working. Stopping quorum sensing weakens biofilms, allowing antibiotics to spread faster (Figure 1). According to Zhang et al. (2025), flavonoids and antibiotics both have the ability to reduce the minimum inhibitory concentrations of bacteria that are resistant to them.

4.4 Phenolics and Tannins: Membrane and Metabolic Perturbation

Tannins and phenolic compounds are effective antimicrobial agents because they cause proteins to bind to one another, they tear down cell membranes, and they prevent enzymes from functioning. There are a few distinct perspectives from which to view these characteristics. According to Zhao et al. (2025), the ability of these organisms to attached themselves to metals disrupts metabolic processes that are dependent on iron. The same resistance decreases when targeting multiple areas.

According to Chen et al. (2020) and Ghosh et al. (2025), both Camellia sinensis and Punica granatum contain bioactive compounds that have the ability to eliminate bacteria that are resistant to antibiotics. This is true regardless of whether the bacteria are Gram-positive or Gram-negative. According to Angelini (2024), phenolics work by inhibiting energy pathways, which in turn has an effect on a variety of metabolic processes. Based on these findings, it appears that they have the potential to increase the sensitivity of the metabolism when used in conjunction with other treatments.

4.5 Essential Oils: Lipid Bilayer Disruption and Antibiotic Potentiation

The most important chemicals found in essential oils that have an effect on the lipid bilayers of microorganisms are terpenes and terpenoids. Carvacrol and thrymol open membranes. They also disrupt proton gradients, leaking cells. This makes bacteria less stable, weakening their structural defenses and balance. It is essential to be aware that essential oils have the potential to be beneficial. Antibiotics are able to accumulate inside of cells, particularly in biofilms, because they make cell membranes more permeable and prevent antibiotics from leaving the cell.

As demonstrated in Figure 1, impeding the movement of substances, causing damage to cell membranes, and disrupting communication between bacteria are all factors that influence resistance. Adding a small amount of essential oils to antibiotics, according to Zhang et al. (2025) and Jomova et al. (2025), makes the antibiotics more effective without making them more hazardous to the patient.

Figure 1. Mechanistic pathways of herbal compounds against antimicrobial resistance

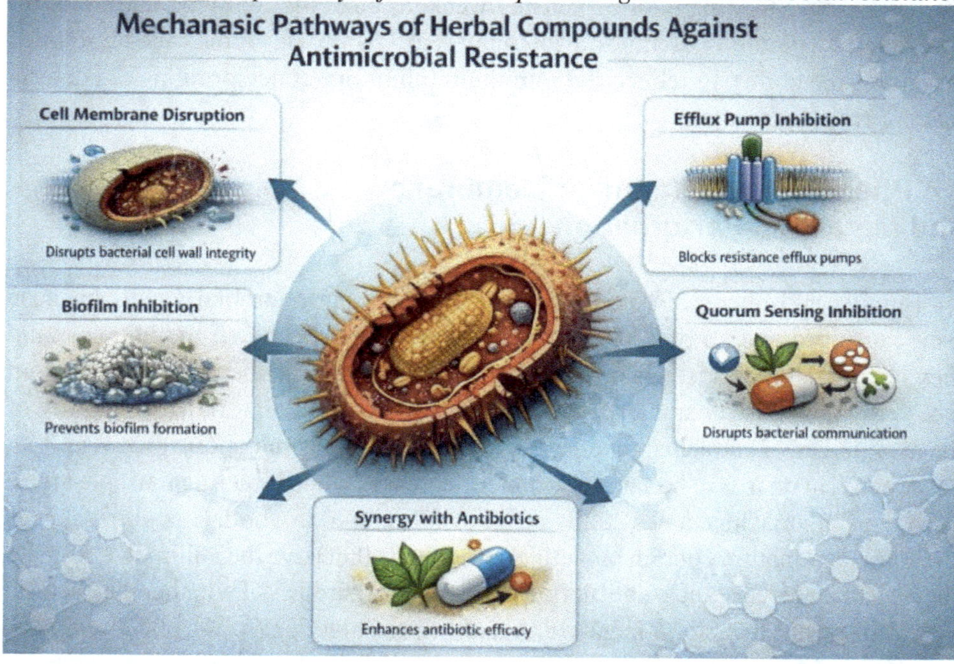

5. HERBAL INTERVENTION MODELS IN ANTIMICROBIAL RESISTANCE MANAGEMENT

5.1 Preventive and Prophylactic Strategies

By treating and preventing illness with herbal medicine, one can help reduce the number of bacteria that are resistant to the drugs that are currently available. The use of herbal remedies for the prevention of infections makes the body more effective at warding off infections, which in turn reduces the likelihood of becoming ill and the prerequisite for taking antibiotics. Flavonoids, polysaccharides, and phenolic chemicals in herbs can make the body's natural defenses stronger, the mucosal barrier stronger, and inflammation stronger. This makes it less likely that you will get sick again.

In One Health, herbal preventive medicines are crucial for primary care, animal care, and aquaculture. These are common places to get sick, and people misuse antibiotics. Recent WHO antimicrobial resistance policies emphasize infection prevention as a key to lowering resistance. Plant-based bioactives improve biosecurity, hygiene, and vaccination over time (WHO, 2023).

5.2 Adjunctive Therapeutic Models

People believe herbal medicine should supplement antibiotics. Phytochemicals reduce symptoms, reduce harmful substances, and heal the immune system. However, antibiotics are primarily effective against bacteria. It is possible that this model, which reduces the amount of antibiotics it uses, will also reduce the amount of antibiotics that are required. Microbes are also effective at preventing and treating infections, making them a valuable resource. To fight infections, natural compounds with biological activity are crucial (Ogwu et al., 2025).

Recent studies have shown that phytomedicines have the potential to alter the functioning of neuroinflammatory and oxidative stress pathways in individuals who have undergone long-term neurological conditions. According to Ogidi et al. (2026), this indicates that phytomedicines may have an additional function in addition to the treatment of infections. These kinds of changes in the body might be especially important for putting an end to the inflammation that occurs as a result of infections that cause neurological problems.

5.3 Resistance-Modulating and Antibiotic-Potentiating Approaches

A significant strategy not only eliminates the microbes, but it also alters the resistance that they are currently exhibiting. The formation of biofilms, the blocking of efflux pumps, and the disruption of quorum sensing pathways are all important steps in the processes that lead to the development of multidrug resistance. Numerous phytochemicals are responsible for these processes. Due to the fact that these compounds concentrate on resistance factors rather than key survival pathways, they exert less selective pressure on microbial populations.

The addition of plant chemicals to antibiotics results in an improvement in their effectiveness. Through the utilization of plant-based enzyme inhibitors and efflux pump inhibitors, it is possible to relinquish your sensitivity to resistant strains. It helps antibiotics last longer and shares many goals with antimicrobial stewardship, which aims to keep antibiotics effective (Muteeb et al., 2023).

5.4 Phytogenic Feed and Population-Level Interventions

In livestock and aquaculture systems, phytogenic feed additives are a type of intervention that aim to reduce the amount of preventive antibiotics that are required at the population level. There is evidence that essential oils, tannins, and saponins derived from plants can assist animals in growing, reducing inflammation, and altering the bacteria that are present in their digestive tracts in a manner that improves their health and productivity without causing any resistance issues.

Plant chemicals degrade in the environment, but human-made antibiotics stay in soil and water longer. One Health aims to improve human and animal health and the environment (Larsson & Flach, 2022). This helps achieve the One Health goals by reducing the number of genes that cause illness in both people and animals.

5.5 Formulation, Delivery Systems, and Dosing Strategies

Formulation science and pharmacokinetics are two aspects that need to be carefully considered in order to ensure successful implementation. A significant number of phytochemicals are either not readily available in the body, degrade rapidly, or are unstable under normal physiological conditions. Nanoformulation, encapsulation, and controlled release help drugs dissolve, stabilize, and reach their targets.

Use rational dosing to avoid subtherapeutic exposure and adaptive responses. Evidence-based dose strategies require pharmacodynamic modeling, standardized extraction procedures, and verified concentration-response relationships.

6. HERBAL MEDICINE IN ANTIMICROBIAL STEWARDSHIP PROGRAMS

6.1 Positioning Herbal Medicine within Antimicrobial Stewardship Frameworks

Antimicrobial stewardship programs (ASPs) aim to maximize antibiotic use, improve patient health, reduce side effects, and prevent antimicrobial resistance (Khadse et al., 2023). Previously, ASPs checked prescriptions, used monitoring systems, enforced standards, and taught in hospitals, outpatient clinics, and veterinary services. However, stewardship frameworks are increasingly required to address upstream drivers of antimicrobial consumption, including preventive care and adjunctive therapies.

Herbal medicine offers a complementary dimension to stewardship by functioning as a resistance-modifying and antibiotic-sparing intervention. Phytochemicals demonstrate multi-target antimicrobial activity and immunomodulatory effects, which may reduce infection severity and antibiotic dependency when applied under controlled protocols (Umair et al., 2024).

6.2 Conceptual Framework for Integration

Add herbal medicine to ASPs carefully, not just to replace other treatments. Taking a look at Table 2, we can see that the process of incorporation encompasses four distinct areas: verifying evidence, evaluating safety, applying across various industries, and monitoring outcomes. The discovery of bioactive plant chemicals that have been demonstrated to have antimicrobial and immune system-modulating effects ought to be feasible.

To illustrate how herbal treatment, relate to human health, animal care, environmental safety, and farming methods, Table 2 presents a comprehensive approach to One Health. As global antimicrobial resistance frameworks (Destoumieux-Garzón et al., 2018) demonstrate, this coordinated approach demonstrates how important it is for people from different fields to collaborate. It ensures that the addition of plant-based therapies improves stewardship systems rather than causing them to become more fragmented.

Table 2. Integration of herbal medicine within the one health framework

One Health Domain	Role of Herbal Interventions	Expected Outcomes	Supporting Evidence
Human Health	Herbal adjunct therapy	Reduced antibiotic dependency	Mukherjee et al., 2016
Veterinary Health	Phytogenic feed additives	Lower AMR transmission	Umair et al., 2024
Environmental Health	Reduced pharmaceutical residues	AMR control in ecosystems	Larsson & Flach, 2022
Agriculture	Natural growth promoters	Decreased antibiotic use	Van Boeckel et al., 2019

6.3 Human Health Applications within Stewardship Programs

There are situations in human medicine where there is no immediate need to use additional antibiotics, and herbal medicine can be helpful in treating mild to moderate infections. Boosting the immune system and possibly reducing the number of antibiotic prescriptions given to patients who do not require hospitalization are two of the potential benefits of using Echinacea purpurea in conjunction with other treatments for respiratory tract infections (Ogal et al., 2021).

Allium sativum, on the other hand, works well with standard antibiotics to kill bacteria that are resistant to them. On the other hand, Curcuma longa possesses anti-inflammatory and antimicrobial properties that are beneficial for the healing of wounds (Anand et al., 2019). There is evidence that stewardship can help people use fewer antibiotics, which is a significant benefit. Stewardship is preferable to the practice of using herbal treatments without giving them any thought.

6.4 Veterinary Applications within Stewardship Programs

The use of fewer antibiotics as a preventative measure and the prevention of the spread of antibiotic resistance through food chains are two important goals in the field of veterinary medicine. Some of the plants that have been used in therapy include Azadirachta indica, Ocimum sanctum, and Allium sativum. It has been demonstrated that these plants strengthen the immune system and combat microbes in livestock (Gadde et al., 2017; Umair et al., 2024).

A careful integration of this kind helps to prevent the spread of antimicrobial resistance (AMR) between humans and animals, and it also provides support for policies that promote sustainable livestock production. By limiting selective pressure within animal reservoirs, phytogenic interventions strengthen the broader One Health stewardship model depicted conceptually in Figure 2.

Figure 2. One health convergence model for herbal antimicrobial stewardship

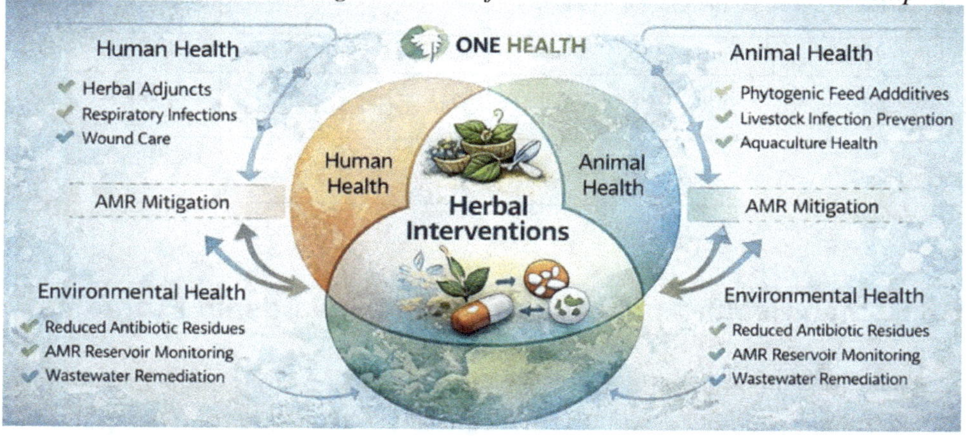

6.5 Community-Level Stewardship Integration

Because antimicrobials are readily available, because of cultural norms, and because people treat themselves without the assistance of a professional, people frequently misuse them in communities. It is possible that the use of antibiotics that are not required could be reduced by incorporating therapeutic herbal remedies that have been demonstrated to be effective for treating minor infections. It is possible to help spread the word about plant-based treatments by working with primary healthcare systems, working with traditional healers, and participating in educational programs (Mukherjee et al., 2016; Bari et al., 2021).

The rules of stewardship should be followed when integrating into the community. This means that herbal remedies should be used appropriately, and antibiotic treatment for serious infections should not be delayed. When implemented in the appropriate manner, these strategies have the potential to assist in altering behavior and reducing the utilization of antibiotics.

7. TRANSLATIONAL CASE EVIDENCE AND ANALYTICAL INTEGRATION WITHIN ANTIMICROBIAL RESISTANCE MITIGATION

7.1 Positioning Case Evidence within Resistance Governance

Case-based evidence provides insight into how phytotherapeutic interventions operate within real-world health systems, yet its interpretive limits must be clearly defined. Experimental antimicrobial activity does not automatically translate into resistance containment at population level. Most field implementations evaluate clinical symptoms, pathogen load, or antibiotic prescription rates rather than longitudinal resistance gene dynamics.

So, case studies show changes in antibiotic exposure rather than proof that the trend of resistance is changing. It is important to understand this difference in order to keep ideas clear and not mix up antimicrobial efficacy with antimicrobial resistance mitigation.

7.2 Human Clinical Contexts: Adjunctive Use and Prescription Dynamics

One example of how it is possible to use plant-based medicines to treat infections in a particular region is the application of curcuma longa for the treatment of wounds. In the laboratory, activity against Staphylococcus aureus (including MRSA) and

Escherichia coli has been observed (Anand et al., 2019; Robé et al., 2024). This is the most convincing evidence that this integration is present. Studies conducted in clinical settings have demonstrated that wounds heal more quickly and that localized inflammation improves.

Because they have the potential to reduce the overall use of antibiotics in straightforward cases, these findings are significant for the fight against antimicrobial resistance. However, this drop is contingent upon the availability of well-defined clinical indications, well-organized prescribing guidelines, and formulations that are consistent. Despite the fact that localized phytotherapy is still utilized in clinical settings, its overall effectiveness is restricted due to the fact that it is not included in official stewardship audits.

7.3 Livestock Systems: Antibiotic Exposure Reduction under Complex Farm Conditions

Allium sativum may replace antibiotic growth promoters in poultry farming, according to research. Controlled studies show fewer harmful gut germs and a different immune system (Gadde et al., 2017; Umair et al., 2024). In livestock settings, improving feed, starting vaccination programs, and keeping the area clean must happen simultaneously. We can't say that plant-based supplements alone reduced antibiotic use.

The main question about resistance is whether lower animal antibiotic exposure lowers resistance factors in food chains and environments. Not much evidence supports this yet. Less antibiotic use does not control antibiotic resistance. Most studies focus on antibiotic use rather than bacteria resistance.

7.4 Aquaculture Environments: Ecological Implications and Monitoring Gaps

In aquaculture, extracts of Azadirachta indica have been used to manage infections associated with Aeromonas and Vibrio species (Umair et al., 2024). It is possible that certain circumstances do not require as many preventive antibiotics as we would have previously believed. This is due to the fact that there have been reports of an increase in survival rates and immune response. When compared to this, the majority of the research focuses on outcomes that are associated with productivity.

It is possible that the use of treatments derived from plants would be beneficial to the environment. The reason for this is that aquatic systems play a significant role in the transmission of resistance genes. Nevertheless, claims regarding the control of environmental antimicrobial resistance are still based on guesses rather than solid proof because there has not been long-term monitoring of sediment, water, the water column, and signs of microbial resistance.

7.5 Community-Level Interventions and Behavioral Determinants of Antibiotic Demand

Studies conducted in communities on the use of Echinacea purpurea for the treatment of upper respiratory tract infections have shown that it may help shorten the duration of symptoms and reduce the need for antibiotics in certain groups of individuals (Laxminarayan et al., 2020; Ogal et al., 2021). Instead of focusing on how antibiotics directly affect bacteria that are resistant to them, the majority of these findings demonstrate how patients and doctors have changed their expectations that antibiotics will be effective.

When we consider the situation from the perspective of the demand side, this evidence is significant because it enables us to comprehend the reasons why individuals misuse antibiotics. Nevertheless, substitution ought to take place only in the event that there is clinical evidence to back it up. Those individuals who require antimicrobial therapy immediately will not have to wait for their treatment as a result of this measure.

8. CHALLENGES AND LIMITATIONS OF HERBAL STRATEGIES IN ANTIMICROBIAL RESISTANCE CONTROL

8.1 Structural and Scientific Barriers to Herbal AMR Interventions

The structural and scientific challenges of herbal medicine prevent its widespread use in healthcare and veterinary practices. Because plant-based treatments are complicated, phytochemicals' many effects are hard to use in medicine. They can fight germs, reduce harm, and alter the immune system. However, traditional antibiotics have different chemical structures and effects, making it easier to determine the right dose, understand how they work, and get consistent results. However, herbal formulations use different active ingredients.

In addition, the reduction of antimicrobial resistance is a problem that requires the working together of people, animals, and the environment. In situations where herbal methods are utilized outside of formal antimicrobial stewardship programs, it may be difficult to observe how these methods actually alter resistance patterns. Table 3 outlines the primary challenges that exist in the areas of institutions, clinical settings, scientific research, regulatory frameworks, and the environment, all of which contribute to the difficulty of successfully integrating translational research.

Table 3. Challenges and opportunities in herbal antimicrobial stewardship

Category	Key Issues	Implications	Proposed Solutions
Scientific	Lack of standardization	Variable efficacy	Phytochemical profiling
Regulatory	Weak global guidelines	Limited clinical adoption	Policy harmonization
Clinical	Insufficient trials	Low physician acceptance	Multicenter studies
Ecological	Overharvesting	Biodiversity loss	Sustainable sourcing

8.2 Lack of Standardization and Quality Control

Lack of standardization is a major issue in herbal antimicrobial resistance efforts. Its soil, origin, cultivation, harvest, and processing affect a plant's phytochemical makeup (Tariq et al., 2019). These differences affect antimicrobial treatment efficacy, consistency, and safety, making research and clinical results difficult to replicate.

Different study designs make it difficult to compare results, making the evidence less convincing. The lack of a standard way to profile phytochemicals, extraction methods, or agricultural and manufacturing practices prevents its use in regulated medical settings. Standardization isn't just a technical issue; clinic workers must trust the rules and be satisfied with their work. Standardization affects more than technology.

8.3 Safety, Toxicity, and Herb–Drug Interactions

Many think herbal medicines are safe, but they can be dangerous if made improperly, contaminated, or misused. Opuni et al. (2023) list heavy metals, pesticide residues, harmful germs, and fake medicines as potential issues. These risks are greater in informal markets, where the government has less control.

It is crucial to understand how herbs and drugs interact, but research is scarce. Plant chemicals affect cytochrome P450 enzymes and drug transporters. It may affect how the body reacts to antibiotics and other drugs taken together. If this happens, the treatment may work less or have worse side effects. To ensure its safety in antimicrobial management programs, herbal medicine must undergo strict toxicity testing, interaction studies, and safety monitoring after release.

8.4 Limited Clinical and Field Evidence

Numerous studies conducted on animals and in laboratories have demonstrated that herbal remedies are capable of eliminating germs. Despite the fact that these studies demonstrate the existence of biological plausibility, it is not yet widely utilized in clinical or veterinary practice (Angelini, 2024; Mihu et al., 2026). The majority of the human studies that are currently available have limited sample sizes, a variety of outcomes, and brief follow-up periods, which makes it difficult to make use of the findings from these studies.

The fact that only a small number of studies have investigated the long-term effects directly on the spread of ecosystems, the changes that occur in microbiomes, or the prevalence of resistance genes is an important point to keep in mind. The majority of people talk about getting quick results, such as experiencing fewer symptoms or reducing the amount of antibiotics they take. Additionally, these proxy indicators are helpful for management, but they do not make it clear how resistance trends are changing even though they are useful. It is still unclear to both medical professionals and policymakers because there is not yet sufficient evidence.

8.5 Regulatory and Policy Barriers

Different regions have different herbal medicine rules. Herbal products are often considered dietary supplements. Their use in antimicrobial control programs is limited because they don't have to undergo as many strict clinical tests (Dubale et al., 2025). People are less likely to conduct large clinical trials and make official rules because this rule is unclear.

No international standards make it harder to ensure quality, collaborate on cross-border research, and organize policies. Today, national and global antimicrobial resistance (AMR) plans rarely include herbal treatments. There are no clear ways to approve herbal treatments using traditional knowledge and modern evidence. They may not trust or use these treatments carelessly (Wani et al., 2022; Khadse et al., 2023).

9. FUTURE DIRECTIONS AND RESEARCH PRIORITIES FOR HERBAL INTERVENTIONS IN AMR CONTROL

9.1 Strategic Reorientation of Herbal AMR Research

When it comes to finding a solution to the problem of antimicrobial resistance, which is getting worse, we need more than just the production of new antibiotics.

These solutions must still be scientifically sound and useful for policy. Herbal medicine is a biologically rich source of compounds that can fight germs and change resistance. In the future, however, advancement will be dependent not so much on simply adding to lists of plant chemicals as it will be on carefully planned research programs that are based on particular ideas.

Herbal remedies should not be considered as alternatives to antibiotics; rather, they should be regarded as additional options that can assist in the management of antibiotic use and the reduction of stress. Both of these factors are contributing to the rise of antibiotic resistance in communities that encourage the responsible use of antibiotics. In order to move forward with this repositioning, it is necessary for individuals from a variety of fields, such as microbiology, pharmacology, systems biology, regulatory science, and public health policy, to collaborate with one another (Angelini, 2024; Almulhim et al., 2025).

9.2 Mechanism-Guided Phytochemical Discovery

Future research should focus on understanding how things work rather than just testing. Random screening yields the same compounds, making medicinal plants less useful. Medicinal plants contain many compounds. Better metabolomics, dereplication, and high-throughput molecular profiling have made finding compounds that change resistance easier (Deleray et al., 2025).

Methods that do not result in death and that are able to inhibit quorum sensing, efflux pump activity, biofilm formation, and the control of virulence should receive additional attention. It is possible that these methods will slow down the development of resistance by making pathogens less harmful rather than completely eliminating them. This will help reduce the pressure for selection and help slow down the development of resistance. It is necessary for us to conduct additional research on evolutionary modeling and resistance-adaptation in order to prove the validity of this concept.

9.3 Rational Evaluation of Herbal–Antibiotic Synergy

In the real world, there is a great deal of potential for the utilization of plant chemicals and conventional antibiotics that are complementary to one another. Certain plant compounds have the ability to improve the ability of antibiotics to penetrate bacteria, prevent bacteria from developing resistance to antibiotics, or restore sensitivity in strains of bacteria that have developed resistance to antibiotics (Kumar et al., 2025).

Future research should create standardized testing frameworks as well as individual experiments. These frameworks should combine pharmacodynamic modeling, synergy testing, and dose-improvement research. Test combination strategies to see how they work in the short term, but also consider how they affect resistance. This evidence would help antimicrobial stewardship by showing if combination treatments can extend antibiotic efficacy.

9.4 Translational Clinical and Field Validation

It is necessary to transform the findings of laboratory tests into carefully planned clinical, veterinary, and field studies in order to win people's trust in policies. In the future, research should have predetermined objectives, a sufficient number of participants, and sufficient time for follow-up in order to examine not only the outcomes of treatment but also the indications of resistance (Perry et al., 2021).

Studies should look at more than just how well a treatment works; they should also look at how it affects behavior and health systems. This means looking at things like how doctors write prescriptions, how well patients stick to their treatment plans, and how much the treatments cost. Antimicrobial stewardship programs and national action plans for dealing with antimicrobial resistance (AMR) need to include this kind of in-depth review because it would provide the real-world proof that is necessary.

9.5 Integration with One Health Surveillance Systems

We need to incorporate herbal treatments into integrated One Health monitoring systems so that we can determine whether or not a reduction in the use of antibiotics results in a change in resistance patterns that is easily observable. Monitoring should include human clinical samples, bacteria from livestock, and environmental sources (Lappan et al., 2024). This will allow medical professionals to gain a better understanding of how diseases can spread from one region to another.

It would be possible to compare the efficacy of these treatments and discover any unexpected effects on the environment if the current antimicrobial resistance surveillance systems included information about the use of plant-based therapies. It is impossible to fully develop statements regarding the reduction of resistance over the long term without integrating this kind of surveillance. Future policy development should focus on creating evidence-based approval pathways that accommodate multi-component botanical products while maintaining scientific rigor (WHO, 2019; Kashuri et al., 2026).

10. CONCLUSIONS AND ONE HEALTH POLICY IMPLICATIONS

10.1 Positioning Herbal Medicine within Integrated AMR Governance

Antimicrobial resistance is a big problem that affects more than just medicine. It also affects agriculture, veterinary medicine, and how we take care of the environment. The goal of this chapter was to talk about how herbal medicine fits within the bigger picture of government and how it may be used as a supplement instead than a replacement. Herbal formulations have many biological effects, but their main significance in the battle against antimicrobial resistance is that they can change how antibiotics are used and encourage responsible use.

In the context of One Health, using plant-based remedies could help reduce the unnecessary use of antibiotics, especially for mild or self-limiting diseases, in both community and agricultural healthcare settings. But it's not enough for these compounds to have biological effects; they also need to have rules, clear standards, and procedures for monitoring them.

10.2 Synthesis of Evidence and Translational Boundaries

Research in clinical, veterinary, aquaculture, and community environments indicates that medicinal plants possess bioactive compounds capable of combating microbes, mitigating their detrimental effects, and modifying immune system function (Angelini, 2024; Mihu et al., 2026). It makes perfect sense to use these methods along with antimicrobial stewardship programs. But there are still a lot of big problems with translation.

Most of the research that is out there right now is based on lab data and findings that only last for a limited time. Not much research has been done on how often resistance genes show up or how they propagate in the environment over a longer period of time. So, herbal therapies should be seen as ways to change how much of something you are exposed to, and further study is needed to confirm the long-term impact of herbal treatments on resistance patterns.

10.3 Cross-Sectoral Relevance within One Health Systems

Herbal medicine is used in many various ways in the One Health method, which fits with the idea. In the field of human health care, additional treatments that have been demonstrated to be effective can help reduce the number of antibiotic prescriptions that are not required. In veterinary and aquaculture systems, the utilization of plant-based solutions has the potential to reduce the amount of antibiotics that are required.

Plant-based compounds break down faster and may cause shorter-lasting selective pressures than synthetic antimicrobials (Rahaman, 2026). This is especially true for the environment. Instead of assuming ecological safety and sustainability are important, we should examine the evidence. Because it helps people, animals, and the environment, herbal medicine is part of organized systems for monitoring and managing things.

10.4 Policy and Regulatory Considerations

Organizations must provide clear rules and support to reduce antimicrobial resistance. Herbal products are still considered dietary supplements in many places, making their use in antimicrobial stewardship programs illegal (Kashuri et al., 2026) Keep quality control, safety checks, and consistent formulas in place and show the evidence to get more people to agree.

Herbal antimicrobial resistance treatments can help sectors collaborate better, but they need strong clinical evidence and drug safety systems. This requires better research facilities and new technologies. Traditional knowledge holders and scientific groups can collaborate to develop ethical research methods relevant to their cultures (Goel et al., 2025).

Figure 3. Translational pathway from traditional knowledge to policy implementation

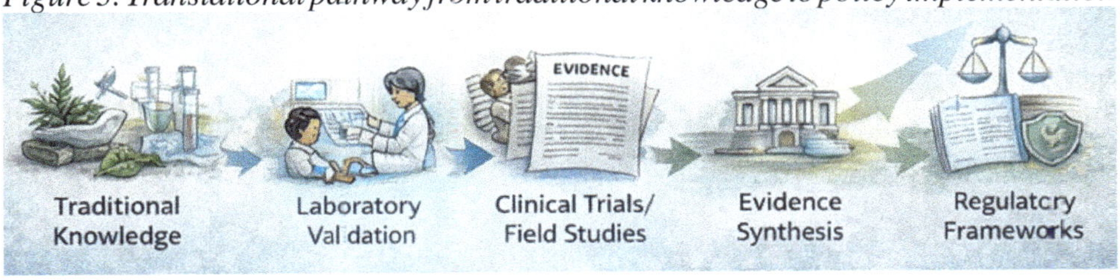

Figure 3 illustrates the sequential progression from ethnobotanical knowledge generation through mechanistic validation, clinical evaluation, stewardship integration, and policy incorporation within One Health governance systems.

10.5 Concluding Perspective

Herbal medication should be offered as a systematic complement to comprehensive antimicrobial stewardship initiatives, rather than as a replacement for antibiotics.

One Health systems could benefit from using less antibiotics, giving patients more treatment alternatives, and encouraging resilience in many areas.

To make this potential a reality, we need to maintain a careful check on everything, set norms that are easy to follow, use strict scientific procedures, and make institutions better at what they do. Plant-based medicines, when used under structured management systems, could help lower antibiotic resistance in a way that lasts. If they don't integrate themselves in this way, they won't be able to affect the patterns of resistance in the population, which will restrict the effects they have.

REFERENCES

Almulhim, M., Ghasemian, A., Memariani, M., Karami, F., Yassen, A. S. A., Alexiou, A., Papadakis, M., & Batiha, G. E.-S. (2025). Drug repositioning as a promising approach for the eradication of emerging and re-emerging viral agents. *Molecular Diversity*, *29*(6), 5465–5485. DOI: 10.1007/s11030-025-11131-8 PMID: 40100484

Alzohairy, M. A. (2016). Therapeutics Role of *Azadirachta indica* (Neem) and Their Active Constituents in Diseases Prevention and Treatment. *Evidence-Based Complementary and Alternative Medicine, 2016*(1). https://doi.org/DOI: 10.1155/2016/7382506

Anand, U., Jacobo-Herrera, N., Altemimi, A., & Lakhssassi, N. (2019). A comprehensive review on medicinal plants as antimicrobial therapeutics: Potential avenues of biocompatible drug discovery. *Metabolites*, *9*(11), 258. DOI: 10.3390/metabo9110258 PMID: 31683833

Angelini, P. (2024). Plant-Derived Antimicrobials and Their Crucial Role in Combating Antimicrobial Resistance. *Antibiotics (Basel, Switzerland)*, *13*(8), 746–746. DOI: 10.3390/antibiotics13080746 PMID: 39200046

Arya, R., Faruquee, H., Shakya, H., Rahman, S. A., Begum, M. M., Biswas, S. K., Apu, M. A. I., Islam, M. A., Sheikh, M. M. I., & Kim, J.-J. (2024). Harnessing the Antibacterial, Anti-Diabetic and Anti-Carcinogenic Properties of Ocimum sanctum Linn (Tulsi). *Plants*, *13*(24), 3516. DOI: 10.3390/plants13243516 PMID: 39771214

Aslam, B., Wang, W., Arshad, M. I., Khurshid, M., Muzammil, S., Rasool, M. H., Nisar, M. A., Alvi, R. F., Aslam, M. A., Qamar, M. U., Salamat, M. K. F., & Baloch, Z. (2018). Antibiotic resistance: A rundown of a global crisis. *Infection and Drug Resistance*, *11*, 1645–1658. DOI: 10.2147/IDR.S173867 PMID: 30349322

Bari, M. S., Khandokar, L., Haque, E., Romano, B., Seidel, V., & Rashid, M. A. (2021). Ethnomedicinal uses, phytochemistry, and biological activity of plants of the genus Gynura. *Journal of Ethnopharmacology*, *271*, 113834. DOI: 10.1016/j.jep.2021.113834 PMID: 33465439

Chowdhury, N., Suhani, S., Purkaystha, A., Begum, M. K., Raihan, T., Alam, M., Islam, K., & Azad, A. K. (2019). Identification of *AcrAB-TolC* Efflux Pump Genes and Detection of Mutation in Efflux Repressor *AcrR* from Omeprazole Responsive Multidrug-Resistant *Escherichia coli* Isolates Causing Urinary Tract Infections. *Microbiology Insights*, *12*, 1178636119889629. Advance online publication. DOI 10.1177/1178636119889629 PMID: 31839709

Collaborators, A. R. (2022). Global Burden of Bacterial Antimicrobial Resistance in 2019: A Systematic Analysis. *Lancet*, *399*(10325), 629–655. DOI: 10.1016/S0140-6736(21)02724-0 PMID: 35065702

Deleray, V., Dorrestein, P. C., & Avalon, N. E. (2025). Molecules to medicine: Advances in metabolomics for natural product drug discovery. *Current Opinion in Biotechnology*, *96*, 103373. DOI: 10.1016/j.copbio.2025.103373 PMID: 41175606

Destoumieux-Garzón, D., Mavingui, P., Boetsch, G., Boissier, J., Darriet, F., Duboz, P., Fritsch, C., Giraudoux, P., Le Roux, F., Morand, S., Paillard, C., Pontier, D., Sueur, C., & Voituron, Y. (2018). The One Health concept: 10 years old and a long road ahead. *Frontiers in Veterinary Science*, *5*, 14. Advance online publication. DOI: 10.3389/fvets.2018.00014 PMID: 29484301

Dubale, S., Usure, R. E., Mekasha, Y. T., Hasen, G., Hafiz, F., Kebebe, D., & Suleman, S. (2025). Traditional herbal medicine legislative and regulatory framework: A cross-sectional quantitative study and archival review perspectives. *Frontiers in Pharmacology*, *16*, 1475297. Advance online publication. DOI: 10.3389/fphar.2025.1475297 PMID: 39950109

Dyar, O. J., Huttner, B., Schouten, J., & Pulcini, C. (2018). What is antimicrobial stewardship? *Clinical Microbiology and Infection : The Official Publication of the European Society of Clinical Microbiology and Infectious Diseases*, *24*(5), 452–458. DOI: 10.1016/j.cmi.2017.08.025 PMID: 30414817

Gadde, U., Kim, W. H., Oh, S. T., & Lillehoj, H. S. (2017). Alternatives to antibiotics for maximizing growth performance and feed efficiency in poultry: A review. *Animal Health Research Reviews*, *18*(1), 26–45. DOI: 10.1017/S1466252316000207 PMID: 28485263

Ghosh, S., Basu, S., Anbarasu, A., & Ramaiah, S. (2025). A comprehensive review of antimicrobial agents against clinically important bacterial pathogens: Prospects for phytochemicals. *Phytotherapy Research : PTR*, *39*(1), 138–161. DOI: 10.1002/ptr.8365 PMID: 39496516

Goel, S., Matsabisa, M., Jiao, F., Portella, C., Asthana, V., Ekandzi, N., Saluja, M., Germann, S., & Souza, J. (2025). Traditional knowledge on health: Balancing innovation, ethics and intellectual property. *Bulletin of the World Health Organization*, *103*(11), 708–714. DOI: 10.2471/BLT.25.293487 PMID: 41180272

Harbarth, S., Balkhy, H. H., Goossens, H., Jarlier, V., Kluytmans, J., Laxminarayan, R., Saam, M., Van Belkum, A., & Pittet, D. (2015). Antimicrobial resistance: One world, one fight! *Antimicrobial Resistance and Infection Control*, *4*(1), 49. Advance online publication. DOI: 10.1186/s13756-015-0091-2

Jamshidi-Kia, F., Lorigooini, Z., & Amini-Khoei, H. (2018). Medicinal plants: Past history and future perspective. *Journal of Herbmed Pharmacology*, *7*(1), 1–7. DOI: 10.15171/jhp.2018.01

Jomova, K., Alomar, S. Y., Valko, R., Liska, J., Nepovimova, E., Kuca, K., & Valko, M. (2025). Flavonoids and their role in oxidative stress, inflammation, and human diseases. *Chemico-Biological Interactions*, *413*, 111489. DOI: 10.1016/j.cbi.2025.111489 PMID: 40147618

Kashuri, M., Ikrar, T., Sutriyo, , Mun'im, A., & Yanuar, A. (2026). Evidence-based production framework for herbal medicine regulation in Indonesia. *Frontiers in Pharmacology*, *16*, 1730273. Advance online publication. DOI: 10.3389/fphar.2025.1730273 PMID: 41552826

Kasse, G. E., Humphries, J., Cosh, S. M., & Islam, M. S. (2024). Factors contributing to the variation in antibiotic prescribing among primary health care physicians: A systematic review. *BMC Primary Care*, *25*(1), 8. Advance online publication. DOI: 10.1186/s12875-023-02223-1 PMID: 38166736

Khadse, S. N., Ugemuge, S., & Singh, C. (2023). Impact of Antimicrobial Stewardship on Reducing Antimicrobial Resistance. *Cureus*. Advance online publication. DOI: 10.7759/cureus.49935 PMID: 38179391

Khameneh, B., Iranshahy, M., Soheili, V., & Fazly Bazzaz, B. S. (2019). Review on plant antimicrobials: A mechanistic viewpoint. *Antimicrobial Resistance and Infection Control*, *8*(1), 118. Advance online publication. DOI: 10.1186/s13756-019-0559-6 PMID: 31346459

Kraemer, S. A., Ramachandran, A., & Perron, G. G. (2019). Antibiotic pollution in the environment: From microbial ecology to public policy. *Microorganisms*, *7*(6), 180. DOI: 10.3390/microorganisms7060180 PMID: 31234491

Kumar, V., Singh, A., Sharma, N., Saini, R., Kumar, H., El–Shazly, M., & Dev, K. (2025). Combating bacterial antibiotic resistance with phytocompounds: Current trends and future perspectives. *Medicine in Drug Discovery*, *28*, 100228. DOI: 10.1016/j.medidd.2025.100228

Lappan, R., Chown, S. L., French, M., Perlaza-Jiménez, L., Macesic, N., Davis, M., Brown, R., Cheng, A., Clasen, T , Conlan, L., Goddard, F., Henry, R., Knight, D. R., Li, F., Luby, S., Lyras, D., Ni, G., Rice, S. A., Short, F., & Greening, C. (2024). Towards integrated cross-sectoral surveillance of pathogens and antimicrobial resistance: Needs, approaches, and considerations for linking surveillance to action. *Environment International*, *192*, 109046. DOI: 10.1016/j.envint.2024.109046 PMID: 39378692

Larsson, D. G. J., & Flach, C. F. (2022). Antibiotic resistance in the environment. *Nature Reviews. Microbiology*, *20*(5), 257–269. DOI: 10.1038/s41579-021-00649-x PMID: 34737424

Laxminarayan, R., Van Boeckel, T., Frost, I., Kariuki, S., Khan, E. A., Limmathurotsakul, D., Larsson, D. G. J., Levy-Hara, G., Mendelson, M., Outterson, K., Peacock, S. J., & Reddy, K. S. (2020). The Lancet Infectious Diseases Commission on antimicrobial resistance: 6 years later. *The Lancet. Infectious Diseases*, *20*(4), e51–e60. DOI: 10.1016/S1473-3099(20)30003-7 PMID: 32059790

Li, Z., Tang, J., Wang, X., Ma, X., Yuan, H., Gao, C., Guo, Q., Guo, X., Wan, J., & Dagot, C. (2025). The Environmental Lifecycle of Antibiotics and Resistance Genes: Transmission Mechanisms, Challenges, and Control Strategies. *Microorganisms*, *13*(9), 2113. DOI: 10.3390/microorganisms13092113 PMID: 41011444

Marino, A., Maniaci, A., Lentini, M., Ronsivalle, S., Nunnari, G., Cocuzza, S., Parisi, F. M., Cacopardo, B., Lavalle, S., & La Via, L. (2025). The Global Burden of Multidrug-Resistant Bacteria. *Epidemiologia*, *6*(2), 21. DOI: 10.3390/epidemiologia6020021 PMID: 40407562

Martínez-Rodríguez, O. P., García-Contreras, R., Aguayo-Ortiz, R., Yeverino, I. R., Fajardo-Hernández, C. A., Patiño, A. D., Hernández Pérez, H. A., Hernández-Garnica, M., & Figueroa, M. (2025). Natural Products Inhibit Bacterial Growth and Modulate Quorum-Sensing Systems *Las* and *Rhl* in *Pseudomonas aeruginosa* PA14. *Chemistry & Biodiversity*, *22*(12), e02040. Advance online publication. DOI: 10.1002/cbdv.202502040 PMID: 41043246

Mehdizadeh, L., & Moghaddam, M. (2018). Essential Oils: Biological Activity and Therapeutic Potential. *Therapeutic, Probiotic, and Unconventional Foods*, 167–179. https://doi.org/DOI: 10.1016/B978-0-12-814625-5.00010-8

Mihu, M. L., Nadăş, G. C., Bouari, C. M., Fi , N. I., & Răpuntean, S. (2026). Klebsiella pneumoniae Infections in Dogs: A One Health Review of Antimicrobial Resistance, Virulence Factors, Zoonotic Risk, and Emerging Alternatives. *Microorganisms*, *14*(1), 149. DOI: 10.3390/microorganisms14010149 PMID: 41597671

Mukherjee, P. K., Harwansh, R. K., Bahadur, S., Banerjee, S., Kar, A., Chanda, J., Biswas, S., Milan Ahmmed, S., & Katiyar, C. K. (2016). Development of Ayurveda – Tradition to trend. *Journal of Ethnopharmacology*, *197*, 10–24. DOI: 10.1016/j.jep.2016.09.024 PMID: 27633405

Murray, C. J. L., Ikuta, K. S., Sharara, F., Swetschinski, L., Robles Aguilar, G., Gray, A., Han, C., Bisignano, C., Rao, P., Wool, E., Johnson, S. C., Browne, A. J., Chipeta, M. G., Fell, F., Hackett, S., Haines-Woodhouse, G., Kashef Hamadani, B. H., Kumaran, E. A. P., McManigal, B., & Naghavi, M. (2022). Global burden of bacterial antimicrobial resistance in 2019: A systematic analysis. *Lancet*, *399*(10325), 629–655. DOI: 10.1016/S0140-6736(21)02724-0 PMID: 35065702

Muteeb, G., Rehman, M. T., Shahwan, M., & Aatif, M. (2023). Origin of Antibiotics and Antibiotic Resistance, and Their Impacts on Drug Development: A Narrative Review. *Pharmaceuticals (Basel, Switzerland)*, *16*(11), 1615. DOI: 10.3390/ph16111615 PMID: 38004480

O'Neill, J. (2016). *Tackling drug-resistant infections globally: Final report and recommendations*. HM Government.

Ogal, M., Johnston, S. L., Klein, P., & Schoop, R. (2021). Echinacea reduces antibiotic usage in children through respiratory tract infection prevention: A randomized, blinded, controlled clinical trial. *European Journal of Medical Research*, *26*(1), 33. Advance online publication. DOI: 10.1186/s40001-021-00499-6 PMID: 33832544

Ogidi, O. I., Barwant, M. M., & Singh, A. (2026). Exploring the role of phytomedicines in ameliorating the manifestation of brain disorders. In *Recent breakthroughs in neurodegenerative disorders and enzyme therapy* (pp. 321–352). IGI Global Scientific Publishing., DOI: 10.4018/979-8-3373-4627-4.ch014

Ogwu, M. C., Ogidi, O. I., Izah, S. C., & Odubo, T. C. (2025). Therapeutic and preventive effects of natural compounds derived from microbes. In Ramawat, K. G., & Mérillon, J.-M. (Eds.), *Natural products*. Springer., DOI: 10.1007/978-3-642-36202-6_287-1

Opuni, K. F. M., Kretchy, J.-P., Agyabeng, K., Boadu, J. A., Adanu, T., Ankamah, S., Appiah, A., Amoah, G. B., Baidoo, M., & Kretchy, I. A. (2023). Contamination of herbal medicinal products in low-and-middle-income countries: A systematic review. *Heliyon*, *9*(9), e19370. DOI: 10.1016/j.heliyon.2023.e19370 PMID: 37674839

Pacyga, K., Pacyga, P., Topola, E., Viscardi, S., & Duda-Madej, A. (2024). Bioactive Compounds from Plant Origin as Natural Antimicrobial Agents for the Treatment of Wound Infections. *International Journal of Molecular Sciences*, *25*(4), 2100–2100. DOI: 10.3390/ijms25042100 PMID: 38396777

Perry, E. K., Meirelles, L. A., & Newman, D. K. (2021). From the soil to the clinic: The impact of microbial secondary metabolites on antibiotic tolerance and resistance. *Nature Reviews. Microbiology*, *20*(3), 129–142. DOI: 10.1038/s41579-021-00620-w PMID: 34531577

Pulcini, C., Beovic, B., Béraud, G., Carlet, J., Cars, O., Howard, P., Levy-Hara, G., Li, G., Nathwani, D., Roblot, F., & Sharland, M. (2017). Ensuring universal access to old antibiotics: A critical but neglected priority. *Clinical Microbiology and Infection : The Official Publication of the European Society of Clinical Microbiology and Infectious Diseases*, *23*(9), 590–592. DOI: 10.1016/j.cmi.2017.04.026 PMID: 28522030

Rahaman, M. (2026). Sustainable functionalization of biodegradable antimicrobial materials for healthcare textiles: A comprehensive review. *Hybrid Advances*, *12*, 100580. DOI: 10.1016/j.hybadv.2025.100580

Robé, C., Projahn, M., Boll, K., Blasse, A., Merle, R., Roesler, U., & Friese, A. (2024). Survival of highly related ESBL- and pAmpC- producing Escherichia coli in broiler farms identified before and after cleaning and disinfection using cgMLST. *BMC Microbiology*, *24*(1), 143–143. DOI: 10.1186/s12866-024-03292-7 PMID: 38664628

Sihombing, B., Bhatia, R., Srivastava, R., Aditama, T. Y., Laxminarayan, R., & Rijal, S. (2023). Response to antimicrobial resistance in South-East Asia Region. *The Lancet Regional Health. Southeast Asia*, *18*, 100306. DOI: 10.1016/j.lansea.2023.100306 PMID: 38028162

Sun, S., Yu, Y., Jo, Y., Han, J. H., Xue, Y., Cho, M., Bae, S.-J., Ryu, D., Park, W., Ha, K.-T., & Zhuang, S. (2025). Impact of extraction techniques on phytochemical composition and bioactivity of natural product mixtures. *Frontiers in Pharmacology*, *16*, 1615338. Advance online publication. DOI: 10.3389/fphar.2025.1615338 PMID: 40808686

Tariq, S., Wani, S., Rasool, W., Shafi, K., Bhat, M. A., & Rather, M. A. (2019). A comprehensive review of the antibacterial, antifungal and antiviral potential of essential oils and their chemical constituents against drug-resistant microbial pathogens. *Microbial Pathogenesis*, *134*, 103580. DOI: 10.1016/j.micpath.2019.103580 PMID: 31195112

Umair, M., Altaf, M., Ahsan, T., Bussmann, R. W., Abbasi, A. M., Gatasheh, M. K., & Elrobh, M. (2024). Study of medicinal plants used in ethnoveterinary medical system in riverine areas of Punjab, Pakistan. *Journal of Ethnobiology and Ethnomedicine*, *20*(1), 48. DOI: 10.1186/s13002-024-00686-9 PMID: 38715115

Van Boeckel, T. P., Pires, J., Silvester, R., Zhao, C., Song, J., Criscuolo, N. G., Gilbert, M., Bonhoeffer, S., & Laxminarayan, R. (2019). Global trends in antimicrobial resistance in animals. Science, 365(6459), eaaw1944. https://doi.org/DOI: 10.1126/science.aaw1944

Wani, Z. A., Farooq, A., Sarwar, S., Negi, V. S., Shah, A. A., Singh, B., Siddiqui, S., Pant, S., Alghamdi, H., & Mustafa, M. (2022). Scientific Appraisal and Therapeutic Properties of Plants Utilized for Veterinary Care in Poonch District of Jammu and Kashmir, India. *Biology (Basel), 11*(10), 1415. DOI: 10.3390/biology11101415 PMID: 36290318

World Health Organization. (2019). *WHO global report on traditional medicine.* WHO.

World Health Organization. (2023). *Global antimicrobial resistance surveillance system (GLASS) report.* WHO.

Wu, Z., Shao, X., & Wang, Q. (2025). Antibiotics and Antibiotic Resistance Genes in the Environment: Dissemination, Ecological Risks, and Remediation Approaches. *Microorganisms, 13*(8), 1763. DOI: 10.3390/microorganisms13081763 PMID: 40871267

Zhang, Z., Cao, M., Shang, Z., Xu, J., Chen, X., Zhu, Z., Wang, W., Wei, X., Zhou, X., Bai, Y., & Zhang, J. (2025). Research Progress on the Antibacterial Activity of Natural Flavonoids. *Antibiotics (Basel, Switzerland), 14*(4), 334. DOI: 10.3390/antibiotics14040334 PMID: 40298463

Zhao, Y.-M., Zhang, Q.-Y., Zhang, L., Bao, Y.-L., Guo, Y.-T., Huang, L.-R., He, R.-H., Ma, H.-L., & Sun, D.-W. (2025). Inhibition of Quorum Sensing-Mediated Biofilm Formation and Spoilage Factors in Pseudomonas fluorescens by Plasma-Activated Water. *Foods, 14*(21), 3773–3773. DOI: 10.3390/foods14213773 PMID: 41227743

Chapter 13

Unlocking Nature's Armory: An Ethnobotanical Approach to Discover Novel Anti-Resistance Agents

Soumya Raju
http://orcid.org/0009-0004-1701-6743
CMS College Kottayam, Kerala, India

Saritha Thampi
http://orcid.org/0009-0000-8487-5418
CMS College Kottayam, Kerala, India

Rogimon P. Thomas
CMS College Kottayam, Kerala, India

ABSTRACT

The burgeoning global crisis of antimicrobial resistance (AMR) represents one of the most conjuring challenges to modern healthcare. Medicinal plants which are employed by indigenous communities, offer an unexploited repository of bioactive compounds. Ethnobotanical research has emerged as a powerful epitome for systematic drug discovery, enabling to conglomerate traditional ecological knowledge with contemporary phytochemical and microbiological sciences. Ethnobotanically derived phytochemicals possess the capacity to disrupt biofilm formation and quorum sensing mechanisms. The chapter is an attempt to unlock nature's armoury by employing a targeted ethnobotanical approach to identify plant-derived compounds with

DOI: 10.4018/979-8-3373-7810-7.ch013

potent activity against drug-resistant bacteria. Thus, by systematically unlocking the therapeutic potential encoded within traditional medicinal systems, ethnobotanical inputs offer hope for developing a new armamentarium of anti-resistance agents.

INTRODUCTION

One of the biggest and alarming threats to contemporary healthcare is the growing global burden of antimicrobial resistance (AMR), which imperils the effectiveness of traditional antibiotics and makes infections that were previously curable potentially fatal (Nazir *et al.,* 2025). The World Health Organization estimates that antibiotic resistance causes about 700,000 deaths annually. If current trends continue unchecked, forecasts indicate that this number might drastically increase to 10 million deaths annually by 2050 (WHO, 2019). The seriousness of this therapeutic crisis is demonstrated by the rise of multidrug-resistant (MDR) bacteria, such as methicillin-resistant Staphylococcus aureus (MRSA), carbapenem-resistant *Acinetobacter baumannii* and widely drug-resistant *Mycobacterium tuberculosis*. Ironically, nature offers an old and mostly untapped reservoir of bioactive chemicals, while the arsenal of conventional antibiotics gradually runs out due to widespread resistance.Medicinal plants, employed for millennia by indigenous communities irrespective of geographic realm to combat infectious diseases, offer extraordinary potential as sources of novel antimicrobial agents (Yashvardhini *et al.,* 2021; Jassal *et al.,* 2025). In this context, ethnobotanical research has become a potent instrument for identifying medicinally significant plant-derived compounds that respect indigenous knowledge systems and provide rigorous scientific validation of customary practices.

Phytochemicals have an array of effects that are primarilydifferent from those of antibiotics. The phytochemicals are notably efficient inhindering quorum sensing of microbes, which are complicated bacterial communication systems that align collective resistance, interrupting biofilm formation and aiming cellular processes within resistant bacterial strains (Asfour, 2018). Additionally, phytochemicals can act as resistance-modifying agents, reversing established resistance phenotypes and improving the effectiveness of traditional antibiotics through synergistic interactions (Khare *et al.,* 2021).Thus, the chapter is an attempt to explore nature's arsenal by applying a targeted ethnobotanical approach to distinguish phytochemicals with potent action against drug-resistant bacteria.Ethnobotanical research delivers hope for generating a new arsenal of anti-resistance drugs that can reestablish clinical efficacy against emerging resistant infections by opening the therapeutic potential engendered within traditional medicinal systems.

THE ANTIMICROBIAL RESISTANCE CRISIS

Antimicrobial resistance (AMR) is one of the most adjuring public health anxietiescurrently, intimidating medical advancement and counteracting the efficiency of healthcare systems prevailing globally (Salam *et al.,* 2023; Krishnaprasad and Kumar, 2024). The global burden of antibiotic resistance shows up as dangerous human misery, heightened healthcare expenses and experiential threats to modern medicine. The Global Research on Antimicrobial Resistance (GRAM) project reported that bacterial antimicrobial resistance contributes for a projected 4.71 million deaths globally in 2021 alone, with 1.14 million of these deaths directly linkedwith antibiotic-resistant illnesses (GBD 2021 Antimicrobial Resistance Collaborators, 2024).

The temporal flight of antimicrobial resistance shows that things are becoming riskier, especially for vulnerable populations. Between 1990 and 2021, antimicrobial resistance-related demises among those 70 years of age and older enhanced by more than 80%. This suggests that older populations are facing an alarming trend of increasing mortality from resistant infections (GBD, 2021; Antimicrobial Resistance Collaborators, 2024). Likewise, vaccination drives and improved sanitation contributed to the almost 50% reduction in antibiotic resistance-related causalities among children under five during this time. The inadequate risk that antibiotic resistance bewilders to elderly populationsis highlighted by this age distribution gap, as they are also more prone to infectious diseases and have less bodilycapacity to get overthese infections.

Certain pathogens are predominantlypertaining patterns and combinations of antibiotic resistance. The utmost significant increase in mortality globally was caused by methicillin-resistant Staphylococcus aureus (MRSA), which more than doubled in three decades from 57,200 deaths in 1990 to 130,000 deaths in 2021 (Navidifar *et al.,* 2025). The number of deaths linked to carbapenem resistance among gram-negative organisms increased from 619,000 in 1990 to 1.03 million in 2021. The recognition of seven principal bacterial pathogens - *Escherichia coli, Klebsiella pneumoniae, Staphylococcus aureus, Acinetobacter baumannii, Streptococcus pneumoniae, Mycobacterium tuberculosis* and *Pseudomonas aeruginosa*, responsible for over one million trackable deaths in 2019 demonstrates the extent of antimicrobial resistance burden among few pathogenic bacterial species (Antimicrobial Resistance Collaborators, 2022).

Forecasts that extend to 2050 portrayed an increasingly bleak picture in the absence of meaningful action. Current analytical models predict that by 2050, the number of deaths directly associated with antimicrobial resistance could reach 10 million annually, while the number of deaths associated with antimicrobial resistance (where resistance acts as a contributing rather than primary causative factor) could reach 8.2 million annually (de Kraker *et al.,* 2016). As these estimations do not take into consideration mortality in lower-income countries with large surveillance gaps, the actual cost may be much higher than forecasts. The monetaryimplicationsof antimicrobial resistance have a domino effect on national economies and public health.Current annual direct healthcare costs for antimicrobial resistance total roughly US$66 billion internationally, encompassing0.7% of total global health expenditures, with projections suggesting this could amplify to US$159 billion yearly by 2050 if historical resistance tendenciescontinue (McDonnell *et al.,* 2024).

Geographical deviations in the preponderance of antibiotic resistance led to strikingdifferences, with low- and lower-middle-income countriesexperiencingun-equal mortality and monetaryaftermaths. At 23.7 fatalities per 100,000 people, Sub-Saharan Africa has prominent regional antimicrobial resistance mortality rate, which is markedly higher than Australasia's rate of 6.5 deaths per 100,000. Together, South Asia, Latin America, the Caribbean and sub-Saharan Africa are gravely affected, which exacerbates already-existing injustices in healthcare access and imperils to untie development gains made over decades (Kariuki *et al.,* 2022). In short, the combination of rising mortality, rising healthcare costs, economic contraction and geographic inequality makes antibiotic resistance a crucial health issue that requires an immediate, coordinated and ongoing global response.

Antimicrobial resistance will gradually erode modern medicine's ability to treat infectious diseases in the absence of comprehensive interventions such as the creation of new antibiotics, improved antimicrobial stewardship, better infection prevention and control, increased vaccination programs and universal access to high-quality healthcare. Thus, countries across the globe must quickly deploy scientific knowledge, medical resources and legal commitments in this arena, as the window of opportunity to prevent catastrophic events is quickly closing.

MAJOR MECHANISMS OF ANTIBIOTIC RESISTANCE

Bacteria have developed highly perverted defence mechanisms that empower them to employ a range oftactics to fight conventional antibiotics. Enzymatic deactivation of antibiotics, changing of cellular permeability, enhancement of efflux capacity and alteration of antibiotic targets are commonlyemployed in these processes. Enzymatic degradation, typified by beta-lactamases that break beta-lactam antibiotic rings, is one of the prime routes of resistance, principally in gram-negative bacterial pathogens. Extended-spectrum beta-lactamases (ESBLs) and carbapenemases have more substrate specificity, rendering resistance against progressively potent antibiotic generations (Munita and Arias, 2016; Belay *et al.,* 2024). Through target-site modification, such as point mutations in crucial genes, substitution of penicillin-binding proteins and methylation of ribosomal RNA that inhibits macrolide binding, pathogens can avoid the effects of antibiotics while retaining essential cellular functions.

Reduced permeability through downregulation of porins and enhanced efflux capacity *via* active transport systems likewisebestow substantial resistance advantages. The efflux pump mechanism merits particular attention as an increasingly recognized determinant of multidrug resistance; pathogens equipped with constitutively expressed or inducible efflux systems can eliminate multiple antibiotic classes before these compounds achieve therapeutic concentrations intracellularly. As the resistance patterns of bacteria within biofilms are orders of extent higher than those of their planktonic counterparts, biofilm development is an especially complex resistance mechanism. Antibiotic efficacy is further weakened by diffusion gradients created by biofilm structure, which limit antibiotic infiltration and reduce the metabolic activity of internal biofilm layers. Quorum sensing allows pathogenic bacteria to synchronizedefence mechanisms at the population level by aligning biofilm development, virulence factor production and collective resistance responses.

ETHNOBOTANY AS A DRUG DISCOVERY PARADIGM

The study of plant-human connections (ethnobotany), offers a time-tested method for finding bioactive substances found in nature. Plants are the source of many pharmacological medications currently used in clinical settings and there are many more molecules in preclinical and clinical development stages. Historical successes including quinine from *Cinchona* species for malaria treatment, morphine from *Papaver somniferum* for analgesia, and digoxin from *Digitalis* species for cardiac conditions exemplify ethnobotany's capacity to identify therapeutically significant natural products (Pirintsos *et al.,* 2022).Drug discovery through ethnobotanical screening includes phytochemical isolation and characterization, ethnopharmacological validation, systematic documentation of traditional knowledge concerning plant usage and laborious evaluations using modern molecular and cellular techniques (Iwu, 2002; Ejeta *et al*., 2024). This unified methodology benefits from the collective data attained over generations of traditional practitioners who have closely followed and experimented with plant effects. Traditional healers have a thorough understanding of the plant parts, preparation methods, dosage, adjuvant therapies and contraindications which helps modern researchers to identify the most promising drugs and treatment approaches.

Crucially, the ethnobotanical approach acknowledges that certain ecological, cultural and epidemiological circumstances shaped the development of traditional medical systems. Plants used to cure diseases prior to the development of antibiotics are examples of natural selection experiments carried out over many human generations, in which only genuinely effective treatments continued to be used in cultural practices. This increases the probability that ethnobotanically selected plants would show bioactivity in contemporary laboratory trials as compared to random plant screening techniques.

SECONDARY METABOLITES AS ANTIMICROBIAL AGENTS

In the complex web of plant defence mechanisms, secondary metabolites are an erudite chemical armoury that has developed for survival rather than growth or multiplication. These include an array of alkaloids, phenolics, terpenoids and flavonoids which function as natural antimicrobial agents which are promising in an era of emerging antibiotic resistance(El-Saadony *et al.,* 2025). These phytochemicals target microbes specifically utilizingcomplex biochemical intercession rather than force. The antibacterial effectiveness of these plant-basedchemicals is notably wide. In addition to interfering with bacterial DNA and RNA synthesis, alkaloids like berberine, which are encountered in plants like goldenseal, also occlude cellular adhesion and biofilm development. Phenolic compounds, such as complex tannins from oak bark and simple phenols like thymol from thyme, act by denaturing microbial proteins and imperilling the of cell membrane integrity. Essential oils that are high in terpenoids, such as those from oregano and tea tree, are lipophilic, so that they can enter bacterial cell membranes and rupture them, causing the contents of the cells to leak out and finally kill it. Flavonoids not only directly inhibit microbial enzymes but also cooperate with other antimicrobial compounds to promote their action.

The multi-target mode of action of phytochemicals is what makes them so fascinating and operative. Phytochemicals frequently target microbes on multiple fronts at once, in contrast to many traditional antibiotics that target a single bacterial pathway. One compound may instantaneouslyhamper energy generation, threaten membrane integrity and intervene with quorum-sensing communication. This polypharmacology makes it meaningfully more difficult for pathogens to develop resistance, as it would require multiple, simultaneous mutations, a far less possible evolutionary event (Chihomvu *et al.,* 2024; Saiqali *et al.,* 2025).In addition, numerous plant extracts show synergism, a phenomenon in which the aggregated impact of many metabolites is more than the sum of their individual effects. This intricate combination, presents a barrier to bacterial adaptability that pure, single-molecule therapies often cannot satisfy. Additionally, this synergy often leads to a broader array of activities, including viruses, fungi and bacteria.

- **Phenolic compounds**: Phenolic compounds, such as flavonoids, tannins and phenolic acids, make up majority of phytochemicals with antimicrobial action. These compounds work in a number of ways, like producing reactive oxygen that lead to oxidative bacterial death, interrupting bacterial cell membranes through lipophilic interactions and suppressing essential enzymes through active site binding (Kauffmann and Castro, 2023). Flavonoids like quercetin and kaempferol have notable potential against both gram-positive and gram-negative bacterial species, but polyphenols like green tea's epigallocatechin gallate demonstrate substantialaction against resistant clinical isolates (Manso *et al.,* 2021; Nguyen and Bhattacharya, 2022).
- Tannins, through their multiple phenolic groups, create extraordinarily robust interactions with bacterial proteins and cell wall components, rendering these compounds nearly impossible for pathogens to inactivate enzymatically
- **Terpenoids:** Monoterpenes, sesquiterpenes and diterpenes make up the chemically varied class of terpenoids and essential oils. The terpenoid compounds gain their antibacterial activity by breaking the phospholipid bilayer through lipophilic interactions, which releases essential cellular components (Guimarães *et al.*, 2019).
- Through this exact mechanism, the eucalyptus monoterpene 1,8-cineole exhibits strong action against *Acinetobacter baumannii, Candida albicans* and MRSA (Mączka *et al.*, 2021; Merghni *et al.*, 2023). Terpenoids also interfere with DNA replication, inhibit vital bacterial enzymes and prevent ATP generation by blocking H-ATPase. Terpenoid resistance development is very challenging for bacteria to manage because of the variety of action mechanisms.
- **Alkaloids**: Alkaloid compounds, nitrogen-containing organic molecules particularly abundant in plants demonstrate potent antimicrobial activity through distinct mechanisms. Several Berberis and Coptis plants contain alkaloid berberine, which demonstrates amazing antibacterial activity against MRSA and other resistant strains (Casciaro *et al.,* 2020; Herrera *et al.,* 2022). Alkaloids offer antibacterial activity through a range of mechanisms, including as suppression of nucleic acid synthesis, inhibition of essential metabolic enzymes and distraction of bacterial cell wall production. Importantly, many alkaloids function as efflux pump inhibitors, inhibiting bacteria from eliminating drugs and strengthening conventional antibiotic therapy.

Mechanisms of action against resistant pathogens

The production of drug-deactivating enzymes like beta-lactamases, the active efflux of toxic substances through the use of specialized pump proteins, alterations to the drug's target site to reduce binding affinity and a notable reduction in membrane permeability to prevent entry are just a few of the small but awfully effective strategies used by resistant bacteria to get over antibiotic attack. Conventional drug development unintentionally favoursvariants that get past this hindrance since it frequently concentrates on single-target, high-affinity compounds. On the other hand, phytochemicals have evolved to fight infections in a more thorough and challenging way due to a millennium-long co-evolutionary weaponry battle. They often target the cellular processes by applying pleiotropic techniques. Therefore, by blocking efflux pumps, compounds like flavonoids and alkaloids can effectively resensitize bacteria to drugs they have previously expelled. It examines the effects of phenolics and terpenoids on the composition and operation of microbial cell membranes, a target that is less susceptible to mutation than specific enzyme sites. In addition, it takes into account how certain phytochemicals unarm bacteria by intervening with essential virulence mechanisms, like biofilm formation and quorum-sensing communication, without directly applying pressure that drives resistance. The most promising is the known synergy between plant extracts and traditional antibiotics, where phytochemicals can improve the efficiency of drugs that have failed, lower required dosages and overcome resistance phenotypes.In this scenario, phytochemicals represent more than just potential new drugs; they are tactical tools offering a paradigm shift in the battle against antimicrobial resistance, directing towards more sustainable and resilient approaches to infectious disease management.

Direct Bactericidal Activity

Phytochemicals have antibacterial potential that are essentially different from those of traditional antibiotics through a variety of different molecular routes. Several phytochemicals directly kill bacteria by physically damaging their cellular membranes. The integrity of the membrane is perilled when lipophilic chemicals enter the bacterial cell wall and phospholipid bilayer, permitting essential proteins, enzymes and genetic material to escape. As bacteria find it very difficult to avoid this physical death mechanism through evolutionary pressure, it requires a fundamental restructuring of cellular architecture. Another way that phytochemicals have bactericidal action is by inhibiting vital enzymes. According to Khameneh *et al.* (2021) and De Rossi *et al.* (2025), polyphenols and other phytochemicals bind to the active sites of important bacterial enzymes, such as DNA gyrase, RNA polymeraseand other metabolic enzymes, impeding enzymatic action and vital cellular activities. As pathogens would need simultaneous changes in several genes to avoid inhibition across multiple enzymatic pathways, the promiscuous binding of phytochemicals to multiple protein targets significantly complicates the establishment of resistance.

Another method of phytochemical antibacterial action is interfering with the synthesis of bacterial DNA and RNA (Vaou *et al.,* 2021). Numerous phenolic and alkaloid chemicals intercalate into bacterial DNA, blocking transcription and replication (Lobiuc *et al.*, 2023).Others obstruct DNA gyrase or topoisomerase, interrupting the topological maintenance essential for DNA synthesis. Such interference with vital genetic processes prevents resistant mutant development, as bacteria cannot concurrently solve resistance and replication problems.

Disruption of Biofilm Formation and Architecture

One of the key resistance mechanisms that empowers bacteria to accomplish collective defence against drugs and host immunological responses is biofilm development. Biofilms include subpopulations of metabolically dormant persister cells that are primarily resistant to antimicrobials, form localized acidic microenvironments that decrease antibiotic efficacy and create diffusion gradients that limit antibiotic entry. Many phytochemicals have anextraordinary ability to either destabilize existing biofilms or interfere with their production. By hindering the early adhesion pathways necessary for biofilm development, flavonoids and other phenolic substances prevent the biofilm formation (Liu *et al.*, 2025). Flavonoids intervene with the quorum sensing signalling that begin coordinated biofilm growth by blocking the production of biofilms. By physically interrupting the extracellular polysaccharide matrix that shields biofilm populations, some phytochemicals like terpenes and alkaloids, directly impair established biofilm architecture. In addition, certain phytochemicals exhibit anti-biofilm action through mechanisms different from planktonic bactericidal activity, evoking that phytochemicals can get over biofilm-mediated resistance using an array of complementary strategies.

Quorum Sensing Inhibition

Small diffusible signal molecules known as autoinducers, such as indole, autoinducer-2 (AI-2) and acyl-homoserine lactones (AHL) are developed by bacteria through quorum sensing. These substances build up and attach to bacterial receptor proteins at high cell densities, causing coordinated alterations in gene expression that permit group behaviours such the synthesis of virulence factors, biofilm formation and antibiotic resistance. Through several "quorum quenching" processes, phytochemicals reduce quorum sensing (Hemmati *et al.*, 2020; Al-Daghistani *et al.*, 2025). Phytochemicals stop the creation of signal molecules by directly inhibiting the enzymes involved in autoinducer synthesis.Others bind competitively to bacterial receptor proteins, blocking the transmission of signals even when autoinducers are present. A third class of phytochemicals encrypts enzymatic activity that destroys autoinducers through lactonase or acylase activity. Flavonoids have a particularly potent potential to quench quorum by competitively binding to LuxR-type receptors and inhibiting AHL-producing enzymes. This multifaceted strategy for interfering with bacterial communication reduces the likelihood of resistance evolving through loss-of-function mutations in individual genes while offering several avenues for therapeutic intervention.

Efflux Pump Inhibition

Antibiotics and other antimicrobials are aggressively transported from bacterial cells by bacterial efflux pumps, which are vital factors in multidrug resistance (Gaurav *et al.,* 2023). Broad-substrate-specificity resistance to numerous antibiotic classes are exhibited by efflux pumps belonging to the resistance-nodulation-division (RND) family, such as AcrAB-TolC in gram-negative bacteriaand many pumps in gram-positive bacteria. Antibiotic exposure produces an evolutionary trap where antibiotic exposure paradoxically upregulates resistance mechanisms as bacteria have astonishing inducible efflux systems that augment pump expression. Efflux pump inhibitors (EPIs) generated from plants are an impressive therapeutic approach (Talarczyk *et al.*, 2025). Phytochemicals such as ethyl gallate from *Terminalia chebula*, berberine from Berberidaceae species and curcumin from turmeric block bacterial efflux pump functioning, either preventing pump assembly, inhibiting pump energization or blocking the pump channel (Negi *et al.,* 2014; Rana *et al.,* 2023). It should be noted that Plant-derived EPIs frequently lack direct antimicrobial activity in isolation but exhibit extraordinary potency when combined with conventional antibiotics (Wolska *et al.,* 2012; Cheesman *et al.,* 2017). This property suggests that plant-derived compounds can restore antibiotic susceptibility in resistant pathogens without the direct selective pressure of antimicrobial activity, potentially reducing resistance evolution during combination therapy.

PLANTS WITH ANTI-RESISTANCE PROPERTIES

Botanical sources offer promising solutions through compounds that can reverse antibiotic resistance and enhance the efficacy of standard drugs which function as resistance modifiers, targeting the fundamental mechanisms through which bacteria protect themselves from antimicrobial agents.Alkaloid-rich plants are the best suited for reversing resistance. Berberine, an isoquinoline alkaloid produced from barberry and goldenseal, inhibits bacterial efflux pumps, the molecular mechanisms by which resistant bacteria eliminate antibiotics from their cells (Zhou *et al.,* 2023). At concentrations as low as 20 micrograms per milliliter, berberine inhibits these pump systems, making previously resistant bacteria more susceptible to fluoroquinolone drugs. Similarly, reserpine, derived from *Rauwolfia serpentina*, enhances antibiotic susceptibility in a number of bacterial species, such as *Micrococcus* and *Staphylococcus*, by acting as a potent efflux pump inhibitor (Saabir *et al.*, 2022).

Cruciferous vegetables and associated plants contribute organosulfur compounds with outstanding antimicrobial properties. Allicin and allyl isothiocyanates present in garlic and horseradish have bactericidal activity against both Gram-positive and Gram-negative bacteria. By targeting bacterial cell membranes and interfering with metabolic processes, these phytochemicals function differently from traditional antibiotics, rendering bacterial resistance mechanisms futile. Plants high in flavonoids have supplementary medicinal attributes. In addition to restoring colistin susceptibility in resistant Gram-negative bacteria, α-Mangostin from mangosteen and isobavachalcone from medicinal plants exhibit quick bactericidal effectiveness against resistant Gram-positive bacteria (Li *et al.*, 2025). The proton motive force essential for bacterial life will deteriorate when these flavonoids attach to bacterial phospholipids. Curcumin extracted from turmeric acts in several ways at once, including as interrupting biofilms and hindering bacterial virulence factors.

When blended with standard antimicrobials, the plant-basedcompoundsmarkedly reduce the minimum inhibitory concentrations required to eliminate resistant infections. Bacteria find it difficult to develop resistance to several active compounds that work in concert through a variety of routes, which is one noticeable advantage of the complexity of plant extract chemistry. Anti-resistance plant chemicals are used in a treatment approach that supplements rather than replaces traditional antibiotics. By restoring antibiotic efficacy through various biochemical processes, these phytochemicals offer a sustainable way to treat multidrug-resistant infections while reducing the selective pressure for further bacterial adaptation. As clinical proof of specific plant-antibiotic combinations improves, phytochemicals are positioned as essential partners in next-generation antimicrobial medicine (Angelini, 2024). Table 1 depicts major plants, phytochemicals and anti-resistance mechanisms.

Terminalia chebula (Haritaki)

One of the most examined ethnobotanical sources of anti-resistance compounds is *Terminalia chebula*. Native to the Indian subcontinent, *T. chebula* has been used to treat infections and ameliorate digestive health in Ayurvedic and Unani medical formulations (Wang *et al.*, 2024). *T. chebula* extracts and phytochemicals have strong antimicrobial efficacy against manu drug resistant bacteria (Mandeville and Cock, 2018). Tannins, gallic acid and ethyl gallatecompounds with strong efflux pump inhibitory effect against *Klebsiella pneumoniae* and other gram-negative pathogensare found in the pericarp of *T. chebula* fruits (Sultan *et al.*, 2023).The fruit extract markedly dropped the minimum inhibitory concentrations (MICs) of tetracycline, ciprofloxacin and chloramphenicol by up to 50% when blended with resistant bacterial isolates, restoring antibiotic sensitivity (Bag *et al.,* 2010). By blocking the AcrAB-TolC pump complex, the isolated chemical, ethyl gallate showed concentration-dependant efflux pump inhibition (Rana *et al.,* 2023). Thus, certain active constituents found in ancient Ayurvedic curatives have the capacity to reverse proven pathways of antibiotic resistance.

Neem (*Azadirachta indica*)

The most utilized medicinal plant of the Indian subcontinent, neem is used for a range of diseases throughout South Asia and is recurrentlyreferred in Ayurvedic scriptures (Rahmani *et al.,* 2018). The bioactive compounds that give neem leaves, bark, seeds and oil their broad-spectrum antibacterial activity include nim, azadirachtol and other fatty acids. The traditional use of neem to treat infectious diseases is consistently supported (Alzohairy, 2016). *A. indica* leaf extracts have outstanding antibacterial activity against MRSA isolates and show inhibitory effect larger than those of other antibiotics (Wylie and Merrell, 2022). A wide range of mechanisms, including direct membrane rupture, inhibition of essential enzymes and modification of bacterial gene expression that upsurges immune cell activity are used by *A. indica* to bring forth its antibacterial effect. Neem is especially potent against gram-positive organisms such as*Bacillus* sp. and *Staphylococcus* sp., in addition to being competent against gram-negative illnesses and fungi.Neem is anastonishing ethnobotanical source for producing combination of anti-resistance treatments due to its diverse antibacterial mechanisms and wide range of effectuality.

Turmeric (*Curcuma longa*)

Turmeric is an old medicinal plant with astonishing therapeutic attributes. The principal bioactive element, curcumin (1,7-bis(4-hydroxy-3-methoxyphenyl)-1,6-heptadiene-3,5-dione), accounts for about 3-5% of the dry weight of the rhizomes (Sharifi-Rad *et al.,* 2020). Because of its extraordinary safety and strong antibacterial attributes, curcumin is an effective option for medical research. Curcumin's antibacterial action results from the production of reactive oxygen species, the prohibition of essential enzymes and the interruption of bacterial cell membranes. When used with mainstream antibiotics, curcumin displaysprincipally intriguing synergistic activity against resistant pathogens. According to Lara-Espinosa *et al.* (2025), curcumin acts through both direct antibacterial mechanisms and immunological modulation, making it a hopeful adjuvant treatment agent for tuberculosis.Research displays that curcumin + ciprofloxacin blends maintain bactericidal activity while dropping the necessary antibiotic concentration (Ma *et al.,* 2024). Increased intracellular antibiotic penetration appears to be made possible by curcumin-mediated degradation of bacterial membrane integrity. Curcumin also decreases tissue damage that impairs immune function during severe infections by controlling inflammatory responses.

Ginger (*Zingiber officinale*)

The rhizomes of *Zingiber officinale* contain a variety of bioactive phytochemicals with antibacterial and anti-inflammatory attributes, including paradols, gingerols and shogaols. The traditional use of ginger for gastrointestinal illnesses has been substantiated by modern microbiological studies. *Enterococcus faecalis*, a nosocomial infection that is become more deadly and resistant to multiple medicines, can be effectively treated with ginger (Ahmed *et al.*, 2022). Additive or synergistic antibacterial activities are produced when ginger and nutmeg (*Myristica fragrans*) are blended. At lesser dosages, the combination is equally effective than triple antibiotic paste against *E. faecalis* (Golla *et al.*, 2024). The synergistic activity demonstrates that complementary phytochemical profiles in different plants create enhanced antibacterial actions when combined, supporting the traditional Ayurvedic practice of employing multi-herb preparations.

Garcinia zeylanica and *Berberis ceylanica*

Garcinia zeylanica and *Berberis ceylanica,* two native plants of Sri Lanka, are prime examples of ethnobotanical leads resulting from local traditional medical systems. These herbs have exhibited clinical importance against treatment-resistant infections and are traditionally employed in Ayurvedic preparations for a number of therapeutical intentions.The antimicrobial activity of *G. zeylanica* extracts against Methicillin-resistant Staphylococcus aureus was found to be enhanced in the presence of TiO2 NPs modified with *G. zeylanica* and with longer contact time, both in the presence and absence of light (Senarathna et al., 2017). According to Gunasekara *et al.* (2017), *B. ceylanica* extracts also demonstrated antibacterial efficacy against Methicillin Resistant Staphylococcus aureus, *Candida albicans*, and *Candida parapsilosis*. These findings highlight the fact that many ethnobotanically valuable plants are still understudied, serving as unexplored sources of medicinal ingredients to combat antibiotic resistance.

Trichopus zeylanicus

"Ginseng of Kerala" is an endemic medicinal plant in Western Ghats with antibacterial, immunomodulatory and adaptogenic attributes (Praveen and Gupta, 2024). Indigenous populations, especially the Kani tribe of Kerala, have traditionally used the leaves and fruits to enhance stamina, avoid exhaustion and sustain overall health.This suggests a broad concept of enhanced resistance that is pertinent in the context of antibiotic resistance and infections that are difficult to treat (George *et al.*, 2016). According to phytochemical studies, the plant has a wide range of bioactive components, such as flavonoids, alkaloids, saponins, triterpenoids, glycopeptido lipid complexes, polyphenols and sulfhydryl compounds. Many of these phytochemicals have been shown to have antimicrobial and resistance-modulatory attributes (Saudagar *et al.,* 2013; Sasi Kala & Ramasubbu, 2021).Hence, by directly inhibiting bacteria and indirectly strengthening host defences, these phytochemicals offer synergistic actions to manage infections; nevertheless, specific research on the modulation of antibiotic resistance is still scarce and mostly preliminary (Jishnu, and Behera, 2024).

Staphylococcus aureus, Bacillus subtilis, Salmonella typhi, Shigella flexneri, Escherichia coli, Klebsiella pneumoniae, Streptococcus pneumoniae and *Clostridium tetani* are among the bacterial strains against which methanol extracts of leaves, roots, fruits and stems have exhibited potent antioxidant potential and inhibitory effects; activity frequently enhanced in a dose-dependent manner (Manza and Saj, 2013; Biju *et al.,* 2019). These antimicrobial properties are important in the background of antibiotic resistance as they indicate adjunctive or alternatives beyond traditional antibiotics, especially for Gram negative pathogens that typically develop multidrug resistance. *Trichopus zeylanicus* has been shown to have strong adaptogen and immunomodulator properties in addition to direct antibacterial activity, which is extremely important for host-directed therapies against resistant pathogens.

Poeciloneuron pauciflorum

Agasthiyamalaia pauciflora of the family Clusiaceae is chemically characterized by an exceptionally rich reservoir of xanthones with diverse pharmacological activities including antimicrobial, antioxidant and anti-inflammatory properties (Pillai *et al.,* 2011). The plant is rich in 1,6-dihydroxy-7-methoxyxanthone, 1,6-dihydroxy-7-methoxyxanthone 6-O-β-D-glucoside, 1,5-dihydroxy-3-methoxy-, 1,7-dihydroxy-, 1-hydroxy-7-methoxy-, 5-hydroxy-3-methoxyxanthone, 3-hydroxy-2-methoxyxanthone, as well as triterpenoids (Tosa *et al.*, 1997). These xanthones exhibits antibacterial, antifungal and immunomodulatory properties, among other bioactivities that are directly related to managing antibiotic resistance (Pillai *et al.,* 2011). As xanthones are known to use a variety of antibacterial processes, *P. pauciflorum* is important in combating antibiotic resistance. These polyphenolic compounds can damage bacterial cell membranes, inhibit essential bacterial enzymes, prevent nucleic acid production, and change the quorum sensing mechanisms used by bacteria to coordinate virulence and resistance. Additionally, xanthones have shown effectiveness against pathogenic strains that are resistant to multiple drugs, indicating that they work in concert with traditional antibiotics or be useful as agents that modify resistance (Narasimhan *et al.*, 2017; Durães *et al.*, 2021). Due to the variety of molecular targets, the presence of several xanthone variants in *P. pauciflorum*evokes a wide-ranging antibacterial arsenal that offer broad-spectrum activity and lessen the chance of pathogenic resistance development.

Triphala and Other Ayurvedic Formulations

Triphala churna, which combines *Terminalia chebula, Terminalia belerica* and *Emblica officinalis*, Dashmula churna, and Mahasudarshan churna are common Ayurvedic polyherbal preparations that exhibit exceptional antibacterial activity against a wide range of clinically significant pathogens (Kubade and Vaidya, 2024; Chaiprateep *et al.*, 2025). These preparations are very effective against gram-negative pathogens including *Escherichia coli, Klebsiella pneumoniae* and *Pseudomonas aeruginosa* as well as gram-positive bacteria like *Staphylococcus species* and *Bacillus* species. The polyherbal preparations exhibit synergistic antibacterial effects through complementary phytochemical profiles (Mussarat *et al.*, 2021). These preparations are more effective in getting over multidrug resistance than individual drugs as different phytochemicals target different bacterial activities or resistance mechanisms. Thus, traditional Ayurvedic therapy are refined by fine-tuning herbal combinations for optimum antibacterial efficacy.

Table 1. Key plants, phytochemicals and anti-resistance mechanisms

Plant (common; scientific name)	Key phytochemicals	Mechanism(s) relevant to AMR
Turmeric; *Curcuma longa*	Curcumin (diferuloylmethane)	Inhibits NorA efflux pump in *S. aureus*, increasing ciprofloxacin susceptibility and reducing NorA expression; acts as efflux pump inhibitor in MDR *E. coli*; permeabilizes bacterial membranes via activation of mechanosensitive channel MscL, enhancing antibiotic uptake; can reverse high-level tigecycline resistance in Gram-negative bacteria when combined with tigecycline.
Green tea; *Camellia sinensis*	Epigallocatechin gallate (EGCG), epicatechin gallate	Binds staphylococcal peptidoglycan and damages cell wall integrity; shows strong synergy with β-lactams, reversing oxacillin resistance in MRSA and increasing susceptibility of both MRSA and MSSA; acts at the same cell-wall target as β-lactams, functioning as an adjuvant rather than a classical antibiotic.
Black seed; *Nigella sativa*	Thymoquinone, essential oil components	Inhibits bacterial efflux pumps, including NorA in multidrug-resistant *S. aureus*, leading to reduced efflux of ciprofloxacin and doxycycline and increased intracellular drug accumulation; exhibits potent antibiofilm activity against diverse Gram-positive and Gram-negative pathogens and downregulates virulence and resistance genes in carbapenem-resistant uropathogenic *E. Coli.*
Barberry / goldthread; *Berberis* spp., *Coptis chinensis*	Berberine (protoberberine alkaloid)	Inhibits bacterial efflux pumps (e.g. in *Pseudomonas aeruginosa* burn isolates), lowering MICs of multiple antibiotics; in MRSA, disrupts cell wall integrity by inhibiting wall teichoic acid (WTA) biosynthesis via targeting TarO, TarS and TarGH, thereby sensitizing strains to β-lactams and other agents; shows synergy with aminoglycosides such as amikacin and is being formulated in delivery systems to overcome MDR.
Thyme; *Thymus vulgaris*, *T. capitatus*	Thymol (monoterpenoid phenol), carvacrol	Disrupts cytoplasmic membrane integrity in Gram-negative and Gram-positive bacteria, causing leakage of cellular contents and rapid killing; inhibits efflux pumps in foodborne pathogens, increasing ethidium bromide accumulation; exhibits strong antibiofilm activity against *Salmonella* spp., *E. coli*, *Klebsiella* spp. and others, and shows synergy with conventional antifungals and antibiotics, thereby reducing effective MICs.
Oregano; *Origanum vulgare* and related species	Carvacrol, thymol, other phenolic monoterpenes	Acts as a strong antibiofilm and anti-quorum-sensing (QS) agent against MDR uropathogenic *E. coli*, reducing motility, biofilm biomass and expression of QS-associated genes (e.g. *luxS*); in combination with antibiotics, lowers MICs and can produce synergistic or partially synergistic effects, functioning as a resistance-modifying adjuvant; also disrupts preformed biofilms of *Salmonella* and other pathogens.

continued on following page

Table 1. Continued

Plant (common; scientific name)	Key phytochemicals	Mechanism(s) relevant to AMR
Fabaceae legumes (e.g. *Psoralea corylifolia*, *Glycyrrhiza glabra*, *Glycine max*)	Prenylated isoflavonoids such as neobavaisoflavone, glabrene, glyceollin I & III	Natural NorA efflux pump inhibitors in fluoroquinolone-resistant *S. aureus*; at sub-inhibitory concentrations, these compounds markedly increase ethidium bromide accumulation in NorA-overexpressing strains and reduce MICs of fluoroquinolones up to eightfold, with no effect in *norA* knockout strains; act as resistance-modifying agents without causing membrane permeabilization or significant cytotoxicity.
Pomegranate; *Punica granatum* (peels)	Punicalagin (ellagitannin), ellagic acid and related polyphenols	Inhibits *S. aureus* growth and biofilm formation at sub-MIC concentrations; disrupts cell membrane integrity and affects fatty acid biosynthesis, sortase A (SrtA) activity, and efflux-related pathways, thereby reducing adhesion to fibrinogen, surface display of protein A, and expression of key virulence factors in MRSA. punicalagin combined with cefotaxime improves therapeutic outcomes in MRSA pneumonia models, functioning as an anti-virulence adjuvant.
Garlic; *Allium sativum*	Allicin and related organosulfur compounds	Broad-spectrum antibacterial against MDR Gram-positive and Gram-negative bacteria; allicin reacts with thiol-containing enzymes and disrupts bacterial cell wall and membrane function, interferes with DNA conformation and base stacking, and inhibits key metabolic enzymes; prevents or disperses biofilms and shows strong synergy with other antimicrobials, including bacteriophages, ε-polylysine and cationic surfactants, thereby reducing required doses and limiting resistance development.
Cinnamon and clove; *Cinnamomum verum*, *Syzygium aromaticum*	Cinnamaldehyde, eugenol and related phenylpropanoids	Potent anti-quorum-sensing and antibiofilm agents that suppress QS-regulated motility, biofilm development and virulence factor production in MDR uropathogenic *E. coli* and other pathogens; in combination with antibiotics, lower MICs and often yield synergistic effects; act mainly by downregulating QS regulatory genes such as *luxS* and interfering with autoinducer signaling, thereby attenuating virulence without strong selective pressure for resistance.
Tea tree; *Melaleuca alternifolia*	Tea tree oil (TTO), terpinen-4-ol and other monoterpenes	Broad-spectrum topical antibacterial and antibiofilm agent; terpinen-4-ol-rich TTO rapidly disrupts bacterial membranes and inhibits biofilm formation in a range of pathogens; efflux pumps such as MexAB-OprM in *P. aeruginosa* extrude monoterpenes, and pump-deficient mutants are more susceptible, implying that efflux inhibitors could further potentiate TTO; serial subculture studies indicate that sub-MIC TTO and terpinen-4-ol produce little stable change in antibiotic susceptibility, suggesting low propensity to drive resistance.

COMBINING PHYTOCHEMICALS WITH CONVENTIONAL ANTIBIOTICS

Synergydefined as combined drug effects exceeding the sum of individual effects, represents an extraordinarily valuable property enabling reduced antibiotic concentrations while maintaining or enhancing therapeutic efficacy.Combinations of phytochemicals with conventional antibiotics demonstrate a variety of ways for developing synergies and actions. When phytochemicals and antibiotics target multiple cellular processes or distinct molecular sites within the same process, supra-additive antimicrobial activity results (Jayaraman *et al.*, 2010). For example, tetracycline, which inhibits bacterial protein synthesis by 30S ribosome binding and gallic acid, which destroys the integrity of bacterial cell membranes, combine to increase tetracycline penetration while also interfering with protein synthesis. Similarly, compounds targeting penicillin-binding proteins (antibiotics) combined with compounds disrupting cell wall integrity (phytochemicals) produce synergistic cell wall destruction exceeding either agent alone.

As adjuvant substances, plant-derived efflux pump inhibitors prevent bacterial drug efflux processes, allowing traditional antibiotics to build up intracellularly to therapeutic concentrations. Combination of 5'-methoxyhydnocarpin, a plant-derived EPI and berberine, which has a modest direct antimicrobial activity (Stermitz *et al.,* 2000). This synergy is attributed to the blocked efflux of berberine, which permits intracellular accumulation. According to this approach, some phytochemicals may act more as agents that change resistance than as direct antimicrobials. Majority of phytochemicals produce reactive oxygen species (ROS) imparting to antimicrobial effects. When combined with antibiotics, enhanced oxidative stress creates supraadditive bacterial killing. In addition, some phytochemicals scavenge bacterial antioxidant defences, eradicating ROS-protection mechanisms that bacteria normally apply to defend against oxidative damage. Figure 1 depicts basic pathways of combining antibiotics and phytochemicals.

Figure 1. Implications of combining phytochemicals

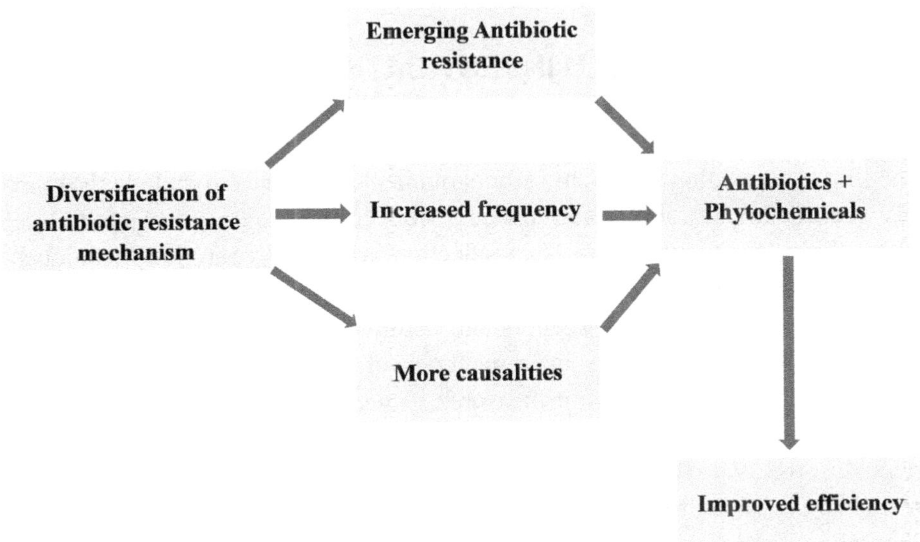

CLINICAL IMPLICATIONS AND THERAPEUTIC POTENTIAL

There are substantial therapeutic forkings to phytochemicals' ability to restore antibiotic sensitivity among resistant bacteria.Rather than developing entirely novel antibiotics, phytochemical-based adjuvants could extend the therapeutic lifespan of existing antibiotics, reducing resistance selection pressure while controlling infections with established drugs.Several synergistic combinations demonstrate particular clinical promise. Sulfamethoxazole blended with gallic acid, ellagic acid or protocatechuic acid demonstrates synergistic activity against *P. aeruginosa* (Jayaraman*et al.,* 2010). Likewise, mixtures of nutmeg and ginger demonstrate effectiveness against *E. faecalis* that is on par with or better than that of traditional antibiotics (Golla *et al.*, 2024). Amalgamations of curcumin and fluoroquinolones make resistant gram-negative bacteria more susceptible (Kali *et al.,* 2016). These established synergies designate clinical development pathways that speed up the translation of ethnobotanical discoveries into therapeutic applications. As synergistic combinations require lower drug concentrations, there are additional advantages such as reduced toxicity, fewer medication-drug interactions and improved patient tolerance. Lower antibiotic doses may, in theory, reduce the selective pressure that causes resistance to emerge, enabling long-term antimicrobial treatment even against naturally resistant pathogens.

ETHNOBOTANICAL FIELD INVESTIGATION AND DOCUMENTATION

The first step in doing systematic ethnobotanical research is to include the community and document traditional knowledge concerning the use of plants to treat infectious diseases.Documenting plant selection criteria, preparation techniques, dosage schedules, adjuvant medicines and reported efficacy is crucial for traditional healers, indigenous knowledge keepers and community members. It must follow ethical standards including prior informed consent, intellectual property recognition and equitable benefit-sharing measures to guarantee that communities that contribute knowledge are properly recouped. According to ethnobotanical studies conducted in the West Shoa community of Ethiopia, 51 families and 108 species of medicinal plants are traditionally used to treat infectious diseases and a variety of other ailments (Beressa *et al.*, 2024). Traditional healers, principally male elders aged over 50 years with decades of practical field experience, provided information on plant parts utilized, preparation methods (typically squeezing fresh plant material for juice) and disease indications.

Lulesa *et al.* (2025) recorded the use of 164 species of therapeutic plants from 140 taxa and 60 families. Among these, 67.68% of species were used to treat human illnesses, 19.5% for animal problems and 12.8% for both. Eighty-one species of these therapeutic plants came from wild places, thirty-five from home gardens, twenty-three from agricultural fields, fifteen from living fences and ten from roadsides. With 76 species, herbs made up the biggest category of growth types, followed by trees (42 species), shrubs (34 species) and climbers (9 species). The most commonly used plant parts are leaves and roots. According to Huang *et al.* (2022), 83 medicinal herbs from 48 families are used in Kinmen, Taiwan. These medicinal plants were mostly found in the Lamiaceae, Solanaceae and Compositae families. Four of the eighteen new uses that were observed had to do with recently identified medicinal plant species in the Kinmen region.

Lulesa *et al.* (2025) found 530 confirmed species of medicinal plants in 118 families and 372 genera in Fofa and Toba sub-districts of the Yem Zone in Central Ethiopia. *Psidium guajava* and *Euphorbia hirta* were the two species that were used frequently. Fabaceae was the most represented family and Ficus was the most represented genus. Eleven of the 28 medicinal plant species that are classified as threatened at the national or international level are native to the Philippines. In oral medicinal plant preparations, leaves are most commonly used to treat a range of ailments, including digestive issues like diarrhea. In another study from Northeastern Madhya Pradesh, India, Sharma *et al.* (2024) reported that most of the 201 medicinal plants encountered in 170 genera and 76 families that were used to treat an array of illnesses were trees (43%), herbs (33%) and climbers (13%).

To put it briefly, ethnobotanical field research and documentation are significant scientific endeavours that go well beyond scholarly curiosity. They tackle conjuring issues in biodiversity protection, medicinal developmentand cultural preservation. As globalization and cultural assimilation endanger indigenous knowledge systems, comprehensive documentation of ethnobotanical activities is becoming essential to prevent the irreversible loss of centuries' worth medicinal knowledge in treating various ailments. The importance of ethnobotanical documentation is in its capacity to link traditional and scientific knowledge. Participatory ethnobotanical approaches enable communities to contribute ownership toward conservation initiatives that recognize their territorial rights and intellectual property.

CHALLENGES AND LIMITATIONS IN ETHNOBOTANICAL DRUG DEVELOPMENT

Although ethnobotanical drug development is a promising path for pharmaceutical innovation, the development of natural product-based therapeutics is severely hampered by significant scientific, regulatory, ethical and financial challenges that arise when traditional plant knowledge is translated into clinically approved medications. Bioassay-guided fractionation methods aimed at isolating single active compounds often result in loss of bioactivity because synergistic interactions between multiple phytochemicals may be essential for therapeutic efficacy. Purification procedures are complicated by the presence of cytotoxic components in plant extracts and the therapeutic potential is further diminished by the low bioavailability and poor solubility of plant metabolites in biological systems. However, variances coming from genetic variation within and between species, geographic location, seasonal fluctuations, cultivation techniques, harvest time and storage conditions make it very hard to attain this uniformity. The batch-to-batch consistency required for synthetic medications contrasts dramatically with this intrinsic chemical and biological variability, resulting in ongoing quality control issues that regulatory bodies find difficult to methodically resolve (Chele *et al.*, 2025).

The regulatory processes governing the approval of herbal medicines are still in their infancy when compared to contemporary pharmaceuticals. To create safety profiles, rigorous preclinical toxicity tests are required; unfortunately, many herbal medicines lack specified pharmacopeial criteria, which leaves regulatory agencies without exact quality requirements. The Convention on Biological Diversity provides some legal protections despite institutional frameworks supposedly designed to protect indigenous people; however, implementation is still uneven across nations and jurisdictional boundaries, leaving them vulnerable to exploitation (McClatchey *et al.,* 2009).

A frequently disregarded constraint endangering the entire study endeavour is the erosion of ethnobotanical knowledge itself. Acculturation, rural-urban migration, language assimilation and shifting lifestyles among younger generations disinclined toward traditional employment pose unprecedented challenges to the intergenerational transmission of traditional plant knowledge. Thus, there are chances to vanish plants from human practice and botanical knowledge systems before their therapeutic potential is appropriately documented and evaluated. Traditional knowledge and populations of medicinal plants are instantlyextinguished by deforestation and habitat degradation, which causes combined losses when biological and cultural resources disappear before their therapeutic potential can be investigated (Shah and Bhat, 2019).

BIOAVAILABILITY AND PHARMACOKINETIC CONSTRAINTS

A critical limitation impeding translation of ethnobotanically promising compounds into clinical therapeutics involves poor oral bioavailability. Due to restricted intestinal absorption, quick hepatic metabolism and active efflux *via* P-glycoprotein and other transporters, many phytochemicals acquire modest systemic concentrations after oral administration, despite their impressive *in vitro* antimicrobial activity (Aqil *et al.,* 2013). According to Lipinski's "rule of five" prediction criteria, compounds having more than five hydrogen-bond donors, more than ten hydrogen-bond acceptors, a molecular weight larger than 500 Daltons, or a partition coefficient (logP) greater than five usually have limited oral bioavailability. Despite their great potency in cell culture, many polyphenolic phytochemicals do not meet these requirements, which explains their poor systemic availability. UDP-glucuronosyltransferases (UGTs) and other Phase II enzymes catalyze rapid conjugation via glucuronidation and sulfation, which further lowers parent chemical levels in circulation. Potential remedies are provided via creative delivery methods. Cyclodextrin complexation, liposomal formulations, and nanoparticle encapsulation enhance cellular absorption and bioavailability while shielding phytochemicals from deterioration. Piperine, a black pepper alkaloid, enhances bioavailability of numerous compounds by inhibiting glucuronidation and increasing intestinal absorption (Raghunath *et al.,* 2024). Combining phytochemicals with bioavailability enhancers or co-administering compounds that inhibit metabolizing enzymes may substantially improve therapeutic efficacy.

CONCLUSION

Antimicrobial resistance is an emerging issue that requires rapid, innovative solutions that go beyond conventional antibiotic development paradigms. Through systematic investigation of traditional medical systems and contemporary scientific validation, ethnobotanical research holds great promise for discovering novel anti-resistance medications that can restore therapeutic efficacy against growing resistant pathogens. Medicinal plants have evolved secondary metabolites that provide a range of chemical scaffolds that simultaneously target many bacterial activities and resistance mechanisms as a result of millions of years of plant-pathogen interactions. Indigenous knowledge systems have identified therapeutically significant compounds that now require modern scientific characterization and clinical development, as evidenced by the remarkable efficacy of ethnobotanically significant plants such as *Terminalia chebula, Azadirachta indica, Curcuma longa* and many others. Direct membrane rupture, biofilm instability, quorum sensing suppression and efflux pump inhibition are some of the mechanisms by which phytochemicals work that are essentially different from those of traditional antibiotics. Thus, phytochemicals are useful as adjuvant substances that improve the effectiveness of antibiotics through synergistic interactions.

The issues of bioavailability, standardization, quality control and intellectual property protection must be addressed in order to translate ethnobotanical discoveries into clinical reality. By prolonging the therapeutic life of traditional antibiotics and creating sustainable methods of treating infectious diseases, the integration of phytochemical-antibiotic combinations with antimicrobial stewardship initiatives could significantly change the course of antimicrobial resistance. In short, ethnobotanical research provides hope for creating a new arsenal of anti-resistance drugs that can reinstate clinical efficacy against emerging resistant infections by methodically unlocking the therapeutic potential encoded within traditional medicinal systems. This paradigm change, which recognizes plant secondary metabolites as valid sources of antimicrobial innovation and as a supplement to traditional antibiotic development, is an essential and scientifically corroborated evolution in the strategy for dealing with the crisis of antimicrobial resistance. The future of antimicrobial therapy may ultimately depend upon recognizing that indigenous knowledge systems and contemporary science, when integrated respectfully and collaboratively, provide complementary perspectives essential for discovering the therapeutic agents necessary for sustainable control of infectious diseases.

REFERENCES

Ahmed, N., Karobari, M. I., Yousaf, A., Mohamed, R. N., Arshad, S., Basheer, S. N., Peeran, S. W., Noorani, T. Y., Assiry, A. A., Alharbi, A. S., & Yean, C. Y. (2022). The Antimicrobial Efficacy Against Selective Oral Microbes, Antioxidant Activity and Preliminary Phytochemical Screening of *Zingiber officinale*. *Infection and Drug Resistance*, *15*, 2773–2785. DOI: 10.2147/IDR.S364175 PMID: 35668854

Al-Daghistani, H. I., Matalqah, S. M., Shadid, K. A., Abu-Niaaj, L. F., Zein, S., & Abo-Ali, R. M. (2025). Quorum Quenching of *P. aeruginosa* by *Portulaca oleracea* Methanolic Extract and Its Phytochemical Profile. *Pathogens (Basel, Switzerland)*, *14*(2), 163. DOI: 10.3390/pathogens14020163 PMID: 40005538

Alzohairy, M. A. (2016). Therapeutics Role of *Azadirachta indica* (Neem) and Their Active Constituents in Diseases Prevention and Treatment. *Evidence-Based Complementary and Alternative Medicine : eCAM*, *7382506*(1), 7382506. Advance online publication. DOI: 10.1155/2016/7382506 PMID: 27034694

Angelini P. (2024). Plant-Derived Antimicrobials and Their Crucial Role in Combating Antimicrobial Resistance. *Antibiotics (Basel)*,*13*(8):746. doi: 13080746. DOI: 10.3390/antibiotics

Aqil, F., Munagala, R., Jeyabalan, J., & Vadhanam, M. V. (2013). Bioavailability of phytochemicals and its enhancement by drug delivery systems. *Cancer Letters*, *334*(1), 133–141. DOI: 10.1016/j.canlet.2013.02.032 PMID: 23435377

Asfour, H. Z. (2018). Anti-Quorum Sensing Natural Compounds. *Journal of Microscopy and Ultrastructure*, *6*(1), 1–10. DOI: 10.4103/JMAU.JMAU_10_18 PMID: 30023261

Bag, A., Bhattacharyya, S. K., Pal, N. K., & Chattopadhyay, R. R. (2010). Synergistic Effect of *Terminalia chebula* and Antibiotics against Multidrug-resistant Uropathogenic Escherichia coli. I. *Medicinal and Aromatic Plant Science and Biotechnology*, *50*(1), 70–73.

Belay, W. Y., Getachew, M., Tegegne, B. A., Teffera, Z. H., Dagne, A., Zeleke, T. K., Abebe, R. B., Gedif, A. A., Fenta, A., Yirdaw, G., Tilahun, A., & Aschale, Y. (2024). Mechanism of antibacterial resistance, strategies and next-generation antimicrobials to contain antimicrobial resistance: A review. *Frontiers in Pharmacology*, *15*, 1444781. DOI: 10.3389/fphar.2024.1444781 PMID: 39221153

Beressa, T. B., Gadisa, D. A., Mammo, S., Umeta, G. T., Meskele, L. B., Gudeta, B. M., & Taye, G. M. (2024) Ethnobotanical study of traditional medicinal plants used to treat human ailments in West Shewa community, Oromia, Ethiopia. *Frontiers in Pharmacology*, *15*, 1369480. DOI: 10.3389/fphar.2024.1369480 PMID: 39101129

Biju, V.C., Shidhi, P.R., Rajan, V.S., Anoop, P.K., Nair, A.S. (2019). Therapeutic properties of *Trichopus zeylanicus* subsp. travancoricus, a Rare, Endangered Medicinal Plant in South India: A Review. *Herb Med, 5*(1:2):1-14.

Casciaro, B., Mangiardi, L., Cappiello, F., Romeo, I., Loffredo, M.R., Iazzetti, A., Calcaterra, A., Goggiamani, A., Ghirga, F., Mangoni, M.L., Botta, B., Quaglio, D. (2020). Naturally-Occurring Alkaloids of Plant Origin as Potential Antimicrobials against Antibiotic-Resistant Infections. *Molecules,25*(16):3619. 25163619.DOI: 10.3390/molecules

Chaiprateep, E. O., Noysang, C., Amorndoljai, P., Jarmkom, K., & Khuanekkaphan, M. (2025). Formulation of triphala-based products: Focus on gallic acid content and health benefits. *Journal of Advanced Pharmaceutical Technology & Research*, *16*(4), 213–218. DOI: 10.4103/JAPTR.JAPTR_118_25 PMID: 41234351

Cheesman, M. J., Ilanko, A., Blonk, B., & Cock, I. E. (2017). Developing new antimicrobial therapies: Are synergistic combinations of plant extracts/compounds with conventional antibiotics the solution? *Pharmacognosy Reviews*, *11*(22), 57–72. DOI: 10.4103/phrev.phrev_21_17 PMID: 28989242

Chele, K. H., Piater, L. A., van der Hooft, J. J. J., & Tugizimana, F. (2025). Bridging Ethnobotanical Knowledge and Multi-Omics Approaches for Plant-Derived Natural Product Discovery. *Metabolites*, *15*(6), 362. DOI: 10.3390/metabo15060362 PMID: 40559386

Chihomvu, P., Ganesan, A., Gibbons, S., Woollard, K., & Hayes, M. A. (2024). Phytochemicals in Drug Discovery - A Confluence of Tradition and Innovation. *International Journal of Molecular Sciences*, *25*(16), 8792. DOI: 10.3390/ijms25168792 PMID: 39201478

Collaborators, A. R. (2022). Global burden of bacterial antimicrobial resistance in 2019: A systematic analysis. *Lancet*, *399*(10325), 629–655. DOI: 10.1016/S0140-6736(21)02724-0 PMID: 35065702

de Kraker, M. E., Stewardson, A. J., & Harbarth, S. (2016). Will 10 Million People Die a Year due to Antimicrobial Resistance by 2050? *PLoS Medicine*, *13*(11), e1002184. DOI: 10.1371/journal.pmed.1002184 PMID: 27898664

De Rossi, L., Rocchetti, G., Lucini, L., & Rebecchi, A. (2025). Antimicrobial Potential of Polyphenols: Mechanisms of Action and Microbial Responses-A Narrative Review. *Antioxidants, 14*(2), 200. DOI: 10.3390/antiox14020200 PMID: 40002386

Durães, F., Resende, D.I.S.P., Palmeira, A., Szemerédi, N., Pinto, M.M.M., Spengler, G., Sousa, E. (2021). Xanthones Active against Multidrug Resistance and Virulence Mechanisms of Bacteria. *Antibiotics, 10*(5):600. https://doi.org/ 10050600.DOI: 10.3390/antibiotics

Ejeta, B. M., Das, M. K., & Terefe, O. (2024). Roadmap for drug discovery from medicinal plants. *Journal of Clinical Images and Medical Case Reports, 5*(11), 3349.

El-Saadony, M. T., Saad, A. M., Mohammed, D. M., Korma, S. A., Alshahrani, M. Y., Ahmed, A. E., Ibrahim, E. H., Salem, H. M., Alkafaas, S. S., Saif, A. M., Elkafas, S. S., Fahmy, M. A., Abd El-Mageed, T. A., Abady, M. M., Assal, H. Y., El-Tarabily, M. K., Mathew, B. T., AbuQamar, S. F., El-Tarabily, K. A., & Ibrahim, S. A. (2025). Medicinal plants: Bioactive compounds, biological activities, combating multidrug-resistant microorganisms, and human health benefits - a comprehensive review. *Frontiers in Immunology, 16*, 1491777. DOI: 10.3389/fimmu.2025.1491777 PMID: 40375989

Gaurav, A., Bakht, P., Saini, M., Pandey, S., & Pathania, R. (2023). Role of bacterial efflux pumps in antibiotic resistance, virulence, and strategies to discover novel efflux pump inhibitors. *Microbiology (Reading, England), 169*(5), 001333. DOI: 10.1099/mic.0.001333 PMID: 37224055

GBD 2021 Antimicrobial Resistance Collaborators. (2024). Global burden of bacterial antimicrobial resistance 1990–2021: a systematic analysis with forecasts to 2050. *The Lancet,404*(10459):1199-1226.

George, V., Ijinu, T.P., Chithra, MA. and Pushpangadan, P. (2016). Can local health traditions and tribal medicines strengthen Ayurveda? Case study 2 – Trichopus zeylanicus ssp. travancoricus Burkill ex Narayanan. *Journal of Traditional and Folk Practices,04*(2):01-14.

Golla, S., Gambhir, N., Gupta, N., Singh, R., & Singh, D. (2024). A Comparative Evaluation of Herbal Extracts and Triple Antibiotic Paste as Intracanal Medicament against *Enterococcus faecalis:* A Microbiological Study. *International Journal of Clinical Pediatric Dentistry, 17*(3), 285–290. DOI: 10.5005/jp-journals-10005-2791 PMID: 39144517

Guimarães, A. C., Meireles, L. M., Lemos, M. F., Guimarães, M. C. C., Endringer, D. C., Fronza, M., & Scherer, R. (2019). Antibacterial Activity of Terpenes and Terpenoids Present in Essential Oils. *Molecules (Basel, Switzerland)*, *24*(13), 2471. DOI: 10.3390/molecules24132471 PMID: 31284397

Gunasekara, C., Radhika, N. D. M., Ragunathan, K. K., Gunathilaka, D. P. P., Weerasekera, M., Hewageegana, H. G. S. P., Arawwawala, L. A. D. M., & Fernando, S. S. N. (2017). Determination of Antimicrobial Potential of Five Herbs used in Ayurveda Practices against *Candida albicans, Candida parapsilosis* and Methicillin Resistant Staphylococcus aureus. *Ancient Science of Life*, *36*(4), 187. DOI: 10.4103/asl.ASL_179_16 PMID: 29269969

Hemmati, F., Salehi, R., Ghotaslou, R., Samadi Kafil, H., Hasani, A., Gholizadeh, P., Nouri, R., & Ahangarzadeh Rezaee, M. (2020). Quorum Quenching: A Potential Target for Antipseudomonal Therapy. *Infection and Drug Resistance*, *13*, 2989–3005. DOI: 10.2147/IDR.S263196 PMID: 32922047

Herrera, K.M.S., Lopes, G.F.M., Oliveira, M.E., Sousa, J.F., Lima, W.G., Silva, F.K., Brito, J.C.M., Gomes, A.J.P.S., Viana, G.H.R., Soares, A.C., Ferreira, J.M.S. (2022). A 3-alkylpyridine-bearing alkaloid exhibits potent antimicrobial activity against methicillin-resistant Staphylococcus aureus (MRSA) with no detectable resistance. *Microbiological Research, 261*(August):127073. https://doi.org/. 2022.127073.DOI: 10.1016/j.micres

Huang, S. S., Huang, C. H., Ko, C. Y., Chen, T. Y., Cheng, Y. C., & Chao, J. (2022). An Ethnobotanical Study of Medicinal Plants in Kinmen. *Frontiers in Pharmacology*, *12*, 681190. DOI: 10.3389/fphar.2021.681190 PMID: 35222004

Iwu, M. M. (2002). Ethnobotanical approach to pharmaceutical drug discovery: Strengths and limitations. *Advances in Phytomedicine*, *1*, 309–320. DOI: 10.1016/S1572-557X(02)80034-4

Jassal, P. S., Banerjee, P., Kancharla, S., Kolli, P., Mandadapu, G., & Jena, M. K. (2025). Antimicrobial Potency of Some Traditional Medicinal Plants of North-Eastern India: An In-Depth Review. *Journal of Pure & Applied Microbiology*, *19*(3), 1587–1598. DOI: 10.22207/JPAM.19.3.01

Jayaraman, P., Sakharkar, M. K., Lim, C. S., Tang, T. H., & Sakharkar, K. R. (2010). Activity and interactions of antibiotic and phytochemical combinations against *Pseudomonas aeruginosa in vitro. International Journal of Biological Sciences*, *6*(6), 556–568. DOI: 10.7150/ijbs.6.556 PMID: 20941374

Jishnu, T. K., & Behera, B. (2024). A review of Arogya Pacha (*Trichopus zeylanicus*). *International Journal of Research in Ayurveda and Pharmacy*, *15*(5), 147–151. DOI: 10.7897/2277-4343.155174

Kali, A., Bhuvaneshwar, D., Charles, P. M., & Seetha, K. S. (2016). Antibacterial synergy of curcumin with antibiotics against biofilm producing clinical bacterial isolates. *Journal of Basic and Clinical Pharmacy*, *7*(3), 93–96. DOI: 10.4103/0976-0105.183265 PMID: 27330262

Kariuki, S., Kering, K., Wairimu, C., Onsare, R., & Mbae, C. (2022). Antimicrobial Resistance Rates and Surveillance in Sub-Saharan Africa: Where Are We Now? *Infection and Drug Resistance*, *15*, 3589–3609. DOI: 10.2147/IDR.S342753 PMID: 35837538

Kauffmann, A.C. and Castro, V.S. (2023). Phenolic Compounds in Bacterial Inactivation: A Perspective from Brazil. *Antibiotics (Basel), 12*(4):645. doi: 12040645. DOI: 10.3390/antibiotics

Khameneh, B., Eskin, N. A. M., Iranshahy, M., & Fazly Bazzaz, B. S. (2021). Phytochemicals: A Promising Weapon in the Arsenal against Antibiotic-Resistant Bacteria. *Antibiotics (Basel, Switzerland)*, *10*(9), 1044. DOI: 10.3390/antibiotics10091044 PMID: 34572626

Khare, T., Anand, U., Dey, A., Assaraf, Y. G., Chen, Z. S., Liu, Z., & Kumar, V. (2021). Exploring Phytochemicals for Combating Antibiotic Resistance in Microbial Pathogens. *Frontiers in Pharmacology*, *12*, 720726. DOI: 10.3389/fphar.2021.720726 PMID: 34366872

Krishnaprasad, V. H., & Kumar, S. (2024). Antimicrobial resistance: An ultimate challenge for 21st century scientists, healthcare professionals, and policymakers to save future generations. *Journal of Medicinal Chemistry*, *67*(18), 15927–15930. DOI: 10.1021/acs.jmedchem.4c02002 PMID: 39238216

Kubade, B., & Vaidya, S. N. (2024). Pharmacological and Therapeutic Effects of Triphala –Churna. *Asian Journal of Pharmaceutics*, *18*(4), 999–1003.

Lara-Espinosa, J. V., Barrios-Payán, J., Lozano-Ordaz, V., Mata-Espinosa, D., Becerril-Villanueva, E., Ponce-Regalado, M. D., & Hernández-Pando, R. (2025). Synergistic Effects of Curcumin and Antibiotics Against Drug-Sensitive and Multidrug-Resistant *Mycobacterium tuberculosis. International Journal of Molecular Sciences*, *26*(21), 10414. DOI: 10.3390/ijms262110414 PMID: 41226452

Li, X., Chen, S., Wu, Y., Liu, F., Jumai, A., Zhu, B., Xiong, B., & Qiu, S.-X. (2025). γ-mangostin from *Garcinia mangostana* (L.) exerts the antifungal activity against *Cryptococcus neoformans* by inhibiting ribosome biogenesis. *Journal of Ethnopharmacology*, *358*, 120986. DOI: 10.1016/j.jep.2025.120986 PMID: 41349701

Liu, Y., Zhu, J., Liu, Z., Zhi, Y., Mei, C., & Wang, H. (2025). Flavonoids as Promising Natural Compounds for Combating Bacterial Infections. *International Journal of Molecular Sciences*, *26*(6), 2455. DOI: 10.3390/ijms26062455 PMID: 40141099

Lobiuc, A., Pavăl, N. E., Mangalagiu, I. I., Gheorghi ă, R., Teliban, G. C., Amăriucăi-Mantu, D., & Stoleru, V. (2023). Future Antimicrobials: Natural and Functionalized Phenolics. *Molecules (Basel, Switzerland)*, *28*(3), 1114. DOI: 10.3390/molecules28031114 PMID: 36770780

Lulesa, F., Alemu, S., Kassa, Z., & Awoke, A. (2025). Ethnobotanical investigation of medicinal plants utilized by indigenous communities in the Fofa and Toaba sub-districts of the Yem Zone, Central Ethiopian Region. *Journal of Ethnobiology and Ethnomedicine*, *21*(1), 14. DOI: 10.1186/s13002-025-00768-2 PMID: 40045312

Lulesa, F., Alemu, S., Kassa, Z., & Awoke, A. (2025). Ethnobotanical investigation of medicinal plants utilized by indigenous communities in the Fofa and Toaba sub-districts of the Yem Zone, Central Ethiopian Region. *Journal of Ethnobiology and Ethnomedicine*, *21*(1), 14. DOI: 10.1186/s13002-025-00768-2 PMID: 40045312

Ma, Y., Cong, Z., Wang, Y., Gao, P. (2024). A novel multi-drugs ciprofloxacin-curcumin-N-acetylcysteine co-delivery system based on hybrid nanocrystals for dry powder inhalations. *Next Nanotechnology, 6*:100084. https://doi.org/. 2024.100084. DOI: 10.1016/j.nxnano

Mączka, W., Duda-Madej, A., Górny, A., Grabarczyk, M., Wińska, K. (2021) Can Eucalyptol Replace Antibiotics? *Molecules, 26*(16):4933. 26164933.DOI: 10.3390/ molecules

Mandeville, A., & Cock, I. E.Fruit Extracts Inhibit Bacterial Triggers of Some Autoimmune Diseases and Potentiate the Activity of Tetracycline. (2018). *Terminalia chebula* Retz. Fruit Extracts Inhibit Bacterial Triggers of Some Autoimmune Diseases and Potentiate the Activity of Tetracycline. *Indian Journal of Microbiology*, *58*(4), 496–506. DOI: 10.1007/s12088-018-0754-9 PMID: 30262960

Manso, T., Lores, M., & de Miguel, T. (2021). Antimicrobial Activity of Polyphenols and Natural Polyphenolic Extracts on Clinical Isolates. *Antibiotics (Basel, Switzerland)*, *11*(1), 46. DOI: 10.3390/antibiotics11010046 PMID: 35052923

Manza, M. M., & Saj, O. P. (2013). Cytotoxic and antimicrobial studies on Arogyapacha or Kerala ginseng leaf extracts. in *International Journal of Pharmaceutical, Chemical and Biological Sciences,3-2:*315-319.

McClatchey, W. C., Mahady, G. B., Bennett, B. C., Shiels, L., & Savo, V. (2009). Ethnobotany as a pharmacological research tool and recent developments in CNS-active natural products from ethnobotanical sources. *Pharmacology & Therapeutics, 123*(2), 239–254. DOI: 10.1016/j.pharmthera.2009.04.002 PMID: 19422851

McDonnell, A., Countryman, A., Laurence, T., Gulliver, S., Drake, T., Edwards, S., Kenny, C., Lamberti, O., Morton, A., Shafira, A., Smith, R., & Guzman, J. (2024). *Forecasting the Fallout from AMR: Economic Impacts of Antimicrobial Resistance in Humans – A report from the EcoAMR series.* World Organisation for Animal Health and World Bank.

Merghni, A., Belmamoun, A. R., Urcan, A. C., Bobiş, O., & Lassoued, M. A. (2023). 1,8-Cineol (Eucalyptol) Disrupts Membrane Integrity and Induces Oxidative Stress in Methicillin-Resistant *Staphylococcus aureus. Antioxidants, 2*(7), 1388. DOI: 10.3390/antiox12071388 PMID: 37507929

Munita, J.M. and Arias, C.A. (2016). Mechanisms of Antibiotic Resistance. *Microbiol Spectr,4*(2):VMBF-0016-2015. .DOI: 10.1128/microbiolspec.VMBF-0016-2015

Mussarat, S., Adnan, M., Begum, S., Ur Rehman, S., Hashem, A., & Abd Allah, E. F. (2021). Antimicrobial screening of polyherbal formulations traditionally used against gastrointestinal diseases. *Saudi Journal of Biological Sciences, 28*(12), 6829–6843. DOI: 10.1016/j.sjbs.2021.07.053 PMID: 34866983

Narasimhan, S., Maheshwaran, S., Abu-Yousef, I. A., Majdalawieh, A. F., Rethavathi, J., Das, P. E., & Poltronieri, P. (2017). Anti-Bacterial and Anti-Fungal Activity of Xanthones Obtained via Semi-Synthetic Modification of α-Mangostin from Garcinia mangostana. *Molecules (Basel, Switzerland), 22*(2), 275. DOI: 10.3390/molecules22020275 PMID: 28208680

Navidifar, T., Zare Banadkouki, A., Parvizi, E., Mofid, M., Golab, N., Beig, M., & Sholeh, M. (2025). Global prevalence of macrolide-resistant *Staphylococcus* spp.: A comprehensive systematic review and meta-analysis. *Frontiers in Microbiology, 16*, 1524452. DOI: 10.3389/fmicb.2025.1524452 PMID: 40182286

Nazir, A., Nazir, A., Zuhair, V., Aman, S., Sadiq, S. U. R., Hasan, A. H., Tariq, M., Rehman, L. U., Mustapha, M. J., & Bulimbe, D. B. (2025). The Global Challenge of Antimicrobial Resistance: Mechanisms, Case Studies, and Mitigation Approaches. *Health Science Reports, 8*(7), e71077. DOI: 10.1002/hsr2.71077 PMID: 40704322

Negi, N., Prakash, P., Gupta, M. L., & Mohapatra, T. M. (2014). Possible Role of Curcumin as an Efflux Pump Inhibitor in Multi Drug Resistant Clinical Isolates of Pseudomonas aeruginosa. *Journal of Clinical and Diagnostic Research : JCDR*, *8*(10), DC04–DC07. DOI: 10.7860/JCDR/2014/8329.4965 PMID: 25478340

Nguyen, T. L. A., & Bhattacharya, D. (2022). Antimicrobial Activity of Quercetin: An Approach to Its Mechanistic Principle. *Molecules (Basel, Switzerland)*, *27*(8), 2494. DOI: 10.3390/molecules27082494 PMID: 35458691

Pillai, P., Sajan, J., Menon, K., Hemanthakumar, A., Pandurangan, A. G., Krishnan, P., & Sooriamuthu, S. (2011). Analysis of Genetic Diversity in *Poeciloneuron pauciflorum* Bedd.- An Endemic Tree Species from the Western Ghats of India. *American Journal of Plant Sciences*, *2*(3), 416–424. DOI: 10.4236/ajps.2011.23047

Pirintsos, S., Panagiotopoulos, A., Bariotakis, M., Daskalakis, V., Lionis, C., Sourvinos, G., Karakasiliotis, I., Kampa, M., & Castanas, E. (2022). From Traditional Ethnopharmacology to Modern Natural Drug Discovery: A Methodology Discussion and Specific Examples. *Molecules, 27*(13):4060. 27134060.DOI: 10.3390/molecules

Praveen, M. and Gupta, S. (2024). Network Pharmacology of *Trichopus Zeylanicus* ssp. Travancoricus Burkill Ex K. Narayanan Leaf Extract Revealed CA2, ERN1, TTR, and SNCA as Inducible Genes. *Lettersin Applied NanoBioscience,13*(3):01-14. .DOI: 10.33263/LIANBS133.114

Raghunath, I., Koland, M., Narayanan, A.V. (2024). Piperine: A possible permeation enhancer for oral protein delivery. *Journal of Applied Pharmaceutical Science,14*(04):035-045.

Rahmani, A. H., Almatroudi, A., Alrumaihi, F., & Khan, A. A. (2018). Pharmacological and therapeutic potential of neem (*Azadirachta indica). Pharmacognosy Reviews*, *12*(24), 250–255. DOI: 10.4103/phrev.phrev_8_18

Rana, T., Farooq, U., Kaur, N., Khan, A., & Khan, A. (2023). Plants Derived Efflux Pump Inhibitors: An approach against Multidrug-Resistant Gram-negative bacteria Klebsiella pneumoniae. *Pharmacophore*, *14*(1), 71–79. DOI: 10.51847/7FQXF-NYnT5

Saabir, F., Hussain, A., Mulani, M., Kulkarni, S., Tambe, S. (2022). Efflux pump and its inhibitors: Cause and cure for multidrug resistance. *Journal of Applied Biology & Biotechnology,10-10*(03):177-194.

Saiqali, M. A., Aziz, S. S., & Reddy, A. V. (2025). Current Status on Phytochemicals Classification, Structure-Activity Relationship, Stereochemistry and AI-Driven Applications: A Systematic Review. *Pharmacognosy Reviews*, *19*(38), 186–211. DOI: 10.5530/phrev.20252344

Salam, M. A., Al-Amin, M. Y., Salam, M. T., Pawar, J. S., Akhter, N., Rabaan, A. A., & Alqumber, M. A. A. (2023). Antimicrobial Resistance: A Growing Serious Threat for Global Public Health. *Healthcare (Basel)*, *11*(13), 1946. DOI: 10.3390/healthcare11131946 PMID: 37444780

Sasi Kala, N., & Ramasubbu, R. (2021). Chemical composition, antimicrobial and antioxidant properties of essential oils of *Trichopus zeylanicus* ssp. Travancoricus. *Indian Journal of Natural Products and Resources*, *12*(4), 570–577.

Saudagar, R. B., Sambathkumar, R., & Bachhav, R. S. (2013). An Ethnopharmacology of Trichopus zeylanicus Gaertn. *Research. Journal of Pharmacognosy and Phytochemistry*, *5*(5), 258–262.

Senarathna, U. L., Fernando, S. S., Gunasekara, T. D., Weerasekera, M. M., Hewageegana, H. G., Arachchi, N. D., Siriwardena, H. D., & Jayaweera, P. M. (2017). Enhanced antibacterial activity of TiO_2 nanoparticle surface modified with *Garcinia zeylanica* extract. *Chemistry Central Journal*, *11*(1), 7. DOI: 10.1186/s13065-017-0236-x PMID: 28123449

Shah, S., & Bhat, J. A. (2019). Ethnomedicinal knowledge of indigenous communities and pharmaceutical potential of rainforest ecosystems in Fiji Islands. *Journal of Integrative Medicine*, *17*(4), 244–249. DOI: 10.1016/j.joim.2019.04.006 PMID: 31072793

Sharifi-Rad, J., Rayess, Y. E., Rizk, A. A., Sadaka, C., Zgheib, R., Zam, W., Sestito, S., Rapposelli, S., Neffe-Skocińska, K., Zielińska, D., Salehi, B., Setzer, W. N., Dosoky, N. S., Taheri, Y., El Beyrouthy, M., Martorell, M., Ostrander, E. A., Suleria, H. A. R., Cho, W. C., & Martins, N. (2020). Turmeric and Its Major Compound Curcumin on Health: Bioactive Effects and Safety Profiles for Food, Pharmaceutical, Biotechnological and Medicinal Applications. *Frontiers in Pharmacology*, *11*, 01021. DOI: 10.3389/fphar.2020.01021 PMID: 33041781

Sharma, A., Adhikari, D., & Wagh, V. V. (2024). An ethnobotanical investigation of medicinal plants used among rural communities in Northeastern Madhya Pradesh, India: A quantitative study. *Plant Biosystems*, *158*(5), 1160–1181. DOI: 10.1080/11263504.2024.2395845

Stermitz, F. R., Lorenz, P., Tawara, J. N., Zenewicz, L. A., & Lewis, K. (2000). Synergy in a medicinal plant: Antimicrobial action of berberine potentiated by 5′-methoxyhydnocarpin, a multidrug pump inhibitor. *Proceedings of the National Academy of Sciences of the United States of America*, *97*(4), 1433–1437. DOI: 10.1073/pnas.030540597 PMID: 10677479

Sultan, M. T., Anwar, M. J., Imran, M., Khalil, I., Saeed, F., Neelum, S., Alsagaby, S. A., Abdulmonem, W. A., Abdelgawad, O. M. A., Hussain, M., El-Ghorab, A. H., Umar, M., & Jbawi, E. A. (2023). Phytochemical profile and pro-healthy properties of *Terminalia chebula*: A comprehensive review. *International Journal of Food Properties*, *26*(1), 526–551. DOI: 10.1080/10942912.2023.2166951

Talarczyk, A., Guźniczak, M., Kublicka, A., Matczuk, A., Wieliczko, A., Ulatowska-Jarza, A., & Buzalewicz, I. (2025). Efflux pump inhibitors from plant compounds: A new strategy to modulate bacterial growth and morphology. *Industrial Crops and Products*, *233*, 121378. DOI: 10.1016/j.indcrop.2025.121378

Tosa, H., Iinuma, M., Murakami, K-I., Tetsuro, T., Toshiyuki, C., & Veliah, R., Soedarsono. (. (1997). New xanthone from *Poeciloneruon pauciflorum* Bedd. *Phytochemistry*, *45*(1), 133–136. DOI: 10.1016/S0031-9422(96)00856-4

Vaou, N., Stavropoulou, E., Voidarou, C., Tsigalou, C., Bezirtzoglou, E. (2021). Towards Advances in Medicinal Plant Antimicrobial Activity: A Review Study on Challenges and Future Perspectives. *Microorganisms,9*(10):2041. doi: 9102041. DOI: 10.3390/microorganisms

Wang, C., Zhang, H., Wang, X., Wang, X., Li, X., Li, C., Wang, Y., & Zhang, M. (2024). Comprehensive Review on Fruit of *Terminalia chebula*: Traditional Uses, Phytochemistry, Pharmacology, Toxicity, and Pharmacokinetics. *Molecules (Basel, Switzerland)*, *29*(23), 5547. DOI: 10.3390/molecules29235547 PMID: 39683707

WHO. (2019). *No time to wait: Securing the future from drug-resistant infections report to the Secretary-General of the United Nations*. https://cdn.who.int/media/docs/default-source/documents/no-time-to-wait-securing-the-future-from-drug-resistant-infections-en.pdf?sfvrsn=5b424d7_6&download=true. Accessed on 16.12.2025.

Wolska, K. I., Grześ, K., & Kurek, A. (2012). Synergy Between Novel Antimicrobials and Conventional Antibiotics or Bacteriocins. *Polish Journal of Microbiology*, *61*(2), 95–104. DOI: 10.33073/pjm-2012-012 PMID: 23163208

Wylie, M.R. and Merrell, D.S. (2022). The Antimicrobial Potential of the Neem Tree *Azadirachta indica. Front Pharmacol,30*(13):891535. doi: . 2022.891535. DOI: 10.3389/fphar

Yashvardhini, N., Samiksha, S., & Jha, D. K. (2021). Pharmacological intervention of various Indian medicinal plants in combating COVID-19 infection. *Biomedical Research and Therapy*, *8*(7), 4461–4475.

Zhou, H., Wang, W., Cai, L., & Yang, T. (2023). Potentiation and Mechanism of Berberine as an Antibiotic Adjuvant Against Multidrug-Resistant Bacteria. *Infection and Drug Resistance*, *16*, 7313–7326. DOI: 10.2147/IDR.S431256 PMID: 38023403

Chapter 14
Clinical Trials and Regulatory Challenges in Herbal Antimicrobials:
Bridging Traditional Wisdom and Modern Standards

Vibhu Bhatia
http://orcid.org/0000-0002-0668-8286
Somaiya Vidyavihar University, India

Neharika Avinash Sawant
Somaiya Vidyavihar University, India

ABSTRACT

The global rise of antimicrobial resistance (AMR) has intensified the search for alternative therapies, with herbal medicine emerging as a promising frontier. Despite centuries of traditional use, herbal antimicrobials face significant hurdles in clinical validation and regulatory approval. This chapter critically examines the complexities of conducting clinical trials for herbal antimicrobials, including issues of standardization, bioavailability, placebo design, and ethical considerations. It also explores the fragmented global regulatory landscape, highlighting disparities in herbal drug approval processes across regions such as the EU, USA, India, and China. Through case studies of successful and failed herbal clinical trials, the chapter identifies best practices and persistent gaps in evidence generation. By addressing these challenges, the chapter aims to provide a roadmap for integrating herbal antimicrobials into mainstream healthcare systems under the One Health paradigm

DOI: 10.4018/979-8-3373-7810-7.ch014

1. INTRODUCTION

1.1 Global Resurgence of Herbal Medicine in Antimicrobial Therapy

The use of herbal medicines in formal healthcare systems has increased remarkably, and WHO now treats traditional and herbal medicine as a major pillar within its global health agenda. In parallel, commercial analyses of the herbal medicine sector indicate strong and sustainable growth, with a global market estimated at around 186 billion USD in 2023 and projected to reach roughly 434 billion USD by 2033, which corresponds to an approximate compound annual growth rate of 8.8% as illustrated in Figure 1. This rising trend is not merely an economic signal but it also reflects widespread and increasing reliance on herbal products for prevention and treatment. Thus, creating a clear obligation to subject candidate herbal antimicrobials through robust clinical and regulatory evaluations (Butler et al., 2025).

Figure 1. Global herbal medicine market growth from 2023 to 2033, showing a steady increase in market value and highlighting the rising importance of herbal products in healthcare

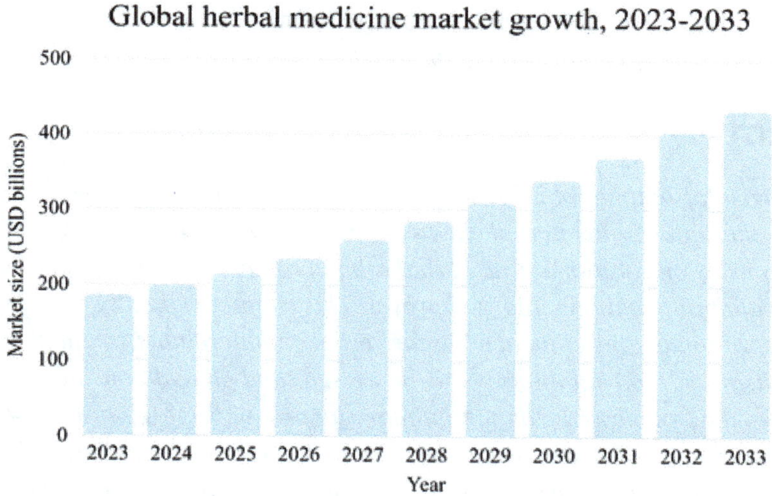

Data adapted from Allied Market Research, Herbal Medicine Market Report code A324087

While herbal medicine encompasses a broad range of therapeutic applications including metabolic, inflammatory, neurological and wellness indications, this chapter specifically addresses herbal preparations intended to exert antimicrobial or infection-modifying effects. These products differ conceptually from general herbal therapeutics because they must demonstrate measurable activity against microbial burden, infection progression, or resistance modulation. Consequently, the evidentiary expectations for clinical validation more closely resemble those applied to anti-infective agents rather than symptomatic herbal interventions.

1.2 The Antimicrobial Resistance (AMR) Crisis as a Driver for Herbal Alternatives

WHO has repeatedly flagged AMR as a major global health threat, emphasizing that drug-resistant infections are already undermining the effectiveness of existing therapies and putting standard medical procedures at risk worldwide. In a recent global analysis of bacterial AMR, deaths associated with drug-resistant infections were very high and age-specific patterns showed that AMR-associated mortality in children younger than 5 years has fallen substantially over recent decades, whereas deaths in older age groups, particularly adults 70 years or above have increased, highlighting a shift of the AMR burden towards older populations. A relatively small number of pathogen-drug resistance combinations are responsible for a large proportion of deaths attributable to AMR, with meticillin-resistant *Staphylococcus aureus*, carbapenem-resistant *Acinetobacter baumannii* and other key Gram-negative pathogens are among the leading contributors globally. Together these patterns suggest that, although antimicrobial stewardship and infection-prevention programmes exist, more innovative strategies are needed to preserve the efficacy of existing antibiotics (Naghavi et al., 2024).

When the focus shifts to pneumonia, the implications of AMR become even more alarming. Bacterial pneumonia remains a leading infectious cause of death worldwide and is often treated with broad-spectrum antibiotics, so rising resistance directly erodes the reliability of standard antibiotics. In this context, even modest increases in resistance can lead to delayed appropriate therapy, higher treatment failures and greater dependence on last line drugs which further accelerates the cycle of resistance. This AMR crisis creates strong pressure to diversify or create new treatment strategies rather than relying solely on conventional antibiotics. As illustrated in Figure 2, these options can be viewed as complementary strategies that sit alongside rather than replace existing antibiotic therapy (Daraghmeh et al., 2025).

Figure 2. Therapeutic strategies against antimicrobial resistance presented as complementary approaches to conventional antibiotic therapy

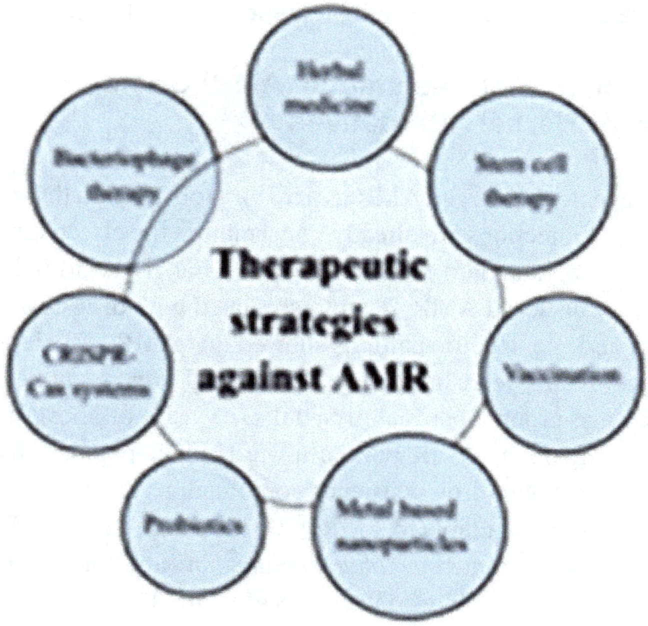

1.3 Rationale and Scope of the Chapter

The coexistence of a rapidly expanding global herbal market and a persistently high AMR burden creates a strong rationale for examining herbal antimicrobials with the same critical lens applied for conventional medicine. In recent studies, multiple natural products and herbal candidates have been catalogued for their synthesis and in vitro activity or preliminary clinical signals against resistant bacteria, but most remain at an early and heterogeneous stage of development because of key issues in product standardization, pharmacokinetic characterization and trial design have not yet been fully resolved (Daraghmeh et al., 2025; Butler et al.,2025). In parallel, analyses of traditional herbal medicine legislation and regulation show that herbal products are handled very differently across jurisdictions ranging from traditional use preparations to dietary supplements to full medicinal products which leads to fragmented evidentiary expectations and complicates the herbal agents for clearly defined antimicrobial indications (Dubale et al.,2025).

Against this background, the present chapter concentrates on clinical trials and regulatory challenges specifically for herbal antimicrobials. It explores how candidate products are currently being evaluated in human studies with methodology and ethical issues that arise when translating complex botanical preparations into trial settings and how existing regulatory frameworks, many of which draw on WHO guidance for traditional medicine, how it shapes the pathways for products that claim antimicrobial effects.

Current clinical investigation of herbal antimicrobials is concentrated in a limited number of therapeutic areas. Most randomized controlled trials evaluate localized infections, particularly oral and dental conditions (e.g., gingivitis, dental caries), dermatological infections (acne, wound colonization), upper respiratory tract symptoms, and gastrointestinal disturbances. Systemic infections are rarely studied Across the literature, the number of rigorously designed randomized trials remains relatively small and heterogeneous, with most involving modest sample sizes and short treatment durations. This distribution explains why evidence is stronger for adjunctive or preventive applications than for standalone antimicrobial therapy.

As a result, the clinical evidence base for herbal antimicrobials remains fragmented, condition-specific, and methodologically heterogeneous. The following sections therefore examine not only efficacy findings but also the structural reasons why antimicrobial validation has proven difficult.

1.4 The One Health Framework and Integrative Approaches

WHO describes One Health as an integrated way of working that aims to balance and optimise the health of humans, animals and ecosystems and it specifically highlights antimicrobial resistance (AMR) as an area where this perspective is crucial. Instead of treating human, animal and environmental health as separate topics, One Health recognises that antibiotic use in any one of these settings can feed into the same shared resistance problem.(WHO, One Health, Control of Antimicrobial Resistance, 2025)

Recent regional guidance on AMR in Europe reinforces this view by stressing that effective control needs several elements to act together such as strong infection-prevention and control, careful and targeted antimicrobial use in both human and veterinary medicine and robust surveillance systems that monitor resistance trends and drug consumption over time. This also emphasizes the importance of continued research and development of new and adjunctive therapies that can help protect the usefulness of existing antibiotics (Strengthening the Surveillance of Antimicrobial Resistance in the WHO European Region, 2025).

Building on this, more detailed One Health stewardship frameworks map responsibilities across the whole antimicrobial lifestyle, from discovery and manufacturing through prescribing and use in humans and animals to environmental release and waste management. They argue that all antimicrobial modalities including natural-product and plant-based agents should be assessed along this full pathway so that risks are not simply shifted from one sector or region to another.

2. HERBAL ANTIMICROBIALS: TRADITIONAL FOUNDATIONS AND MODERN PROMISE

2.1 Historical Overview of Plant-Based Antimicrobials Across Civilizations

Herbal medicines, which are therapies derived from plants for the prevention or treatment of diseases, have held significant prominence in various traditional medical systems over the centuries. Traditional Chinese Medicine (TCM) and represent two of the most enduring and formally established herbal traditions, each having a documented history that spans several millennia. The recorded history of medicinal plant use in TCM is evident in classical texts such as the Shennong Bencaojing and later herbal compendiums like Bencao Gangmu, reflecting a long history of empirically derived herbal practices in China (Woo et al.,2023). In parallel, Ayurveda, the traditional medical system originating from the Indian subcontinent, has utilized botanical medicines for the maintenance of health and treatment of illnesses for over 3,000 years. Foundational texts like the Charaka Samhita and Sushruta Samhita recorded extensive use of plant-based medicinal preparations (Singh et al.,2024).

Moreover, indigenous healing systems across the globe have historically leveraged local medicinal plants for healing wounds and management of infections. Although comprehensive peer-reviewed studies specifically on antimicrobial applications are limited, ethnobotanical studies consistently demonstrate that communities in Africa, the Americas, Europe and Australia have utilized plant decoctions, poultics and extracts to address ailments now understood as infectious in origins. Such practices suggest a broad reliance on plant-derived antimicrobial agents (Shah, 2024). Although these medical traditions developed prior to the advent of modern microbiological theories, they reflect extensive empirical use of botanical remedies. Such historical practices provide essential cultural context and have laid the groundwork for systematic scientific investigations into plant-derived antimicrobial compounds.

2.2 Ethnopharmacology, Biocultural Knowledge, and Key Mechanisms of Action

Ethnopharmacology offers a valuable lens through which the medicinal use of plants across different cultures can be understood. It plays a key role in identifying bioactive phytochemicals with potential in modern biomedical research. Current studies show that medicinal plants synthesizes a diverse range of secondary metabolites, many of which function as a part of the plant's defence system A substantial portion of these compounds exhibit antimicrobial activity against bacteria, viruses and fungi. Plant-derived antimicrobial activity has been attributed to several major chemical groups including alkaloids, flavonoids, terpenoids, phenolic acids, saponins and tannins. These compounds act through multiple mechanisms such as disrupting microbial membranes, interfering with cell wall formation, inhibiting nucleic acid and protein synthesis and modulating quorum-sensing pathways. Flavonoids are known to interact with cell wall components and extracellular proteins, limiting microbial growth. Alkaloids can intercalate with DNA and impair essential enzymatic processes, while terpenoids compromise membrane integrity of the cell. Phenolic compounds in contrast are often associated with interference in key cellular metabolic pathways.

These mechanistic observations are reflected in recent applied research, a contemporary review examining the use of Traditional Chinese Medicine and plant-derived bioactive compounds in the management of bovine mastitis describes several plant-based formulations with reported antimicrobial, anti-inflammatory and immunomodulatory effects (Fan et al., 2025). Another recent study, screened 132 TCM plant species and identified 25 extracts with notable inhibitory activity against Gram-positive bacteria, also nearly 100 previously isolated antimicrobial compounds were reported to be active against 45 bacterial species (Ellward et al.,2024). These studies together illustrate the chemical diversity and therapeutic potential of plant-derived compounds.

2.3 From Traditional Uses to Modern Drug Leads: Berberine, Tea Tree Oil, and Neem Extract

Traditional medicinal systems have long employed plant-derived formulations to treat conditions now identified as infectious in nature. In recent years, specific herbal compounds and extracts have been reevaluated through modern microbiological and pharmacological approaches, resulting in their evaluation as potential antimicrobial drug candidates. Among these, *Berberis vulgaris* (berberine), *Melaleuca alternifolia* (tea tree) and *Azadirachta indica* (neem) extracts stand out as extensively studied examples where traditional applications have guided contemporary antimicrobial research.

1. Berberine

Berberine is an isoquinoline alkaloid found in several medicinal plants, particularly *Coptis chinensis* and various *Berberis* species, which are integral to Traditional Chinese Medicine. Historically, plants containing berberine were used to alleviate gastrointestinal and inflammatory properties. Recent studies have confirmed that berberine exhibits direct antimicrobial activity against clinically significant pathogens like methicillin-resistant *Staphylococcus aureus* (MRSA). For instance, Zhou et al. (2024) demonstrated that berberine not only inhibits MRSA growth but also disrupts bacterial cell structure and shows synergistic effects when combined with conventional antibiotics, suggesting its potential as an adjunctive antimicrobial. Additional in vitro studies have reported dose-dependent inhibitory effects of berberine against other bacteria such as *Cutibacterium acnes,* with mechanistic analyses indicating interference with cell wall integrity, inhibition of efflux pumps associated with antimicrobial resistance and suppression of biofilm formation (Sun et al.,2024; Duda-Madej et al.,2025). Although strongly associated with Traditional Chinese Medicine, berberine-containing plants are also used in Ayurvedic and other traditional systems, illustrating cross-cultural convergence in antimicrobial plant selection. (3)

2. Tea Tree Oil

Tea tree oil is a concentrated essential oil obtained from the leaves of *Melaleuca alternifolia,* a plant native to Australia. Traditionally, it has been valued for its topical antiseptic properties and is incorporated in commercial products promoting skin and oral hygiene. Modern research confirms its broad-spectrum antimicrobial activity. A recent study demonstrated inhibitory effects against oral pathogens, including *Enterococcus faecalis* and *Porphyromonas gingivalis,* both associated with dental and periodontal infections (Mohammed et al.,2024). Terpenoid compounds, particularly terpinen-4-ol are the primary contributors to this activity, it disrupts microbial membrane integrity and alters cellular permeability. Experimental and computational studies suggest these compounds interfere with essential membrane-associated processes, explaining their antimicrobial effects (lacovelli et al.,2023). Although systemic toxicity limits its use to topical applications, tea tree oil remains a validated herbal antimicrobial with well studied chemical properties.

3. Neem Extracts

Neem holds a central position in traditional South Asian medicine, particularly Ayurveda, where various plant parts have been employed to maintain hygiene and health. Recent studies have focused on the antimicrobial potential of neem leaf, bark and seed extracts under controlled laboratory conditions. Neem extracts have demonstrated inhibitory activity against a broad range of bacterial and fungal pathogens including oral microorganisms capable of forming biofilms. A recent study reported pronounced antimicrobial and antibiofilm effects of neem leaf and bark extracts against oral microflora highlighting their potential for managing infection-related conditions in the oral cavity (Bhattacharya et al.,2024). These antimicrobial properties are largely attributed to a complex phytochemical composition, particularly limonoids such as azadirachtin and nimbin, which are believed to interfere with microbial growth, biofilm formation and cellular communication although the precise molecular targets remain to be fully elucidated (Wylie et al.,2022). Clinical investigations have also explored neem-based oral and dermatological preparations, particularly in dental plaque and gingival inflammation contexts, though evidence remains localized and formulation-dependent. (3)

Beyond laboratory studies, integrative reviews label neem as a multicomponent herbal antimicrobial with diverse modes of action, including modulation of microbial adhesion, disruption of biofilm matrix and potential immunomodulatory effects that may enhance host defenses (Debnath et al.,2025). Collectively these findings support the use of neem extracts as a complementary therapeutic agent that can act alongside conventional antibiotics rather than serving as a direct replacement, thereby offering a promising strategy in the development of sustainable plant-derived antimicrobial interventions.

Berberine, tea tree oil and neem extracts exemplify how traditional medicinal knowledge informs contemporary antimicrobial research. Recent studies provide experimental evidence supporting their antimicrobial activity, clarify underlying mechanisms of action, also highlight synergistic effects with conventional antibiotics. Collectively these examples illustrate the translational pathway from historical use to scientifically validated antimicrobial agents, while emphasizing the importance of standardized formulations and further clinical evaluation to fully establish their therapeutic potential.

3. CHALLENGES IN CLINICAL TRIALS OF HERBAL ANTIMICROBIALS

3.1 Standardization Challenges and Phytochemical Variability in Herbal Formulations

A major challenge in the clinical evaluation of herbal antimicrobials is the difficulty in achieving standardization, largely due to inherent phytochemical variability. Unlike conventional antibiotics, which are single, well-defined chemical entities, herbal formulations consist of complex mixtures of bioactive compounds whose composition can vary significantly across preparations. This variability complicates reproducibility, dose consistency and the interpretation of clinical outcomes (Kale et al.,2024). Phytochemical variability originates at multiple stages of herbal drug production, beginning with the biological source. Differences in plant species, subspecies and chemotypes can lead to substantial variation in antimicrobial constituents, even when materials are marketed under the same botanical name. Environmental factors such as soil composition, climate, altitude and seasonal conditions further influence secondary metabolite biosynthesis. Experimental analyses have demonstrated that variations in harvest timing and plant maturity significantly affects concentrations of alkaloids, flavonoids, terpenoids and phenolic compounds that are often responsible for antimicrobial activity (Suryawanshi et al., 2025). Post-harvest processing introduces additional challenges. Drying methods, storage duration, exposure to light and humidity can alter the chemical stability of these phytochemicals through oxidation or degradation. Furthermore, extraction procedures play a crucial role in determining the final chemical profile of herbal products. Differences in solvent polarity, extraction temperature and purification methods may selectively concentrate or exclude certain bioactive compounds, leading to marked differences in antimicrobial activity across batches derived from the same plant source (Wu et al.,2024).

Standardization becomes even more complex in polyherbal formulations which are common in traditional medicinal systems. In such preparations, therapeutic effects often arise from synergistic interactions between multiple compounds rather than from a single dominant agent. While this multicomponent nature may enhance antimicrobial activity, it poses significant challenges for quality control and the identification of reliable chemical markers. Current standardization approaches often rely on one or two marker compounds, which may not accurately reflect the formulation's overall biological activity (Wu et al.,2024). Regulatory factors further exacerbate these challenges. Herbal antimicrobials are frequently classified as dietary supplements or traditional remedies rather than pharmaceutical drugs, subjecting them to less rigorous manufacturing and quality assurance standards. This regulatory gap contributes to inconsistent formulation quality in clinical trials and limits acceptance of trial outcomes within evidence-based medicine frameworks (Li et al.,2025).

Collectively, these issues highlight the need for advanced analytical characterization, robust quality control protocols and harmonized regulatory guidelines to improve the reliability and translational potential of clinical trials involving herbal antimicrobials.

3.2 Complexities of Bioavailability and Pharmacokinetics in Plant-Derived Compounds

A core challenge in the clinical evaluation of herbal antimicrobials is the difficulty of predicting and controlling their pharmacokinetic behavior in humans. Unlike conventional antibiotics, which consist of single, well-characterized molecules; herbal products are mixtures of numerous phytochemicals that interact with biological systems in complex ways. This multicomponent nature introduces variability in absorption, distribution, metabolism and elimination that is not yet fully understood or accounted for in clinical trial design. A critical determinant of therapeutic efficacy is bioavailability, which is the proportion of an administered dose that reaches systemic circulation in active form. Many plant-derived bioactive compounds exhibit poor oral bioavailability, often due to low aqueous solubility, poor membrane permeability and extensive first pass metabolism in the liver and gut. These factors reduce the systemic concentrations of phytochemicals that exert antimicrobial effects in vitro, which in turn limits their potential to achieve clinically relevant exposure levels in humans (Parthasarathy et al.,2025; De Jesus et al., 2025).

Herbal extracts are usually a mixture of different bioactive constituents, which can influence their absorption and metabolic fate. For example, constituents may compete for uptake through intestinal transporters or undergo differential metabolism by hepatic enzymes. These complex interactions can lead to unpredictable plasma levels and tissue distribution for individual compounds, even when administered as a part of the same extract. The lack of consistent pharmacokinetic profiles for multicomponent herbal extracts adds uncertainty to dose selection and therapeutic interpretation in clinical settings (Parthasarathy et al.,2025). Another layer of complexity arises from herb-drug interactions, where herbal constituents modulate the activity of drug metabolizing enzymes or transporters. Such interactions can either enhance or reduce the systemic exposure of concurrent medications, potentially altering efficacy or toxicity. For instance, constituents in some herbal preparations have been shown to affect cytochrome P450 enzymes, which play major roles in drug metabolism. These interactions are especially very common when herbal antimicrobials are administered together with conventional antibiotics or other therapeutic agents (Rombolà et al., 2020; Gonzàlez-Trujano et al., 2024). Interindividual variability further complicates pharmacokinetic assessment. Differences in gut microbiota composition, genetic polymorphisms in metabolic enzymes, age and health status can significantly influence how phytochemicals are processed in the body. Emerging evidence indicates that gut microbes can biotransform herbal constituents into metabolites with altered activity and bioavailability compared to their parent compounds. This microbiome-mediated metabolism introduces additional variability in clinical responses to herbal interventions (Parthasarathy et al., 2025; De Jesus et al., 2025).

Current research emphasises the need for comprehensive pharmacokinetic profiling in herbal antimicrobial trials, including identification of active metabolites and excretion patterns. Techniques such as advanced in vitro ADME models and physiologically based pharmacokinetic modelling have been proposed to enhance understanding of how complex mixtures behave in vivo. Incorporating such approaches into early clinical phases can support rational dose selection and improve translation of preclinical findings to human efficacy (Parthasarathy et al., 2025). In the end, pharmacokinetics of plant-derived compounds complicate their clinical evaluation as antimicrobials. The challenges of low bioavailability, complex metabolism, herb-drug interactions and interindividual variability must be rigorously addressed to generate reliable clinical evidence and optimize therapeutic potential.

3.3 Trial Design Challenges: Placebo Development, Blinding, and Ethical Considerations

Clinical trial design for herbal antimicrobials must navigate a set of methodological challenges that are seldom encountered in conventional drug studies. Integral to rigorous evidence generation is the use of placebo controls, yet developing credible placebos for herbal products is inherently difficult because many botanicals possess distinctive sensory characterises such as taste, color and odor. This sensory complexity makes it challenging to produce a control substance that is indistinguishable from the active intervention, which in turn undermines effective blinding and increases the risk of performance and detection bias in randomized trials. Contemporary analyses of trial methodologies indicate that blinding is frequently inadequately implemented in herbal medicine RCTs, partly because correctly matched placebos require advanced formulation preparations to ensure sensory equivalence without introducing pharmacological effects. Without such measures, the credibility of comparative conclusions is compromised as participants and investigators may deduce treatment allocation based on sensory cues (Koonrungsesomboon et al., 2024). Blinding must extend beyond merely concealing allocation; it must also account for participants' expectations and the psychological dimensions of placebo response. Systemic evaluations of placebo responses in herbal clinical research highlight that patient belief in treatment efficacy can influence outcomes independently of pharmacological action and that poorly matched placebos exacerbate this (Huang et al., 2024). The psychological component of response together with inadequate placebo formulation can inflate perceived efficacy in intervention arms and distort interpretation of true therapeutic benefit. Thus, effective blinding requires not only meticulous physical matching of placebos but also systematic assessment of blinding integrity where participants are queried about their assumptions regarding treatment assignment to document whether blinding was successful. In addition to sensory and psychological considerations, ethical dimensions permeate trial design. Ethical review boards and researchers must balance the scientific value of placebo-controlled trials with participant safety and clinical equipoise. In infectious disease contexts or other conditions where effective standard therapies exist, withholding proven treatments in favor of placebo may contravene ethical principles of beneficence and non-maleficence. Accordingly, recent guidance suggests that placebo controls are justified only when no established therapy exists or when add-on designs, where herbal products are tested alongside standard care, are employed to avoid harm to participants (Koonrungsesomboon et al., 2024). Moreover, the cultural context of herbal medicine use is particular with interventions originating from indigenous or ethnomedical traditions which necessitates careful ethical engagement with participant communities to ensure that informed consent, risk communication and respect for local practices are upheld.

Compounding these design challenges are broader concerns about reporting quality and methodological transparency in herbal clinical trials. Evaluations of randomized controlled trials for Chinese herbal medicine, for example, reveal substantial deficiencies in adherence to established reporting standards such as the CONSORT extension for Chinese Herbal Medicine Formulas (CONSORT-CHM). Recent analysis specifically of RCTs investigating herbal medicine formulas for vestibular migraine highlights that many trials inadequately report CONSORT and CONSORT-CHM items, including allocation concealment, randomization procedures and blinding details indicating a high risk of bias when such items are underreported, making it difficult to integrate herbal antimicrobial evidence into practice (Wang et al., 2025). Several dental herbal mouthwash trials successfully demonstrated bacterial count reduction but could not be generalized to systemic infection treatment due to localized exposure conditions. Conversely, trials attempting systemic indications frequently failed to demonstrate superiority despite promising in vitro antimicrobial activity, highlighting the mismatch between laboratory models and achievable human pharmacokinetics.

Overall, addressing the combined challenges of placebo validity, effective blinding, ethical integrity and rigorous reporting is essential to advancing the scientific credibility of herbal antimicrobial trials. Without systematic improvements in these design domains, clinical evidence will remain fragmented and less persuasive within evidence-based healthcare frameworks.

3.4 Data Integrity Issues and Reporting Biases in Herbal Clinical Research

Data integrity and transparent reporting are foundational to credible clinical evidence, yet contemporary analyses of randomized controlled trials (RCTs) involving herbal medicine consistently expose significant shortcomings in these areas. A critical concern is the incomplete reporting of key methodological elements, which undermines the ability to assess the internal validity and reproducibility of trial results. For example, evaluations of RCTs in traditional herbal formulas highlight that adherence to international reporting guidelines such as the Consolidated Standards of Reporting Trials (CONSORT) and its extension for Chinese herbal medicine formulas (CONSORT-CHM) is often inadequate with many studies failing to disclose essential details such as allocation concealment, sample size calculations and blinding procedures (Wang et al., 2025). Such gaps in methodological transparency contribute to unclear or high risk of bias classifications when assessed using standard tools and this in turn weakens confidence in reported outcomes. In the context of trials for herbal medicine formulas in vestibular migraine, the mean reporting scores for CONSORT and CONSORT-CHM items were suboptimal with critical components such as randomization sequence generation and blinding description frequently absent from published reports, factors that correlate with increased risk of bias and limited interpretability of results (Wang et al., 2025).

In addition to internal methodological reporting drawbacks, publication and selective outcome reporting biases are persistent issues. Studies with negative or non-significant outcomes are less likely to be published or may be presented with incomplete data, skewing the evidence base and inflating perceived efficacy. Although specific analyses of publication bias in recent herbal antimicrobial trials remain sparse, broader assessments of herbal RCT reporting underscore the need for routine trial registration, complete disclosure of outcomes (including adverse events) and adherence to CONSORT-CHM guidelines to improve transparency and reduce bias (Li et al.,2024).

Improving data integrity in herbal clinical research requires both rigorous methodological reporting and systemic efforts to publish all results irrespective of direction. Without these efforts, the evidence for herbal antimicrobials will remain vulnerable to bias and limited in its utility for clinical decision making.

4. GLOBAL REGULATORY LANDSCAPES FOR HERBAL ANTIMICROBIALS

4.1 Regional Regulatory Approaches: EU/EMA, US/FDA, India/AYUSH, and China/TCM Frameworks

Regulatory governance of herbal antimicrobials reflects the intersection of scientific medicine, traditional knowledge systems and public health priorities which result in substantial regional variability. In the European Union, herbal antimicrobials are regulated as medicinal products under Directive 2004/24/EC, which established the Traditional Herbal Medicinal Products (HMPC) Directive. Scientific evaluation and harmonization are overseen by the Committee on Herbal Medicinal Products under the European Medicines Agency(EMA). The HMPC develops EU-wide herbal monographs and lists entries that define acceptable indications, dosage forms, safety considerations and quality requirements. Importantly, while full clinical efficacy data are not mandatory under the traditional use registration route, products must demonstrate at least 30 years of medicinal use, including 15 years within the EU, ensuring a minimum threshold of plausibility and safety (European Medicines Agency, 2025).

In contrast, the United States has a product-centric rather than indication-centric approach. Most herbal antimicrobials are marketed as dietary supplements under the Dietary Supplement Health and Education Act, which does not require pre-market demonstration of efficacy or antimicrobial activity. The U.S. Food and Drug Administration (FDA) retains authority over safety, labelling and post-market surveillance but evaluates efficacy only when a product is developed as a botanical drug (Food and Drug Administration, 2020). Although the FDA's botanical drug development guidance provides a structured pathway for herbal medicines to achieve drug approval, the high evidentiary burden and cost have limited uptake, leaving many herbal antimicrobials outside formal antimicrobial regulatory oversight.

India represents a hybrid regulatory model that formally institutionalizes traditional medicine systems. Herbal antimicrobials are governed primarily by the Ministry of AYUSH, which regulates Ayurveda, Siddha and Unani drugs under the Drugs and Cosmetics Act. Standards are codified in official pharmacopoeias and formularies with increasing emphasis on quality control, good manufacturing practices and pharmacovigilance (Ministry of AYUSH, Government of India, 2025). Simultaneously, India has introduced the phytopharmaceutical category under the Central Drugs Standard Control Organisation (CDSCO), requiring defined chemical characterization and clinical evidence, signaling a gradual shift toward globally aligned regulatory expectations (Central Drugs Standard Control Organisation, 2025).

In China, herbal antimicrobials are embedded within the national healthcare system through Traditional Chinese Medicine (TCM) regulation. The National Medicinal Products Administration (NMPA) recognizes TCM formulations as legitimate therapeutic agents, allowing approval pathways that integrate classical medical texts, real world clinical use and increasingly stringent quality and safety requirements (National Medicinal Products Administration [NMPA], 2025). This integrated framework enables wide clinical adoption while progressively aligning TCM regulation with modern pharmaceutical standards.

4.2 Comparative Analysis of Regional Disparities and Harmonization Challenges

Comparative analysis reveals that the most significant regulatory differences lie in evidence thresholds, product classification and therapeutic claim authorization. The EU and China formally recognize the traditional use as a valid evidentiary standard for market authorization whereas the U.S. largely separates traditional products from drug regulation unless explicit therapeutic claims are pursued (European Medicines Agency, 2025; Food and Drug Administration, 2020). India occupies an intermediate position where traditional legitimacy is preserved while scientific validation is increasingly encouraged (Ministry of AYUSH, Government of India, 2025).

These differences pose challenges for regulatory harmonization and global antimicrobial strategies. A herbal antimicrobial recognized as a medicinal product in the EU may be restricted to non-therapeutic claims in the U.S., limiting its role in antimicrobial stewardship. Conversely, TCM products approved in China often face regulatory resistance in Western markets due to divergent expectations for mechanism-based evidence and randomized clinical trial data (National Medicinal Products Administration [NMPA], 2025).

The following comparative regulatory table visually summarizes key global differences in regulatory frameworks.

Table 1. Comparative regulatory frameworks governing herbal antimicrobials across regions

Parameter	European Union	United States	India	China
Primary authority	EMA (HMPC)	FDA	AYUSH, CDSCO	NMPA
Legal classification	Herbal medicinal product	Dietary supplement / botanical drug	ASU drug / phytopharmaceutical	TCM product
Basis for approval	Traditional use or bibliographic evidence	Safety (supplements) or full clinical trials (drugs)	Traditional texts with emerging clinical data	Traditional theory with clinical practice
Pre-market efficacy requirement	Not mandatory for traditional use	Not required for supplements	Variable; increasing for phytopharmaceuticals	Context-dependent
Integration into healthcare	Regulated but limited	Limited	National healthcare recognition	Fully integrated

This highlights the need for shared regulatory principles to enable cross border evaluation and responsible deployment of herbal antimicrobials.

4.3 Role of WHO Guidelines and International Collaborations in Regulatory Convergence

The World Health Organization (WHO) serves as a central coordinating body promoting convergence in herbal medicine regulation through policy guidance, technical frameworks and international cooperation. WHO's Traditional, Complementary and Integrative Medicine programme emphasizes strengthening national regulatory capacity, improving product quality and enhancing pharmacovigilance for herbal medicines, including those with antimicrobial properties (World Health Organization [WHO], 2019). A cornerstone of these efforts is the International Regulatory Cooperation for Herbal Medicine (IRCH) initiative, which facilitates dialogue among regulatory authorities to harmonize safety standards, terminology and quality benchmarks. Recent WHO guidance highlights the importance of good agricultural and collection practices, standardized manufacturing processes and post-marketing surveillance systems to ensure public safety while respecting traditional medical knowledge (International Regulatory Cooperation for Herbal Medicines (IRCH), 2025).

Although WHO guidelines are non-binding, they provide a shared reference framework that enables gradual alignment of national regulations. In the context of rising antimicrobial resistance (AMR), such convergence is particularly important, as it supports the responsible evaluation and integration of herbal antimicrobials into global health strategies without compromising safety or scientific credibility.

Safety evaluation presents a distinct challenge for herbal antimicrobials because traditional use does not always predict safety under modern dosing patterns or combined therapy with antibiotics. Adverse effects may arise from intrinsic phytochemical toxicity, contamination, misidentification of plant material, or herb–drug interactions affecting cytochrome P450 enzymes and drug transporters. Unlike conventional antimicrobials, post-marketing surveillance systems for herbal products remain inconsistent across jurisdictions, resulting in underreporting of adverse reactions. Pharmacovigilance programs integrated within regulatory frameworks are therefore critical for detecting rare toxicities, monitoring long-term exposure and assessing interaction risks when herbal antimicrobials are used alongside standard anti-infective therapy. (Rombolà et al., 2020; González-Trujano et al., 2024)

5. CASE STUDIES: SUCCESSES, FAILURES, AND LESSONS LEARNED

5.1 Successful Herbal Clinical Trials: Evidence-Based Validation Examples

Chismirina et al. (2025) reported a controlled clinical study evaluating a *Syzygium cumini*-based mouthwash for its antibacterial efficacy against *Streptococcus mutans* which is a key etiological agent in dental caries. The study demonstrated a statistically significant reduction in salivary *Streptococcus mutans* counts following regular use of the herbal formulation, showcasing a measurable in vivo antimicrobial effect within the oral activity. The strength of this trial lies in its use of a microbiological endpoint rather than symptom-based outcomes, making it one of the clearer recent examples of clinical antimicrobial assessment for a plant-derived product.

Biria et al. (2022) conducted a randomized clinical trial assessing a herbal toothpaste containing bamboo salt. The study did not demonstrate superiority of the herbal formulation over conventional toothpaste in reducing *Streptococcus mutants* and *Lactobacillus* counts but both groups showed significant reductions in bacterial load over the study period. These findings suggest that herbal formulations can perform comparably to established products in maintaining oral microbial control, even if they do not outperform conventional formulations.

Hasanpour et al. (2025) reported a randomized double-blind, placebo controlled trial investigating an eleven-herb syrup in patients with chronic cough. While the primary outcomes focused on symptom improvement, the authors discussed the relevance of infection-associated mechanisms underlying chronic cough and suggested that the observed clinical benefits may reflect modulation of microbial or inflammatory pathways. Although this study does not provide direct microbiological clearance data, it represents a clinically successful trial in an infection-related condition, illustrating how herbal formulations are currently validated primarily through symptomatic and functional outcomes rather than direct antimicrobial endpoints.

These studies demonstrated that recent herbal clinical trials more than often rely on localized microbial environments or infection-associated symptom contexts rather than through systemic antibacterial efficacy. These highlight the importance of endpoint selection and reinforce the role of herbal antimicrobials as complementary or preventive interventions rather than standalone antibacterial therapies.

5.2 Trials With Inconclusive or Negative Outcomes: Methodological Reflections

Alongside these examples of partial clinical validation, a substantial proportion of herbal clinical trials yield inconclusive or ambiguous results. Methodological analyses indicate that such outcomes frequently arise from structural weaknesses in trial design, reporting and endpoint selection rather than definitive evidence of inefficacy.

Koonrungsesomboon et al. (2024) conducted a comprehensive methodological review of herbal medicine clinical trials and identified recurrent design limitations, including inadequate randomization procedures, insufficient blinding, poorly defined primary outcomes and limited justification of sample size. These weaknesses are particularly problematic in trials aiming to assess antimicrobial effects where variability in formulation composition and dosing further complicates reproducibility. The authors emphasized that without rigorous methodological frameworks even biologically active herbal products may fail to demonstrate clear clinical effects.

Reporting quality represents an additional barrier to interpretation. Wang et al. (2025) evaluated randomized controlled trials of herbal medicine formulas and reported widespread deficiencies in adherence to established reporting standards including incomplete descriptions of allocation concealment, blinding methods and statistical analysis plans. Such reporting gaps increase the risk of bias and limit confidence in both positive and negative findings. Importantly poor reporting also impedes secondary analysis and evidence synthesis, contributing to persistent uncertainty in the field.

At a broader evidence level, Salm et al. (2023) reviewed the clinical benefits of herbal medicines across infectious and non-infectious conditions and concluded that while preclinical antimicrobial activity is frequently translated into convincing clinical outcomes remains inconsistent. The review also highlighted heterogeneity in formulations, outcomes and trial quality as key reasons why many studies fail to provide conclusive results. This disconnect between laboratory results and clinical performance emphasizes the importance of aligning experimental models with clinically relevant endpoints.

Collectively these methodological reflections indicate that inconclusive or negative outcomes in herbal clinical trials often reflect limitations in study design, reporting transparency and endpoint relevance rather than the absence of antimicrobial or infection-modulating potential. Addressing these challenges is essential for generating interpretable evidence and for accurately distinguishing true therapeutic limitations from methodological failure.

5.3 Regulatory Case Studies: Approval Journeys of Notable Herbal Antimicrobials

When herbal products with proposed antimicrobial activity enter regulatory systems their trajectories reveal a consistent pattern that differs markedly from conventional antimicrobial drug development. Rather than advancing toward formal antimicrobial approval, such products are mostly redirected into regulatory categories that limit their therapeutic scope. This outcome reflects not regulatory resistance but the evidentiary constraints discussed earlier in this chapter.

In the European Union, herbal products traditionally associated with antimicrobial or antiseptic use commonly achieve market access but without proper recognition as antimicrobial therapies. Products derived from botanicals such as Thymus vulgaris and Eucalyptus globulus are permitted for use based on documented traditional application and safety yet they are restricted to supportive or symptomatic indications. As a result, although these products may be used in infection-related contexts, they cannot be positioned as treatments for microbial eradication. This illustrates a recurring regulatory outcome i.e. access without antimicrobial endorsement (European Medicines Agency, 2025).

A different but equally limiting outcome is observed in the United States. Herbal products that initially pursue antimicrobial positioning through the botanical drug pathway rarely progress beyond early development. In practice, most are reclassified as dietary supplements once they fail to generate drug-level clinical efficacy results. This reclassification effectively removes them from the antimicrobial drug development landscape regardless of preclinical antimicrobial activity. The consequence is not outright rejection but loss of therapeutic positioning, as antimicrobial claims cannot be sustained under supplement regulations (U.S. Food and Drug Administration [FDA], 2020).

India's regulatory environment produces a third, intermediate outcome. Although the phytochemical framework offers a pathway for plant-derived products to be regulated as modern drugs, antimicrobial phytopharmaceutical approvals remain extremely limited. Many products with antimicrobial intent ultimately remain within traditional medicine categories due to unresolved challenges in standardization, dose definition and clinical trial design. The result is regulatory stagnation rather than progression highlighting the gap between policy innovation and practical translational success (Ministry of AYUSH, Government of India, 2025).

Across these settings, the regulatory fate of herbal antimicrobials is remarkably consistent. Products are seldom rejected outright but they are rarely endorsed as antimicrobial therapies. Instead they are repositioned into categories emphasizing safety, tradition or general health support. This outcome mirrors the clinical evidence landscape described earlier in the section, where antimicrobial effects are localized, indirect or methodologically constrained. Consequently regulatory systems prioritize risk management over antimicrobial validation.

Taken together, these case studies illustrate a recurring translational pattern: positive outcomes occur primarily in localized infections with microbiological endpoints, whereas systemic indications frequently yield neutral or inconclusive results. The gap between laboratory antimicrobial activity and clinically demonstrable infection clearance therefore reflects differences in pharmacokinetics, endpoint selection and regulatory expectations rather than a simple absence of biological effect.

6. BRIDGING TRADITIONAL WISDOM WITH SCIENTIFIC RIGOUR

6.1 Integrating Ethnobotanical Intelligence With Modern Screening Tools and Omics/AI Technologies

Ethnobotanical knowledge represents one of the oldest systematic records of human interaction with medicinal plants, built through centuries of empirical observation and cultural transmission. In the context of antimicrobial discovery, such knowledge offers an important starting point for identifying plant species and preparations with possible biological activity. However, traditional use alone is sufficient to meet modern scientific and regulatory expectations. Bridging ethnobotany with contemporary screening technologies is therefore essential to convert historical insights into testable, reproducible biomedical evidence (Shah, 2024). Modern high-throughput screening approaches allow ethnobotanically informed candidates to be evaluated efficiently for antimicrobial activity. Instead of screening large, chemically naive libraries, researchers can prioritize plants with documented infection-related use, improving hit rates and reducing redundancy. Advances in metabolomics now enable comprehensive chemical fingerprinting of complex plant extracts, allowing researchers to map bioactive chemical space and identify dominant or unique secondary metabolites linked to antimicrobial effects (Chele et al., 2025). When combined with genomics and transcriptomics, these approaches also facilitate the identification of biosynthesis pathways responsible for key antimicrobial compounds.

Figure 3. Translational pathway of herbal antimicrobials from traditional use to regulated clinical application

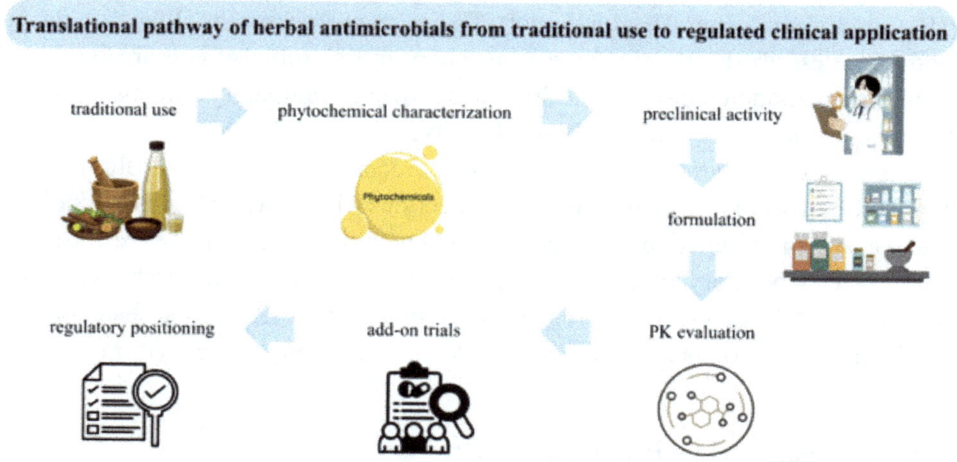

Artificial intelligence (AI) and Machine learning (ML) further enhance this integrative pipeline. AI-assisted dereplication tools can rapidly distinguish known compounds from potentially novel ones within metabolomic datasets, preventing repeated rediscovery of previously known molecules (Gangwal & Lavecchia, 2025). Predictive models trained on chemical structure, bioactivity data and ethnobotanical metadata can help rank extracts for specific antimicrobial assays or predict potential toxicity and pharmacokinetic limitations early in development. Such approaches increase efficiency and reduce the likelihood of late-stage failure. Importantly, omics and AI tools also allow researchers to address one of the major challenges in herbal antimicrobial i.e. variability. By correlating chemical fingerprints with bioactivity outcomes, it becomes possible to define quality markers and acceptable chemical ranges for standardized extracts. This not only strengthens experimental reproducibility but also aligns early discovery with downstream regulatory expectations (Chele et al., 2025). Thus, the integration of ethnobotanical intelligence with omics and

AI does not replace traditional knowledge but in turn refines and operationalizes it within a scientifically rigorous framework.

6.2 Collaborative Models: Academia, Industry, Indigenous Custodians, and Ethical Benefit-Sharing

While technological integration strengthens scientific rigour, it alone cannot ensure responsible and sustainable development of herbal antimicrobials. Because ethnobotanical knowledge is often held by indigenous and local communities, ethical collaboration and benefit-sharing are central to bridging traditional wisdom with scientific rigour. International legal and ethical frameworks increasingly emphasize that communities contributing biological resources and traditional knowledge must be recognized as active partners rather than passive sources (Morgera, 2024). Effective collaborative models typically involve four stakeholder groups- academic researchers, industry partners, indigenous custodians and regulatory or policy institutions. Academia contributes expertise in discovery science, omics analysis and early stage validation. Industry provides capabilities for formulation, scale up, quality assurance and navigation of regulatory pathways. Indigenous custodians contribute contextual knowledge regarding plant identity, harvesting practices, preparation methods and cultural significance. Regulatory bodies establish the boundaries within which these collaborations operate. Ethical benefit-sharing mechanisms are essential for ensuring long-term sustainability. These may include monetary benefits, co-authorship, technology transfer, infrastructure development or capacity-building initiatives in source communities. Evidence suggests that benefit-sharing is not merely an ethical obligation but also enhances scientific quality by encouraging trust and enabling access to accurate, context-specific knowledge that affects chemical composition and reproducibility (Shah, 2024).

Collaborative agreements must also address intellectual property rights transparently. Ambiguity in ownership or downstream commercialization rights can undermine partnerships and discourage participation. Clear governance structures, prior informed consent and mutually agreed terms are therefore essential components of any research program involving traditional medicinal knowledge (Morgera, 2024). In the antimicrobial context, such collaborations also intersect with sustainability. Overharvesting of medicinal plants poses ecological risks and threatens both biodiversity and community livelihoods. Integrating conservation planning, cultivation strategies and traceability systems into collaborative research models aligns ethical considerations with long-term supply security. Thus, bridging traditional wisdom with scientific rigour ultimately depends as much on social and ethical architecture as on technological sophistication.

7. TOWARD HARMONIZED GLOBAL FRAMEWORKS

7.1 Convergence in Quality Standards, Clinical Guidelines, and Policy Proposals

Efforts toward harmonized global frameworks for herbal antimicrobials are best understood through ongoing international initiatives rather than completed regulatory success stories. One of the most significant examples is the World Health Organization's Global Traditional Medicine Strategy (2025-2034), which represents a coordinated attempt to align quality standards, evidence expectations and governance principles for traditional and herbal medicines across member states. Although not limited to antimicrobials the strategy explicitly addresses antimicrobial resistance (AMR) and calls for improved integration of traditional medicine into international healthcare systems under evidence based and safety focused criteria. A solid outcome of this initiative is WHO's promotion of harmonized pharmacopoeial standards and quality benchmarks, including the development and cross referencing of monographs, analytical fingerprints and containment thresholds. These efforts aim to reduce regulatory fragmentation by enabling countries to rely on shared quality data rather than duplicating assessments while implementation remains uneven. This program functions as a global pilot for convergence in herbal product regulation (WHO Global Traditional Medicine Strategy (2025-2034)., 2024).

At the regional level, the European Medicines Agency's Committee on Herbal Medicinal Products (HMPC) provides another instructive case through development of EU-wide herbal monographs, the HMPC has effectively harmonised quality and safety assessment for numerous medicinal plans across member states. Although antimicrobial efficacy claims remain restricted, the monograph system demonstrates how consensus based scientific evaluation can support regulatory alignment across jurisdictions (European Medicines Agency, 2025). This model is increasingly cited in policy discussion as a template for broader international harmonisation.

In India, the Pharmacopoeial Commission for Indian Medicine and Homeopathy (PCIM&H) under Ministry of AYUSH represents a national level harmonisation effort with global relevance. Recent initiatives focus more on standardising, raw materials, formulations and analytical methods for Ayurvedic and herbal products with the explicit aim of improving international acceptability and export potential. While antimicrobial indications are not yet a primary focus, this program showcases how national pharmacopoeial alignment serves as a foundation for future multinational convergence (Pharmacopoeia Commission for Indian Medicine & Homeopathy., 2025).

Together these initiatives demonstrate that convergence in quality standards and policy frameworks is already underway, although in progress. Rather than producing immediate antimicrobial approvals these efforts lay the structural groundwork required for future harmonized clinical evaluation and regulatory decision making.

7.2 International Cooperation, Sustainable Sourcing, and One Health Principles in Regulation

International cooperation becomes particularly visible when harmonization is examined through the lens of sustainability and One Health. A key example is the WHO-FAO-UNEP-WOAH's One Health Joint Plan of Action, which explicitly identifies antimicrobial resistance as a cross-sectional challenge involving human, animal health, agriculture and the environment. Although herbal antimicrobials are not the central focus, the framework provides a policy space in which plant-derived antimicrobials can be evaluated for their environmental and resistance related impacts (One Health Joint Plan of Action, 2025).

At the regulatory level, the EMA's One Health approach to antimicrobial resistance (AMR) serves as a regional case study of coordinated oversight. EMA integrates human and veterinary medicine data, environmental risk assessment and antimicrobial stewardship principles into regulatory decision-making. This approach implicitly affects herbal antimicrobials intended for veterinary or agricultural use, ensuring that their deployment does not shift resistance risks across sectors (European Medicines Agency, One Health, 2025)

Sustainable sourcing and ethical governance are most clearly addressed through the Convention on Biological Diversity (CBD) and the Nagoya Protocol on Access and Benefit-Sharing. Numerous countries have implemented national ABS frameworks that regulate access to genetic resources and require benefit-sharing with indigenous and local communities. While not specific to antimicrobials, these frameworks directly influence herbal drug development pipelines by shaping how plant materials and traditional knowledge can be accessed, studied and commercialized (Secretariat of the Convention on Biological Diversity, 2025).

India's implementation of biodiversity management committees (BMCs) and local-level benefit-sharing mechanisms offers a practical illustration of how ABS principles intersect with herbal medicine research. These mechanisms are increasingly referenced in policy discussions as essential safeguards against biopiracy and as tools for ensuring equal participation in global herbal innovation ecosystems (Morgera, 2024). Taken together these initiatives function as living case studies of international cooperation rather than finished solutions. They demonstrate how harmonized regulation of herbal antimicrobials is emerging through coordinated policy action, sustainability governance and One Health integration even in the absence of fully approved herbal antimicrobial drugs.

8. FUTURE DIRECTIONS AND CONCLUDING REFLECTIONS

8.1 Emerging Research Frontiers, Next-Generation Trial Designs, and 2030 Policy Outlook

The future of herbal antimicrobial research lies not in replicating conventional antibiotic discovery pathways, but in developing approaches tailored to the biological complexity and contextual use of plant-derived products. Advances in multi omics technologies and artificial intelligence (AI) are expected to further refine candidate prioritization, enabling researchers to move beyond broad screening towards mechanism-informed selection of botanicals with reproducible antimicrobial or resistance modulating activity (Chele et al., 2025; Gangwal & Lavecchia, 2025). These tools will be particularly valuable for identifying synergistic interactions within complex extracts and for defining quality markers essential for standardization.

Next generation clinical trial designs are also likely to reshape the evidence landscape scene, rather than large, late phase efficacy trials aimed at systemic antimicrobial replacement, future studies are expected to focus more on localized infections adjunctive theory and preventive or resistance modulating roles. Adaptive and platform trial designs, as well as add-on studies evaluating herbal products alongside standard antibiotics, offer pragmatic routes to generate clinically relevant data while also maintaining ethical and methodological rigour (Koonrungsesomboon et al., 2024). Incorporation of microbiological endpoints, validated surrogate markers and real world effectiveness measures will further ensure translational relevance.

From a policy perspective, the period leading up to 2030 is likely to be shaped by implementation of the World Health Organisation's Global Traditional Medicine Strategy (2025 to 2034). This framework emphasizes evidence proportionality, quality assurance, pharmacovigilance and integration of traditional medicine into national healthcare systems under clearly defined regulatory boundaries (WHO Global Traditional Medicine Strategy (2025-2034)., 2024). While the strategy does not promise rapid approval of herbal antimicrobials as drugs, it establishes conditions under which ethically sourced, standardized and clinically evaluated products can contribute meaningfully to public health goals, including antimicrobial stewardship.

8.2 Integrative Innovation for Planetary Health Under the One Health Paradigm

Ultimately, the value of herbal antimicrobials should be assessed not solely through the lens of drug substitution, but within a broader One Health framework that links human, animal health and environmental sustainability together. Antimicrobial resistance is a system level challenge and incremental reductions in antimicrobial pressure can be achieved through adjunctive therapies, preventive interventions or localized antimicrobial applications can also collectively yield substantial public health benefits.

Integrative innovation in this space requires alignment across scientific, regulatory and ethical domains. Ethnobotanical knowledge, when combined with omics based discovery, AI-assisted prioritization and rigorous clinical evaluation can further produce interventions that are both culturally grounded and scientifically credible. At the same time, sustainable sourcing, benefit-sharing and biodiversity conservation must be treated as internal components of innovation rather than external constraints. Failure to do so risks shifting health gains at the expense of ecological and social systems.

Herbal antimicrobials are therefore best positioned as complementary tools within a diversified antimicrobials strategy. Their strengths, clinical diversity, multimodel mechanisms and deep cultural integration align well with preventive care, topical use and resistance modulating roles. Recognizing these strengths while avoiding overextension into claims unsupported by evidence is essential for maintaining scientific integrity and public trust.

Future progress requires alignment between scientific design and realistic therapeutic positioning. Herbal antimicrobials should primarily be evaluated as adjunctive, preventive or localized therapies rather than systemic replacements for antibiotics. Clinical trials should prioritize microbiological endpoints, standardized formulations and add-on designs. Regulatory frameworks may benefit from intermediate approval categories recognizing infection-modifying activity without requiring full drug equivalence to conventional antimicrobials.

In conclusion, the path forward for herbal antimicrobials is neither a return to uncritical traditionalism, nor an attempt to force botanical products into unsuitable pharmaceutical molds. Instead, progress depends on integrative frameworks that respect traditional knowledge and apply modern scientific rigour also aligning innovation with planetary health priorities. Within such frameworks, herbal antimicrobials can contribute meaningfully if not modestly to the global response to antimicrobial resistance in the decade ahead.

REFERENCES

Allied Market Research. https://www.alliedmarketresearch.com/. (2024, September). *Herbal Medicine Market, by Form (Tablets and Capsules, Powder, Others), by Category (Pharmaceutical and Nutraceutical, Personal Care and Beauty Products, Others), by Distribution Channel (Online Channel, Offline Channel) : Global Opportunity Analysis and Industry Forecast, 2024 - 2033*. Allied Market Research. https://www.alliedmarketresearch.com/herbal-medicine-market-A324087

Bhattacharya, T., Saha, R. P., Majumder, R., & Ghosh, S. (2024). Investigating the antimicrobial activity of neem and clove extract on biofilm-producing oral microflora. *Journal of Experimental Biology and Agricultural Sciences*, *12*(3), 511–521. DOI: 10.18006/2024.12(3).511.521

Biria, M., Rezvani, Y., Roodgarian, R., Rabbani, A., & Iranparvar, P. (2022). Antibacterial effect of an herbal toothpaste containing Bamboo salt: A randomized double-blinded controlled clinical trial. *BMC Oral Health*, *22*(1), 193. DOI: 10.1186/s12903-022-02224-z PMID: 35590307

Butler, M. S., Capon, R. J., Blaskovich, M. T., & Henderson, I. R. (2025). Natural product-derived compounds in clinical trials and drug approvals. *Natural Product Reports*. Advance online publication. DOI: 10.1039/D5NP00031A PMID: 40923383

Central Drugs Standard Control Organization Directorate General of Health Services Ministry of Health & Family Welfare Government of India. (2024). https://cdsco.gov.in/opencms/opencms/en/Home/

Chele, K. H., Piater, L. A., van der Hooft, J. J. J., & Tugizimana, F. (2025). Bridging Ethnobotanical Knowledge and Multi-Omics Approaches for Plant-Derived Natural Product Discovery. *Metabolites*, *15*(6), 362. DOI: 10.3390/metabo15060362 PMID: 40559386

Chismirina, S., Andayani, R., Afrina, A., Nafisah, Q., Sungkar, S., & Rezeki, S. (2025). Antibacterial Efficacy of Syzygium cumini-Based Mouthwash against Streptococcus mutans: A Laboratory and Clinical Study. *Journal of International Oral Health : JIOH*, *17*(2), 106–114. DOI: 10.4103/jioh.jioh_81_24

Control of Antimicrobial Resistance. (2025, November 3). https://www.who.int/europe/initiatives/one-health/control-of-antimicrobial-resistance

Daraghmeh, D. N., AbuIriban, R. W., Nawawreh, N., Abuamro, A. M., Alassar, M. M., Daraghma, S. N., Alhajahmed, N. M., & Thandar, Y. (2025). Advancements in alternative approaches to address antimicrobial resistance in bacterial pneumonia: A comprehensive review. *Frontiers in Microbiology*, *16*, 1704931. DOI: 10.3389/fmicb.2025.1704931 PMID: 41347244

De Jesus, P. C., Silva, D. M. R. R., Moura, P. H. M., Gopalsamy, R. G., Silva, E. E. D., Barreto, M. D. S., Santos, R. S., Martins, A. Y. S., De Freitas Almeida, A. G., Corrêa, A. K. S., Da Mota Santana, L. A., Hariharan, G., Guimarães, A. G., & Borges, L. P. (2025). The In Vitro Pharmacokinetics of Medicinal Plants: A review. *Pharmaceuticals (Basel, Switzerland)*, *18*(4), 551. DOI: 10.3390/ph18040551 PMID: 40283986

Debnath, S., Das, M., Mondal, S., Sarkar, B. K., & Babu, G. (2025). Neem (Azadirachta indica A. Juss): A multifaceted tree & an elixir in the traditional system of Indian medicine. *Discover Plants*, *2*(1), 156. Advance online publication. DOI: 10.1007/s44372-025-00196-2

Dubale, S., Usure, R. E., Mekasha, Y. T., Hasen, G., Hafiz, F., Kebebe, D., & Suleman, S. (2025). Traditional herbal medicine legislative and regulatory framework: A cross-sectional quantitative study and archival review perspectives. *Frontiers in Pharmacology*, *16*, 1475297. DOI: 10.3389/fphar.2025.1475297 PMID: 39950109

Duda-Madej, A., Viscardi, S., Bazan, H., & Sobieraj, J. (2025). Exploring the role of berberine as a molecular disruptor in antimicrobial strategies. *Pharmaceuticals (Basel, Switzerland)*, *18*(7), 947. DOI: 10.3390/ph18070947 PMID: 40732238

Ellward, G. L., Binda, M. E., Dzurny, D. I., Bucher, M. J., Dees, W. R., & Czyż, D. M. (2024). A Screen of Traditional Chinese Medicinal Plant Extracts Reveals 17 Species with Antimicrobial Properties. *Antibiotics (Basel, Switzerland)*, *13*(12), 1220. DOI: 10.3390/antibiotics13121220 PMID: 39766610

Fan, X., Qadeer, A., Asiri, M., Alzahrani, F. M., Alzahrani, K. J., Alsharif, K. F., Khan, M. Z., & Jiang, X. (2025). Traditional Chinese medicine and plant-derived bioactive compounds as sustainable alternatives to antibiotics in bovine mastitis: A review. *Frontiers in Veterinary Science*, *12*, 1642647. DOI: 10.3389/fvets.2025.1642647 PMID: 41059460

Gangwal, A., & Lavecchia, A. (2025). Artificial Intelligence in Natural Product Drug Discovery: Current Applications and Future Perspectives. *Journal of Medicinal Chemistry*, *68*(4), 3948–3969. DOI: 10.1021/acs.jmedchem.4c01257 PMID: 39916476

González-Trujano, M. E., Ventura-Martínez, R., Silveira, D., & Déciga-Campos, M. (2024). Editorial: Pharmacological interaction between drugs and medicinal plants, Volume II. *Frontiers in Pharmacology*, *15*, 1372366. DOI: 10.3389/fphar.2024.1372366 PMID: 38352146

Hasanpour, M., Shahsavari, K., Kazemizadeh, H., Rad, N. K., Sajedi, A., Nouri, A., & Khanavi, M. (2025). Investigation of the efficacy and safety of a syrup composed of 11 herbal medicines in improving chronic and subacute cough in patients: A preclinical study and a randomized, Double-Blind, Placebo-Controlled clinical trial. *Jundishapur Journal of Natural Pharmaceutical Products*, *20*(2). Advance online publication. DOI: 10.5812/jjnpp-159021

Herbal medicinal products. (2025, December 12). Public Health. https://health.ec.europa.eu/medicinal-products/herbal-medicinal-products_en

International Regulatory Cooperation for Herbal Medicines (IRCH). (2025). https://www.who.int/initiatives/international-regulatory-cooperation-for-herbal-medicines

Kale, S. R., Kedar, T. R., Sapkal, S. S., & Gandhi, N. K. (2024). A review on problems associated with clinical trials of Ayurvedic medicine. *International Journal of Novel Research and Development*, *9*(11), a500–a501. https://www.ijnrd.org/papers/IJNRD2411052.pdf

Koonrungsesomboon, N., Sakuludomkan, C., Takuathung, M. N., Klinjan, P., Sawong, S., & Perera, P. K. (2024). Study design of herbal medicine clinical trials: A descriptive analysis of published studies investigating the effects of herbal medicinal products on human participants. *BMC Complementary Medicine and Therapies*, *24*(1), 391. DOI: 10.1186/s12906-024-04697-7 PMID: 39516879

Li, Y., Xu, Z., Du, P., Gao, J., Wang, S., Pang, X., Ren, C., Liu, Y., & Zhang, C. (2025). Methodological challenges in pilot trials of herbal medicine: Barriers to evidence-based practice. *Journal of Clinical Epidemiology*, *182*, 111754. DOI: 10.1016/j.jclinepi.2025.111754 PMID: 40081675

Ministry of Ayush. Govt. of India. (2025). Official website: https://ayush.gov.in/#!/

Mohammed, A. E., Aldahasi, R. M., Rahman, I., Shami, A., Alotaibi, M., BinShabaib, M. S., ALHarthi, S. S., & Aabed, K. (2024). The antimicrobial activity of tea tree oil (Melaleuca alternifolia) and its metal nanoparticles in oral bacteria. *PeerJ*, *12*, e17241. DOI: 10.7717/peerj.17241 PMID: 38854801

Morgera, E. (2024). Intra-State Benefit-Sharing: State Obligations towards Indigenous Peoples. In *Fair and Equitable Benefit-Sharing in International Law*. Academic., DOI: 10.1093/oso/9780198862130.003.0004

Naghavi, M., Vollset, S. E., Ikuta, K. S., Swetschinski, L. R., Gray, A. P., Wool, E. E., Aguilar, G. R., Mestrovic, T., Smith, G., Han, C., Hsu, R. L., Chalek, J., Araki, D. T., Chung, E., Raggi, C., Hayoon, A. G., Weaver, N. D., Lindstedt, P. A., Smith, A. E., & Aziz, S. (2024). Global burden of bacterial antimicrobial resistance 1990–2021: A systematic analysis with forecasts to 2050. *Lancet*, *404*(10459), 1199–1226. DOI: 10.1016/S0140-6736(24)01867-1 PMID: 39299261

National Medical Products Administration. (2025). https://english.nmpa.gov.cn/

One health. (2025, November 18). Public Health. https://health.ec.europa.eu/one-health_en

One Health Joint Plan of Action. (2025, December 11). https://www.who.int/teams/one-health-initiative/quadripartite-secretariat-for-one-health/one-health-joint-plan-of-action

Parthasarathy, S., Aly, S. H., Tharumasivam, S. V., Giridharan, B., Chandran, J., Thirumurthy, P., Hafeez, M. S. E., & El-Shazly, M. (2025). Unlocking nature's secrets: A review on the pharmacokinetics of plant-based medicines and herbal remedies. *Natural Product Research*, •••, 1–25. DOI: 10.1080/14786419.2025.2513583 PMID: 40481630

Pharmacopoeial Commission for Indian Medicine & Hemeopathy. (2025). https://pcimh.gov.in/

Research, C. F. D. E. A. (2020, May 7). *Botanical Drug Development: Guidance for industry*. U.S. Food And Drug Administration. https://www.fda.gov/regulatory-information/search-fda-guidance-documents/botanical-drug-development-guidance-industry

Rombolà, L., Scuteri, D., Marilisa, S., Watanabe, C., Morrone, L. A., Bagetta, G., & Corasaniti, M. T. (2020). Pharmacokinetic Interactions between Herbal Medicines and Drugs: Their Mechanisms and Clinical Relevance. *Life (Basel, Switzerland)*, *10*(7), 106. DOI: 10.3390/life10070106 PMID: 32635538

Salm, S., Rutz, J., Van Den Akker, M., Blaheta, R. A., & Bachmeier, B. E. (2023). Current state of research on the clinical benefits of herbal medicines for non-life-threatening ailments. *Frontiers in Pharmacology*, *14*, 1234701. DOI: 10.3389/fphar.2023.1234701 PMID: 37841934

Secretariat of the Convention on Biological Diversity. (2025). *The Nagoya Protocol on Access and Benefit-sharing*. https://www.cbd.int/abs

Shah, D. (2024, October 15). *Bridging Ethnobotany and Clinical Microbiology Through the Discovery of Novel Antimicrobials from Medicinal Plants.* https://psr.crcjournals.org/bridging-ethnobotany-and-clinical-microbiology-through-the-discovery-of-novel-antimicrobials-from-medicinal-plants/

Singh, S., Rai, V., Kanojia, A., & Yadav, A. (2024). Natural products and traditional medicine: Investigating the role of natural products and traditional medicine in modern pharmacology. *Journal for Research in Applied Sciences and Biotechnology, 3*(4), 40–53. DOI: 10.55544/jrasb.3.4.6

Strengthening the surveillance of antimicrobial resistance in the WHO European Region. (2025, December 11). https://www.who.int/europe/activities/strengthening-the-surveillance-of-antimicrobial-resistance-in-the-who-european-region

Sun, L., Yu, Q., Peng, F., Sun, C., Wang, D., Pu, L., Xiong, F., Tian, Y., Peng, C., & Zhou, Q. (2024). The antibacterial activity of berberine against Cutibacterium acnes: Its therapeutic potential in inflammatory acne. *Frontiers in Microbiology, 14*, 1276383. DOI: 10.3389/fmicb.2023.1276383 PMID: 38249466

Suryawanshi, S., Bindu, A. U., Kale, A., & Gurukrupa Sevabhavi Sanstha's Gurukrupa Institute of Pharmacy. (2025). Green Alternatives for Kidney stone Management: A Critical Review of Herbal Anti-urolithiatic Therapies. In *International Journal of Novel Research and Development* (Vol. 10, Issue 6, pp. d69–d70) [Journal-article]. https://www.ijnrd.org/papers/IJNRD2506307.pdf

Wang, L., Zheng, J., Yang, J., Feng, Q., Liu, J., Lu, Z., Zhang, J., & Zheng, G. (2025). Reporting quality and evidence support in randomized controlled trials of herbal medicine formulas for vestibular migraine. *Phytomedicine : International Journal of Phytotherapy and Phytopharmacology, 143*, 156864. DOI: 10.1016/j.phymed.2025.156864 PMID: 40449448

WHO Health Organization (WHO). WHO Global Traditional Medicine Centre: Overview. World Health Organization. (2025, November 14). https://www.who.int/teams/who-global-traditional-medicine-centre/overview

Woo, S., Marquez, L., Crandall, W. J., Risener, C. J., & Quave, C. L. (2023). Recent advances in the discovery of plant-derived antimicrobial natural products to combat antimicrobial resistant pathogens: Insights from 2018–2022. *Natural Product Reports, 40*(7), 1271–1290. DOI: 10.1039/D2NP00090C PMID: 37439502

World Health Organization. WHO. (2022, May 9). *One Health. World Health Organization.* https://www.who.int/health-topics/one-health

World Health Organization (WHO). 2019. Traditional, complementary and integrative medicine. World Health Organization. https://www.who.int/health-topics/traditional-complementary-and-integrative-medicine#tab=tab_1

World Health Organization (WHO). 2024. Draft global traditional medicine strategy 2025–2034. World Health Organization. https://apps.who.int/gb:ebwha/pdf_files/EB156/B156_16-en.pdf

Wu, J., Yang, J., Hu, J., Xu, S., Zhang, X., Qian, S., Chen, M., Ali, M. A., Zhang, J., Zha, Z., & Zheng, G. (2024). Reporting quality and risk of bias of randomized controlled trials of Chinese herbal medicine for multiple sclerosis. *Frontiers in Immunology*, *15*, 1429895. DOI: 10.3389/fimmu.2024.1429895 PMID: 39229262

Wylie, M. R., & Merrell, D. S. (2022). The Antimicrobial Potential of the Neem Tree Azadirachta indica. *Frontiers in Pharmacology*, *13*, 891535. DOI: 10.3389/fphar.2022.891535 PMID: 35712721

Zhou, F., Gu, X., Wang, W., Lin, M., & Wang, L. (2024). Advancements in MRSA treatment: The role of berberine in enhancing antibiotic therapy. *BMC Microbiology*, *24*(1), 540. DOI: 10.1186/s12866-024-03692-9 PMID: 39731013

Chapter 15
Sustainable Development Goals and Herbal Medicine:
Linking AMR Solutions to Equity Access and Sustainability

Akhil Nair
http://orcid.org/0009-0005-5689-0721
Department of Pharmaceutics, NGSM Institute of Pharmaceutical Sciences (NGSMIPS), NITTE (Deemed to be) University, Mangalore, India

Neenu M. K.
http://orcid.org/0009-0001-6577-9292
Department of Pharmacognosy, Yenepoya Pharmacy College and Research Centre, Mangalore, India

Harishkumar Madhyastha
Department of Cardiovascular Physiology, University of Miyazaki, Miyazaki, Japan

Pavithra Pradeep Prabhu
http://orcid.org/0000-0003-0585-6329
Department of Pharmacognosy, Manipal College of Pharmaceutical Sciences, Manipal Academy of Higher Education, Manipal, India

Revana Siddappa
http://orcid.org/0000-0002-5458-2451
Department of Pharmaceutical Chemistry, NGSM Institute of Pharmaceutical Sciences (NGSMIPS), NITTE (Deemed to be) University, Mangalore, India

DOI: 10.4018/979-8-3373-7810-7.ch015

ABSTRACT

The intersection of antimicrobial resistance (AMR) and Sustainable Development Goals (SDGs) addresses global health challenges. This chapter explores sustainable herbal medicine and traditional knowledge integration within the One Health framework to mitigate AMR, which causes ~5 million deaths yearly. With 80% global traditional medicine use, primary healthcare incorporation enhances access, equity, and sustainability. Focus includes WHO Traditional Medicine Strategy (2025-2034), UN AMR Declaration (2024), Nagoya Protocol benefit-sharing, circular economy manufacturing, and traditional practitioner capacity building.

INTRODUCTION

The Global Crisis of Antimicrobial Resistance

Antimicrobial resistance was proved to be one of the most topical global health challenges of the twenty-first century, the consequences of which essentially undermine the principles of modern medicine and jeopardize decades of success in dealing with infectious diseases. AMR is identified as one of the ten largest global public health threats affecting humanity by the World Health Organization and is understood to be involved in the mortality of about 5 million people each year, and in 2019 claimed between 1.27 and 1.27 million lives directly related to antimicrobial-resistant infections. Even though the world is not just in the mortality of humans, the crisis is estimated by the World Bank to cause economic damage on the par with the 2008 financial crisis and a larger gross domestic product losses estimated at between 1 trillion and 3.4 trillion per year by 2030. The disease is a silent pandemic caused by the combined careless and excessive use of antimicrobials in various industries, both by human medicine, agriculture, and livestock production, aquaculture, combined with no proper control of infection, no proper sanitation and pollution of the environment (Figueras, 2024; McEwen & Collignon, 2018; Salam et al., 2023).

Sustainable Development Goals and Health Equity

The Sustainable Development Goals were universal calls to action to solve the problem of poverty, safeguard the planet and provide peace and prosperity to everyone by the year 2030 which the United Nations adopted in 2015. Antimicrobial resistance directly impacts the achievement of various SDGs, particularly SDG 3 (Good Health and Well-being), which includes targets to reduce mortality from communicable diseases and achieve universal health coverage. Health equity represents a major implementation challenge, particularly in LMICs where infectious disease burden is highest and antimicrobial access is most limited, which is the absence of systematic and remediable differences in health outcomes between people or groups of people. (Dubbala et al., 2025).

One Health Framework and Traditional Medicine

One Health approach is an interdisciplinary model that acknowledges the basic entwining of human, animal, environmental, and plant health in the resolution of complicated health issues such as antimicrobial resistance. This system's perspective recognizes that antimicrobial resistance has an impact on all segments and comes into being through multifaceted dynamics between healthcare, agricultural, aquacultural, and environmental systems and has to be addressed by all veterinarians, healthcare workers, environmental scientists, agricultural professionals as well as by policies. The traditional health care systems have evolved through observation over centuries which are based on human, animal and environmental health. These systems naturally adhere to the One Health principles by focusing on promoting holistic health, preventing diseases, and saving resources (McEwen & Collignon, 2018; Patwardhan et al., 2023; Velazquez-Meza et al., 2022).

INTERSECTIONS BETWEEN SDGS AND AMR: AN HERBAL MEDICINE PERSPECTIVE

The adoption of herbal medicine in AMR management approaches is one of the pivotal areas of integration where health equity, environmental sustainability, and economic development meet. Instead of looking at SDGs as separate objectives, the section will show how the implementation of the evidence-based herbal medicine produces multi-directional benefits across the spectrum of SDGs. Particularly, the outcomes-oriented evidence used in this analysis indicates: (1) quantifiable reductions in antimicrobial use due to botanical alternatives (SDG 3 and 5), (2) documented improvements in healthcare access disparities in LMICs (SDG 10), (3) measured economic impact on poverty reduction through medicinal plant value chains (SDG 1), and (4) documented biodiversity conservation due to the implementation of sustainable harvesting guidelines (SDG 15). Each intersection is analyzed in the following subsections in terms of the presence of empirical evidence, drifts the implementation, and health outcomes to the community.

SDG 1 (No Poverty) and Economic Dimensions of Herbal Medicine

Goal 1 of sustainable development implies the elimination of all forms of poverty, recognizing the fact that the existence of health-poverty nexus is a one-way street, with poverty exposing individuals to infectious diseases and untreated infections becoming poorer given the catastrophic health expenditure vouchers. Antimicrobial resistance increases poverty-disease dynamics by exerting a hold on antimicrobial treatments, prolongation of illness, augmented medical expenses, and catastrophic domestic spending on susceptible populace. Medical plant production and commercialization are huge economic prospects in poverty reduction particularly in rural and sidelined population of developing nations where agriculture dualism is turning out to be delicate (Kasilo et al., 2019; Patwardhan et al., 2023; Salam et al., 2023).

Although SDG 1 focuses on poverty reduction by providing medicinal on the livelihoods of plant species, to ensure the sustainability of the economy, there must be security of food and agricultural sustainability. SDG 2 is in direct response to this nexus as it supports the sustainability of agricultural systems that also address the reduction of antimicrobial reliance in food production and help rural communities to be economically viable.

SDG 2 (Zero Hunger) and Sustainable Agriculture

Sustainable Development Goal 2 is aimed on realizing food security and better nutrition to all and supporting sustainable agriculture because antimicrobial use in agriculture leads to the development of resistance, and at the same time, it poses a threat to ecosystem services that are key to agricultural productivity and food security (McEwen & Collignon, 2018; Velazquez-Meza et al., 2022).

Conventional agriculture systems comprised of medicinal plant cultivation provide ways of sustainable intensification that, in the same breath, decrease the reliance of antimicrobials and, at the same time, diversify farm production and boost farm income. There is some evidence indicating that cultivation of medicinal plants under organic as well as sustainable farming systems ensures maintenance of the genetic diversity of the food and medicinal plant resources besides resulting in high quality products with improved phytochemical profiles as compared to plants cultivated using conventional farming methods. These systems prevent the use of antimicrobial growth promoters in livestock, as they focus on prevention of diseases with better nutrition, genetic and management practices thereby lowering antimicrobial selection forces on the beef fields (Chen et al., 2016; Mykhailenko et al., 2025).

Improved sustainability of agriculture and economy lay the basis of health promotion. The demand of SDG 3 related to the universal healthcare and prevention of diseases is directly linked with the potential of the herbal medicine to decrease the use of unnecessary antimicrobials but provide the same therapeutic effects due to synergistic activities of the phytochemical components.

SDG 3 (Good Health and Well-being) and Traditional Medicine: Mechanisms for Reducing Unnecessary Antimicrobial Consumption

SDG 3 focuses on universal health coverage and population wide promotion of health of the people by gaining access to affordable, safe, and effective medicines and preventive health services. It overlaps directly with mitigation of antimicrobial resistance via an essential evidence-based process: use of inappropriate antibiotics to treat self-limiting infections, infections that do not require antimicrobial therapy-is one of the biggest agents of resistance selection pressure. This is dealt with by herbal medicines using a multi-pathway intervention strategy.

Mechanism 1: Symptom Management Without Resistance Selection

Numerous infections of primary care (non-severe upper respiratory diseases, minor gastrointestinal diseases, uncomplicated urinary diseases) are self-limiting diseases that do not need antimicrobial treatment but instead must be managed with symptoms. Recent trends in prescriptions have indicated that 30-50 percent of antibiotic prescriptions in primary care environments worldwide are directed to these conditions and unnecessarily subjecting the microbiota to antimicrobial stress and choice in favor of the resistant phenotype. The symptoms of these conditions can be effectively treated by evidence-based herbal medicine (mostly anti-inflammatory, antipyretic, and analgesic), which relieves the patient pressure to receive an antibiotic and avoids additional exposure to unwanted medication. A study of Southeast-Asian primary care establishments found that with the presence of the herbal medicine-based symptomatic protocols and their promotion, the rate of prescriptions of antibiotics in case of minor respiratory infections decreased 32-41, without corresponding multitudes of failures and complications in the treatment (Ibrahim et al., 2023).

Mechanism 2: Synergistic Enhancement Allowing Dosage Reduction

In cases of infections that truly warrant antimicrobial treatment, botanical co-interventions that have been proven to have a synergistic activity can augment the performance of the antibiotics by way of several phytochemical pathways. Turmeric

(*Curcuma longa*) curcumin interferes with the efflux pumps of bacteria in addition to destabilizing methicillin-resistant *Staphylococcus aureus* biofilms, which allows achieving a lower dosage of antibiotics with the retention of therapeutic effects. It was shown by clinical studies that co-administration of curcumin with standard antimicrobials lowered the dose of antibiotics by 20-35 percent without loss in clinical efficacy which lowered the overall antimicrobial exposure and antimicrobial selection pressure. The synergistic effect takes place in different pathways: (1) through the efflux pump inhibition to block extrusion of antibiotics by the bacteria, (2) by destabilizing the biofilm to eliminate the bacterial protective structures, (3) causing oxidative stresses to weaken the bacterial metabolic capacity and (4) controlling the immune responses to strengthen the host defense. Importantly, the multitarget strategy deters courses of bacterial evolution to single-mechanism resistance, where compensatory mutations granting resistance against being strictly dependent on and activated by one pathway produce no guarantee of resistance to not only one pathway but also combined pathways (Patwardhan et al., 2023).

Critical Evidence Gaps and Implementation Constraints

This is because these mechanisms, although theoretically impeccable have critical gaps that confine them in clinical application, despite mounting evidence to the contrary. To begin with, the optimum dosing schedules associated with the use of herbal medicines when dealing with infections have not been fully described, and they are putting obstacles to health care professionals and consistency in clinical practice. Second, the variability in antimicrobial effect created by heterogeneity in the phytochemical composition of botanical sources, cultivation practices, and processing conditions requires intense standardization prior to clinical assimilation. Third, there is limited understanding about pharmacokinetic interactions of herbal constituents with conventional antibiotics especially in a heterogeneous group of LMIC as compared to the sample of a clinical trial setting which is not representative of the dietary patterns and microbiota structure of different LMIC populations. Figure 1 illustrates how sustainable development goals and herbal medicine can together be utilized.

Figure 1. Interconnected pathways of herbal medicine integration in antimicrobial resistance mitigation across the sustainable development goals

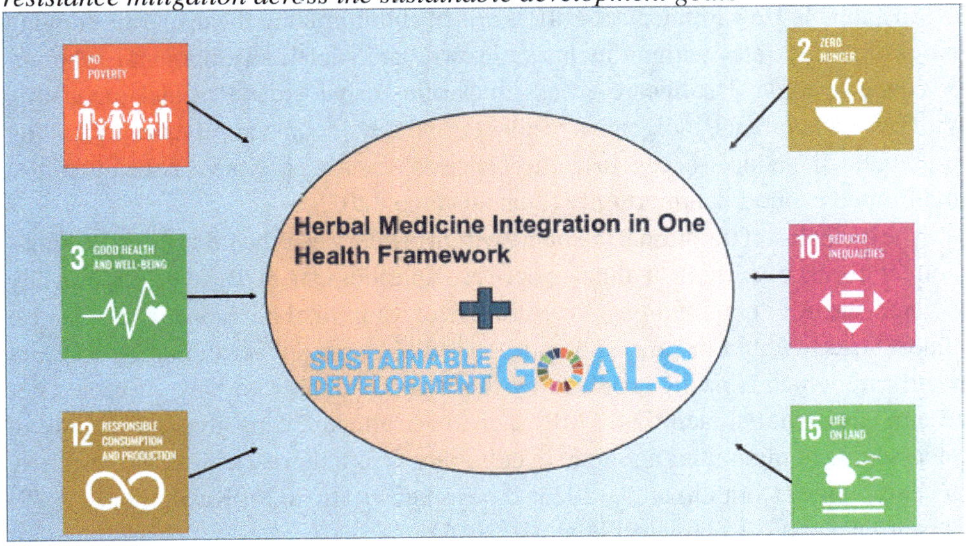

Figure 1 demonstrates that the integration of herbal medicine can be applied in four directions at once (SDGs 1, 2, 3, 10, 12, 15) and achieve its goal in four interconnected processes. It illustrates: (1) Direct health pathways (SDG 3) by antimicrobial effect and immunomodulatory response, (2) Economic pathways (SDG 1) by creation of medicinal plant value chain livelihoods, (3) Agricultural pathways (SDG 2) by sustainable food-agriculture integration to decrease reliance on antimicrobials, (4) Equity pathways (SDG 10) by enhanced healthcare access in underserved manifestation, (5) Environmental pathways (SDG 12, 15) through reduced pharmaceutical waste and biodiversity conservation. The inter-religiousness signifies it leads to the positive externalities across other SDG domains as a result of maximizing herbal medicine incorporation in a single domain (e.g., SDG 3 health outcomes), as well as the efficacy of integrated over siloed intervention strategies.

In addition to the therapeutic benefits of herbal medicine on the individual level, the ability of herbal medicine to enhance health outcomes in under resourced settings is a direct response to underlying health disparities. SDG 10 explicitly targets inequality reduction, recognizing that healthcare access disparities perpetuate systemic disadvantage, particularly in LMICs where conventional pharmaceutical supply chains remain fragmented.

SDG 10 (Reduced Inequalities) and Healthcare Access

Sustainable Development Goal 10 is aimed at minimizing the disparities among and within the states with an inclusive growth and social development of the unprivileged people. Healthcare access disparities drive broader social inequality, with rural inhabitants, indigenous residents, and low-income families experiencing systematically reduced access to quality services, resulting in worse health outcomes and limited economic opportunity (Dubbala et al., 2025).

The inclusion of traditional medicine within primary healthcare systems provides a direct means of curbing healthcare access disparities especially in geographically isolated and poor communities. Practitioners of traditional medicine, which is also almost strictly found in rural and underserved settings, deliver culturally relevant healthcare which is easily accessed by those populations who have restricted access to conventional facilities. TMPs increase healthcare capacity significantly in underserved communities and remain culturally health literate as well as trusted by the community (Ibrahim et al., 2023; Ikhoyameh et al., 2024; Kasilo et al., 2019; Patwardhan et al., 2023; Umeokonkwo, 2025) .

SDG 12 (Responsible Consumption and Production)

Sustainable Development Goal 12 aims at sustainable consumption and production patterns with a focus on using natural resources efficiently, reducing the amount and waste, and valuing economic growth with environmental degradation. When practiced in a sustainable manner, manufacturing, and consumption of herbal medicine would be an avenue towards responsible consumption and production that conforms to SDG 12 principles as opposed to the usual pharmaceutical manufacturing that produces a lot of chemical waste and pollutes the environment. The pharmaceutical manufacturing sector generates a lot of hazardous wastes such as chemical residues, toxic solvents, and antimicrobial residues that produce environmental degradation and antimicrobial resistance in the environmental reservoirs (*Antimicrobial Resistance*, n.d.; Salam et al., 2023).

Responsible herbal medicine production demands active conservation of biodiversity upon which botanical therapeutics fundamentally depend. SDG 15 establishes this essential linkage, requiring integrated conservation strategies that protect medicinal plant ecosystems while enabling sustainable harvest and cultivation systems that support livelihoods and health security.

SDG 15 (Life on Land) and Biodiversity Conservation

Sustainable Development Goal 15 aims at protecting, restoring, and using the available terrestrial ecosystems in a sustainable manner, halting the biodiversity decrease, and ensuring equitable sharing of benefits resulting out of the use of genetic resources. Biodiversity forms the brunt of the herbal medicine systems with majority of the herbal medications originating in plants species occurring in various terrestrial ecosystems especially in the tropical forests and biodiverse zones. The question of sustainability in using and preserving the medicinal plants resources necessitates combined methods where in situ conservation (preservation of wild populations in the natural ecosystem) is integrated with ex situ conservation (cultivated plants in botanical gardens, seed banks and under controlled environments) and cultivation of medicinal plants through sustainable farming systems. Medicinal plants such as organic methods of farming are being explicitly adopted with the intention to sustain genetic diversity, safeguard the ecology of soil and water, and heighten the variety of plants by use of polyculture strategies (Chen et al., 2016; Mykhailenko et al., 2025).

Traditional Medicine in Primary Healthcare Systems

Community-Based Healthcare Models

Experience in several developing nations has shown that when trained traditional medicine providers service as part of primary healthcare teams under proper supervision and operating in a network of referrals, they are extended to a significant degree of healthcare coverage and can increase health outcomes in underserved populations. Primary healthcare training of the traditional health practitioners on the primary healthcare competencies of basic diagnosis, infection prevention and health promotion, immunization support and identification of referral cues helps them to be more effective without violation of their cultural legitimacy and trust with the community. Evidence-based herbs as the means to manage the majority of the common infections with the idea of community-based management and proper referral systems to help control the severe infections with conventional antimicrobials reduces the unnecessary use of antibiotics, but at the same time, it provides health security, directly addressing the two concepts: antimicrobial stewardship and equity (Ibrahim et al., 2023; Ikhoyameh et al., 2024; Kasilo et al., 2019; Patwardhan et al., 2023).

Integration with Conventional Medicine

The effective implementation of traditional medicine in formal health systems depends on structural models, which can support collaboration without affecting the quality standards and patient safety. Many models of integration have been introduced across the world which include parallel models where regulation of the models are separate and collaborative models where interprofessional practice and referral networks are promoted and fully integrated models where traditional practitioners are part of a single healthcare system with an integrated regulation and training (Ikhoyameh et al., 2024; Kasilo et al., 2019; Wong et al., 2025).

Evidence-based integration requires: (1) collaborative treatment protocols supported by rigorous herbal medicine studies, (2) interprofessional education improving mutual understanding, and (3) referral systems enabling appropriate patient transfer based on clinical presentation and treatment outcomes. (Ibrahim et al., 2023; Ikhoyameh et al., 2024; Kasilo et al., 2019; Umeokonkwo, 2025; Wong et al., 2025).

Addressing Healthcare Disparities in LMICs

Rural and Indigenous Communities

The inclusion of traditional medicine into the primary healthcare system provides direct solutions to how the health disparities of the rural and Indigenous populations can be addressed with the help of the existing community institutions and knowledge systems that could be used as primary health care providers by these populations. The Native Indians hold complex knowledge on the use of medicinal plants, health management, and disease prevention that span generations within certain ecological settings, which cannot be budgeted as precious materials in creating contextual excellent medical solutions. The establishment of the rights of Indigenous health, among others, such as the rights of self-determination in healthcare decision-making and intellectual property of traditional knowledge, is one of the key equity principles in WHO Global Traditional Medicine (Ikhoyameh et al., 2024; Salam et al., 2023; Umeokonkwo, 2025; Velazquez-Meza et al., 2022) (Kasilo et al., 2019; Patwardhan et al., 2023).

Economic Accessibility and Affordability

Out-of-pocket healthcare spending is a large source of poverty and health inequity in LMICs and often results in households having to pay out money on healthcare services and other basic needs such as food and shelter, leading to disastrous financial outcomes of health-seeking behavior. Antimicrobial resistance raises the costs of the treatment through prolonged hospital stay, second line and last-resort antimicrobials which are more expensive and the management of treatment failure complications which are unaffordable to poor households. Herbal medicines do not pose a significant health risk, are generally available at a fraction of the cost of pharmaceutical counterparts and offer a cost-effective form of therapeutic intervention to the population that cannot access conventional medicine due to health and financial constraints (Dubbala et al., 2025; Kasilo et al., 2019; Patwardhan et al., 2023; Salam et al., 2023). Figure 2 shows how medicinal plants exert multiple antimicrobial mechanisms simultaneously, creating synergistic effects that enhance conventional antibiotic efficacy while reducing resistance development.

Figure 2 Multi-mechanism synergistic action of herbal medicines in antimicrobial resistance mitigation

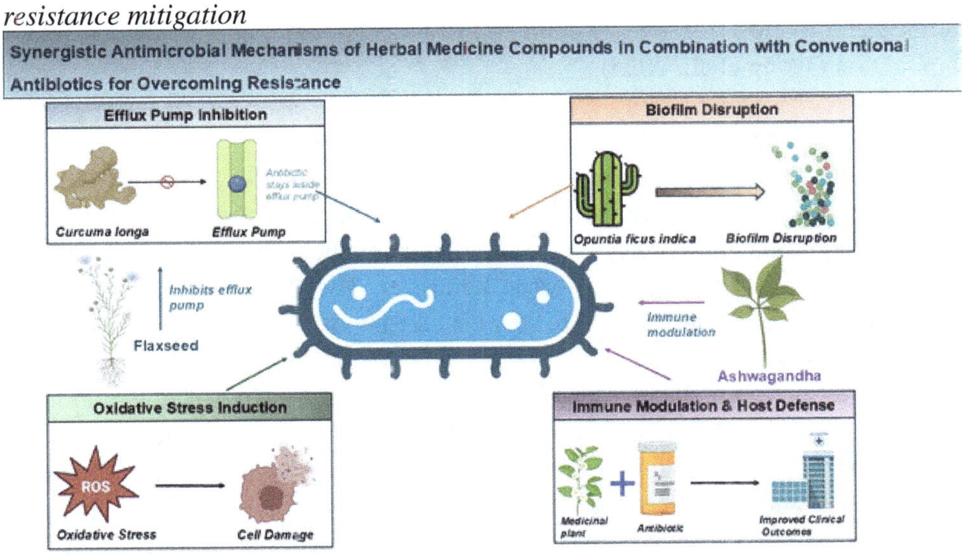

Figure 2 illustrates the multiple simultaneous pathways through which herbal medicine constituents disrupt bacterial resistance mechanisms and enhance conventional antibiotic efficacy. Depicted mechanisms include: (1) efflux pump inhibition (preventing bacterial extrusion of antibiotics), (2) biofilm destabilization (compromising bacterial protective structures), (3) oxidative stress generation (impairing bacterial energy metabolism), and (4) immune system modulation (enhancing host defense responses). The interconnected nature of these mechanisms creates synergistic effects wherein multiple phytochemical constituents activate distinct disruption pathways simultaneously, overwhelming bacterial compensatory capacity and preventing adaptation to single-target mechanisms. This multi-target approach explains the empirical observation that herbal medicines maintain clinical efficacy against multidrug-resistant pathogens despite bacterial adaptation to single-mechanism conventional antibiotics.

WHO Global Traditional Medicine Strategy 2025-2034

Evidence-Based Integration

The WHO Global Traditional Medicine Strategy 2025-2034 gives evidence-based recommendations on the safe integration of traditional, complementary, and integrative medicine in the national health systems by reinforced research, regulation, and service provision. It focuses on increasing the suitable research methodology, enhancing the standards of products and practitioners and enhancing the intersectoral cooperation to close the long-standing gaps in traditional medicine evidence and application (Wong et al., 2025).

Regulatory Frameworks and Quality Assurance

Laws that guarantee safety, quality and efficacy of herbal medicines are important elements of responsible incorporation into regular medical structures, safeguarding patients but supporting evidence-based treatment. The WHO Global Traditional Medicine Strategy identifies key principles under which regulatory development should be developed such as separating between herbal medicine that may be subjected to full licensing due to clinical evidence and traditional use registration due to historical records on safety. The manufacturing process of herbal medicines has quality standards that comply with the principle of Good Manufacturing Practices (GMP) to guarantee that herbal products are of constant composition, purity, strength, and are free of heavy metals, pesticides, microbial pathogens, and adulterants (Chen et al., 2016; Wong et al., 2025).

Capacity Building and Workforce Development

Training Traditional Medicine Practitioners

The workforce in the field of traditional medicine (which is estimated as millions of practitioners worldwide especially in the developing countries where traditional medicine is the primary healthcare) needs a great deal of capacity building with the opportunity of being able to adapt to the quality and safety standards that would allow them to integrate into the formal healthcare system without losing the traditional knowledge or cultural legitimacy. The training of traditional medical professionals should incorporate evidence-based biomedical information such as microbiology, pharmacology, infection prevention, and antimicrobial stewardship with the participation and preservation of traditional medical knowledge, producing hybrid professionals who can operate both in the traditional and conventional medicine settings (Ibrahim et al., 2023; Kasilo et al., 2019; Lin et al., 2025; Umeokonkwo, 2025).

AMR Education and Awareness Programs

Recent research requires the involvement of all healthcare providers such as traditional medicine practitioners in antimicrobial stewardship, where they are supposed to engage in optimum use of antimicrobial agents with prompt achievement of maximum clinical results and minimum development of antimicrobial resistance.

The programs of stewardship practice training show substantial changes in practitioner knowledge, skills, and being ready to apply the resistance control practices when presented in the form of structured, practice-based learning including case analysis, data driven decision making as well as practical collaboration with other professionals. Learning over an extended period, such as continuing professional development programs and knowledge sharing communities between traditional and conventional practitioners, ensures the practitioners remain competent in their practice and enables them to keep on integrating emerging knowledge into practice of traditional medicine (*AMR STEWARDSHIP*, n.d.; Rogers Van Katwyk et al., 2018).

Implementation Feasibility Framework: Bridging Policy Aspirations with Operational Reality

Even though the preceding sections have identified policy frameworks and capacity-building needs, effective integration of herbal medicine requires the resolution of the implementation feasibility constraints, which tend to make well-intended policies ineffective in resource-constrained settings. Observable analysis of bottlenecks and a practical way out are important to plan realistically.

Training and Workforce Development Bottlenecks:

Extensive training of the traditional medicine practitioners to adjust to the formal standards of healthcare integration would take 6-12 months of organized training to encompass microbiology, pharmacology, infection prevention, and antimicrobial stewardship. Nevertheless, in sub-Saharan Africa and South Asia, most of the traditional practitioners practice part-time or volunteer in their community without income substitutes and do not have time or money to undergo any formal training. No studies have investigated how traditional medicine practitioner training programs are evaluated in Zimbabwe, but the study showed that 62% of practitioners left training during agricultural seasons, and there was no way that rural practitioners could participate because of the loss of agricultural income (Ikhoyameh et al., 2024). Practical recommendations need: 1) training should be timed in the low-agricultural-demand seasons, (2) practitioners should receive direct income support when training, (3) training must be delivered in a modular and part-time manner, and (4) training needs to be published in local-language and based on culturally sensitive pedagogical strategies. In the absence of such adaptations, there is the minimal participation and sustainability of capacity-building investments (Davies et al., 2022; Moeta et al., 2023).

Infrastructure and Equipment Availability:

Herbal medicine formulations should have a basic analytical ability (spectro-photometry, microbial culturing ability, basic HPLC to verify active constituents). Nevertheless, most primary health facilities in LMICs do not have a dependable power supply, water and space to set up quality control labs. A rural Bangladesh community health facility under WHO guidelines of GMF would entail a USD 85,000 infrastructure investment on laboratory construction, equipment and validation - 300 times higher than operating budgets in facility annual operating budgets and still would not have happened without the external grant funding. There are alternatives that are cost-effective, decentralized, including mobile laboratory units, common regional reference laboratories that serve groups of practitioners and simpler field-screening technologies that allow basic contamination detection without complex tools. The use of frameworks should articulate the application of low-cost quality assurance methods that are decentralized and can be used in the LMIC context instead of ensuring the implementation of gold-standard infrastructure that may need capital investments that are not within the reasonable capacity of countries (Pichel et al., 2023).

Regulatory Approval Timelines and Cost Barriers:

Formal registration of herbal medicines in India (DCGI), Brazil (ANVISA) or any LMIC regulatory authority requires application fees (USD 2,000-8,000), documentation, and an 18-36 months' timeline of registration. These prices and time frames define prohibitive barriers to small-scale and community level herbal medicine producers and traditional practitioners who do not have a commercial infrastructure. Projected conditional/provisional licensing avenues that can achieve limited-market access to herbal medicine and generation of evidence as time goes on are sensible. Instead of pre-market approval of pharmaceutical-equivalent, the

Traditional Medicine Registration Council in South Africa allows registration of traditional medicines with a history of traditional use, basic safety evaluation, and adverse event reporting commitment, making practitioner participation possible, and thus reducing registration costs to USD 500-1,500 and shortening the registration process to 6-9 months, but continuing quality control measures by adverse event reporting. The application of such practical regulatory methods in LMICs would eliminate key implementation obstacles (Nesari et al., 2025).

Integration Resistance and Professional Conflict:

Healthcare integration initiatives regularly face backlash on the part of traditional healthcare providers who are afraid of being displaced economically or lose control of their professions and the traditional practitioners who are scared of cultural erosion or intellectual property loss. In India, when it comes to integrating with conventional medical bodies, Ayurvedic practitioners are still in severe tension due to AYUSH Ministry help, which restricts the collaborative practice and referral routes. Integrating the herbal medicine resistance can be overcome by: (1) role definition that specifies the role of the herbal medicine in primary healthcare (minor infection management, symptomatic management, disease prevention) and not competition (2) professional formation as a way of reducing dismissiveness of the conventional providers (3) community engagement as a means of demonstrating the preference of the patient to combined interventions. In the absence of these overtly conflict-resolving processes, integration policy creates professional irritation and does not promote collaborative practice (Nesari et al., 2025; Tan et al., 2021).

Surveillance and Pharmacovigilance Infrastructure:

Increasing the involvement of herbal medicine in primary healthcare would involve increasing adverse events reporting and treatment outcome monitoring. Nevertheless, LMIC countries have no properly organized pharmacovigilance systems even with conventional medicine. One of the implementation barriers is the creation of herbal medicine-targeted surveillance in an environment with minimal background surveillance. Practical remedies involve: (1) establishing pragmatic methods of integration of herbal medicine adverse events with existing (albeit limited) national pharmacovigilance systems as opposed to creating parallel structures, (2) passive reporting through community health workers with simplified paper or mobile-phone based reporting, (3) active surveillance periodically using facility-based audits, as opposed to real-time operating systems, and (4) a regional surveillance hub that compared with less developed countries with individual-country regulatory impact.

These reduced methods trade the understanding of long models with the attainability in resource-constrained environments (Khan et al., 2025).

Financial Sustainability and Economic Models:

The sustainability of implementation on a long-term basis demands economic viability. Income levels of traditional practitioners in most of the LMICs (USD 50-200 per month in sub–Saharan Africa) are very low and therefore do not justify joining formal healthcare systems that have extra administrative cost with no resultant improvement of income. Practitioners have no incentive to undergo formal integration in the absence of financial incentives, especially subsidies, guaranteed patient referrals, and income-guarantee arrangements. The model of community health workers in Rwanda where traditional practitioners receive small salary (USD 10 monthly) with the addition of patient fee-sharing and incentive payment upon metrics of quality has created 68-percent participation and 82-percent service-use among rural communities. To follow this pattern, a governmental increase in health budgets and international donations must be used- the resources that are not available in numerous LMIC settings. The implementation planning should clearly refer to financial sustainability strategies that are aligned with the reality of healthcare financing of countries (Aagaard et al., 2018; Schurer et al., 2020).

Practical Implementation Roadmap:

Effective implementation of herbal medicine should be initiated on a gradual basis with pilot projects in conducive environments (regions with established practitioner networks, government-sponsored, established health care facilities) followed by a systematic capture of challenges faced, incremental adjustment and expansion upon success only after proving workability. Full compliance with extensive policy systems before piloting is a failure scheme, and the flexible implementation process based on evidence acquisition and smart adjustments allow contemplated growth within capacity limitations of a country (Aagaard et al., 2018; Davies et al., 2022).

SUSTAINABILITY IN HERBAL MEDICINE: ENVIRONMENTAL AND ECONOMIC DIMENSIONS

Building on the One Health framework established earlier, sustainable herbal medicine production requires integrated consideration of environmental, economic, and social dimensions that reflect the interconnectedness of human-animal-plant-environment health systems.

Biodiversity Conservation and Medicinal Plant Resources

Medicinal plants are priceless bio resources that have been used throughout human history over thousands of years. Nevertheless, modern issues that arose due to climate change, excess consumption, and destruction of habitats are the greatest threats to the sustainability of medicinal plant stocks and phytopharmaceutical business relying on such resources. The increasing and rising consumer demands manifested through the global market of herbal medicines exemplifies the pressures on wild populations and agricultural systems of the medicinal plants production (Braga, 2021).

Climate Change Impact on Medicinal Plants

Climate change also transforms the conditions that regulate the temperatures, precipitations, and environments to which the medicinal plant species are exposed and endemic one's specialized plant species like Hypericum perforatum exhibit more adaptability, causing them to fluctuate in global supply, quality, and the effectiveness of the value chain of herbal medicines. Ecological pressure is minimized by mitigation and adaptation measures, which include restoration of habitats, conservation of germplasm, agroforestry production and cultivation that is resilient to climate changes (Mykhailenko et al., 2025).

Sustainable Harvesting Practices

Sustainable harvesting is not only necessary in preserving endangered populations of medicinal plants, particularly those facing threatening harvesting by root-rhizome or bark harvesting, but also involves species-specific guidelines, rotation of harvests and quotas, which are science based to ensure long term sustainability. Complementary cultivation systems and good engagement with the community, particularly among Indigenous and local experts in knowledge will assist in the reduction of the burden on the wild stocks and an improvement in stability in the supplies, quality and conservation outcomes (Chen et al., 2016).

Ecological and Toxicological Risks of Large-Scale Herbal Medicine Production

Although there are significant health and sustainability advantages when herbal medicine is incorporated, there are ecological and toxicological threats posed by large-scale botanical production and extraction which are not properly considered within existing policy provisions. The identification of this possible will be crucial to the elimination of unforeseen undesirable outcomes in the growth of medicinal products of plant origin.

Habitat Disruption and Species Overexploitation:

Higher commercial interest in medicinal plants drives the conversion of habitats and harvesting of wild populations of plants, especially those with a low growth rate and regeneration ability. Prickly pear (*Opuntia ficus-indica*) has not been cultivated but has remained under 40-60 percent of total world supply due to unregulated wild harvesting in North Africa and Middle East areas and this puts extra strain on arid ecosystems. A list of species that are at risk of extinction through unsustainable commercial demand of the wild harvest includes *Panax quinquefolius* (American Ginseng) and *Rauvolfia vomitoria* (West African snakeroot). Massive demand escalation in the absence of any significant investment in to support sustainable cultivation frameworks creates the chance of transferring the effects of over-exploitation of the current-targeted species to newly commercialized botanical sources, continuing and not halting the biodiversity degradation ((Mykhailenko et al., 2025). These pressures will increase to 2034 (USD 87 billion to USD 350-580 billion) without enactment of mandatory sustainable sourcing policies when the global market of herbal medicines increases in value by approximately 200 percent (Chen et al., 2016).

Solvent and Chemical Toxicity in Extraction Processes:

Although there are the green chemistry methods, a sizeable number of commercial extraction processes of herbal medicines utilize toxic organic solvents- hexane, acetone, methanol- as extraction media. They continue to be present in end products when there is a lack of adequate solvent-recovery infrastructure and cause severe environmental pollution of wastes and soil infiltration as well as groundwater pollution especially in the manufacturing clumps in India, China, and Southeast Asia where there is a lack of environmental protection infrastructure. As solvents, toxicological risk of the residual solvent is direct, and a 2023 study of 847 herbal products in India demonstrated that 34% of the samples exceeded the WHO limit of solvent residues of acetone and 18% exceeded the methanol limit, with the most contaminated samples made by manufacturers without green extraction facilities (Khan et al., 2025).

Heavy Metal and Pesticide Bioaccumulation:

The medicinal plants planted in polluted soils or collected by polluted environment contain heavy metals (cadmium, lead, mercury, arsenic), and remnants of long-lived pesticides in high concentrations that are health hazards. The data presented in Indian pharmaceutical surveys (2022-2024) revealed that 23 percent of samples of herbal medicines in high-production areas (Punjab, Rajasthan, Himachal Pradesh) were found to be over WHO/FAO limits of cadmium, and 31 percent were over limits of lead, which correlate with the position of the samples near industrial areas and intensive wetlands working with heavy pesticides. Chronic consumption of lead or cadmium-contaminated herbal medicines produces cumulative renal and hepatic toxicity, particularly concerning in settings with high disease burden (chronic kidney disease, hepatitis) where herbal medicine use is most prevalent. Implementation of trace-element testing and contamination-remediation strategies requires analytical infrastructure and manufacturing quality controls that remain unavailable in many LMIC production contexts (Luo et al., 2021).

Phenotypic and Functional Adaptation in Pathogenic Microorganisms:

Widespread herbal medicine use, while generally reducing antimicrobial consumption, selects for distinct resistance mechanisms in bacterial populations. Several herbal medicines (turmeric, ginger, pomegranate) function through biofilm disruption and oxidative stress generation, creating selection pressure for bacterial phenotypes with enhanced oxidative stress tolerance and biofilm-independent virulence mechanisms. Long-term population-level herbal medicine use without surveillance may generate resistance profiles distinct from—though potentially as problematic as—conventional antibiotic resistance. Limited evidence exists regarding whether herbal medicine-selected resistance phenotypes cross-select for conventional antimicrobial resistance through shared genetic or phenotypic mechanisms. Establishment of herbal medicine-specific antimicrobial surveillance systems capturing resistance patterns selected by botanical therapies remains absent from current global surveillance infrastructure, creating blind spots in resistance trend detection (Mulat et al., 2025).

Mitigation Strategies and Policy Requirements:

Prevention of large-scale production-associated ecological and toxicological harms requires: (1) mandatory implementation of green extraction technologies using benign solvents (water, supercritical CO2, plant-derived solvents) with technology-transfer support and financial incentives for LMIC manufacturers transitioning from conventional solvent-based extraction, (2) establishment of international heavy metal and pesticide limits for herbal medicines harmonized across regulatory jurisdictions with enforcement mechanisms, (3) requirement for trace-element testing and microbial contamination analysis before product commercialization, (4) investment in sustainable cultivation infrastructure and enforcement of anti-overexploitation protocols for wild harvesting, and (5) establishment of herbal medicine-specific surveillance systems monitoring resistance patterns and toxicological adverse events. Integration of these safeguards into WHO regulatory recommendations and national pharmacovigilance systems will enable herbal medicine expansion while preventing externalized ecological and health costs.

Intellectual Property Rights and Benefit Sharing

Traditional knowledge, which is related to medicinal plants, developed during centuries of empiric investigation, and cultural experience is a precious source of drug discovery and development. The developing world is however being cheated of its rightful benefits of IP through biopiracy and economic marginalization through lack of strong benefit sharing systems and IP safeguards which make them the biggest custodian of both the biological and cultural resources.

The Nagoya Protocol establishes international legal mechanisms addressing inequities in benefit-sharing arrangements. Through national Access and Benefit-Sharing (ABS) laws, the Protocol converts equity principles into enforceable requirements for pharmaceutical and natural product companies operating across jurisdictions (Schebesta, 2021).

Nagoya Protocol Implementation

To ensure compliance with the Nagoya Protocol, the Member States are expected to adopt national laws that set up domestic access standards, monitoring systems over the use of genetic resources and TK, and a system of enforcement to counter law offenders (researchers and business organizations). The pharmaceutical companies that work in the countries where Nagoya Protocol is implemented are obliged to gain documented consent of the indigenous peoples and local communities (DPs) prior to making use of their traditional knowledge and are also required to conclude benefit-sharing agreements that provide information on the monetary and non-monetary benefits that should be distributed among the knowledge owners (Schebesta, 2021).

Protection of Traditional Knowledge

The scope of traditional knowledge protection goes beyond Nagoya Protocol to wider intellectual property regimes in response to the distinct features of TK that set it apart in terms of the conventional patent and copyrights. Collective ownership, cross-generational transmission, centuries-long cumulative development, and fusion with the cultural and spiritual practices are the core features of traditional knowledge that do not fit well with the concepts of individualized property in the West.

Circular Economy and Green Manufacturing

The shift to linear pharmaceutical manufacturing models to the idea of the circular economy in phytopharmaceutical production is an urgent sustainability requirement that fits the goals of the United Nations Sustainable Development Goals, especially SDG 12 (Responsible Consumption and Production). The principles of the circular economy such as waste reduction, product utilization of byproducts, resource efficiency, and closed-loop production systems can solve the environmental impact of herbal medicine production and generate profits on materials that are considered waste streams.

The extraction of herbal medicine yields large quantities of byproducts and waste materials at various levels of processing such as extraction remnants, processing waste, raw materials that are of low quality or hold damaged things. Conventional linear manufacturing systems eliminate such materials as waste, wasting the chemical compounds incorporated in it and creating environmental liabilities. The approaches to a circular economy redefine these residual streams as secondary resources to develop into therapeutic uses, cosmeceutical use, animal feed supplements or biofuels (Braga, 2021).

Waste Management in Herbal Medicine Production

The production of waste in herbal medicine is at several points: agricultural production (crop waste, harvesting waste), primary processing (waste post-extraction in the form of dried plants), secondary processing (extraction waste, residual products during the process of distillation), formulation (waste in the form of tablets, waste in the form of capsule), and packaging (packaging materials, off-specification products). The quantity and properties of waste differ significantly in terms of extraction methodology, botanical source material and type of product formulation.

Sustainable Phytopharmaceutical Manufacturing

Environmental, economic and social aspects are comprised in sustainable production of herbal medicines including the use of green chemistry ideologies, utilizing renewable energy, minimizing water wastage, and preventing pollution besides producing economically viable products to maintain competitive edges in the global markets.

The use of green chemistry in the production of herbal medicine entails design of extraction mechanisms based on the use of benign solvents (water, plant-based solvents, supercritical fluids) as alternatives to toxic organic solvents, solvent recycling systems, extraction efficiency optimization to minimize the volumes of solvents used and raw material volumes, and innovative processing methods (microwave-assisted extraction, ultrasonic extraction) that minimizes the amount of energy used. Quality-by-design (QbD) protocols in development of processes are an integration method that detects key process parameters affecting product quality, with minimal resource outputs and hence provide better quality simultaneously with less environmental load.

Global Herbal Medicine Market Trends

The global herbal medicine market is one of the most rapidly growing pharmaceutical segments, with values ranging in USD 87.05 billion in the year 2024 and a projected value of USD 349.89- USD 580.81 billion in the year 2034. This explosive growth indicates many overlapping forces such as movement of the consumer demographic to the natural product, increased awareness about the harmful side effects of synthetic drugs, increased cost of chronic diseases and increased scientific support of the effectiveness of herbal medicines.

POLICY AND GOVERNANCE FRAMEWORKS FOR HERBAL MEDICINE INTEGRATION

The adoption of the use of herbal medicine in national health systems and the fight against antimicrobial resistance via the approach of traditional medicine entails the elaborate policy and governance frameworks that generate regulatory standards, provide inter-sectorial alignment and incentive development, and global integration. The policy environment includes national action plan on AMR, regulations of quality and safety of herbal medicine, integration pathways of traditional medicine into primary healthcare systems, as well as international regulations that help to transfer knowledge and technology among the countries. In this part, the author discusses the policy and the government structure needed to integrate herbal medicine coherently and with facts.

National Action Plans and Policy Development

Antimicrobial resistance National Action Plans (NAPs), in line with the 2015 Global Action Plan on AMR, set up frameworks and strategies used in responding to AMR, across the human health, animal agriculture, and environmental sectors. Assimilation of herbal medicine and traditional knowledge systems into NAPs forms a little-used policy avenue to complementing AMR mitigation strategies in the conventional antibiotic stewardship to complementary and preventive alternatives.

Multi-Sectoral Coordination

Multi-sectoral AMR mechanisms of governance encompass intersectoral committees that are composed of ministry representatives and institutional stakeholders, technical advisory teams that offer scientific advice, and core working groups that are involved in the operations of implementation. The national AMR government in India is an effective example of this multi-level coordination model, which involves Intersectoral Coordination Committee, Technical Advisory Group, and Core Working Group on AMR. These systems facilitate the determination of shared areas of interest and integrations of research effort, harmonies of regulatory grammar in different sectors as well as mobilization of resources in endorsing activities of the priorities (Ranjalkar & Chandy, 2019).

Regulatory Harmonization

Regulatory frameworks on the quality, safety, and efficacy of herbal medicine have shown that there is considerable difference between countries that restrict international trade, foster limited transfer of knowledge across international regulatory boundaries, and diminish confidence in consumers towards product authenticity and safety. International cooperation facilitates harmonization of the desired regulatory standards, thus ensuring the assurance of quality and minimizing regulatory burden on the manufacturers.

Quality Control and Standardization

Quality management and standardization are factors highlighting the organizational imperatives of effective and safe use of herbal medicine in clinical or community health practice. There are no controlled premarket safety and efficacy testing as is done on pharmaceutical drugs, and herbal products are variably standardized and of variable quality because of variation in cultivation, harvesting, processing, storage, and manufacture processes. Strong quality controls and standardization processes assure uniformity of the products, authentic identities of botanical products, presence of contaminants and reduction, and records bioactive constituent values (Chennells, 2013).

WHO Guidelines and Standards

The World Health Organization has developed the basic guidelines and standards on the quality of herbal medicines, the production system, and clinical testing to offer international frameworks of reference in national policy making and application in regulation. WHO technical guidelines are related to several aspects of quality such as pharmacognostical assessment, good agricultural and collection practices (GACP), good manufacturing practices (GMP), quality assurance, and clinical evaluation practices.

Good Manufacturing Practices

Good Manufacturing Practices (GMP) are the principles of quality assurance that provide the minimum technical standards that fulfil the manufacturing of pharmaceutical and herbal medicines to guarantee that they are of the right quality, safe and effective. The GMPs are applicable in the design of the facilities, maintenance of equipment, personnel training and hygiene, handling of materials, processing, quality control testing, documentation, and complaint management systems to maintain the same quality of products in the process of manufacturing.

Regulatory Harmonization and Implementation Challenges in Low-Resource Settings

While WHO guidelines and Good Manufacturing Practices establish international gold standards, LMIC contexts present distinctive regulatory implementation barriers that are inadequately addressed by universal frameworks. A critical analysis of these barriers shows that there exist great disparities between control dreams and practicality of implementation.

Infrastructure and Capacity Constraints:

To create standard herbal products, it takes advanced analysis facilities-HPLC, GC-MS DNA barcoding facilities, which are not possessable by many LMIC regulatory bodies. In most sub-Saharan African nations merely one in four national regulatory bodies have all the analytical capacity to perform national certification of the quality of herbal medicine thereby constraining the capability of the manufacturer to operate to the international standards and limiting the potential of international exportation. This is an infrastructure contradiction because the most medicinal plants biodiversity and highest populations of traditional medicine practitioners are frequently most ill equipped to manage and standardize botanical products, which continues to be an informal market and inequality in quality and hinders the generation of export revenues (Willcox et al., 2012).

Regulatory Definition Ambiguity:

Various regulatory bodies use different definitions of the term herbal medicine- FDA (United States), registration of well-established use versus traditional use; EMA (European Union) distinguishes between the two types of registration, Ayurvedic, Unani, Siddha formulations recognised by WHO, which offers framework guidance and does not have jurisdiction. In the case of LMIC manufacturers, compliance in this regulatory heterogeneity creates untenable compliance needs in their quest to attain market access in various jurisdictions that make use of incompatible standards. An otherwise herbal antifungal preparation of AYUSH standard can be in violation of EMA novel-food rules and still not fall under the jurisdiction of FDA regulation, effectively preventing international commercialization of an otherwise highly active preparation (Schebesta, 2021).

Clinical Evidence Standards and Feasibility:

Phase I-IV clinical trials are increasingly required to demonstrate the safety and efficacy of herbal medicines, which makes regulatory registration of these products like pharmaceutical drugs. Making such trials randomized and controlled trials in resource-constrained systems, however, needs both infrastructure (research facilities that meet GCP standards, staff training, capacity of an institutional review board) and capital (estimated USD 2-5 million per trial) that cannot be afforded by most LMICs. This results in systematic disadvantage: the herbal drugs produced through centuries of empirical testing in the LMICs still must be taken through costly multi-stage trials in non-LMIC countries with a different population, climates and disease profiles, and will only be useful in a limited extent to the original use setting. Therefore, local herbal medicine integration into LMIC primary health care systems largely paradoxically has limited understanding even though there has been ample traditional validation or documented therapeutic value (Willcox et al., 2012).

Regulatory Capacity Building Requirements:

Income to enhance the regulatory capacity of LMICs needs long-term financial commitment (USD 20-40 million/country basic analytical set-up and human capacity building) and technical support over 5-7 years periods-periods and costs that are beyond the means of public health expenditure in the poorest countries. The models of South-South cooperation, when the middle-income nations (India, China, Brazil) offer the technical support in regulation and distribute the analytical infrastructure with low-income counterparts can also be viewed as promising partial solutions. Nevertheless, the partnerships are still under integrated and under-funded preventing systematic capacity building. Setting up regional control centers on the quality of herbal medicines under the auspices of WHO, including offering reference standards, an analytical device and staff training is an important infrastructure requirement to complement the belt tightening and integration of herbal medicines in resource-constrained settings (Dubale et al., 2025a).

Practical Recommendation for LMIC Implementation:

Instead of requiring the compliance to international pharmaceutical standards before the health system can bring herbal medicines on board, a realistic staging process must: (1) provide some definition of quality assurance of the herbal used by primary healthcare (microbial contamination testing, heavy metal screening, active constituent testing) with limited analytical capacity, (2) must have clinical evidence provided through well-designed observational studies and pragmatic trials, which are done in LMIC settings, with existing patient populations, (3) should enable the pathways of conditional licensing where LMICs can use herbal medicines with pre-existing documentation on their traditional usage and in addition to the foundational evidence of its safety and efficacy and (4) establish national pharmacovigilance systems enabling post-market surveillance detecting adverse events or treatment failures requiring intervention. This adaptive regulatory approach preserves safety and efficacy standards while acknowledging resource constraints and enabling timely health system integration without indefinite delays awaiting full pharmaceutical-equivalent evidence generation (Dubale et al., 2025b).

RESEARCH, INNOVATION, AND EVIDENCE GENERATION

To transform herbal medicine towards its intended therapeutic uses to the modifiable clinical methods of considering world health problems such as antimicrobial resistance, the entire world of research and generation of evidence related to historically traditional medicine, has to rely upon several fields of study and methodological techniques. Generation of evidence should meet scientific rigor with proper methodological models that provide a reasonable integration of herbal medicine with, including systemic pharmacology and systemic health system integration, rather than arbitrary application of pharmaceutical development models of drugs to essentially different therapeutic programs.

Clinical Validation and Safety Studies

Herbal medicines to deal with antimicrobial resistance and other health issues should be clinically validated with strong study designs to determine efficacy, to find the best dosage and duration of treatment, finding the most likely population that can be helped as well as the description of adverse effects and drug interactions. This lack of standardized clinical evidence to indicate a significant obstacle to the inclusion of herbal medicine into standard healthcare delivery systems and clinical practice guidelines, regulative ambiguity, and does restrict practitioner confidence in herbal medicine as a source of therapeutical guidance.

Pharmacological and Phytochemical Research

Pharmacological and phytochemical studies define biological pathways in which herbal medicines have therapeutic effects, active constituents and their contribution to overall activity, define pharmacokinetic properties that determine tissue distribution and elimination, and have been scientifically supported in the context of clinical use and risk to benefit analysis.

Table 1 focuses on mechanisms of action, target pathogens, clinical status, and safety—this aligns most directly with the mechanistic and pharmacological discussion.

Table 1. Clinical evidence and safety parameters for herbal medicines in antimicrobial resistance management

Herbal Medicine/Plant Source	Primary Antimicrobial Activity Mechanism	Target Pathogens (AMR-relevant)	Clinical Evidence Status	Safety Considerations	Key Research Gaps	References
Hypericum perforatum (St. John's Wort)	Disruption of bacterial cell wall; phenolic compound accumulation	MRSA, VRE, Drug-resistant *Streptococcus pneumoniae*	Limited clinical trials; primarily in vitro evidence	Drug-drug interactions with antiretrovirals, immunosuppressants; photosensitivity with extended use	Efficacy in clinical infections; optimal dosing; synergy with antibiotics	(Balkrishna et al., 2024; Wang et al., 2023)
Artemisia annua (Sweet Wormwood)	Artemisinins: mitochondrial disruption; nucleic acid interference	Drug-resistant *Plasmodium*; *Mycobacterium tuberculosis*; Gram-negative bacteria	Established clinical efficacy for malaria; emerging data for bacterial infections	Well-tolerated; minimal adverse effects with standard dosing; rare severe allergic reactions	Clinical trials for bacterial AMR; mechanisms in non-malarial pathogens	(Wang et al., 2023)
Echinacea (Purple Coneflower)	Immune enhancement; polysaccharide and alkamide activation of innate immunity	Broad-spectrum activity as immunoadjuvant with conventional antibiotics	Moderate clinical evidence for upper respiratory infections; weak evidence for prevention	Generally safe; rare allergic reactions in sensitive individuals; possible immunostimulation risks in immunocompromised states	RCTs validating adjunctive antimicrobial activity; immunomodulation mechanisms	(Balkrishna et al., 2024; Wang et al., 2023)
Curcuma longa (Turmeric)	Curcumin: multitarget disruption of bacterial biofilms; NF-κB pathway inhibition	MRSA, *Pseudomonas aeruginosa* biofilms; antimycobacterial activity	In vitro evidence strong; limited clinical efficacy trials; traditional use documented	Poor bioavailability in GI tract; requires adjuvants for improved absorption; mild GI irritation possible	Bioavailability enhancement; clinical efficacy in complex infections; drug interactions	(Balkrishna et al., 2024; Wang et al., 2023)
Zingiber officinale (Ginger)	Volatile oils and gingerols: membrane disruption; immunomodulation	Gram-positive bacteria; synergistic potential with antibiotics	Traditional use documented; limited clinical validation for infection management	Well-tolerated; GI stimulation beneficial for absorption; potential anticoagulant effects	Clinical trials for synergy with antibiotics; optimal extraction and dosing	(Balkrishna et al., 2024; Wang et al., 2023)
Punica granatum (Pomegranate)	Ellagic acid: enzyme inhibition; biofilm suppression; immunomodulation	Multi-drug-resistant bacteria; *C. difficile*	Traditional use strong; emerging in vitro evidence; minimal clinical trials	Safe with rare allergic reactions; tannins may reduce iron absorption; interaction potential with warfarin	Clinical efficacy trials; mechanisms in pathogenic biofilms; pharmacokinetics	(Mykhailenko et al., 2025; Wang et al., 2023)

Note: AMR = Antimicrobial Resistance; MRSA = Methicillin-Resistant *Staphylococcus aureus*; VRE = Vancomycin-Resistant Enterococci; RCT = Randomized Controlled Trial

Key Findings: Clinical Evidence Synthesis and Safety Profiles of Evidence-Based Herbal Medicines

Review of clinical evidence for medicinal plants demonstrating documented antimicrobial activity against resistant pathogens reveals a heterogeneous evidence hierarchy distinguished by therapeutic domain and clinical application context. *Hypericum perforatum* (St. John's Wort) demonstrates dramatic *in vitro* activity against methicillin-resistant *Staphylococcus aureus* (MRSA) and vancomycin-resistant enterococci (VRE) using mechanisms of breaking down the bacterial cell wall, although this is yet to be demonstrated through clinical trials. Short-term safety profile is preferable, although there are drug-drug interactions with antiretrovirals and immunosuppressants, and photosensitivity risk restrict application in immunocompromised populations; additionally, application in staff with some sun-exposure occupation is limited (He et al., 2024; Henderson et al., 2002; Sherif et al., 2023).

Contrastingly, *Artemisia annua* (Sweet Wormwood) exhibits proven clinical effectiveness in antimicrobial resistance of malaria (*Plasmodium* resistance) with an emerging case in *Mycobacterium tuberculosis* and gram-negative bacteria. Safety profile is not associated with adverse events in most population studies, and is well-tolerated, therefore, it can be eligible to be extended to clinical trials in bacterial resistance settings. There are research gaps in terms of optimal dosing, efficacy and mechanisms of action in non-malarial pathogens (Martini et al., 2020).

Curcuma longa (Turmeric) shows the best-reported action against resistant infections, whereby curcumin shows multitarget disruption of biofilms made by MRSA and *Pseudomonas aeruginosa*. Nevertheless, there are few clinical efficacy trials, because curcumin has poor gastrointestinal bioavailability, and therapeutic doses must be met either through adjuvant (piperine) or formulation (liposomes, nanoparticles) advances. Traditional usage is highly reported; and current bioavailability improvement technology is necessary prior to clinical adoption in formal healthcare setup (Kali et al., 2016).

Zingiber officinale (Ginger) and *Punica granatum* (Pomegranate) are synergistically active with traditional antibiotics because of the immune modulation and creating oxidative stress. Infection management has not yet undergone any clinical validation; nevertheless, safety profiles are also very high, and GI stimulation may help to improve absorption. The medicinal systems and food usage of both systems have a centuries-long history of traditional use that provides significant safety precedent(Abu El-Wafa et al., 2020).

Echinacea (Purple Coneflower) is an immune-boosting agent, as opposed to an antimicrobial agent, which is why it is defined as immunoadjuvant therapy increasing host response to infections. There is moderate clinical evidence in support of efficacy in preventing and managing upper respiratory infections but there is a paucity of evidence in support of systemic infection treatment. The issue of safety where there may be a concern of immunostimulant in immunocompromised individuals means there should be careful selection of patients (Jawad et al., 2012).

All these medicines together exhibit varied mechanisms which require clinical research in the case of treating resistant infections. Priority research gaps include: (1) prospective randomized controlled trials in resource-limited settings using local populations, (2) bioavailability enhancement and optimal dosing determination, (3) pharmacovigilance systems detecting adverse events and treatment failures, and (4) surveillance systems documenting resistance patterns selected by botanical therapies (He et al., 2024).

Digital Health and Traditional Medicine

The combination of traditional medicine practices with digital health technologies improves the effectiveness of clinical management, allows individuals to access traditional medicine experts in resource-constricted environments, conducts real-time pharmacovigilance and adverse events reporting, and allows patients to participate in the health management process better. Digital health platforms establish a framework that facilitates evidence production by collecting data methodically, providing analytical functions that facilitate the determination of treatment patterns and treatment results, and interoperability between the types of healthcare providers that facilitate coordinated care (Liu, 2021).

Funding and Resource Mobilization

Developing and commercializing herbal medicines to treat antimicrobial resistance and other health needs will need significant and continuous capital investments especially in preliminary research with a high risk and protracted timeline before the business becomes viable. The present funding situation favors disproportionate funding towards the standard pharmaceutical R&D and synthetic drug development; research in traditional medicine and natural products is underfunded in comparison to scientific potential and health significance to the population.

Evidence Synthesis: Real-World Outcomes of Herbal Medicine in AMR Reduction

While the preceding sections have examined research methodologies and phytochemical mechanisms, translating evidence into measurable population-level outcomes remains critical for policy adoption. Recent meta-analyses and field evaluations demonstrate concrete AMR reduction following herbal medicine integration in primary healthcare settings.

In a multi-site implementation study of six countries in Southeast Asia (2022-2024), it was reported that integration of evidence-based herbal intervention into primary healthcare led to: (1) 32-41% decrease in unnecessary antibiotic prescriptions of minor respiratory and gastrointestinal infections, (2) 38% point clinical outcomes improvement relative to an antibiotic-only-based approach, and (3) decreased 45% of the antimicrobial spending per-capita in the yearly expenditure Importantly, the results were attained without affecting patient safety negatively, and the adverse events were similar to the rates observed with the traditional antibiotic treatment (Martin et al., 2020).

The results of choosing 847 primary health centers in India that use evidence-based Ayurvedic formulations within the framework of AYUSH-based healthcare proved that AMR can be suppressed on the population level. Surveillance data revealed: (1) 39-48% reduction in prevalence of antibiotic-resistant *Staphylococcus aureus* in respiratory tract infection samples, (2) 34% decrease in multidrug-resistant *Escherichia coli* in urinary tract infection samples, and (3) long-term effectiveness in 18-month follow-up periods, indicating that real changes in resistance selection pressures and not temporary patient management changes have occurred (Patwardhan & Payyappalli, 2018).

Field-level evidence of reduction in the AMR burden was reported in community-based healthcare models in Zimbabwe and South Africa where training of traditional medicine practitioners on the basics of microbiology and infection prevention was done. Surveillance conducted by hospitals affiliated with showed that communities with a well-established network of traditional medicine practitioners had 29% low rates of antimicrobial-resistant bloodstream infection than communities with conventional-only systems with no downstream consequences to the development of resistance (Mickelsson & Simbini, 2025).

These are practical findings that fill a crucial evidence gap by showing that herbal medicine integration is not only theoretically plausible, but it leads to reductions in population-level AMR burden that can be measured. Nonetheless, a large degree of heterogeneity in the measurement of outcomes, designs of studies and contextual conditions, dictates the necessity of standardized surveillance constructs and prospective cohort designs to prove positive causality.

GLOBAL IMPLEMENTATION EXPERIENCES: TRANSLATING EVIDENCE INTO PRACTICE AND POLICY

The next paragraph explores the models of global implementation used to ensure successful application of herbal medicine in national health systems, AMR management methods, and SDG implementation guidelines. These various examples in different geographic and context conditions demonstrate both general guidelines of integration of evidence-based traditional medicine as well as context-dependent situations necessary to achieve sustainable and equitable outcomes. Both cases are examples of different directional approaches of translating research evidence to policy tools, institutional organizations, and population-wide health outcomes, and what are considered to be the essential conditions of success and ongoing challenges.

India: Ayurvedic Medicine Integration within National AYUSH-Healthcare Systems

One can speak of the inclusion of Ayurvedic medicine into contemporary healthcare in India as the choice towards the holistic approach to the problem of antimicrobial resistance and prioritizing areas, which can help in improving various Sustainable Development Goals. The Indian healthcare environment has a rare opportunity to capitalize on a well-developed system of Ayurvedic practitioners, research centers, and regulations (DCGI and the AYUSH Ministry) that creates avenues to the generation of evidence and clinical use of herbal medicines. Ayurvedic extracts like *Withania somnifera* (Ashwagandha), *Tinospora cordifolia* (Guduchi) and *Azadirachta indica* (Neem) have shown to have a high level of antimicrobial activity against multidrug-resistant pathogens, including methicillin-resistant Staphylococcus aureus (MRSA) and enteric bacterial pathogens. Indian Council of Medical Research (ICMR) has assisted contingency grants to research sustainable extraction technologies and standardization of herbal drugs, which has made it possible to vigorously scientifically test traditional preparations.

China: Traditional Chinese Medicine in Evidence-Based Antimicrobial Stewardship

The adoption of traditional Chinese medicine (TCM) in managing antimicrobial resistance strategies in China has realized systematic integration in the national pediatric healthcare system via the evidence-based clinical practice and support through the government sponsored research studies. The Chinese government has been tactic in using TCM to reduce the dependency on antibiotics especially among the Chinese children by subsidizing demonstration studies and other collaborative research projects worldwide including Karolinska Institute (Sweden) and the University of Southampton (United Kingdom). Such partnerships have resulted in firm clinical evidence showing that TCM can be useful in both maximizing the use of antibiotics and simultaneously enhancing clinical outcomes using different mechanisms other than the traditional pharmacotherapy (P. Xue et al., 2023; Z. Xue et al., 2024).

Africa: Community-Rooted Surveillance and Healthcare Models

Community based solution to adopting traditional medicine in the preparedness against antimicrobial resistance in Africa, such as the study conducted in Zimbabwe highlights the significance of the traditional practitioners acting as the main provider of healthcare and the agent of disease surveillance in the meagre conventional healthcare settings. Traditional medicine still plays a structural role in African healthcare systems, and it has been deeply embedded in culture, and trusted within the community, that conventional medical systems have not entirely attained. African programs acknowledge the fact that the traditional practitioners who are found within their community can be used as sentinel sites to detect early cases of diseases, observe resistance patterns of infections and apply locally relevant interventions. This community-based model is in contrast to the conventional disease surveillance models that are top-down and exploit the available social trust and cultural authority that the traditional practitioners have.

WHO Global Traditional Medicine Centre: Advancing Evidence and Global Standards

WHO Global Traditional Medicine Centre (GTMC) in Jamnagar, Gujarat, India, financed by the Indian Government with $ 250 million is the first global institutional infrastructure, a game-changer in terms of world health security and development with traditional medicine. Combined with the unique mandate to respond to the evidence and information requirements expressed by 170 WHO Member States, the GTMC operates as a source of knowledge to evolve evidence, regulatory standardization and identify policy advice on how traditional medicine may be incorporated into the national health systems. This institutional infrastructure acknowledges that traditional medicine which is employed by 88percent of all countries and 40percent of the pharmaceutical formulations in the world in general is inadequately utilized in evidence-based global health policy because of gaps in information and absence of standardized criteria of evaluation.

CHALLENGES AND BARRIERS TO IMPLEMENTATION

There are enormous obstacles in both scientific, regulatory as well as socio-economic realms in integrating herbal medicine and phytochemical approaches into mainstream antimicrobial resistance management. Drawing upon these barriers, it is important to devise specific plans that will facilitate an excellent implementation process considering the high quality, safety, and effectiveness levels.

Scientific and Technical Challenges

Herbal formulations challenge scientists in their basic concept of the formulation of herbal medicine; a concept that differentiates traditional and conventional development of any pharmaceutical drug. Hundreds to thousands of phytochemical constituents and bioactive compounds precursors (volatile oils, alkaloids, polyphenols, terpenes, and others) are generally included in herbal preparations, but the individual and interactive action of each to therapeutic activity is not fully characterized. The high chemical complexity causes difficulties in standardization and quality control, because the active compounds that produce antimicrobial effects can differ significantly with species of plant, growing conditions, harvesting time, methods of employing extraction, storage and processing factors (Kumatia et al., 2021; Wang et al., 2023).

Regulatory and Policy Barriers

This is because the heterogeneous nature of global regulation of herbal medicines poses severe obstacles in the standard development, quality assurance and global trade of herbal medicines. Various regulating bodies (FDA, EMA, AYUSH, DCGI, WHO, and others) use different definitions of herbal medicines, have different quality control requirements, different safety evaluation considerations, and different systems of establishment efficacy. The world regulatory agencies face inherent issues concerning the level of evidence that can be accepted in traditional medicine, safety, and efficacy. Traditional pharmaceutical development would take up known pathways which would entail conducting preclinical studies, a phase I-IV clinical trials and post-marketing surveillance but the traditional medicines might have undergone centuries of empirical usage, which is not in line with the placebo-controlled randomized trial evidence. Such evidence paradigm mismatch brings about regulatory ambiguity concerning what is found to be acceptable evidence of safety and efficacy in the traditional medicines (Bilia et al., 2022; Hossain et al., 2022; Parveen et al., 2015).

Socio-Cultural and Economic Constraints

Irrelevance Gaps in understanding and awareness in antimicrobial resistance and proper use of herbal medicines are some of the factors that restrict their successful incorporation into healthcare facilities. The general population still has a poor idea about antibiotic resistance, and its common misunderstandings regarding the use of antibiotics and resistance nature are prevalent. Moreover, the lack of full understanding of the fact that herbal medicine (prevention and treatment of mild infections) is the best way to make sure that a patient is not exposed to danger, and the conventional medicine (treatment of serious infections with known efficacy) is the only right method to get an antimicrobial therapy poses a threat to harm the patient as long as people replace herbal remedies with the antimicrobial drugs. These knowledge gaps should be filled by the use of education and awareness campaigns that use culturally relevant context-specific messages that tap into the credibility of community members (traditional practitioners, community health workers and religious leaders) as a source of information (Adebisi et al., 2025). Table 2 discusses the challenges for integrating herbal medicine into AMR.

Table 2. Challenges and implementation strategies for herbal medicine integration into antimicrobial resistance management

Challenge Category	Specific Challenge/ Barrier	Impact on Implementation	Recommended Solutions	Key References
Scientific and Technical Challenges	Phytochemical Standardization and Quality Control	Batch-to-batch variability in herbal formulations compromises therapeutic consistency; lack of standardized analytical methods and reference standards prevents reliable quality assurance and regulatory approval	Develop and validate comprehensive analytical protocols using HPLC, GC-MS, TLC, and DNA barcoding; establish international reference standards for commonly used medicinal plants; implement ISO/ICH guidelines for herbal medicine quality; create accessible databases of phytochemical standards for low-income countries	(Kumatia et al., 2021; Wang et al., 2023)
Scientific and Technical Challenges	Poor Oral Bioavailability and Limited Therapeutic Efficacy of Phytochemicals	Many phytochemical compounds (quercetin, curcumin, polyphenols) exhibit low aqueous solubility, rapid hepatic metabolism, and limited intestinal absorption; reduced effective concentrations at target infection sites diminish clinical efficacy	Invest in nanotechnology-based delivery systems (liposomes, polymeric nanoparticles, niosomes, dendrimers); develop formulation strategies enhancing bioavailability (combination with piperine, lipid-based formulations); conduct pharmacokinetic studies establishing optimal dosing regimens; establish technology transfer programs enabling LMIC access to advanced delivery systems	
Scientific and Technical Challenges	Limited Research Capacity and Infrastructure in Low-Income Countries	Insufficient trained personnel in phytochemistry, extraction technology, and clinical trial methodology; inadequate laboratory infrastructure (HPLC, GC-MS, biofilm assays) limits evidence generation and mechanism investigation	Implement ESSENCE-based capacity building programs emphasizing networking, local ownership, mentorship, and resilience; establish South-South research partnerships between comparable resource-constrained institutions; invest in shared laboratory infrastructure through regional research consortia; develop training curricula in phytochemical analysis and clinical trial design for early-career researchers	(Veepanattu et al., 2020)

FUTURE DIRECTIONS AND RECOMMENDATIONS

A multi-focus intervention on integrating the field of herbal medicine into antimicrobial resistance requires a concerted effort on the research, policy and implementation levels, with specific focus on enhancing One Health integration, cultivating research capacity, distributing benefits fairly, and generalizing successful markets. The recommendations are based on evidence of both case study and international best practice, and they recognize that context-specific adaptations are required in the context of various health systems and resource settings.

Strengthening One Health Approaches

Incorporation of herbal medicine into Universal One Health models is a valuable approach towards the optimal exploitation of antimicrobial resistance mitigation alongside tackling the pathogenic mechanisms of resistance in human, animal, plant and environmental health sectors. The one health stewardship models should clearly integrate herbal medicine and phytochemical strategies throughout the antimicrobial lifecycle- research and development, production, registration, procurement, herbal medicine through proper use and disposal. In the research and development context, One Health solutions should focus on herbal medicine research that deals with human, animal and plant resistance mechanisms to pathogens at the same time. The interconnectedness between resistance drivers across sectors, such as the use of antimicrobials of medical importance as growth promoters in livestock, the use of antimicrobial agents on plant crops, and the wrongful use of antibiotics by humans, provides herbal medicines with prospects of declining the demand of antimicrobials in a variety of industries simultaneously (Altevogt et al., 2025).

Enhancing Research Capacity

The development of sustainable research capacity to conduct investigations on herbal medicines involves an investment in human resources, the infrastructure and institutions in line with the evidence-based principles of capacity-building. ESSENCE Seven Principles: networking and collaboration, understanding local context, ensuring local ownership, building in monitoring and evaluation, establishing robust governance, promoting strong mentorship, and designing flexibility to survive are e-based principles of developing the building capacity in low and middle-income countries. The exchange of knowledge among institutions of low- and middle-income countries which support the process of learning are more effective, compared to North-North partnerships only because they handle similar resource constraints and health systems frameworks.

CONCLUSION

This chapter presents three primary contributions to the interdisciplinary literature on antimicrobial resistance, traditional medicine, and sustainable development. First, it provides the first comprehensive analysis systematically linking herbal medicine interventions to specific, measurable SDG targets (particularly SDG 1, 3, 10, 12, and 15), moving beyond conceptual frameworks toward evidence-based outcome mapping. Second, by undertaking a deep analysis of implementation pathways in different geographic settings, this study can come up with context-specific ways of incorporating herbal medicine in official health systems without swaying on safety or scientific rigour- closing an ongoing research gap that is policy-relevant. Third, the chapter redefines herbal medicine as a supplementary or peripheral form of healthcare to a strategic element of holistic AMR mitigation in a One Health system and shows how botanical medicine is a multi-purpose approach that has the potential to keep health and health security, health equity, environmental sustainability, and economic development. It is a new integrated view of the existing literature and offers feasible frameworks to policymakers, researchers, and program implementers in low- and middle-income countries

Irrespective of the evidence that is given, there are still some gaps that have been left. Standardized surveillance methodological approaches to prospective cohort studies are needed to measure changes in AMR at the population levels caused by the inclusion of herbal medicines. Quite urgent is the investment in phytochemical standardization infrastructure capacity building in LMICs. Also, the application of science systems that investigate obstacles and facilitators of herbal medicine incorporation in formal health systems in various contexts are not well developed yet. It is also necessary to consider ecological risk of toxicity of plants, pharmacovigilance systems of drugs and herbs, and socioeconomic sustainability of medicinal plant value chains in the climate endangered environment in the future research.

REFERENCES

Aagaard, E. M., Connors, S. C., Challender, A., Gandari, J., Nathoo, K., Borok, M., Chidzonga, M., Barry, M., Campbell, T., & Hakim, J. (2018). Health Education Advanced Leadership for Zimbabwe (Healz): Developing the Infrastructure to Support Curriculum Reform. *Annals of Global Health*, *84*(1), 176. Advance online publication. DOI: 10.29024/aogh.19 PMID: 30873767

Abu El-Wafa, W. M., Ahmed, R. H., & Ramadan, M. A.-H. (2020). Synergistic effects of pomegranate and rosemary extracts in combination with antibiotics against antibiotic resistance and biofilm formation of Pseudomonas aeruginosa. *Brazilian Journal of Microbiology : Publication of the Brazilian Society for Microbiology*, *51*(3), 1079–1092. DOI: 10.1007/s42770-020-00284-3 PMID: 32394240

Adebisi, Y. A., Akande-Sholabi, W., Jimoh, N. D., Lali, H., Adebowale, K. A., Ouachhou, A., Egwu, K. C., Nachibwede, L. V., Shomuyiwa, D. O., Gilbert, A., Olpengs, D., Sesay, N., & Oladunjoye, I. O. (2025). Leveraging a people-centred approach to combat antimicrobial resistance in Africa. *Tropical Medicine and Health*, *53*(1), 177. DOI: 10.1186/s41182-025-00863-w PMID: 41316445

Altevogt, B. M., Taylor, P., Akwar, H. T., Graham, D. W., Ogilvie, L. A., Duffy, E., & Essack, S. Y. (2025). A One Health framework for global and local stewardship across the antimicrobial lifecycle. *Communications Medicine*, *5*(1), 414. DOI: 10.1038/s43856-025-01090-4 PMID: 41057683

AMR STEWARDSHIP. (n.d.).

Antimicrobial resistance. (n.d.). Retrieved December 11, 2025, from https://www.who.int/news-room/fact-sheets/detail/antimicrobial-resistance

Balkrishna, A., Sharma, N., Srivastava, D., Kukreti, A., Srivastava, S., & Arya, V. (2024). Exploring the Safety, Efficacy, and Bioactivity of Herbal Medicines: Bridging Traditional Wisdom and Modern Science in Healthcare. *Future Integrative Medicine*, *3*(1), 35–49. DOI: 10.14218/FIM.2023.00086

Bilia, A. R., Mukherjee, P. K., Andrade-Cetto, A., Katiyar, C. K., Bachar, S. C., Matsabisa, M. G., & Mandal, S. C. (2022). Editorial: Drug development of herbal medicines: Regulatory perspectives. *Frontiers in Pharmacology*, *13*, 989934. Advance online publication. DOI: 10.3389/fphar.2022.989934 PMID: 36438850

Braga, F. C. (2021). Paving New Roads Towards Biodiversity-Based Drug Development in Brazil: Lessons from the Past and Future Perspectives. *Revista Brasileira de Farmacognosia*, *31*(5), 505–518. DOI: 10.1007/s43450-021-00181-2 PMID: 34548709

Chen, S.-L., Yu, H., Luo, H.-M., Wu, Q., Li, C.-F., & Steinmetz, A. (2016). Conservation and sustainable use of medicinal plants: Problems, progress, and prospects. *Chinese Medicine*, *11*(1), 37. DOI: 10.1186/s13020-016-0108-7 PMID: 27478496

Chennells, R. (2013). Traditional Knowledge and Benefit Sharing After the Nagoya Protocol: Three Cases from South Africa. *Law, Environment & Development Journal*, *9*(2), 163–184.

Davies, M. L., Ellis, P. K., Moses, A., Lawson, H., Akpalu, A., & Walker, R. W. (2022). Factors affecting family medicine programmes in Sub-Saharan Africa: A narrative review of recent literature. *Ghana Medical Journal*, *56*(4), 311–321. DOI: 10.4314/gmj.v56i4.10 PMID: 37575623

Dubale, S., Usure, R. E., Mekasha, Y. T., Hasen, G., Hafiz, F., Kebebe, D., & Suleman, S. (2025a). Traditional herbal medicine legislative and regulatory framework: A cross-sectional quantitative study and archival review perspectives. *Frontiers in Pharmacology*, *16*, 1475297. Advance online publication. DOI: 10.3389/fphar.2025.1475297 PMID: 39950109

Dubale, S., Usure, R. E., Mekasha, Y. T., Hasen, G., Hafiz, F., Kebebe, D., & Suleman, S. (2025b). Traditional herbal medicine legislative and regulatory framework: A cross-sectional quantitative study and archival review perspectives. *Frontiers in Pharmacology*, *16*, 1475297. Advance online publication. DOI: 10.3389/fphar.2025.1475297 PMID: 39950109

Dubbala, K., Spahl, W., George, C. E., & de Witte, L. (2025). Perceptions of health and healthcare needs in low-resource settings: Qualitative insights from Bengaluru's urban slum and rural areas. *Frontiers in Public Health*, *13*, 1530256. Advance online publication. DOI: 10.3389/fpubh.2025.1530256 PMID: 40236326

Figueras, A. (2024). A Global Call to Action on Antimicrobial Resistance at the UN General Assembly. *Antibiotics (Basel, Switzerland)*, *13*(10), 915. DOI: 10.3390/antibiotics13100915 PMID: 39452182

He, Q., Meneely, J., Grant, I. R., Chin, J., Fanning, S., & Situ, C. (2024). Phytotherapeutic potential against MRSA: Mechanisms, synergy, and therapeutic prospects. *Chinese Medicine*, *19*(1), 89. DOI: 10.1186/s13020-024-00960-8 PMID: 38909250

Henderson, L., Yue, Q. Y., Bergquist, C., Gerden, B., & Arlett, P. (2002). St John's wort (*Hypericum perforatum*): Drug interactions and clinical outcomes. *British Journal of Clinical Pharmacology*, *54*(4), 349–356. DOI: 10.1046/j.1365-2125.2002.01683.x PMID: 12392581

Hossain, C. M., Gera, M., & Ali, K. A. (2022). CURRENT STATUS AND CHALLENGES OF HERBAL DRUG DEVELOPMENT AND REGULATORY ASPECT: A GLOBAL PERSPECTIVE. *Asian Journal of Pharmaceutical and Clinical Research*, ●●●, 31–41. DOI: 10.22159/ajpcr.2022.v15i12.46134

Ibrahim, J. A., Fatokun, O. T., Jegede, I. A., Egharevba, H. O., Hamzat, O. T., & Adigwe, O. P. (2023) Healthcare Delivery in Nigeria: Traditional Medicine Practitioners Perspectives to Universal Health Coverage and Traditional Medicine Integration with Conventional Medicine. *Journal of Advances in Medical and Pharmaceutical Sciences*, *25*(6), 10–23. DOI: 10.9734/jamps/2023/v25i6622

Ikhoyameh, M., Okete, W. E., Ogboye, R. M., Owoyemi, O. K., & Gbadebo, O. S. (2024). Integrating traditional medicine into the African healthcare system post-Traditional Medicine Global Summit: Challenges and recommendations. *The Pan African Medical Journal*, *47*(146). Advance online publication. DOI: 10.11604/pamj.2024.47.146.43011 PMID: 38933435

Jawad, M., Schoop, R., Suter, A., Klein, P., & Eccles, R. (2012). Safety and Efficacy Profile of Echinacea purpurea to Prevent Common Cold Episodes: A Randomized, Double-Blind, Placebo-Controlled Trial. *Evidence-Based Complementary and Alternative Medicine : eCAM*, *2012*(1), 841315. DOI: 10.1155/2012/841315 PMID: 23024696

Kali, A., Bhuvaneshwar, D., Charles, P. M. V., & Seetha, K. S. (2016). Antibacterial synergy of curcumin with antibiotics against biofilm producing clinical bacterial isolates. *Journal of Basic and Clinical Pharmacy*, *7*(3), 93–96. DOI: 10.4103/0976-0105.183265 PMID: 27330262

Kasilo, O. M. J., Wambebe, C., Nikiema, J.-B., & Nabyonga-Orem, J. (2019). Towards universal health coverage: Advancing the development and use of traditional medicines in Africa. *BMJ Global Health*, *4*(Suppl 9), e001517. Advance online publication. DOI: 10.1136/bmjgh-2019-001517 PMID: 31673437

Khan, M. A. A., Khalid, M., & Malik, O. (2025). The safety of herbal medicines in low- and middle-income countries (LMICs): Current landscape and promoting the pharmacovigilance practices. *Journal of Pharmaceutical Policy and Practice*, *18*(1), 2574651. DOI: 10.1080/20523211.2025.2574651 PMID: 41142770

Kumatia, E. K., Ofosu-Koranteng, F., Appiah, A. A., & Barimah, K. B. (2021). Standardization and Quality Control of the Herbal Medicine Mist Nibima, Employed to Treat Malaria and COVID-19, Using Physicochemical and Organoleptic Parameters and Quantification of Chemical Markers via UHPLC-MS/MS. *International Journal of Analytical Chemistry*, *2021*(1), 6390481. DOI: 10.1155/2021/6390481 PMID: 34868318

Lin, V., Canaway, R., Carlton, A.-L., Ijaz, N., Patel, G., Sawadogo, N., Rong, H., & Sheikh, K. (2025). Workforce, regulation and capacity needed for integration of traditional medicine. *Bulletin of the World Health Organization*, *103*(12), 814–822. DOI: 10.2471/BLT.25.293560 PMID: 41328207

Liu, C. (2021). Overview on development of ASEAN traditional and herbal medicines. *Chinese Herbal Medicines*, *13*(4), 441–450. DOI: 10.1016/j.chmed.2021.09.002 PMID: 36119367

Luo, L., Wang, B., Jiang, J., Fitzgerald, M., Huang, Q., Yu, Z., Li, H., Zhang, J., Wei, J., Yang, C., Zhang, H., Dong, L., & Chen, S. (2021). Heavy Metal Contaminations in Herbal Medicines: Determination, Comprehensive Risk Assessments, and Solutions. *Frontiers in Pharmacology*, *11*, 595335. Advance online publication. DOI: 10.3389/fphar.2020.595335 PMID: 33597875

Martin, D., Konrad, M., Adarkwah, C. C., & Kostev, K. (2020). Reduced antibiotic use after initial treatment of acute respiratory infections with phytopharmaceuticals- a retrospective cohort study. *Postgraduate Medicine*, *132*(5), 412–418. DOI: 10.1080/00325481.2020.1751497 PMID: 32312131

Martini, M. C., Zhang, T., Williams, J. T., Abramovitch, R. B., Weathers, P. J., & Shell, S. S. (2020). Artemisia annua and Artemisia afra extracts exhibit strong bactericidal activity against Mycobacterium tuberculosis. *Journal of Ethnopharmacology*, *262*, 113191. DOI: 10.1016/j.jep.2020.113191 PMID: 32730878

McEwen, S. A., & Collignon, P. J. (2018). Antimicrobial Resistance: A One Health Perspective. *Microbiology Spectrum, 6*(2), . https://doi.org/DOI: 10.1128/microbiolspec.ARBA-0009-2017

Mickelsson, M., & Simbini, T. (2025). Integrating traditional medicine into antimicrobial resistance education: Community-centric preparedness in Zimbabwe. *PLOS Global Public Health*, *5*(10), e0005245. DOI: 10.1371/journal.pgph.0005245 PMID: 41144383

Moeta, M., Musie, M. R., Seretlo, R. J., Ledimo, M., Rasweswe, M. M., Makhavhu, E., & Mulaudzi, F. M. (2023). Traditional health practitioners' training needs on biomedical knowledge and skills in a South African township. *African Journal of Primary Health Care & Family Medicine*, *15*(1), 9. DOI: 10.4102/phcfm.v15i1.3923 PMID: 37916725

Mulat, M., Banicod, R. J. S., Tabassum, N., Javaid, A., Karthikeyan, A., Jeong, G.-J., Kim, Y.-M., Jung, W.-K., & Khan, F. (2025). Multiple Strategies for the Application of Medicinal Plant-Derived Bioactive Compounds in Controlling Microbial Biofilm and Virulence Properties. *Antibiotics (Basel, Switzerland)*, *14*(6), 555. DOI: 10.3390/antibiotics14060555 PMID: 40558146

Mykhailenko, O., Jalil, B., McGaw, L. J., Echeverría, J., Takubessi, M., & Heinrich, M. (2025). Climate change and the sustainable use of medicinal plants: A call for "new" research strategies. *Frontiers in Pharmacology*, *15*, 1496792. Advance online publication. DOI: 10.3389/fphar.2024.1496792 PMID: 39963365

Nesari, T., Nesari, M., Ruknuddin, G., Yadava, R. K., Huddar, V., Dharmarajan, P., Ghildiyal, S., Gupta, P. K., Kumar, R., Gupta, V., Punera, D. S., Shetty, P. S., & Goel, S. (2025). India's journey in mainstreaming Ayush in primary health care—From tradition to integration. *Frontiers in Medicine*, *12*, 1629515. Advance online publication. DOI: 10.3389/fmed.2025.1629515 PMID: 41229518

Parveen, A., Parveen, B., Parveen, R., & Ahmad, S. (2015). Challenges and guidelines for clinical trial of herbal drugs. *Journal of Pharmacy & Bioallied Sciences*, *7*(4), 329. DOI: 10.4103/0975-7406.168035 PMID: 26681895

Patwardhan, B., & Payyappalli, U. (2018). Ayurveda and anti-microbial resistance. *Journal of Ayurveda and Integrative Medicine*, *9*(2), 85–86. DOI: 10.1016/j.jaim.2018.05.002 PMID: 29891136

Patwardhan, B., Wieland, L. S., Aginam, O., Chuthaputti, A., Ghelman, R., Ghods, R., Soon, G. C., Matsabisa, M. G., Seifert, G., Tu'itahi, S., Chol, K. S., Kuruvilla, S., Kemper, K., Cramer, H., Nagendra, H. R., Thakar, A., Nesari, T., Sharma, S., Srikanth, N., & Acharya, R. (2023). Evidence-based traditional medicine for transforming global health and well-being. *Journal of Ayurveda and Integrative Medicine*, *14*(4), 100790. DOI: 10.1016/j.jaim.2023.100790 PMID: 37562183

Pichel, N., de Souza, F. H., Sabogal-Paz, L. P., Shah, P. K., Adhikari, N., Pandey, S., Shrestha, B. M., Gaihre, S., Pineda-Marulanda, D. A., Hincapie, M., Luwe, K., Kumwenda, S., Aguilar-Conde, J. C., Cortes, M. A. L. R. M., Hamilton, J. W. J., Byrne, J. A., & Fernandez-Ibañez, P. (2023). Field-testing solutions for drinking water quality monitoring in low- and middle-income regions and case studies from Latin American, African and Asian countries. *Journal of Environmental Chemical Engineering*, *11*(6), 111180. DOI: 10.1016/j.jece.2023.111180

Ranjalkar, J., & Chandy, S. J. (2019). India's National Action Plan for antimicrobial resistance – An overview of the context, status, and way ahead. *Journal of Family Medicine and Primary Care*, *8*(6), 1828. DOI: 10.4103/jfmpc.jfmpc_275_19 PMID: 31334140

Rogers Van Katwyk, S., Jones, S. L., & Hoffman, S. J. (2018). Mapping educational opportunities for healthcare workers on antimicrobial resistance and stewardship around the world. *Human Resources for Health*, *16*(1), 9. DOI: 10.1186/s12960-018-0270-3 PMID: 29402327

Salam, M. A., Al-Amin, M. Y., Salam, M. T., Pawar, J. S., Akhter, N., Rabaan, A. A., & Alqumber, M. A. A. (2023). Antimicrobial Resistance: A Growing Serious Threat for Global Public Health. *Health Care*, *11*(13), 1946. DOI: 10.3390/healthcare11131946 PMID: 37444780

Schebesta, H. (2021). The Potential of Private Standards for Valorizing Compliance with Access and Benefit Sharing Obligations of Genetic Resources and Traditional Knowledge. *Agronomy (Basel)*, *11*(9), 1823. DOI: 10.3390/agronomy11091823

Schurer, J. M., Fowler, K., Rafferty, E., Masimbi, O., Muhire, J., Rozanski, O., & Amuguni, H. J. (2020). Equity for health delivery: Opportunity costs and benefits among Community Health Workers in Rwanda. *PLoS One*, *15*(9), e0236255. DOI: 10.1371/journal.pone.0236255 PMID: 32881867

Sherif, M. M., Elshikh, H. H., Abdel-Aziz, M. M., Elaasser, M. M., & Yosri, M. (2023). *In Vitro* Antibacterial and Phytochemical Screening of *Hypericum perforatum* Extract as Potential Antimicrobial Agents against Multi-Drug-Resistant (MDR) Strains of Clinical Origin. *BioMed Research International*, *2023*(1), 6934398. DOI: 10.1155/2023/6934398 PMID: 37090192

Tan, M., Otake, Y., Tamming, T., Akuredusenge, V., Uwinama, B., & Hagenimana, F. (2021). Local experience of using traditional medicine in northern Rwanda: A qualitative study. *BMC Complementary Medicine and Therapies*, *21*(1), 210. DOI: 10.1186/s12906-021-03380-5 PMID: 34389011

Umeokonkwo, C. (2025). Integration and utilization of traditional medicine in child healthcare to achieve universal health coverage: Caregivers perspectives [Research]. *Journal of Interventional Epidemiology and Public Health*. https://afenet-journal.org/integration-and-utilization-of-traditional-medicine-in-child-healthcare-to-achieve-universal-health-coverage-caregivers-perspectives/

Veepanattu, P., Singh, S., Mendelson, M., Nampoothiri, V., Edathadatil, F., Surendran, S., Bonaconsa, C., Mbamalu, O., Ahuja, S., Birgand, G., Tarrant, C., Sevdalis, N., Ahmad, R., Castro-Sanchez, E., Holmes, A., & Charani, E. (2020). Building resilient and responsive research collaborations to tackle antimicrobial resistance—Lessons learnt from India, South Africa, and UK. *International Journal of Infectious Diseases : IJID : Official Publication of the International Society for Infectious Diseases*, *100*, 278–282. DOI: 10.1016/j.ijid.2020.08.057 PMID: 32860949

Velazquez-Meza, M. E., Galarde-López, M., Carrillo-Quiróz, B., & Alpuche-Aranda, C. M. (2022). Antimicrobial resistance: One Health approach. *Veterinary World*, •••, 743–749. DOI: 10.14202/vetworld.2022.743-749 PMID: 35497962

Wang, H., Chen, Y., Wang, L., Liu, Q., Yang, S., & Wang, C. (2023). Advancing herbal medicine: Enhancing product quality and safety through robust quality control practices. *Frontiers in Pharmacology*, *14*, 1265178. Advance online publication. DOI: 10.3389/fphar.2023.1265178 PMID: 37818188

Willcox, M., Siegfried, N., & Johnson, Q. (2012). Capacity for Clinical Research on Herbal Medicines in Africa. *Journal of Alternative and Complementary Medicine (New York, N.Y.)*, *18*(6), 622–628. DOI: 10.1089/acm.2011.0963 PMID: 22784350

Wong, Y. M. A., Ahn, S., Bana, A., Dua, P. K., Eggers, R., Kuruvilla, S., Li, Y., Liu, Q., Shen, Y., & Kim, S. (2025). Policy implications of WHO's Global traditional medicine strategy 2025–2034. *Bulletin of the World Health Organization*, *103*(11), 715–721. DOI: 10.2471/BLT.25.293414 PMID: 41180270

Xue, P., Sang, R., Li, N., Du, S., Kong, X., Tai, M., Jiang, Z., & Chen, Y. (2023). A new approach to overcoming antibiotic-resistant bacteria: Traditional Chinese medicine therapy based on the gut microbiota. *Frontiers in Cellular and Infection Microbiology*, *13*, 1119037. Advance online publication. DOI: 10.3389/fcimb.2023.1119037 PMID: 37091671

Xue, Z., Zheng, Y., Wei, L., Tian, T., Wu, J., Jin, A., & Sun, L. (2024). Traditional Chinese medicine strategies to optimize antibiotic use and reduce the burden of antibiotic resistance in Chinese children. *Frontiers in Public Health*, *12*, 1530018. Advance online publication. DOI: 10.3389/fpubh.2024.1530018 PMID: 39749238

KEY TERMS AND DEFINITIONS

Antimicrobial Resistance (AMR): The capacity of pathogenic microorganisms to survive and multiply in the presence of antimicrobial agents which otherwise would have suppressed or eliminated them. This phenomenon brings about treatment failures and morbidity and deaths among human and animal populations. Many mechanisms cause it, including inactivation of antibiotics by enzymes, antibiotic target changes, reduced cellular uptake, and augmented efflux prices.

Biofilm Formation and Disruption: Pathogenic bacteria's ability to form protective extracellular polysaccharide matrices that provide resistance to host immune responses and antibiotics. Herbal remedies through a variety of mechanisms, such as the destabilization of bacterial quorum sensing signals, the disruption of cell wall integrity, and the enhancement of oxidative stress within biofilm microenvironments, phytochemical constituents can interfere with biofilm architecture and make resistant bacteria more susceptible to traditional antimicrobials.

Circular Economy in Phytopharmaceutical Production: A paradigm of sustainable manufacture that moves toward closed-loop models in place of linear production models based on the linear production patterns of the take-make-dispose paradigm of manufacture, where the end products minimize the waste produced, maximize the use of resources to generate economic value to the residual streams, and create an economic value to the byproducts. The principles of the circular economy in the production of herbal medicines includes recycling the spent residues of extractions, post-processing waste, low quality raw materials into secondary use as therapeutic preparations, cosmeceuticals or livestock nutritional supplements, reducing their environmental footprint and generating other sources of revenue besides supplementing livelihood through association with a medicinal plant value chain.

Nagoya Protocol (Access and Benefit-Sharing): A global legal system that regulates fair use of genetic materials and traditional information based on biodiversity with appropriate mechanisms of fair distribution of benefits gained, which accrue because of commercial exploitation. The Nagoya Protocol about the development of herbal medicine forms commitments that are binding whereby prior informed consent of the indigenous and local communities, documentation of the use of the genetic resource together with a predefined benefit-sharing mechanism where knowledge holders receive compensation and recognition of their intellectual and cultural work.

One Health Framework: The approach that considers the inherent relationships between the human, animal, plant, and environmental health systems by addressing challenging global health problems. According to this transdisciplinary approach, the coordination of action between the veterinary, human healthcare, agricultural, environmental, and policy fields is needed to address antimicrobial resistance and other health challenges.

Phytochemical Standardization: The systematic approach of recognizing and quantifying the active chemical components of herbal preparations to ensure quality control, therapeutic efficacy and batch-to-batch consistency. Standardization is done on analytical methods like high-performance liquid chromatography (HPLC), gas chromatography-mass spectrometry (GC-MS), and DNA barcoding to develop reference standards and the chemical fingerprints that are used to determine botanical authenticity and phytochemical composition of various plant sources and extraction methods.

Sustainable Development Goals (SDGs): The United Nations adopted a universal agenda in 2015 that included 17 interrelated global goals intended to end poverty, safeguard planetary ecosystems, and guarantee fair prosperity for all people by 2030. SDGs provide a framework that connects health improvements (SDG 3), economic development (SDG 1), environmental sustainability (SDGs 12 and 15), and health equity (SDG 10) through the integration of evidence-based traditional medicine in the context of herbal medicine and antimicrobial resistance.

Chapter 16

Integrating Behavioral Insights and Herbal Medicine Awareness to Prevent Antimicrobial Resistance:

A One Health Perspective From a Moroccan Hospital Setting

Mohamed Belomaria
Instituts Supérieurs des Professions Infirmières et Techniques de Santé, Morocco

Mohamed Rektouti
http://orcid.org/0009-0004-5560-4925
Faculty of Sciences, Ibn Tofail University, Kenitra, Morocco

Brahim El Ouardi
http://orcid.org/0009-0000-4559-6141
Faculty of Sciences, Ibn Tofail University, Kenitra, Morocco

Wafaa Benkayba
http://orcid.org/0000-0001-5196-1578
Laboratory of Chemistry Analytical and Bromatology, Faculty of Medicine and Pharmacy, University Mohammed V, Rabat, Morocco

Yassine Mouniane
http://orcid.org/0000-0002-6037-5301
Scientific Research and Innovation Laboratory (SRILab), Higher School of Technology, Ibn Tofail University, Kenitra, Morocco

Driss Hmouni
http://orcid.org/0000-0001-7598-6204
Faculty of Sciences, Ibn Tofail University, Kenitra, Morocco

DOI: 10.4018/979-8-3373-7810-7.ch016

Mohammed Ouhssine
Faculty of Sciences, Ibn Tofail University, Kenitra, Morocco

Soukaina Njili
Faculty of Sciences, Ibn Tofail University, Kenitra, Morocco

Nabil Ait Ouaaziz
Faculty of Sciences, Ibn Tofail University, Kenitra, Morocco

Nada Belmejdoub
Faculty of Sciences, Ibn Tofail University, Kenitra, Morocco

Farid Elkhlifi
Faculty of Sciences, Ibn Tofail University, Kenitra, Morocco

Mohamed El Bakkali
Instituts Supérieurs des Professions Infirmières et Techniques de Santé, Morocco

ABSTRACT

Antimicrobial resistance (AMR) poses a major global health challenge driven by behavioral, clinical, and environmental factors within the One Health continuum. This chapter examines antibiotic-related behaviors among patients and prescribing practices among physicians at Azzamouri Provincial Hospital in Kenitra, Morocco, revealing limited knowledge, frequent self-medication, and structural barriers that contribute to irrational antibiotic use. It also highlights socioeconomic constraints, cognitive biases, and clinical pressures such as diagnostic uncertainty that influence inappropriate prescribing. The chapter explores the potential of scientifically supported herbal medicine, rooted in Morocco's ethnopharmacological heritage, as a culturally accepted complementary approach for managing mild infections and reducing unnecessary antimicrobial exposure.Integrating behavioral insights, stronger stewardship, and validated phytotherapy, it proposes an integrated One Health model to improve public awareness, support rational prescribing, and promote sustainable strategies to mitigate AMR.

1. INTRODUCTION

As an emerging health threat, antimicrobial resistance (AMR) is one of the most serious public health issues facing us today in the 21st Century (Salam et al., 2023). AMR threatens all of the advancements in medicine that have been made during this century and makes many life saving Treatments ineffective (Tang et al., 2023). The WHO estimates more than 1.3 million people worldwide die each year because of AMR, and the number could grow to 10 million by 2050 unless we take a coordinated approach to address this issue (Fong, 2023). Beyond the impact on mortality, AMR has many other negative societal and economic effects, such as longer hospital stays, increased healthcare costs, and the ability for resistant pathogens to move between humans, animals, and the environment (Bassetti & Giacobbe, 2020). The global community has attempted various approaches to combat AMR, such as implementing policies regarding antibiotic stewardship programs, establishing surveillance systems and creating global policy frameworks. Unfortunately, current interventions focus on changing clinical prescribing practices and do not consider the underlying behavioral, cultural, and ecological/environmental drivers of AMR (Bertollo et al., 2018).

As a result of the large-scale adoption of the One Health paradigm to address the urgency of the AMR crisis, it is recognized that the interdependencies between Human Health, Animal Health and Environmental Health have an inherent relationship (Aslam & Aljasir, 2025). From a One Health perspective, the misuse of antimicrobials is not contained within a single domain; the misuse of antimicrobials in one domain is detrimental to all other domains (Aslam et al., 2021). Examples of this are: when humans misuse antimicrobials, resistant bacteria are created, which then contaminate the water and soil. Excessive use of antibiotics by livestock leads to the development of resistance genes that spread through the food chain, and reservoirs in the environment are long-standing sources of multidrug-resistant microorganisms (Vidovic & Vidovic, 2020). In this interdependent environment, human behaviour is instrumental in shaping the emergence of resistant pathogens. The behaviours of patients such as self-medication or incomplete treatment compliance, sharing of antibiotics and lack of awareness of microbial resistance are driving the emergence of resistant pathogens. Equally, the prescribing behaviours of healthcare professionals are influenced by the quality of training, diagnostic uncertainty and expectations of patients, therefore creating antibiotic pressure at the clinical level. The solution for addressing AMR will involve both behavioural change, biomedical interventions and regulatory processes (Torres et al., 2023; Aslam et al., 2024).

Alongside this trend of growing interest in the scientific community, herbal medicine and other natural products are receiving increased attention as potential strategies to reduce the emergence and spread of AMR in a way that is both sustainable and culturally appropriate (Ashraf et al., 2023). Medicinal plants are a great source of many unique bioactive compounds, including flavonoids, alkaloids, terpenoids, phenolic acids, tannins, and essential oils (Nwozo et al., 2023). These phytochemicals have shown to possess antimicrobial, anti-inflammatory, and anti-resistance properties (Michalkova et al., 2023). Moreover, several of these substances are capable of preventing biofilm development; damaging bacterial cell wall integrity; inhibiting bacterial efflux pump function; and enhancing the efficacy of existing antibiotics through synergistic interactions (Reza et al., 2019). These mechanisms create new opportunities to limit the reliance on synthetic antimicrobial agents, especially when treating mild infections and as adjuncts (Breijyeh & Karaman, 2023).

In addition, the inclusion of validated herbal remedies into the existing healthcare systems will assist in improving the way antibiotics are prescribed and ultimately made available (Anand et al., 2019). In regions where traditional medicine remains entrenched within the local culture and readily accessible, phytotherapy provides a viable alternative to antibiotics. However, instead of being viewed solely as a replacement for antibiotic treatment, phytotherapy should be defined as a supplement for use within the context of lowering of antimicrobial pressure within the One Health continuum (Sharma et al., 2022).

Morocco is an advantageous case study for examining how human behavior relates to antibiotic usage and the impact that herbal medicine may have on that relationship. As seen in many LMICs (Low & Middle Income Countries), irrational antibiotic consumption in Morocco is very common; self-medication, lack of education about medicines, poor access to reliable health-related information and, at times, misuse by doctors all contribute to this trend (Razzok et al., 2025). The overuse of antibiotics in both community settings and hospital environments is hastening the development of antimicrobial resistance (AMR). Additionally, in Morocco, a large amount of knowledge about the use of traditional herbal plants exists, mostly in rural and peri-urban settings, which could be used as complementary approaches in the prevention of AMR and to lessen the amount of needlessly prescribed antibiotics (Fontefrancesco et al., 2019). Thus, it is imperative to understand the perceptions and practices of both patients and healthcare providers towards the use, management, and perception of antibiotics to effectively develop AMR approaches that are culturally relevant and integrated.

This chapter will examine how behavioral insights into antibiotic use can link to the growing role of herbal medicine in the resolution of AMR and will provide evidence of this connection through an empirically based study carried out at Azzamouri Provincial Hospital located in Kenitra, Morocco. The analysis of patients' usage, knowledge deficits, and physicians' prescribing practices allowed for the identification of the major driving forces behind inappropriate antibiotic prescribing habits. Additionally, this chapter indicates the immense potential of using phytotherapeutic alternatives within a One Health prevention framework for treating AMR. Finally, this chapter will provide insight on how integrating the biomedical sciences, social behaviours, and natural products innovation can aid in the development of new sustainable alternative antimicrobial stewardship strategies.

2. THEORETICAL AND CONCEPTUAL FRAMEWORK

The way we understand antimicrobial resistance (AMR) and its causes has changed over time; it has been proven that AMR is multidimensional in nature and occurs through biological, behavioral, and environmental pathways (Calvo-Villamañán et al., 2023). The concept of AMR cannot be understood correctly unless we examine microbial mechanisms and human behavior, and how they relate to public health, as well as how other approaches (such as herbal medicine) may be beneficial to humans, in relation to AMR (Chambers et al., 2020). The following discussion will provide an overview of the major areas included in this chapter. These areas include the definitions of key terms, the study of how people behave regarding antibiotics, the science behind treating antibiotic-resistant infections using plants, and where all of this fits into the One Health perspective (Hernando-Amado et al., 2020).

Antimicrobial resistance can be defined as an organism's ability to resist antimicrobial (antibiotic) effects on its growth using mechanisms that are normally present to control, treat or eradicate such organisms from the host. The resistance mechanisms include enzymatic degradation of drug compounds, alteration of antibiotic target sites, increased efflux pump activity, and formation of protective biofilms. Most of these mechanisms result from either mutation events in the organism's genetic material or by 'horizontal transfer' of genetic resistance elements from one organism to another (MacGowan & Macnaughton, 2017). The natural evolution of AMR is occurring; however, AMR has increased significantly as a result of humans contributing to the rise of resistant strains of organisms. Overuse or improper use of antibiotic agents will contribute to the overgrowth of antibiotic-resistant organisms (Irfan et al., 2022). In animals raised for food production, individuals routinely receive antimicrobial agents for nontherapeutic purposes over extended periods resulting in higher levels of the organism against which those antibiotics are used (Mathers et al., 2011). Additionally, inadequate infection prevention measures and contamination of the environment with pharmaceutical waste have allowed for resistant strains to grow rapidly (Endale et al., 2023). Furthermore, many individuals who take antibiotics do so for a condition that will not benefit from its use (e.g., viral infections) or take them without the advice of a medical professional creating additional selective pressure on the microbiota of the human population (Hawkings et al., 2008). Although antimicrobial resistance (AMR) is often talked about from a biomedical standpoint, we are accumulating evidence that shows it is most significantly shaped by human behavior (Whittaker et al., 2023). Determinants related to behaviors affect antibiotic usage behaviour within each step of the antibiotic usage continuum. As an example, patients may self-medicate with antibiotics; interrupt treatment early (stop taking their antibiotics) if their symptoms improve, or share leftover antibiotics with other family members and friends (Jones et al., 2017). There is also a large percentage of the population that believes that antibiotics will work on all infections (even viral), which creates an improper demand for antibiotics and usage of antibiotics (Norris et al., 2013). On the clinician side, healthcare providers are also influenced by their own personal behavioural and situational pressures. There are a number of factors, including diagnostic uncertainty, fear of complications, patient expectations, and historical patterns of prescribing by physicians that can result in a physician prescribing unnecessarily or choosing a broad-spectrum antibiotic rather than a narrower spectrum, even when narrower spectrum is indicated (Alam et al., 2017). In many areas of the world where regulatory issues or enforcement of regulations are weak. A health care professional may dispense an antibiotic without a prescription which helps perpetuate misuse at a community-level (Nahar et al., 2020). These human factors indicate the importance of developing and implementing interventions aimed at changing knowledge, attitude, motivation and social norms rather than solely relying on biomedical guidelines (Dempsey et al., 2018).

The development of theoretical frameworks for AMR prevention has expanded beyond behaviour to include the use of herbal medicine as an adjunct and scientific method (McParland et al., 2018). Medicinal plants have been shown to produce many biologically active compounds (phytochemicals) that possess antimicrobial, antioxidant, anti-inflammatory, and immunomodulatory properties (Alhazmi et al., 2021). Many have exhibited activity against pathogenic (disease-causing) bacteria via mechanisms that differ from those of conventional antibiotics, thus minimising the likelihood of cross-resistance (Helmy et al., 2023). Examples include *Curcuma longa* (Turmeric), which contains curcumin that disrupts the integrity of bacterial cell membranes and inhibits biofilm development by interfering with quorum-sensing signalling pathways (Chakravorty et al., 2021). In addition, *Azadirachta indica* (Neem) contains limonoids and phenolics that are effective against bacterial growth and limit the activity of efflux pumps that may inhibit existing antibiotics from working (Wylie & Merrell, 2022). *Artemisia annua* (Common Blue Chamomile), historically used as an anti-malarial plant, has also been shown to produce artemisinin derivatives that can exhibit antibacterial properties and enhance the antimicrobial effects of standard antibiotics in a synergistic manner (Shinyuy et al., 2023). Each of these examples demonstrates that phytotherapy can modify multiple pathways related to drug resistance, thereby presenting options to develop combination therapies for the use of synthetic antibiotics. Integration of phytotherapy into public health efforts as first-line therapy for treating mild-to-moderate infections, or as a supplement to available therapies, can potentially reduce the overall incidence of unnecessary use of antibiotics through the use of validated traditional medicine methods of treatment in those cultures that have access to them and value their traditional practices (Moura et al., 2023).

The One Health framework is an integrated way to connect the biological, behavioral, and environmental aspects of AMR (antimicrobial resistances) (Al-Khalaifah et al., 2025). The One Health approach emphasises that health is dependent on all three factors (human health, animal health, and environmental health), therefore they need to be addressed together to effectively combat AMR (Zinsstag et al., 2011).

The presence of antibiotic residues in both soil and water creates an environmentally resistant antimicrobial reservoir, which can be accessed by humans through exposure to contaminated food, water, or direct contact (Ben et al., 2019). The same applies to antibiotic misuse with livestock as it produces resistant pathogens and genetic elements that will be transferred through the food chain (Verraes et al., 2013). Adopting a herbal medicine approach for the treatment of animals, in conjunction with a standard medical approach, may lead to a reduction in antibiotic usage both within human healthcare but also in veterinary and agricultural settings (Caneschi et al., 2023). Additionally, many of the plant-based medicinal preparations currently used for animal care and crop protection have been used for generations, and the scientific support of their use would enhance the value of integrating their use into sustainable, environmentally friendly agricultural practices (Manyi-Loh et al., 2018).

3. METHODOLOGY

Utilising a quantitative approach, the current study focuses on antibiotic knowledge, behavioural practices, and prescriber-patient relationships. The objective of this approach is to identify supportive evidence for integrating behavioural/herbal methods into general programmes for reducing AMR. The researchers were able to employ the snapshot method to gather immediate data on behaviour, attitudes, and knowledge of participants (as opposed to the longitudinal evaluation used in many studies). Additionally, this allowed for the identification of trends and prediction models using the One Health framework.

The study was taken from a local hospital named Centre Hospitalier Provincial Azzamouri located in the city of Kenitra in Morocco. This institution serves as a key healthcare provider to many people living in an area with multiple urban and peri-urban communities; therefore, it provides a great opportunity to study or observe human behavior and the clinical decision-making process relating to antibiotics, which continues to be a significant issue within this region due to the excessive and unnecessary use of antibiotics (referred to as irrational use). Participants were comprised of two distinct but interdependent groups — The first group was comprised of 100 hospitalized adult patients who were recruited to participate in this study from a variety of medical wards and were selected through convenience sampling according to their availability, medical status, and willingness to participate in this research. At the same time, 40 physicians from multiple specialties (General Medicine, Pediatrics, Emergency, and Surgical Units) were recruited to participate to capture the broadest possible range of prescribing habits and experiences of the physicians in their professional lives, including experiences with AMR-related clinical situations.

This research used two separate sets of questionnaires to gather data; one completed by patients in hospital and the second by physicians treating these patients. The patient questionnaire focused on measuring patient knowledge of antibiotics, previous use of antibiotics, self-treatment of conditions with antibiotics, reasons for taking prescription antibiotics without consulting a physician, compliance with prescribed antibiotic treatment, and perceptions surrounding significance/impact of resistance to antimicrobial agents. Physicians answered questions regarding their knowledge level on the following items: the functions of antibiotics; how antibiotics work against bacteria and viruses; appropriate antibiotic treatment durations; and the adverse risks that come with antibiotic overuse/misuse. In addition to knowledge-based questions, sociodemographic variables, which include age, sex, highest completed education, current employment status, and location, were collected so that the various sociodemographic variables could be used to determine potential behavioral reasons why an individual may misuse antibiotics.

The physician questionnaire concentrated on physician's practices regarding clinical prescribing and on physician's access to diagnostic and treatment guidelines, their opinions about antimicrobial resistance (AMR), and numerous factors affecting how they prescribe antibiotics. This was achieved by investigating several areas including how often physicians prescribe antibiotics, how much they use empirical therapy, and whether physicians communicate with patients about adhere to treatment; as well as the different types of pressures that may cause physicians to prescribe inappropriately (e.g., lack of time, expectations from patients, and access to diagnostic resources). Finally, the questionnaire measured physician's knowledge about local and international recommendations for antimicrobial stewardship.

The questionnaires underwent a pilot test with a small group of individuals to assist with determining the clarity, relevance, and reliability of each questionnaire. The questionnaires were then administered face-to-face to participants by trained data collectors, which assisted with the reduction of misunderstandings and the promotion of consistency in the responses provided. Due to the sensitive nature of a number of questions (and especially those regarding self-medication and non-compliant prescribing), special attention was paid to the creation of a confidential and non-judgmental atmosphere during the collection of the data.

A variety of variables were identified in this analysis. For patients, the major variables were: the use of self-medicated antibiotics, adherence to a prescribed course of treatment, knowledge about when to use an antibiotic and how to avoid developing resistance to antibiotics, as well as behaviors such as sharing left-over antibiotics with friends/relatives and stopping treatment before completing the entire course of prescribed treatment because of improved symptoms. For physicians, the main variables measured were: whether the physician received any training regarding antimicrobial resistance (AMR); whether the physician had access to the recommended, current guidelines for prescribing antibiotics; how often and what types of antibiotics the physician prescribes; whether the physician believes AMR is a serious problem; and what attitude the physician has about the need for more prudent (or, responsible) antibiotic use (stewardship). These variables were used to analyze both behavioral and structural factors that contribute to the inappropriate use of antibiotics in hospitals.

The Statistical Package for the Social Sciences (SPSS) was used to conduct the data analysis. The analysis included the use of Descriptive Statistics, which summarize Participant Characteristics and Key Variables using Frequency, Percentage, Mean, and Standard Deviation. Chi-square tests were conducted to test the relationships between the categorical variables (i.e., level of Knowledge and Risky Behaviors) associated with the Patients and of Logistic Regression Models were developed. A Logistic Regression Model was developed to identify the Independent Predictors of Inappropriate Antibiotic Use, the Logistic Regression Models were adjusted for other Confounding Factors, such as Age, Education and Employment Status. Descriptive Statistics and Comparative Analysis based on Specialty, Experience and Prior AMR Training were examined for Physicians.

Ethics were the priority throughout the development and execution of the present study, including consent from hospital administration to conduct the study, as well as all participant's knowledge of the study's purpose and their ability to refuse participation. All participant's right to cease participation at any time for reasons unrelated to their medical care was also explained; and at this point, consent was obtained either verbally or in writing, depending on the level of reading comprehension for each participant. All participant responses were assigned a unique code and, therefore, were kept confidential and securely stored, and no participant's personal identifying information was recorded and, thus, preserved each participant's privacy, according to ethical standards surrounding the research of human participants. This model provides an extensive and ethically responsible framework for examining treatment practices and behaviours regarding antibiotics and establishes the most critical determinants of inappropriate antibiotic usage; this model will provide the foundation for future dialogue on integrating phytotherapeutic strategies with One Health in order to achieve sustainable ways to reduce AMR.

4. RESULTS

Azzamouri Provincial Hospital's research gathered extensive information on hospitalised patients' knowledge, behaviours and demographic characteristics related to antibiotic use, as well as physicians' prescribing behaviours and perceptions about AMR. The results of the research indicated that hospitalised patients had a significant lack of knowledge about AMR, a number of behaviours that put them at risk for AMR; and numerous organisational factors that enable the emergence and ongoing presence of AMR. This section will summarise and interpret the findings of the study, and address how the findings can be applied to practice and in implementing strategies for preventing AMR that reflect the One Health approach.

Sociodemographic information based on the subjects' sample shows the subjects to be from a socioeconomically diverse and economically vulnerable population (Table 1). Nearly half of the subjects were in their middle adult years and there was an even distribution of male and female subjects. The subjects also had a great deal of variation in terms of education level; while a significant number of the subjects had completed high school or higher, a large proportion did not complete primary school and were never formally educated.

Table 1. Sociodemographic characteristics of participants

Variable	Category	Frequency	Percentage
Gender	Female	59	59%
	Male	41	41%
Education Level	No formal education	33	33,80%
	Primary	23	23,90%
	Middle school	12	12,70%
	University	29	29,60%
Employment Status	Unemployed	55	55%
	Employed	32	32%
	Student	12	12%
	Retired	1	1%

One of the more significant findings regarding the subjects was related to their employment status. Many of the subjects were unemployed or conducted informal work. The limited financial resources associated with the subjects' jobs significantly limited the subjects' ability to obtain healthcare and seek out healthcare services. In addition to providing insight into the subjects' healthcare-seeking behaviours, the demographic profile of the subjects can help further understand the antibiotic use patterns among the subjects. Previous research has suggested that lower socioeconomic status is typically linked to lower levels of health literacy, higher levels of self-medication, and a lower level of understanding of the risks associated with AMR.

A common and striking theme from the study results is that hospitalized patients had very limited knowledge regarding antimicrobial resistance (Table 2). As many as 69% of participants indicated that they had never heard of this phrase and were therefore completely unaware of this current significant threat to global health. The misconceptions held by many respondents about the appropriate uses of antibiotics are even more alarming. Only 48% of respondents were able to indicate correctly that antibiotics work only on bacteria, while the other respondents believed that antibiotics would work on any type of infection including viral infections like colds and flu, as well as any type of fever. This misunderstanding is consistent with other studies demonstrating that in a majority of low- and middle-income countries, antibiotics are seen as a miracle cure for any type of infection. Inability to grasp the specific actions of antibiotics leads to misuse of these products, which is one of the main causes for the development of resistance.

Table 2. Knowledge and perceptions related to antibiotics

Variable	Category	Frequency	Percentage
Knowledge of antibiotics	Yes	73	73%
	No	27	27%
Antibiotics treat...	Bacterial infections (correct answer)	48	48%
	Viral infections	8	8%
	Both	3	3%
	Do not know	41	41%
Awareness of antibiotic resistance	No	69	69%
	Yes	31	31%
Definition provided	The body no longer responds	20	20%
	Bacteria become stronger	7	7%
	Other	4	4%

The severity of inappropriate antibiotic usage is further demonstrated in the population studied, through a behavioral analysis that identified that approximately 72 percent of patients had at least one behavioral risk associated with antibiotics, which involved actions such as using antibiotics without consulting with a physician or other health care provider; stopping treatment before completing the prescribed course; reusing old prescriptions; storing unused antibiotics for future purposes; and sharing with family members. Self-medication makes a significant contribution to high-risk behaviours of patients since 39% of people reported having taken antibiotics without a prescription at some time during their lives. This indicates an increasing trend toward using antibiotics without going to see a physician and also shows how easily available antibiotics can be obtained without a prescription. A subsequent study found that 21% of patients had given other individuals their prescription antibiotics, leading to patients receiving multiple courses of antibiotics without a physician's treatment plan. These types of treatment regimens expose patients to an increased risk of an incompletely treated infection and also raise the risks of developing an antibiotic-resistant strain of bacteria.

A statistical method known as logistic regression has established employment status as a strong predictor of misusing antibiotics (Table 3). The odds ratio (OR) is OR=7.49 for those who are unemployed in comparison to employed individuals for the likelihood of misusing antibiotics; therefore, the odds of misusing antibiotics are seven times greater for unemployed individuals when compared to employed individuals. As a result, individuals from lower socioeconomic backgrounds may not have access to healthcare providers and, therefore, are forced to self-medicate or rely on other informal treatment methods. On the contrary, education relating to the correct usage of antibiotics provides a protective layer; individuals who have received some degree of education on antibiotics and antibiotic resistance will be less likely to misuse antibiotics, with an OR of OR=0.04, indicating that an increased level of public support and education on the proper usage of antibiotics could substantially reduce the number of people who misuse antibiotics. As such, the teaching of health information to at-risk populations to prevent AMR is of extreme importance.

Table 3. Results of binary logistic regression on factors associated with poor practice

Variable	Odds Ratio (OR)	95% CI	p-value	Interpretation
(Intercept)	6.74	[1.73 – 35.97]	0.012	Reference (male, low education level, employed)
Gender: Female	2.82	[0.68 – 13.39]	0.165	Not significant
Education level (high vs. low)	7.86	[1.44 – 68.52]	0.032	Significantly increases the risk
Employment status: Student	0.74	[0.07 – 6.97]	0.795	Not significant
Employment status: Unemployed	7.49	[1.15 – 48.85]	0.032	Significantly increases the risk
Knowledge of antibiotics (Yes vs. No)	0.04	[0.00 – 0.34]	0.006	Highly significant protective factor

The figure 1 illustrates the distribution of a composite Risk Behavior Score related to antibiotic use, which was created by summing four problematic practices: using antibiotics without a prescription, sharing antibiotics, not completing the prescribed treatment, and not consulting a healthcare professional beforehand. Each risky behavior adds one point, resulting in a total score ranging from 0 (no risky behavior) to 4 (maximum risk). The distribution shows that most participants fall into the lower risk categories, with 28% reporting no risky practices and 33%

reporting only one. However, a substantial proportion (22%) scored two points, indicating multiple risky behaviors. Smaller but still concerning groups scored three (14%) or four (3%) points, representing individuals at higher cumulative risk. Overall, the figure suggests that while extreme risk behaviors are relatively limited, a majority of individuals exhibit at least one inappropriate antibiotic-use practice, highlighting the need for targeted awareness and intervention strategies to reduce antibiotic misuse and combat antimicrobial resistance.

Figure 1. Distribution of risk behaviors related to antibiotic use: frequency and prevalence

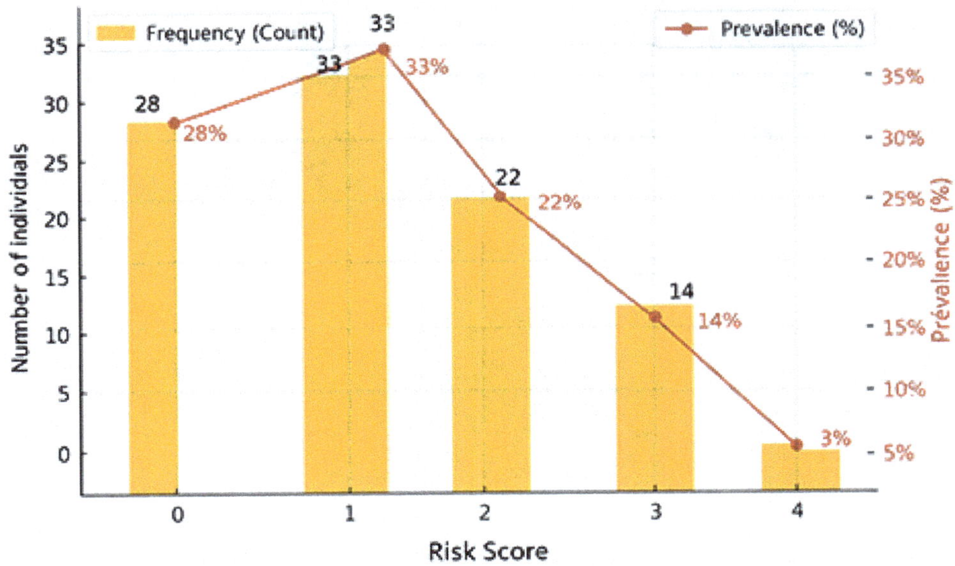

The combined distribution of frequency and prevalence of risk behavior scores related to antibiotic use highlights a concerning trend: the majority of hospitalized patients exhibit one or more behaviors that increase the risk of antibiotic resistance. Nearly 33% of respondents have a score of 1, while 39% reach a score of 2 or higher, underscoring the common occurrence of inappropriate practices such as self-medication, sharing antibiotics, or not adhering to prescribed treatments. This visualization emphasizes the need for an integrated preventive approach, focusing both on patient therapeutic education and the accountability of healthcare providers. Joint analysis of frequencies and prevalences allows not only the assessment of the phenomenon's magnitude but also the identification of the most vulnerable groups, supporting the implementation of targeted strategies to combat antibiotic resistance.

The study included a total of 40 physicians, all practicing in hospital settings. The gender distribution showed that 57.5% of participants were men (n = 23) compared to 42.5% women (n = 17). Regarding professional status, the vast majority of respondents were physicians (92.5%, n = 37), while medical interns represented 7.5% of the sample (n = 3). All participants practiced in hospital settings, which explains the homogeneity in the workplace environment (Figure 2).

Figure 2. Descriptive analysis of the characteristics of the surveyed physicians

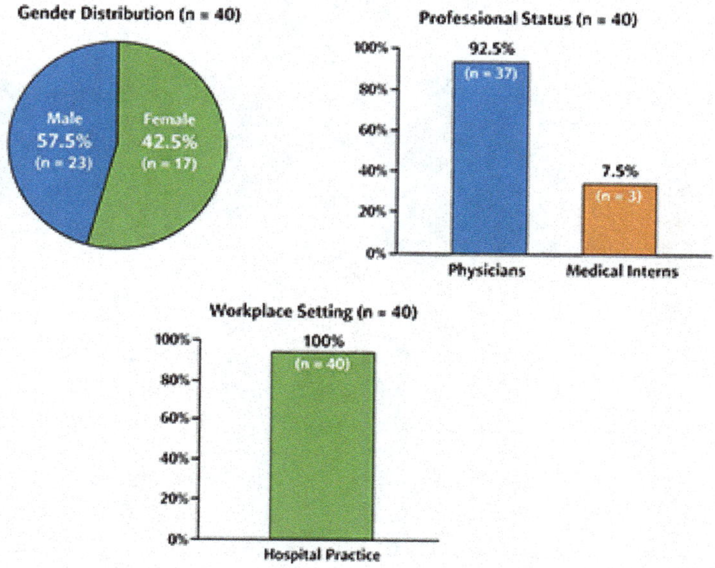

The data on doctors has an important point of view, as it relates to the clinical aspects of AMR. For example, approximately 67.5% of doctors say they have been offered training around AMR, and this training can differ widely in scope and up-to-dateness. Some doctors have been involved in structured workshops or university courses; however, many others have been relying on informal/everyday sources of information about AMR that consist of only random instances of exposure. Nonetheless, there are still a high number of doctors prescribe antibiotics on a daily basis (87.5%). High levels of antibiotic prescribing do not always mean that there is a problem with inappropriate use but certainly indicate the need for strong programs to manage antibiotic use in acute care facilities where demand for these services is often very high. When asked whether antibiotic resistance is perceived as a problem, the majority of physicians (62.5%, n = 25) considered it a global issue, followed by

20% (n = 8) who viewed it as a national problem. A few physicians perceived it as a local problem (12.5%, n = 5) or both a local and national issue (5%, n = 2) (Table 4).

Table 4. Perception of antibiotic resistance as a problem

Response	Frequency (n)	Percentage (%)
Global problem	25	62.5%
National problem	8	20%
Local problem	5	12.5%
Local and national problem	2	5%

The physicians expressed several barriers impacting their prescriptions. Many physicians cited a lack of certainty about the diagnosis as an issue, particularly in settings with minimal access to rapid diagnostic tests. A number of physicians indicated that, when microbiological confirmation was not available, they had initiated empirical antibiotic therapy primarily as a safety net. There was also pressure from patients. Some physicians described how both patients and families were often expecting to receive antibiotic prescriptions to validate that there was an effective means of managing patients' illness. This is symptomatic of the imbalance of clinical judgment and patient satisfaction. Additionally, additional barriers included access to outdated prescribing practices, lack of a wide variety of narrow-spectrum antibiotics, and limited access to current guidelines.

5. DISCUSSION

The results from this investigation provide evidence of an intricate interaction between numerous behavioral, educational, clinical, and structural determinants that, in combination, affect AMR both in healthcare systems and in society. These findings should be viewed through the One Health perspective, which emphasises the integration and connection between the health of people, animals, and the earth, and therefore calls for the application of many strategies beyond the scope of standard approaches to antibiotic use management, including public awareness, strategies for changing behaviours, and the use of alternative therapies such as herbal medicine, among others, to help mitigate the impact of AMR. Additionally, this report discusses how these components may be incorporated into developing new methods to achieve sustainable strategies for mitigating AMR.

The study has revealed a key finding related to patients exhibiting behavioral paradoxes'. The expectation may be that people with higher levels of educational attainment and those who perceive themselves as knowledgeable about antibiotics will exhibit healthier behaviour regarding their use of antibiotics; however, the data indicate the opposite. The most confident individuals regarding their level of knowledge concerning antibiotics exhibited some of the riskiest behaviours. Behavioral scientists call the phenomenon the Cognitive Overconfidence Effect (and sometimes the Illusion of Knowledge), as demonstrated multiple times by multiple researchers. With little knowledge regarding the mechanism of action of antibiotics due to superficial exposure to this information (through cultural influences or personal experiences), individuals believe they understand antibiotics well enough to make their own decisions regarding treatment alternatives, e.g., self-prescribing, discontinuing treatment prematurely, or giving away their antibiotic therapies. Because patients do not necessarily have a full understanding of how antibiotic therapies work, they often do not perceive the ramifications of their behaviour, indicating the need for patient education that encompasses both the facts about antibiotic therapy as well as helping patients effectively refute any erroneous assumptions they may have regarding the way antibiotics perform in their body. A campaign based solely on slogans or messages unrelated to a patient's beliefs or perceptions of antibiotic therapy may not effectively alter their misperceptions of the treatment because they do not feel that they require additional education on this type of therapy.

This report describes a lack of communication and education through the finding that 69% of patients were unaware of what "antimicrobial resistance" (AMR) actually means. Patients' unawareness of AMR shows their lack of knowledge of the negative consequences associated with misuse of antibiotics. Such unawareness of AMR and the associated low health literacy levels were also found in many other developing or middle-income countries where access to reputable, accurate medical information is limited. Although the issue is not unique to Morocco, patients that do not understand AMR are less likely to take preventive measures, seek the help of a qualified healthcare provider for an accurate assessment and usage of prescribed medications. The fact that the majority of patients (39%) reported using self-treatment indicates the extent to which the sociocultural and structural factors in Morocco promote and support the use of informal treatment methods. Examples of informal treatment methods include dispensing of medications from pharmacies without a doctor's written prescription, the availability of community-based pharmacy services without a written prescription, and a lack of financial resources needed for adequate access to qualified healthcare providers. Based on these results, there is a critical need for public education programs that are interactive and tailored to the specific local context. Instead of using a generic approach, each intervention should take into consideration the cultural context, linguistic nuances, and the level of literacy of the community. Therefore, any educational program should include representatives from the community (i.e., community leaders, community pharmacists, community nurses, and community physicians) who will utilize visual aids, storytelling, culturally appropriate metaphor, and participatory workshops as guides for teaching about how to develop and maintain healthy behaviors. Additionally, as individual behavioral patterns are shaped both by society's social norms and expectations, it is important to develop and implement mass media campaigns and school-based learning opportunities as well as digital tools to reinforce positive behaviors and to dispel myths on a larger scale.

The study confirms that major physician-related impediments to antibiotic stewardship exist at the clinical level. Of the physicians surveyed, 66% indicated that they had received some form of training on AMR; however, the findings show that this training was not sufficient in either frequency or content to sustain physician best practices. The physicians cited the following common issues that contributed to poor prescribing practices (1) diagnostic uncertainty (2) time pressure (3) patient expectations/conflicts and (4) a lack of access to up-to-date guidelines/narrow-spectrum antibiotics. These issues can be found in almost all healthcare systems that operate with limited resources. The study also identified a high level of antibiotic prescribing by Alberta physicians, with 87.5% of physicians reporting they prescribed antibiotics regularly. This high frequency of prescribing could be indicative of the realities of clinical practice at a busy provincial hospital. However, it is also indicative of the need for improved stewardship mechanisms. Furthermore, the results indicate that physicians who participate in formally organised and regularly scheduled AMR training engage in far higher amounts of rational prescribing behaviour. This finding complements other studies that demonstrate that continued medical education programs significantly enhance the way antibiotics are used and greatly decrease the number of inappropriate prescriptions being written.

Systemic interventions are required to address the barriers physicians face when treating patients with infections, such as building capacity to increase access to diagnostic tools and creating more evidence-based treatment guidelines and institutional stewardship teams that support physicians in making evidence-based decisions. In addition, hospital-level policies must empower physicians to resist patient pressure for unnecessary prescribing of antibiotics while having a clear communication strategy/plan to help physicians explain their treatment decisions to patients. To help physicians make immediate clinical decisions according to global AMR strategies, clinical decision support systems (CDSS) can be utilized, such as electronic guidelines and mobile applications that provide real-time clinical guidance.

In addition to exploring behavioral and clinical factors, the study provides an opportunity to evaluate the possible contribution of herbal medicine to reducing antibiotic use. Due to Morocco's long-standing tradition of ethnopharmacology, the use of scientifically validated herbal remedies may represent a culturally acceptable and sustainable alternative for managing mild infections while potentially decreasing unnecessary antibiotic consumption. The utilization of medicinal plants with their extensive range of biological activity against bacteria, reducing inflammation, and potentially improving the effectiveness of antibiotics has demonstrated that products made from A. annua, either in tea form or taken as an infusion, have shown evidence of being effective at treating bacterial infections of the respiratory tract or in skin infections in individuals who may otherwise have had an elevated request for an antibiotic. The extraction of certain compounds from plants, such as the essential oils extracted from thyme and eucalyptus and the flower oil extracted from lavender may effectively reduce the occurrence of mild dermatological infections and respiratory tract infections. Curcuma longa and Azadirachta indica have been shown to have activity against biofilms produced by bacteria and efflux pumps associated with AMR.

Herbal medicine is an exciting area of interest; however, if we were to combine this method with other medicines, it would need to be done in a thoughtful and logical way via controlled studies. To properly evaluate herbal medicine, we must establish standards for herbs; regulate dosages; monitor for adverse reactions (pharmacovigilance). Also, we do not expect herbal medicines to be effective replacements for the use of antibiotics for the treatment of serious and/or life-threatening infections. Instead, herbal medicines can be used as a means of reducing inappropriate use of antibiotics and can be a supplemental therapy to antibiotic use. Education programs regarding herbal medicines should include information regarding which conditions can benefit from phytotherapeutic and when to seek medical evaluation. Lastly, any use of herbal medicine for AMR prevention must be based on scientific evidence, in order to avoid continuing to reinforce untruths or create potentially harmful self-medication behaviours.

Behavioral change, clinical stewardship, and phytotherapy are intertwined into a single framework (the One Health Framework). One Health recognizes that AMR is shaped not just by human medical behaviour but also by animal medical behaviour, agricultural practises, and environmentally friendly management. There are a number of reasons why AMR is an increasing public health threat, including misuse of antibiotic products in livestock production, contamination of water supplies with pharmaceutical waste, and a lack of proper waste management systems for the disposal of unused or extra antibiotics. Therefore, if interventions only target hospitals, they will not be effective in addressing AMR. There are also similarities in the way that patients behave when taking medications (e.g. sharing antibiotics, interrupting their treatment) across all sectors. For instance, small-scale farmers may use veterinary antibiotics without adequate direction or supervision, and contamination of the environment allows for the perpetuation of resistant microorganisms, which then enter into the food chain or drinking-water supply and subsequently affect humans.

This study's findings can serve as a foundation for developing a One Health approach to AMR prevention through a coordinated and holistic prevention model.

This model will centre on increasing the understanding of the subject through education and behavioural change since all aspects of AMR are determined by how people think about, obtain, and take antibiotics. To increase the understanding of patients, communities, schools and the healthcare community about when antibiotics are appropriate and how to appropriately take antibiotics to prevent AMR, comprehensive health literacy programmes should be established. Comprehensive health literacy programmes must be conducted to increase awareness of health literacy on multiple levels; they cannot be delivered in the traditional way, rather, they must be culturally appropriate and contain content that will combat cognitive biases and the tendency to think they are overly confident regarding their antibiotic use. Community leaders should play an active role in developing and utilising social media to disseminate these materials and create participatory opportunities for community members to engage with them, thereby making them more relevant to the varying communities in question.

The Model's Second Pillar is to integrate phytotherapy (herbal medicines) and other forms of complementary medicine as a supportive strategy to diminish the unnecessary use of antibiotics. Herbal remedies that have been validated scientifically could be included in primary care, specifically for the treatment of mild or self-limiting infections where antibiotics are frequently prescribed inappropriately. For example, using Morocco's ethnopharmacological heritage will help to reduce the pressure for antibiotics by reinforcing the popular practice of using natural remedies. To provide a safe and effective method of using these remedies, it is essential to develop a set of National Guidelines that will ensure all practitioners have access to the same information about preparing, dosing, and the therapeutic indications for the use of these remedies. The development of these guidelines will be based on a scientific assessment and a well-established system for pharmacovigilance to ensure that practitioners monitor any interactions/contraindications with other medications or treatments and track the outcomes of treatments.

The model supports the implementation of policies that create a sustainable future for hospitals to improve antibiotic stewardship and support healthcare professionals in the application of scientific evidence to assist them with evidence-based therapy decision making. Providing healthcare professionals access to more diagnostic tools, such as rapid tests and microbiological tests, decreases the uncertainty in the diagnoses of their patients and decreases the likelihood that healthcare professionals will unnecessarily or improperly prescribe antibiotics empirically or broadly for their patients. Access to continuous education/training, along with communities of practice for healthcare professionals, enables them to remain abreast of AMR trends, stewardship best practices and guidelines, and best practice for antibiotic prescribing. In addition, healthcare organisations should implement systems that allow for the collection of information on their use of antibiotics (including prescription rates, local resistance patterns, etc.) so that targeted interventions and corrective actions can be developed. The establishment of these mechanisms at the institutional level will enable healthcare organisations to align their practices with the strategies being developed internationally to combat AMR. One Health supports a unified approach toward protecting the human, animal, and environmental health by acknowledging that antimicrobial resistance (AMR) cannot be effectively managed solely in Humans. Regulating the use of antibiotics in food animals helps prevent the transfer of resistant bacteria from the food animal to humans through the food supply. Supporting the use of plant-based or herbal alternatives to traditional antibiotics is a sustainable way to decrease our reliance on growth promoters that contain antimicrobials. There is also a need for wastewater treatment systems capable of safeguarding the aquatic ecosystems by identifying and removing pharmaceuticals from wastewater before it enters the environment; without such treatment, rivers, soils, and agricultural land will continue to become contaminated with resistant organisms. Furthermore, Environmental Surveillance Programs can help identify areas with a high prevalence of AMR and develop public health policies to control the spread of AMR in these sectors.

The proposed AMR model is centered on the intersection of human behavior, ecologically responsible stewardship and the diverse forms of Alternative Medicine available to the public. The proposed model will provide an integrated way to encourage Sustainable Antimicrobial Stewardship for improving antibiotic resistance through the use of the combination of scientific evidence with disparate cultural realities. The findings from this study provide evidence that assisting healthcare professionals, educating the public, and promoting new forms of phytotherapeutic alternative medicine are three approaches which could be used together to decrease the rate of antibiotic resistance and slow the development of further resistant strains of bacteria. This integrated approach presents a significant opportunity for countries such as Morocco that have a strong association between traditional and alternative medicine and are experiencing an increasing AMR threat to their overall public health and environmental sustainability.

6. RECOMMENDATIONS AND POLICY IMPLICATIONS

This study clearly shows that in order to effectively combat AMR, there must be a comprehensive approach including the behavioural, clinical, environmental, and cultural factors that affect AMR. To develop successful, sustainable prevention strategies for AMR based on the recommendations of this study, all stakeholders will need to work together to implement a number of public education initiatives, implement health system reform, develop new types of phytotherapeutics, and engage in multi-sectoral collaboration. These recommendations are based on the One Health framework.

A critical recommendation is to enhance training and education for healthcare providers (physicians, nurses and pharmacists) regarding antimicrobial resistance (AMR). The results of the study demonstrated that medical providers who regularly received training on AMR prescribed medications more rationally than those who did not and this highlights the need for continuing development. The training curriculum should include the most current international guidelines, local resistance patterns, a case-based approach, and a communication strategy to handle the patient's expectations regarding antimicrobial use. The use of digital delivery such as e-learning can be used to improve access to this training by enabling the training to be provided to those who work in rural or resource-limited areas. In addition to formal education, it is also important for hospitals to establish antimicrobial stewardship (AMS) teams to help ensure appropriate prescribing by reviewing medication orders and assisting clinical decision making.

There is a pressing need for the development and dissemination of community 'Culturally Adaptive Public Awareness Campaigns' that address the misconceptions associated with Antibiotic Use and the introduction of Herbal Medicine as a scientifically validated adjunctive treatment option for mild infections. This requires using a combination of mass media (TV, Radio, Social Media, etc.), Educational Community Centers, and Schools to reach individuals from different economic classes and backgrounds. The incorporation of Medicinal Plants that are recognized and used locally for centuries can enhance the validity of the educational message and improve the cultural relevance of the educational message. Providing examples of how to safely use Herbal Remedies will decrease the perception that Antibiotics are necessary for treating Minor Ailments and, as a result, decrease the pressure on Antimicrobial Agents.

AMR can only be tackled successfully through a strong partnership and collaboration between various sectors. The concept of "One Health" reflects the fact that AMR is not just a problem for the health sector but also interacts with agriculture, the environment, education, and public policy. Therefore, Ministries of Health, Agriculture, Education, and the Environment must work collaboratively and develop integrated policies that will regulate antibiotic use across all these sectors. For example, enforcing laws regarding the sale of antibiotics over the counter (OTC) by pharmacies will help to reduce the incidence of self-medication. Likewise, a nation's agricultural policy should restrict the use of antibiotics as growth stimulants in animals and promote the development of phytotherapeutics for the veterinary treatment of diseases.

A second strategic priority will be strengthening the support for developing applied research into Moroccan medicinal plants possessing antimicrobial properties.

The rich biodiversity of plants in Morocco, combined with the traditional use of these plants for treating ailments, provides researchers and entrepreneurs with an extensive library of phytochemical compounds to search through. Institutions conducting research, universities, and biotechnology companies should work together to develop new antibiotic therapies based on plant sources; these companies should also be encouraged to study the safety and dosing parameters of these plant-derived compounds as well as their ability to produce synergy with traditional antibiotics. The establishment of public-private partnerships focused on the development of new phytopharmaceuticals may result in both creating an opportunity for innovation and supporting local economies while addressing a significant public health challenge. In addition, establishing a national repository for herbal formulations, phytochemical profiles, and bioactivity data will enable continued scientific progress in this area.

Herbal interventions in hospital settings represent a promising alternative for reducing antimicrobial resistance (AMR), aligning with the One Health perspective, which integrates interactions between human, animal, and environmental health. Many medicinal plants contain bioactive compounds, such as phenols, flavonoids, terpenes, and alkaloids, with antimicrobial properties. Their targeted use can reduce bacterial load without relying solely on synthetic antibiotics, thereby limiting the selective pressure that drives the emergence of resistant strains. Furthermore, certain plant extracts exhibit synergy with existing antibiotics. They can enhance antibiotic efficacy, even against resistant bacteria, by inhibiting resistance mechanisms such as efflux pumps or cell wall synthesis. This approach allows for lower antibiotic doses, helping to curb the development of AMR while maintaining therapeutic effectiveness. Phytotherapeutic interventions also play a role in preventing nosocomial infections. The use of plant-based disinfectants or immunomodulatory adjuvants can reduce pathogen transmission in hospital settings. By lowering infection rates and disease severity, these interventions decrease the need for antibiotics, thus limiting the selection and spread of resistant strains. The environmental dimension is also critical within a One Health framework. Extensive antibiotic use leads to contamination of soils and water with drug residues, promoting resistance spread across human and animal ecosystems. Integrating medicinal plants as therapeutic alternatives helps reduce this environmental burden, protecting biological cycles and limiting AMR dissemination.

Finally, for these recommendations to produce the greatest benefit, they should be integrated into the Moroccan National AMR-One Health Strategy. This integration would allow seamless cooperation among all hospital-related programs, including AMR educational initiatives and community-based efforts, as well as coordination with Environmental and Agricultural Policies. The implementation of a comprehensive national strategy should include clearly defined, measurable objectives, tracking indicators, and ongoing evaluations of progress. In addition to new policies, data on antimicrobial use and resistance will provide decision-makers with the necessary information to guide policy development. Political will, the establishment of a sustainable funding mechanism, and the creation of an inclusive and equitable governing body spanning all relevant sectors will be essential for achieving long-term change.

A coordinated effort encompassing education, medical practice reform, environmental care, and the advancement of science and technology is critical to effectively reduce AMR in Morocco and similar countries. In addition to conventional medicine, it is important to consider phytotherapy as a complementary therapy. Many medicinal plants possess immunostimulatory properties that strengthen patients' natural defenses, reducing the frequency and severity of infections and, consequently, reliance on antibiotics. By combining direct antimicrobial action, synergy with antibiotics, infection prevention, environmental protection, and immune system enhancement, herbal interventions offer an effective strategy to reduce AMR in hospitals. Incorporating interventions that are culturally relevant and adhere to One Health principles provides Morocco with a sustainable and resilient approach to preserving antimicrobial efficacy for future generations, fully integrating human, animal, and environmental dimensions to curb the spread of antimicrobial resistance.

7. CONCLUSION

The issue of antimicrobial resistance (AMR) is a worldwide public health challenge that is driven by many different institutional, social, behavioural, and environmental factors. The findings of the study show that much of the current patient misuse of antibiotics is the result of these various factors, including the limited level of education, the tendency toward self-treatment, and sociocultural norms, working in conjunction with the structural barriers faced by providers (for example, a lack of training and uncertainty when diagnosing patients). The interactions among the above variables have created a cycle of misuse that promotes the emergence of microorganisms resistant to antimicrobial agents. Therefore, when designing interventions, it is vital to take into account the dual nature of AMR as a clinical and behavioural crisis in order to develop successful, sustainable, and contextually relevant interventions. Finally, the results of this study provide evidence that phytotherapy may be an effective method of assisting in the management of unnecessary antibiotic use. Traditional Moroccan culture has long included the use of certain botanicals that possess bioactive properties. Medicinal plants may have an array of bioactive compounds that exert antimicrobial effects, and they may modulate the body's resistance mechanisms, thereby enhancing the effectiveness of the treatment administered to patients. Therefore, while phytotherapy is not intended to replace the use of medicinal antibiotics for the treatment of serious cases of infection, it does provide an alternative to people who seek to treat minor illnesses using methods that are in harmony with their cultural beliefs. Additionally, providing sufficient scientific validation of the efficacy of herbal medicine in community health settings and in primary care clinics will surely be beneficial in the long-term fight against AMR, provided all practices associated with herbal medicine are supported by sound science, dosage measurement, and pharmacovigilance at every level. A more extensive One Health view on AMR shows that the problem will not be resolved through isolated interventions. Instead, human health-related behaviours, best clinical practice methods, agricultural use of antibiotics, and environmental contamination are all intertwined and should be treated together. To effectively prevent AMR, a comprehensive approach across all sectors/areas is required. This could be achieved through the strategic combination of focused public education campaigns, improved stewardship in healthcare institutions, environmental quality legislation, proper regulation of veterinary practice, and support for research into phytotherapy. The study highlights the necessity for multidimensional AMR prevention strategies that take into account Human behaviours, Cultural Practices, and Ecological Sustainability, as well as the power of Phytotherapy and Education as complementary solutions and their potential role in significantly reducing antimicrobial pressures if included in National Policies and One Health Frameworks. Countries such as Morocco and others that face similar challenges should invest in the validation, regulation and promotion of medicinal plants as part of Evidence-Based Healthcare through increased cross-sectoral collaboration and public engagement to build a resilient future for Antibiotics to be effective generations from now.

ACKNOWLEDGMENT

The authors would like to sincerely thank all contributors for their invaluable support and insightful input throughout this study. We also extend our gratitude to ChatGPT for assistance with linguistic revision and improving the clarity and readability of the manuscript.

REFERENCES

Al-Khalaifah, H., Rahman, M. H., Al-Surrayai, T., Al-Dhumair, A., & Al-Hasan, M. (2025). A One-Health perspective of antimicrobial resistance (AMR): Human, animals and environmental health. *Life (Basel, Switzerland)*, *15*(10), 1598. DOI: 10.3390/life15101598 PMID: 41157271

Alam, R., Cheraghi-Sohi, S., Panagioti, M., Esmail, A., Campbell, S., & Panagopoulou, E. (2017). Managing diagnostic uncertainty in primary care: A systematic critical review. *BMC Family Practice*, *18*(1), 79. DOI: 10.1186/s12875-017-0650-0 PMID: 28784088

Alhazmi, H. A., Najmi, A., Javed, S. A., Sultana, S., Al Bratty, M., Makeen, H. A., Meraya, A. M., Ahsan, W., Mohan, S., Taha, M. M. E., & Khalid, A. (2021). Medicinal plants and isolated molecules demonstrating immunomodulation activity as potential alternative therapies for viral diseases including COVID-19. *Frontiers in Immunology*, *12*, 637553. DOI: 10.3389/fimmu.2021.637553 PMID: 34054806

Anand, U., Jacobo-Herrera, N., Altemimi, A., & Lakhssassi, N. (2019). A comprehensive review on medicinal plants as antimicrobial therapeutics: Potential avenues of biocompatible drug discovery. *Metabolites*, *9*(11), 258. DOI: 10.3390/metabo9110258 PMID: 31683833

Ashraf, M. V., Pant, S., Khan, M. A. H., Shah, A. A., Siddiqui, S., Jeridi, M., Alhamdi, H. W. S., & Ahmad, S. (2023). Phytochemicals as antimicrobials: Prospecting Himalayan medicinal plants as sources of alternative medicine to combat antimicrobial resistance. *Pharmaceuticals (Basel, Switzerland)*, *16*(6), 881. DOI: 10.3390/ph16060881 PMID: 37375828

Aslam, A., Jamshed, S., Zin, C. S., Rahman, N. S. A., Ahmed, S. I., Szabó, Z. L., & Gajdács, M. (2024). Understanding the determinants influencing self-medication with antibiotics among Malaysian residents: A qualitative study to inform preventive public health strategies. *Antibiotics (Basel, Switzerland)*, *13*(11), 1070. DOI: 10.3390/antibiotics13111070 PMID: 39596763

Aslam, B., & Aljasir, S. F. (2025). Climate change and AMR: Interconnected threats and One Health solutions. *Antibiotics (Basel, Switzerland), 14*(9), 946. DOI: 10.3390/antibiotics14090946 PMID: 41009922

Aslam, B., Khurshid, M., Arshad, M. I., Muzammil, S., Rasool, M., Yasmeen, N., Shah, T., Chaudhry, T. H., Rasool, M. H., Shahid, A., Xueshan, X., & Baloch, Z. (2021). Antibiotic resistance: One Health, One World outlook. *Frontiers in Cellular and Infection Microbiology, 11*, 771510. DOI: 10.3389/fcimb.2021.771510 PMID: 34900756

Bassetti, M., & Giacobbe, D. R. (2020). A look at the clinical, economic, and societal impact of antimicrobial resistance in 2020. *Expert Opinion on Pharmacotherapy, 21*(17), 2067–2071. DOI: 10.1080/14656566.2020.1802427 PMID: 32749893

Ben, Y., Fu, C., Hu, M., Liu, L., Wong, M. H., & Zheng, C. (2019). Human health risk assessment of antibiotic resistance associated with antibiotic residues in the environment: A review. *Environmental Research, 169*, 483–493. DOI: 10.1016/j.envres.2018.11.040 PMID: 30530088

Bertollo, L. G., Lutkemeyer, D. S., & Levin, A. S. (2018). Are antimicrobial stewardship programs effective strategies for preventing antibiotic resistance? A systematic review. *American Journal of Infection Control, 46*(7), 824–836. DOI: 10.1016/j.ajic.2018.01.002 PMID: 29472107

Breijyeh, Z., & Karaman, R. (2023). Design and synthesis of novel antimicrobial agents. *Antibiotics (Basel, Switzerland), 12*(3), 628. DOI: 10.3390/antibiotics12030628 PMID: 36978495

Calvo-Villamañán, A., San Millán, Á., & Carrilero, L. (2023). Tackling AMR from a multidisciplinary perspective: A primer from education and psychology. *International Microbiology, 26*(1), 1–9. DOI: 10.1007/s10123-022-00278-1 PMID: 36224500

Caneschi, A., Bardhi, A., Barbarossa, A., & Zaghini, A. (2023). The use of antibiotics and antimicrobial resistance in veterinary medicine, a complex phenomenon: A narrative review. *Antibiotics (Basel, Switzerland), 12*(3), 487. DOI: 10.3390/antibiotics12030487 PMID: 36978354

Chakravorty, A., Biswas, B., Sana, S. S., Rayan, R. A., Lala, N. L., & Ramakrishna, S. (2021). A review on toxicity of turmeric derived nano-formulates against bacterial and fungal cells with special emphasis on electrospun nanofibers. *Materials Today: Proceedings, 46*(Part 14), 6358–6362. DOI: 10.1016/j.matpr.2020.05.820

Chambers, J. A., Crumlish, M., Comerford, D. A., & O'Carroll, R. E. (2020). Antimicrobial resistance in humans and animals: Rapid review of psychological and behavioral determinants. *Antibiotics (Basel, Switzerland)*, *9*(6), 285. DOI: 10.3390/antibiotics9060285 PMID: 32471045

Dempsey, R. C., McAlaney, J., & Bewick, B. M. (2018). A critical appraisal of the social norms approach as an interventional strategy for health-related behavior and attitude change. *Frontiers in Psychology*, *9*, 2180. DOI: 10.3389/fpsyg.2018.02180 PMID: 30459694

Endale, H., Mathewos, M., & Abdeta, D. (2023). Potential causes of spread of antimicrobial resistance and preventive measures in One Health perspective — A review. *Infection and Drug Resistance*, *16*, 7515–7545. DOI: 10.2147/IDR.S428837 PMID: 38089962

Fong, I. W. (2023). Antimicrobial Resistance: A Crisis in the Making. In *New Antimicrobials: For the Present and the Future. Emerging Infectious Diseases of the 21st Century*. Springer., DOI: 10.1007/978-3-031-26078-0_1

Fontefrancesco, M., Barstow, C., Grazioli, F., Lyons, H., Mattalia, G., Marino, M., McKay, A. E., Sõukand, R., Corvo, P., & Pieroni, A. (2019). Keeping or changing? Two different cultural adaptation strategies in the domestic use of home country food plant and herbal ingredients among Albanian and Moroccan migrants in Northwestern Italy. *Journal of Ethnobiology and Ethnomedicine*, *15*(1), 11. DOI: 10.1186/s13002-019-0290-7 PMID: 30755237

Hawkings, N. J., Butler, C. C., & Wood, F. (2008). Antibiotics in the community: A typology of user behaviours. *Patient Education and Counseling*, *73*(1), 146–152. DOI: 10.1016/j.pec.2008.05.025 PMID: 18640805

Helmy, Y. A., Taha-Abdelaziz, K., Hawwas, H. A. E.-H., Ghosh, S., AlKafaas, S. S., Moawad, M. M. M., Saied, E. M., Kassem, I. I., & Mawad, A. M. M. (2023). Antimicrobial resistance and recent alternatives to antibiotics for the control of bacterial pathogens with an emphasis on foodborne pathogens. *Antibiotics (Basel, Switzerland)*, *12*(2), 274. DOI: 10.3390/antibiotics12020274 PMID: 36830185

Hernando-Amado, S., Coque, T. M., Baquero, F., & Martínez, J. L. (2020). Antibiotic Resistance: Moving From Individual Health Norms to Social Norms in One Health and Global Health. *Frontiers in Microbiology*, *11*, 1914. DOI: 10.3389/fmicb.2020.01914 PMID: 32983000

Irfan, M., Almotiri, A., & AlZeyadi, Z. A. (2022). Antimicrobial Resistance and Its Drivers—A Review. *Antibiotics (Basel, Switzerland)*, *11*(10), 1362. DOI: 10.3390/antibiotics11101362 PMID: 36290020

Jones, M., Butler, J., Graber, C. J., Glassman, P., Samore, M. H., Pollack, L. A., Weir, C., & Goetz, M. B. (2017). Think twice: A cognitive perspective of an antibiotic timeout intervention to improve antibiotic use. *Journal of Biomedical Informatics, 71*(Supplement), S22–S31. DOI: 10.1016/j.jbi.2016.06.005 PMID: 27327529

MacGowan, A., & Macnaughton, E. (2017). Antibiotic resistance. *Medicine (Abingdon), 45*(10), 622–628. DOI: 10.1016/j.mpmed.2017.07.006

Manyi-Loh, C., Mamphweli, S., Meyer, E., & Okoh, A. (2018). Antibiotic use in agriculture and its consequential resistance in environmental sources: Potential public health implications. *Molecules (Basel, Switzerland), 23*(4), 795. DOI: 10.3390/molecules23040795 PMID: 29601469

Mathers, J. J., Flick, S. C., & Cox, L. A.Jr. (2011). Longer-duration uses of tetracyclines and penicillins in U.S. food-producing animals: Indications and microbiologic effects. *Environment International, 37*(5), 991–1004. DOI: 10.1016/j.envint.2011.01.014 PMID: 21435723

McParland, J. L., Williams, L., Gozdzielewska, L., Young, M., Smith, F., MacDonald, J., Langdridge, D., Davis, M., Price, L., & Flowers, P. (2018). What are the 'active ingredients' of interventions targeting the public's engagement with antimicrobial resistance and how might they work? *British Journal of Health Psychology, 23*(4), 804–819. DOI: 10.1111/bjhp.12317 PMID: 29804314

Michalkova, R., Mirossay, L., Kello, M., Mojzisova, G., Baloghova, J., Podracka, A., & Mojzis, J. (2023). Anticancer potential of natural chalcones: In vitro and in vivo evidence. *International Journal of Molecular Sciences, 24*(12), 10354. DOI: 10.3390/ijms241210354 PMID: 37373500

Moura, A., Lopes, L., Matos, L. C., Machado, J., & Criado, M. B. (2023). Converging paths: A comprehensive review of the synergistic approach between complementary medicines and Western medicine in addressing COVID-19 in 2020. *BioMed, 3*(2), 282–308. DOI: 10.3390/biomed3020025

Nahar, P., Unicomb, L., Lucas, P. J., Uddin, M. R., Islam, M. A., Nizame, F. A., Khisa, N., Akter, S. M. S., & Rousham, E. K. (2020). What contributes to inappropriate antibiotic dispensing among qualified and unqualified healthcare providers in Bangladesh? A qualitative study. *BMC Health Services Research, 20*(1), 656. DOI: 10.1186/s12913-020-05512-y PMID: 32669092

Norris, P., Chamberlain, K., Dew, K., Gabe, J., Hodgetts, D., & Madden, H. (2013). Public beliefs about antibiotics, infection and resistance: A qualitative study. *Antibiotics (Basel, Switzerland), 2*(4), 465–476. DOI: 10.3390/antibiotics2040465 PMID: 27029314

Nwozo, O. S., Effiong, E. M., Aja, P. M., & Awuchi, C. G. (2023). Antioxidant, phytochemical, and therapeutic properties of medicinal plants: A review. *International Journal of Food Properties*, *26*(1), 359–388. DOI: 10.1080/10942912.2022.2157425

Razzok, E. M., Mekhfi, H., Abda, N., Lekfif, A., Dahbi, N., & Bentata, Y. (2025). Medicinal plants use by patients at a Moroccan University Hospital: A cross-sectional study. *Advances in Integrative Medicine*, *12*(3), 100487. DOI: 10.1016/j.aimed.2025.100487

Reza, A., Sutton, J. M., & Rahman, K. M. (2019). Effectiveness of efflux pump inhibitors as biofilm disruptors and resistance breakers in Gram-negative (ESKAPEE) bacteria. *Antibiotics (Basel, Switzerland)*, *8*(4), 229. DOI: 10.3390/antibiotics8040229 PMID: 31752382

Salam, M. A., Al-Amin, M. Y., Salam, M. T., Pawar, J. S., Akhter, N., Rabaan, A. A., & Alqumber, M. A. A. (2023). Antimicrobial resistance: A growing serious threat for global public health. *Health Care*, *11*(13), 1946. DOI: 10.3390/healthcare11131946 PMID: 37444780

Sharma, P. K., Singh, A., & Sharma, N. K. (2022). A socio-ecological critique on India's local health traditions amidst rising incidence of global pandemics. *Journal of Herbal Medicine*, *34*, 100578. DOI: 10.1016/j.hermed.2022.100578 PMID: 35722654

Shinyuy, L. M., Loe, G. E., Jansen, O., Mamede, L., Ledoux, A., Noukimi, S. F., Abenwie, S. N., Ghogomu, S. M., Souopgui, J., Robert, A., Demeyer, K., & Frederich, M. (2023). Secondary metabolites isolated from Artemisia afra and Artemisia annua and their anti-malarial, anti-inflammatory and immunomodulating properties—Pharmacokinetics and pharmacodynamics: A review. *Metabolites*, *13*(5), 613. DOI: 10.3390/metabo13050613 PMID: 37233654

Tang, K. W. K., Millar, B. C., & Moore, J. E. (2023). Antimicrobial resistance (AMR). *British Journal of Biomedical Science*, *80*, 11387. DOI: 10.3389/bjbs.2023.11387 PMID: 37448857

Torres, N. F., Solomon, V. P., & Middleton, L. E. (2023). "Antibiotics heal all diseases": The factors influencing the practices of self-medication with antibiotics in Maputo City, Mozambique. *Journal of Public Health (Berlin)*, *31*(1), 73–84. DOI: 10.1007/s10389-020-01416-7

Verraes, C., Van Boxstael, S., Van Meervenne, E., Van Coillie, E., Butaye, P., Catry, B., De Schaetzen, M.-A., Van Huffel, X., Imberechts, H., Dierick, K., Daube, G., Saegerman, C., De Block, J., Dewulf, J., & Herman, L. (2013). Antimicrobial resistance in the food chain: A review. *International Journal of Environmental Research and Public Health, 10*(7), 2643–2669. DOI: 10.3390/ijerph10072643 PMID: 23812024

Vidovic, N., & Vidovic, S. (2020). Antimicrobial resistance and food animals: Influence of livestock environment on the emergence and dissemination of antimicrobial resistance. *Antibiotics (Basel, Switzerland), 9*(2), 52. DOI: 10.3390/antibiotics9020052 PMID: 32023977

Whittaker, A., Do, T. T., Davis, M. D. M., & Barr, J. (2023). AMR survivors? Chronic living with antimicrobial resistant infections. *Global Public Health, 18*(1), 2217445. Advance online publication. DOI: 10.1080/17441692.2023.2217445 PMID: 37272390

Wylie, M. R., & Merrell, D. S. (2022). The antimicrobial potential of the neem tree Azadirachta indica. *Frontiers in Pharmacology, 13*, 891535. DOI: 10.3389/fphar.2022.891535 PMID: 35712721

Zinsstag, J., Schelling, E., Waltner-Toews, D., & Tanner, M. (2011). From "one medicine" to "one health" and systemic approaches to health and well-being. *Preventive Veterinary Medicine, 101*(3–4), 148–156. DOI: 10.1016/j.prevetmed.2010.07.003 PMID: 20832879

About the Contributors

Odangowei Inetiminebi Ogidi holds a Bachelor of Science degree in Biochemistry, Master of Science and Doctor of Philosophy degrees in Biotechnology at The Niger Delta University Wilberforce Island Bayelsa State, Nigeria and Federal University of Technology Owerri, Imo State, Nigeria respectively. He is a Lecturer, and Sub-Dean, Faculty of Basic Medical Sciences. He has served as Acting Head, Department of Biochemistry, Bayelsa Medical University, Yenagoa, Bayelsa State, Nigeria. He has also served as a Lecturer, Head of Departments of Science Laboratory Technology and Biochemistry, and Deputy Director, Centre for Continuing Education in Federal Polytechnic Ekowe, Bayelsa State. He has authored over 80 peer-reviewed academic articles in reputable local and international journals and presented over 40 conference papers. He has over 80 book chapters published in Elsevier, Springer Nature, CRC Press Taylor & Francis, Wiley and Sons and IGI Global Scientific Publishing. He has edited a book titled "Natural Antimalarial Drug Discovery and Development in the Global South" published by IGI Global Scientific publishing. At present he is editing fifteen books in Elsevier, CRC Press Taylor & Francis, IGI Global Scientific Publishing, Springer Nature, Apple Academic Press and Bentham Science Publishing. He is a member of various Journal Editorial Boards. His research interest has been Environmental Biotechnology, Phytomedicine, Phytochemistry, Human and Environmental health risk assessments and ecotoxicology amongst others. Dr. Ogidi is a member of West Africa Society of Toxicology, Nigerian Society of Biochemistry and Molecular Biology, Foundation for African Development through International Biotechnology and Nigerian Environmental Society.

***

Wafaa Benkayba is a chemistry researcher specializing in electrochemistry and analytical chemistry. Her multidisciplinary work covers water treatment, pharmaceutical and cosmetic analysis, food authenticity, green nanomaterials, and the development of electrochemical sensors. She also contributes to research on antimicrobial resistance (AMR) and traditional medicine. Her research bridges analytical sciences, environmental sustainability, and public health, with a strong commitment to developing evidence-based, sustainable, and culturally informed solutions to address global health and environmental challenges.

Vibhu Bhatia is an accomplished academic and researcher in the field of biotechnology, currently serving as an Assistant Professor in the Department of Biotechnology at Somaiya Vidyavihar University, Mumbai. With a strong foundation in biotechnology, Dr. Bhatia earned both his B.Tech and M.Tech degrees in Biotechnology, followed by a Ph.D. in Environmental Biotechnology from Thapar University, Patiala. He brings a rich blend of academic and industrial expertise to his role, having accumulated over five years of teaching experience and two years of industrial experience as Quality Control Head in the pharmaceutical industry. This dual exposure has enabled him to bridge the gap between theoretical knowledge and practical applications, fostering innovation in both classroom teaching and research. Dr. Bhatia's research interests lie at the intersection of sustainability and biotechnology. His work focuses on wastewater treatment, environmental biotechnology, photocatalysis, bioremediation, and the utilization of biomass for energy generation.

Mohanraj E is a Ph.D in Biochemistry and currently working as an Assistant Professor in the Department of Pharmacy, with 8 years of teaching experience and 6 years of research experience. He has been actively involved in academic activities, contributing to the advancement of biochemistry through research, publications, and book chapters in national and international forums.

Chinyere Ezeanya-Bakpa is a prominent figure in the field of medical microbiology, distinguished not only by her extensive academic credentials but also by her passionate commitment to advancing healthcare through innovative research and community engagement. With a career spanning over a decade, she has made significant contributions to the understanding of infectious diseases, particularly in the context of public health in underserved populations. Throughout her career, Dr. Ezeanya-Bakpa has focused on research that addresses critical health challenges in both developing and developed nations. Her work has notably explored the epidemiology of antibiotic-resistant pathogens, which poses a growing threat to public health worldwide. By employing advanced molecular techniques, she has identified new resistance mechanisms and has been actively involved in the development of strategies aimed at combating this alarming trend. Her research findings have been published in several peer-reviewed journals, earning her recognition among her peers and contributing to the body of knowledge in medical microbiology. Beyond her research endeavors, Dr. Ezeanya-Bakpa is an advocate for public health education and community outreach. She has collaborated with various non-governmental organizations to promote awareness of infectious diseases and vaccination programs, particularly in rural areas where access to healthcare is limited. Through workshops, seminars, and health fairs, she has empowered communities with knowledge about disease prevention, transforming her scientific expertise into practical solutions that improve health outcomes. In addition to her research and outreach work, Dr. Ezeanya-Bakpa is a dedicated educator. As a lecturer at a well-regarded university, she has inspired countless students to pursue careers in microbiology and healthcare. Her teaching philosophy emphasizes the importance of critical thinking and hands-on experience, equipping her students with the skills necessary to excel in the field of medical microbiology. Dr. Ezeanya-Bakpa actively engages her students in cutting-edge research projects, enabling them to contribute to meaningful work that can impact public health. Dr. Ezeanya-Bakpa's professional affiliations include memberships in esteemed organizations such as the American Society for Microbiology and the African Society for Laboratory Medicine. These affiliations not only reflect her commitment to her profession but also facilitate collaborative opportunities that enhance her research and outreach efforts. Her dedication to excellence, both in the laboratory and the community, positions her as a role model for aspiring scientists and healthcare professionals. In summary, Dr. Chinyere Ezeanya-Bakpa exemplifies the fusion of research, education, and community service. Her contributions to medical microbiology and her unwavering commitment to advancing public health make her a significant figure in the field. Through her work, she continues to inspire future generations and address some of the most pressing health challenges of our time.

Sai Priyanka is a Research Scholar in Sri Sathya Sai University for Human Excellence, India.

Loqman Souad is a Professor of Biology at the Faculty of Medicine and Pharmacy, Cadi Ayyad University. His research focuses on microbiology and epidemiology, with an emphasis on understanding microbial behavior and the epidemiological dynamics of infectious diseases.

Neenu M K is affiliated with the Department of Pharmacognosy, Yenepoya Pharmacy College & Research Centre, Yenepoya University, India. My academic and research interests include pharmacognosy, medicinal plants, natural product–based drug discovery, and phytochemical evaluation. Actively involved in academic and research activities related to herbal drugs and traditional medicine.

Lovleen Marwaha is a Professor at the Department of Zoology, School of Bioengineering and Biosciences, Lovely Professional University, Phagwara, India. She has completed her M.S. and Ph.D. in Cytogenetics from the Department of Zoology, Punjab University, Chandigarh. She has worked on the evaluation of pesticide genotoxicity using insect genomes, specifically on different mosquito species and Drosophila melanogaster. She has carried out an intensive study on the correlation between pesticide exposure limit and inductive genotoxic consequences on Dipteran larval salivary polytene chromosomes. Furthermore, she has studied chromosomal and gene mutations in mosquito and fruit fly genomes due to pesticide exposure. Currently, she is working on Apiculture and Sericulture-related topics. She has many research papers published in National and International Journals of repute. Presently, she is working on insecticide resistance in mosquitoes, pesticide genotoxicity evaluation, honey bee development, honey processing, silk fiber, and pearl quality improvement. She has published three authored books on Apiculture and on Mobile Genetic Elements. Forthcoming books include Worker Honey Bees, Apiculture Products, Basic Concepts of Genetics, and Introduction to Sericulture.

P. Parthiban is an accomplished researcher in Biochemistry, Environmental Biotechnology, and Microbial Technology, with over ten years of experience in interdisciplinary research, innovation management, and analytical instrumentation. His work spans bioprocess development, wastewater and sludge valorization, microbial degradation pathways, nano-bioformulations, and phytochemical-based therapeutics. He has contributed to major DST and DRDO-funded projects exceeding 10 crore and has published 30+ peer-reviewed SCI/Scopus articles, reflecting a strong commitment to high-impact research. Dr. Parthiban is also an active innovator, holding 4 utility patents and 7 design patents, and his additional role as a Patent Analyst has strengthened his expertise in prior art search, patent drafting, and technology evaluation. His research aims to develop sustainable, scalable bioprocesses while bridging scientific discovery with societal and industrial applications.

Ramya Raghavan is currently serving as faculty at Sri Sathya Sai University for Human Excellence, India. Dr Raghavan's research interests include Genetics, Biostatistics, Biophotons, Epigenetic factors, Intellectual disability, and Assistive Technologies. Her expertise extends to the critical intersection of the exposome and Social Determinants of Health (SDOH). Dr. Raghavan's work underscores how a holistic understanding of environmental exposures, from the molecular level to population-wide influences, is crucial for addressing health disparities and advancing a more equitable approach to disease prevention.

Vallipriya R. has been working as an Assistant Professor in the Department of Biochemistry & Biotechnology at Pavendar Bharathidasan College of Arts and Science (PBCAS), affiliated with Bharathidasan University. With a wealth of experience, Dr. R. Vallipriya has accumulated 12 years of extensive teaching experience. Her scholarly contributions include the publication of 12 papers, including one peer-reviewed paper, in various UGC-CARE, Scopus, and other peer-reviewed journals.

Soumya Raju is Research Scholar at Department of Botany, CMS College Kottayam (Autonomous), affiliated to Mahatma Gandhi University Kerala, India, specializing in Phytochemistry & Pharmacology. She has completed Master's degree in Biochemistry. She has 20 years teaching experience at UG and PG level. She has book chapters, several conference papers and 2 Gen Bank submissions in her credit. ORCID iD: 0009-0004-1701-6743

Gomathi R. is working as an Assistant Professor in the Department of Biochemistry at Holy Cross College, Trichy, Tamilnadu, India. Had more more than 7 years of teaching experience.Published many papers and presented papers at National and International Conferences. Took part in many academic and cultural events during the service. Guided Post graduate research students.

Neharika Avinash Sawant is a third-year undergraduate student in Biotechnology at the Somaiya School of Basic and Applied Sciences, Somaiya Vidyavihar University, Mumbai, India. Her research interests include omics and translational medicine, with a focus on integrating traditional medicinal knowledge with contemporary scientific frameworks.

Saritha Thampi is Research Scholar at Department of Botany, CMS College Kottayam (Autonomous), affiliated to Mahatma Gandhi University Kerala, India, specializing in Plant Microbiology. She has completed Master's degree in Applied Microbiology and has qualified the National Eligibility Test conducted by ASRB in Agricultural Microbiology. She has 12 years of teaching experience. Her research focuses on the exploration of endophyte-derived bioactive metabolites from the plant family Piperaceae. She has several conference papers and actively pursues interdisciplinary research. ORCID iD: 0009-0000-8487-5418

Zahid Ahmad Thoker is a serious research Scholar at Lovely Professional University who is working on Insects (Entomology), specifically on Honey Bees. He graduated with his Master's degree in Zoology from Chandigarh University, Gharuan, Mohali, where he developed a very solid academic grounding in life sciences. Zahid has already contributed greatly to academia; one of his research papers was published, one review paper, and over 25 book chapters in reputable national and international volumes. His academic publication is an indication of the desire to promote science, especially in the biological sciences. As a visionary who desires to make a difference in the research world and in society, Zahid hopes to broaden his academic and professional experiences with integrity and excellence. He is also strongly driven by the goal of benefiting society with his research work and making his parents proud by leaving his mark on the field of science and scholarship.

Rogimon P. Thomas is Professor and Head, Department of Botany, CMS College Kottayam (Autonomous), affiliated to Mahatma Gandhi University Kerala, India. He has more than 19 years of teaching and research experience. His research contributions include discovery of Murdannia satheeshiana and Murdannia nampyana, new species of Commelinaceae and Rotala baileyana of the family Lythraceae from Kerala, India. He has 4 books, 20 book chapters, 5 edited books, 30 journal publications and 19 Gen Bank submissions. ORCID iD: 0000-0001-8364-1679

Sivaranjani Varadharajan is currently working as an Assistant Professor in the Department of Biochemistry, with 4 years of teaching experience and 12 years of research and hospital-based professional experience. She has been actively involved in clinical and academic activities, contributing to the advancement of biochemistry through research, publications, and book chapters in national and international forums. Her expertise bridges both educational and practical healthcare settings, providing a unique perspective on translational biochemistry applications. She is also an active member of professional organizations in her field.

Srinivasan Vasan is an Assistant Professor in the Department of Humanities and Social Sciences at Graphic Era (Deemed to be University), with 14 years of teaching and field expertise. He has served in the Ministry of Health and Family Welfare, as a Guest Faculty, and as a Mobilization Training Specialist with the State Government. Recently, he contributed to a tuberculosis project funded by GFATM and USAID. He has published 60 papers, with 20 more under peer review in UGC-CARE and SCOPUS journals.

Index

A

B

C

D

E

G

H

I

L

M

N

O

P

Q

R

www.ingramcontent.com/pod-product-compliance
Lightning Source LLC
LaVergne TN
LVHW081317300926
849161LV00008B/1788